T0326395

HETEROGENEITY OF FUNCTION IN NUMERICAL COGNITION

HETEROGENEITY OF FUNCTION IN NUMERICAL COGNITION

Edited by

AVISHAI HENIK

Department of Psychology,
Ben-Gurion University of the Negev,
Beer-Sheva, Israel

WIM FIAS

Ghent University,
Department of Experimental Psychology,
Ghent, Belgium

ACADEMIC PRESS

An imprint of Elsevier

Academic Press is an imprint of Elsevier
125 London Wall, London EC2Y 5AS, United Kingdom
525 B Street, Suite 1650, San Diego, CA 92101-4495, United States
50 Hampshire Street, 5th Floor, Cambridge, MA 02139, United States
The Boulevard, Langford Lane, Kidlington, Oxford OX5 1GB, United Kingdom

Library of Congress Cataloging-in-Publication Data
A catalog record for this book is available from the Library of Congress

British Library Cataloguing-in-Publication Data
A catalogue record for this book is available from the British Library

ISBN: 978-0-12-811529-9

For information on all Academic Press publications visit our
website at https://www.elsevier.com/books-and-journals

  Working together
to grow libraries in
developing countries

www.elsevier.com • www.bookaid.org

Publisher: Nikki Levy
Acquisition Editor: Natalie Farra
Developmental Editor: Kristi Anderson
Production Project Manager: Sujatha Thirugnana Sambandam
Cover Designer: Matthew Limbert

Typeset by TNQ Books and Journals

Contents

MARC BRYSBAERT

II

PERFORMANCE CONTROL AND SELECTIVE ATTENTION

III

SPATIAL PROCESSING AND MENTAL IMAGERY

IV

EXECUTIVE FUNCTIONS

16. (How) Are Executive Functions Actually Related to Arithmetic Abilities?   337

KIM ARCHAMBEAU AND WIM GEVERS

V

MEMORY

17. Numerical Cognition and Memory(ies)   361

PIERRE BARROUILLET

18. Hypersensitivity-to-Interference in Memory as a Possible Cause of Difficulty in Arithmetic Facts Storing   387

MARIE-PASCALE NOËL AND ALICE DE VISSCHER

19. Working Memory for Serial Order and Numerical Cognition: What Kind of Association? 409

STEVE MAJERUS AND LUCIE ATTOUT

20. Do Not Forget Memory to Understand Mathematical Cognition 433

VALÉRIE CAMOS

Index 449

List of Contributors

Kim Archambeau   Center for Research in Cognition and Neurosciences, ULB Neuroscience Institute, Université Libre de Bruxelles, Brussels, Belgium

Lucie Attout   Université de Liège, Liège, Belgium

Lauren S. Aulet   Emory University, Atlanta, GA, United States

Pierre Barrouillet   University of Geneva, Geneva, Switzerland

Mattan S. Ben-Shachar   Ben-Gurion University of the Negev, Beer-Sheva, Israel

Andrea Berger   Ben-Gurion University of the Negev, Beer-Sheva, Israel

Mario Bonato   Ghent University, Ghent, Belgium; University of Padova, Padova, Italy

Marc Brysbaert   Ghent University, Ghent, Belgium

Valérie Camos   Université de Fribourg, Fribourg, Switzerland

Chi-Ngai Cheung   Emory University, Atlanta, GA, United States

Lucy Cragg   The University of Nottingham, Nottingham, United Kingdom

Bert De Smedt   Faculty of Psychology and Educational Sciences, University of Leuven, Leuven, Belgium

Alice De Visscher   Université Catholique de Louvain, Louvain-la-Neuve, Belgium

Wim Fias   Ghent University, Ghent, Belgium

Andrea Frick   University of Fribourg, Fribourg, Switzerland

Wim Gevers   Center for Research in Cognition and Neurosciences, ULB Neuroscience Institute, Université Libre de Bruxelles, Brussels, Belgium

Camilla Gilmore   Loughborough University, Loughborough, United Kingdom

Liat Goldfarb   University of Haifa, Haifa, Israel

Avishai Henik   Ben-Gurion University of the Negev, Beer-Sheva, Israel

Shachar Hochman   Ben-Gurion University of the Negev, Beer-Sheva, Israel

Teresa Iuculano   Stanford University, Stanford, CA, United States

Naama Katzin   Ben-Gurion University of the Negev, Beer-Sheva, Israel

André Knops   Humboldt-Universität zu Berlin, Berlin, Germany

Patrick Lemaire   Aix-Marseille University & CNRS, Marseille, France

Stella F. Lourenco   Emory University, Atlanta, GA, United States

Steve Majerus   Université de Liège, Liège, Belgium; Fund for Scientific Research FNRS, Brussels, Belgium

Pawel J. Matusz   University of Hospital Centre – University of Lausanne, Lausanne, Switzerland

Vinod Menon   Stanford University, Stanford, CA, United States

Rebecca Merkley   University of Western Ontario, London, ON, Canada

Wenke Möhring   Universität Basel, Basel, Switzerland

Nora S. Newcombe   Temple University, Philadelphia, PA, United States

Marie-Pascale Noël   Université Catholique de Louvain, Louvain-la-Neuve, Belgium

Aarthi Padmanabhan   Stanford University, Stanford, CA, United States

Jérôme Prado   Institut des Sciences Cognitives Marc Jeannerod – UMR 5304, Centre National de la Recherche Scientifique (CNRS) & Université de Lyon, Bron, France

Gaia Scerif   University of Oxford, Oxford, United Kingdom

Kim Uittenhove   University of Geneva, Geneva, Switzerland

Klaus Willmes   RWTH Aachen University, Aachen, Germany

Acknowledgments

The field of numerical cognition has made impressive progress in the last three decades. Much research has been devoted to unravel the mental processes involved in numerical cognition and mathematical thinking. It is clear that in addition to unique and specific mental processes, numerical cognition is being contributed to by general domains such as language, attention, and memory. The discussion and exchange of knowledge between these fields are important. We think of the current volume as a contribution to this effort.

To encourage discussions among the various fields of study, we organized a workshop titled "Heterogeneous Contributions to Numerical Cognition" in June 2016 in Ghent, Belgium. When we started to contemplate this workshop, we already thought that the end result would be an edited book with contributions by the participants. This is an occasion to thank all those who contributed to this endeavor.

The symposium was financed in part by the ERC (European Research Council) and by funds coming from the University of Ghent. ERC funds were part of an advanced researcher grant (295644) to AH, and funds from the University of Ghent were provided by the Research Council. We would like to extend our thanks to the ERC and the University of Ghent for supporting the workshop.

We would like to thank Desiree Meloul and Kerensa tiberghien for administrative and logistic support.

Last but not least, we would like to thank our students and colleagues who contributed each one in his/her own way to develop ideas, thoughts, and plans of research throughout the years.

Avishai Henik and Wim Fias

Introduction

Wim Fias[1], Avishai Henik[2]

[1]Ghent University, Ghent, Belgium; [2]Ben-Gurion University of the Negev, Beer-Sheva, Israel

The study of the cognitive basis of the human capacity for arithmetic and mathematics has a fairly long history. The oldest studies using numbers or numerical tasks were mainly conducted to answer general questions in cognitive psychology, such as the structure of memory, without truly considering mathematical cognition as a separate cognitive skill. Neuropsychology described patients with particular numerical or mathematical deficits, but much of the neuropsychology standard testing lacked examination of numerical cognition. Educational science, on the other hand, considered mathematical skill as a separate entity, but it did not concentrate on the underlying cognitive system. With cognitive neuropsychology becoming more dominant than neuropsychology and cognitive psychology, the modular organization of the numerical cognitive system became more emphasized, defining it as a cognitive skill that is partly separated from other cognitive skills.

This brief history culminated in the triple code model proposed by Dehaene in the seminal volume of the journal Cognition in 1992 and later in a revised form by Dehaene and Cohen in the journal of Mathematical Cognition in 1995. It suggests that mathematical cognition is subserved by three types of number representations, each supporting specific aspects of numerical cognition. The Arabic number form supports the visual encoding of Arabic numbers. The verbal number representations are primarily involved in cognitive operations that rely on verbal routines, such as the tables of multiplication, which are learned by verbal drill and are assumed to be processed like nursery rhymes. Finally, there is the type of representation that encodes the numerical magnitudes expressed by numbers. This code is important in tasks in which numerical magnitude is fundamental, like comparison, or during the execution of calculation procedures, like calculating 7+6 and splitting the problem into (7+3) +3. Interestingly, each of these numerical representational codes has been attributed to specific brain structures. The Arabic number form involves the inferior temporal regions, the verbal code involves the left hemisphere's language regions, and the magnitude code involves the intraparietal sulcus.

This model has had a profound impact on how the field of numerical cognition further developed. Driven by some key observations, the focus became more and more on the parietal magnitude representing system. Nieder and Miller (2004) were able to demonstrate the biological reality of magnitude-coding neurons in the intraparietal sulcus of macaque monkeys, which was further supported by similar magnitude-tuned numerical coding in humans with functional magnetic resonance imaging (fMRI) adaptation (Piazza, Izard, Pinel, Le Bihan, & Dehaene, 2004). The fact that the accuracy of this approximate number magnitude-coding system increases with age (Piazza et al., 2010), and the demonstration that the ability to distinguish numerosities following the principles of approximate neural coding correlated with school levels of mathematical skill (Halberda, Mazzocco, & Feigenson, 2008), caused a shift toward assigning a massive amount of explanatory power to this approximate number system (ANS).

Although the ability to accurately represent number magnitude can undoubtedly have an influence on the development of mathematical skill, there are a number of reasons to assume that it cannot be the only critical component that is involved in number processing tasks and that would explain variation in mathematical skill. First, it has been demonstrated that the correlation between the accuracy of the ANS system, as measured by the comparison between two collections of dots, and mathematical skill is primarily driven by trials in which numerosity differences, and differences in terms of nonnumerical aspects of the stimulus are incongruent (for instance, when the more numerous set is presented such that it occupies less space) (Gilmore et al., 2013). Moreover, it has been suggested that numerosity correlates highly with noncountable variables (e.g., density) that modulate comparative judgments (Leibovich, Katzin, Harel, & Henik, 2017). These issues indicate that not only the ability to distinguish the numerosity is important but also the ability to resolve the conflict induced by the distracting information, suggesting cognitive control processes are contributing to mathematical cognition (Bull & Scerif, 2001; Cragg & Gilmore, 2014; Chapters 13 and 14). A second line of evidence comes from developmental disorders. Although cases of isolated dyscalculia have been described, developmental dyscalculia more frequently occurs comorbid with dyslexia (Shalev, Auerbach, Manor, & Gross-Tsur, 2000), suggesting a role for language-related processing. Third, spatial processing is intimately linked with mathematical processing, among other things, as evidenced by fMRI studies showing that the neural regions that are involved in spatially directed eye movements are similarly recruited while performing addition and subtraction (corresponding to right and left eye movements, respectively; see Knops, Thirion, Hubbard, Michel, & Dehaene, 2009). Fourth, memory processing is an important factor as well, as witnessed by different correlational studies that have established that working memory skill correlates with mathematical skill (Geary, 1993). The fact that both working memory tasks and number processing tasks engage overlapping parietal regions (e.g., Attout, Fias, Salmon, & Majerus, 2014) provides additional ground for assuming a strong link between memory and mathematical performance. Not only working memory is important but also the structure of memory is important. Computational work has shown that general principles of activation and competition in associated networks account for many of the behavioral phenomena that are observed during mental arithmetic (Rotem & Henik, 2015; Verguts & Fias, 2005). Similarly, mechanisms of proactive interference may have shaped the nature of the memory representations storing the tables of multiplication (De Visscher & Noel, 2014). Finally, it is obvious that mathematical tasks require planning and supervisory control. This is confirmed by the fact that neuroimaging of number and mathematical tasks shows a strong involvement of prefrontal brain areas (Anderson, Betts, Ferris, & Fincham, 2011; Arsalidou & Taylor, 2011).

Overall, it is thus clear that many cognitive functions are involved in number processing and mathematics. We believe that to advance toward a detailed understanding of the neurocognitive mechanisms of numerical and mathematical cognition, we need to enhance our insight into how language, memory, attentional control, executive function, and spatial cognition relate to mathematical tasks.

From this perspective we organized a workshop for which we invited three speakers for each of those topics, one speaker focusing on the general cognitive psychology, one on the neurocognitive basis, and one on development. We had hoped that the complementary expertise of the speakers, both not only within their domain but also across domains, would lead to stimulating interactions and exchanging of ideas that would ultimately contribute to further progress in the field of numerical and mathematical cognition. The contributions in the present volume showed that our hope was not unrealistic.

References

Anderson, J. R., Betts, S., Ferris, J. L., & Fincham, J. M. (2011). Cognitive and metacognitive activity in mathematical problem solving: Prefrontal and parietal patterns. *Cognitive, Affective, & Behavioral Neuroscience*, 11(1), 52–67. http://doi.org/10.3758/s13415-010-0011-0.

Arsalidou, M., & Taylor, M. J. (2011). Is 2+2=4? Meta-analyses of brain areas needed for numbers and calculations. *Neuroimage*, 54(3), 2382–2393. http://doi.org/10.1016/j.neuroimage.2010.10.009.

Attout, L., Fias, W., Salmon, E., & Majerus, S. (2014). Common neural substrates for ordinal representation in short-term memory, numerical and alphabetical cognition. *PLoS One*, 9(3), e92049. http://doi.org/10.1371/journal.pone.0092049.

Bull, R., & Scerif, G. (2001). Executive functioning as a predictor of children's mathematics ability: Inhibition, switching and working memory. *Developmental Neuropsychology*, 19, 273–293.

Cragg, L., & Gilmore, C. (2014). Skills underlying mathematics: The role of executive function in the development of mathematics proficiency. *Trends in Neuroscience and Education*, *3*, 63–68. https://doi.org/10.1016/j.tine.2013.12.001.

De Visscher, A., & Noël, M.-P. (2014). The detrimental effect of interference in multiplication facts storing: Typical development and individual differences. *Journal of Experimental Psychology: General*, *143*(6), 2380–2400. http://doi.org/10.1037/xge0000029.

Dehaene, S. (1992). Varieties of numerical abilities. *Cognition*, *44*, 1–43.

Dehaene, S., & Cohen, L. (1995). Towards an anatomical and functional model of number processing. *Mathematical Cognition*, *1*(1), 83–120.

Geary, D. C. (1993). Mathematical disabilities: Cognitive, neuropsychological and genetic components. *Psychological Bulletin*, *114*(2), 345–362.

Gilmore, C., Attridge, N. F., Clayton, S., Cragg, L., Johnson, S., Marlow, N., et al. (2013). Individual differences in inhibitory control, not non-verbal number acuity, correlate with mathematics achievement. *PLoS One*, *8*(6), e67374. https://doi.org/10.1371/journal.pone.0067374.

Halberda, J., Mazzocco, M. M. M., & Feigenson, L. (2008). Individual differences in non-verbal number acuity correlate with maths achievement. *Nature*, *455*(7213), 665–668. http://doi.org/10.1038/nature07246.

Knops, A., Thirion, B., Hubbard, E. M., Michel, V., & Dehaene, S. (2009). Recruitment of an area involved in eye movements during mental arithmetic. *Science*, *324*(5934), 1583–1585. http://doi.org/10.1126/science.1171599.

Leibovich, T., Katzin, N., Harel, M., & Henik, A. (2017). From 'sense of number' to 'sense of magnitude' – the role of continuous magnitudes in numerical cognition. *Behavioral and Brain Sciences*, *40*, e164. https://doi.org/10.1017/S0140525X16000960.

Nieder, A., & Miller, E. K. (2004). A parieto-frontal network for visual numerical information in the monkey. *Proceedings of the National Academy of Sciences of the United States of America*, *101*(19), 7457–7462. http://doi.org/10.1073/pnas.0402239101.

Piazza, M., Facoetti, A., Trussardi, A. N., Berteletti, I., Conte, S., Lucangeli, D., … Zorzi, M. (2010). Developmental trajectory of number acuity reveals a severe impairment in developmental dyscalculia. *Cognition*, *116*(1), 33–41. http://doi.org/10.1016/j.cognition.2010.03.012.

Piazza, M., Izard, V., Pinel, P., Le Bihan, D., & Dehaene, S. (2004). Tuning curves for approximate numerosity in the human intraparietal sulcus. *Neuron*, *44*(3), 547–555. http://doi.org/10.1016/j.neuron.2004.10.014.

Rotem, A., & Henik, A. (2015). Development of product relatedness and distance effects in typical achievers and in children with mathematics learning disabilities. *Journal of Learning Disabilities*, *48*, 577–592. https://doi.org/10.1177/0022219413520182.

Shalev, R. S., Auerbach, J., Manor, O., & Gross-Tsur, V. (2000). Developmental dyscalculia: Prevalence and prognosis. *European Child & Adolescent Psychiatry*, *9*(Suppl 2), 1158–1164. Retrieved from http://www.ncbi.nlm.nih.gov/pubmed/11138905.

Verguts, T., & Fias, W. (2005). Interacting neighbours: A connectionist model of retrieval in single-digit multiplication. *Memory and Cognition*, *33*(1), 1–16.

LANGUAGE

1

Numbers and Language: What's New in the Past 25 Years?

Marc Brysbaert

Ghent University, Ghent, Belgium

OUTLINE

The relationship between mathematical and verbal performance is not a clear one. On the one hand, both abilities seem to be related; on the other hand, many school systems offer pupils the opportunity to choose between a language-oriented education and a mathematics-oriented education, suggesting the two types of skills diverge.

THE RELATIONSHIP BETWEEN VERBAL AND ARITHMETICAL PERFORMANCE IN THE WISC INTELLIGENCE TEST

One way to assess the relationship between verbal and arithmetical performance is to see how they correlate in intelligence tests. Wechsler (1949), for instance, made a distinction between verbal intelligence and performance intelligence, and he included a test of arithmetic in the verbal scale. In the Wechsler Intelligence Scale for Children (WISC), the test of arithmetic correlated .6 with the verbal scale and .45 with the performance scale (Seashore, Wesman, & Doppelt, 1950). For comparison, the vocabulary subtest (the most typical verbal test in the WISC) correlated .75 with the verbal scale and .55 with the performance scale. The test with the highest correlation to the performance scale was the object assembly test, correlating .35 with the verbal scale and .6 with the performance scale. The arithmetic test was retained as part of the verbal scale when the WISC was revised for the first time (and called the WISC-R).

Factor analyses indicated, however, that a third factor was present in the WISC and the WISC-R. This factor was difficult to interpret but was called "freedom from distractibility." A new test was added to the WISC-III (the symbol search test) to better measure the elusive third factor, but this did not succeed very well because a new factor analysis hinted at four factors (Keith & Witta, 1997) as shown in Fig. 1.1. Surprisingly, in this analysis, the arithmetic subtest became the best measure of the factor freedom from distractibility. In addition, this factor had the highest correlation with the overall score (considered to be a measure of general intelligence, or g). Both findings suggested that the factor freedom from distractibility was more central to intelligence than its name suggested. Keith and Witta (1997) proposed to rename it as "quantitative reasoning." Unfortunately, only two tests loaded on the factor, which is weak evidence for a factor.

The fourth revision of the WISC included extra tests to better measure the four first-order factors. In particular, it was hypothesized that the

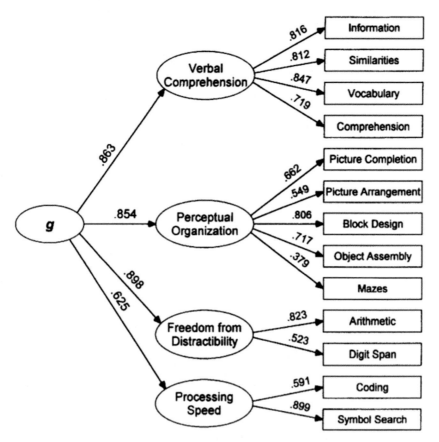

FIGURE 1.1 Outcome of a hierarchical factor analysis of the WISC-III. Subtests are a measure of both general intelligence (g) and a first-order variable. Four first-order variables could be discerned. The test of arithmetic loaded on the first-order variable freedom from distractibility, together with the digit span test. *From Keith, T. Z., & Witta, E. L. (1997). Hierarchical and cross-age confirmatory factor analysis of the WISC-III: What does it measure?* School Psychology Quarterly, 12(2), 89–107.

freedom from distractibility factor actually could be a working memory factor and new tests were added to better capture it. This seemed to work reasonably well (Keith, Fine, Taub, Reynolds, & Kranzler, 2006), as shown in Fig. 1.2 (although a solution with five first-order factors provided a better fit).

All in all, the analyses of the WISC-III and WISC-IV confirm that arithmetic skills are correlated to language skills (via the high correlations with *g*) and at the same time form two different intelligence factors on which individuals can score high or low (see also Reynolds, Keith, Flanagan, & Alfonso, 2013).

A similar conclusion was reached on the basis of an analysis specifically geared toward mathematical knowledge. Taub, Floyd, Keith, and McGrew (2008) predicted mathematics achievement on the basis of IQ subtests. The mathematics tests consisted of a calculation test (ranging from simple addition facts to calculus) and an applied problems test in which the nature of the problem had to be comprehended, relevant information identified, calculations performed, and solutions stated. Data were available from 5-year-old children to 19-year-old children. Fluid reasoning (similar to working memory) had an effect in all age groups (see also Primi, Ferrão, & Almeida, 2010). For the younger pupils, processing speed also contributed to the performance, whereas for the older participants, crystallized intelligence became more important (arguably to retrieve simple solutions from long-term memory; see also Calderón-Tena & Caterino, 2016).

Interestingly, the broader Cattell–Horn–Carroll (CHC) model, on which the analyses of Figs. 1.1 and 1.2 were based and which is currently seen as the best summary of intelligence research, postulates the existence of a separate first-order factor quantitative knowledge (McGrew, 2009), as shown in Fig. 1.3. Therefore, the expectation is that with the right tests included, numerical knowledge will come out as an individual type of intelligence, although so far attempts have not been successful (e.g., Keith, Low, Reynolds, Patel, & Ridley, 2010).

COGNITIVE PROCESSES INVOLVED IN NUMERICAL COGNITION: THE 1990s

The relationship between verbal and mathematical performance in intelligence tests provides an interesting background but is limited by the type of tests used to assess the various skills. In the psychometric tradition, tests have mainly been proposed via trial and error, starting with the first intelligence test published by Binet and Simon (1907). Tests that correlated with school achievement were retained, others were replaced. In addition, there is a strong force not to change existing arrangements too much, as

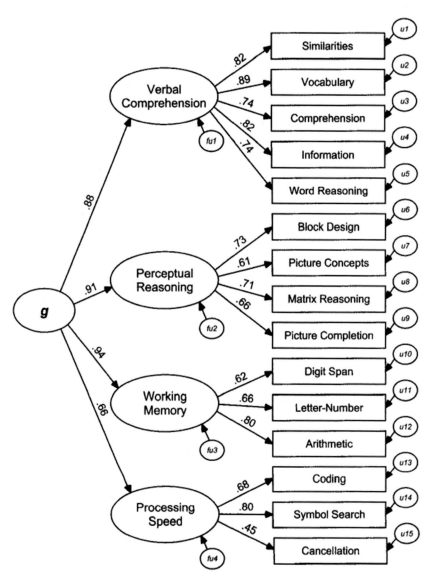

FIGURE 1.2   Outcome of a hierarchical factor analysis of the WISC-IV. Subtests are a measure of both general intelligence (*g*) and a first-order variable. Four first-order variables could be discerned. The test of arithmetic loaded on the first-order variable working memory. *From Keith, T. Z., Fine, J. G., Taub, G. E., Reynolds, M. R., & Kranzler, J. H. (2006). Higher order, multisample, confirmatory factor analysis of the Wechsler Intelligence Scale for Children – fourth edition: What does it measure?* School Psychology Review, 35(1), 108–127.

C. Cattell-Horn-Carroll (CHC) Integrated Model

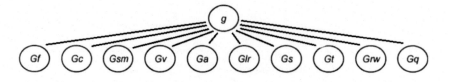

Gf	Fluid reasoning
Gc	Comprehension knowledge
Gsm	Short-term memory
Gv	Visual processing
Ga	Auditory processing
Glr	Long-term storage and retrieval
Gs	Cognitive processing speed
Gt	Decision and reaction speed
Grw	Reading and writing
Gq	Quantitative knowledge

FIGURE 1.3 The Cattell–Horn–Carroll model of intelligence. The model postulates a general intelligence factor and ten first-order factors, of which quantitative knowledge is one. *Based on McGrew, K. S. (2009). CHC theory and the human cognitive abilities project: Standing on the shoulders of the giants of psychometric intelligence research.* Intelligence, 37(1), 1–10. *Used with permission from Elsevier.*

practitioners do not like radically new revisions of existing tests. Therefore, the factors emerging from the factor analyses may to some extent be an artifact of the initial choices made in the design of intelligence tests (but see Jewsbury, Bowden, & Strauss, 2016 for an interesting study about the overlap between the processes postulated in cognitive models of executive function and factors emerging from the CHC model of intelligence).

A different approach is to start from an analysis of the cognitive processes involved in number and word processing, independent of whether these processes are related to individual differences in performance. The 1990s were a particularly fruitful decade in this respect, largely because of the publications of Dehaene and the responses they elicited from other researchers.

According to Dehaene (1992; Dehaene, Dehaene-Lambertz, & Cohen, 1998), a tacit hypothesis in cognitive arithmetic was that numerical abilities derived from human linguistic competence. In his own words (Dehaene, 1992, pp. 2–3): "For the lay person, calculation is the numerical activity par excellence. Calculation in turn rests on the ability to read, write, produce or comprehend numerals […]. Therefore number processing, in its fundamental form, seems intuitively linked to the ability to mentally manipulate sequences of words or symbols according to fixed transcoding or calculation rules."

Based on data from adults, infants, and animals, Dehaene (1992) argued that number processing on the basis of symbols is not the only processing the human brain is capable of. A second pathway makes use of an innate quantity system. The quantity system is based on analog encoding and allows accurate representations for numbers up to 3 or 4 and approximate quantities for larger numbers. Evidence for the involvement of such a quantity-based pathway was found in animals, young children, and in number comparison tasks with adults. When participants are asked to indicate whether a two-digit number is larger or smaller than 65, response times decrease as a function of the logarithm of the distance between the number and 65. Therefore, participants are faster at indicating that 61 is smaller than 65 than that 63 is smaller than 65. Importantly, they are also faster at indicating that 51 is smaller than 65 than that 59 is smaller than 65, a finding that would not be predicted if two-digit numbers were encoded entirely as ordered sequences of two symbols. Both 51 and 59 start with the tens digit 5, which is different from the tens digit of the comparison number 65. Therefore, if the comparison was based on the tens digits alone, no difference would be predicted in deciding that 51 is smaller than 65 than in deciding that 59 is smaller than 65 (as both comparisons would boil down to deciding that 5 is smaller than 6). The metaphor of an analog, compressed *number line* was proposed, with clearer distinctions at the low end than at the high end.

The verbal pathway and the number line pathway were part of the triple-code model Dehaene (1992) proposed for number processing. He argued that the meaning of numbers is encoded in three ways:

1. An auditory-verbal code, similar to the semantic representations of words.
2. A visual-Arabic code, in which numbers are manipulated in Arabic format on a spatially extended representational medium.
3. An analog-magnitude code, in which numerical quantities are represented as inherently variable distributions of activation over a compressed analogical number line.

Dehaene further proposed that the three codes interact with each other and are activated by different types of input, as shown in Fig. 1.4.

WHAT HAVE WE LEARNED ABOUT NUMBERS AND THEIR RELATION TO LANGUAGE SINCE?

Dehaene's (1992) article and the special journal issue, of which it was part, were a catalyst in number processing research. While research before had been scattered, now a sufficiently large group of scholars took up the topic and became organized by arranging symposia and workshops and by publishing special journal issues and edited handbooks (e.g., Campbell,

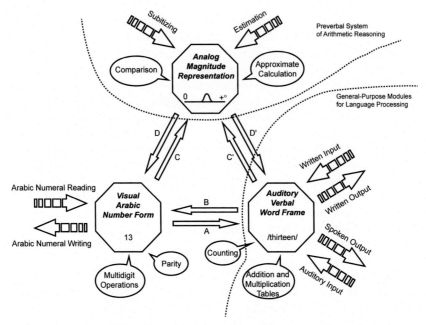

FIGURE 1.4   Dehaene's triple-code model of number processing. The three octagons represent the three codes that together form the meaning of numbers. For each code, the input and output and the main operations involving the code are given. *From Dehaene, S. (1992). Varieties of numerical abilities. Cognition, 44(1), 1–42.*

2005; Kadosh & Dowker, 2015). The number of paper submissions to journals grew so substantial that editors started to appoint dedicated action editors for the topic.

Unfortunately, the large number of publications has not (yet?) led to a flurry of "established findings" related to numerical and verbal performance. As it happens, I could find five. Four other topics are still in full debate. I discuss them successively.

THINGS WE HAVE LEARNED I: SMALL NUMBERS ARE EASIER TO PROCESS THAN LARGE NUMBERS

A consistent finding in number processing is that small numbers are easier to process than large numbers. One of the first robust findings was that people can easily discriminate between one, two, three, and sometimes four elements but require increasingly more time to discern five, six, seven,… elements. The fast perception of small numbers of elements is called *subitizing* (Kaufman, Lord, Reese, & Volkmann, 1949; Taves, 1941).

Small numbers are also easier to compare with each other than large numbers: People are faster to indicate that two is smaller than three than that eight is smaller than nine (Moyer & Landauer, 1967).

Finally, arithmetic operations are easier with small numbers than with large numbers (Knight & Behrens, 1928). The problem $2+3$ is easier to solve than $4+5$; the same is true for 2×3 versus 4×5. This problem size effect is present when the numbers are presented as Arabic digits and when they are presented as words (Noël, Fias, & Brysbaert, 1997).

THINGS WE HAVE LEARNED II: THE ANALOG-MAGNITUDE SYSTEM ACTIVATES A PART OF THE CORTEX THAT IS NOT INVOLVED IN LANGUAGE PROCESSING

As Fig. 1.5 shows, most brain regions of the left cerebral cortex are involved in language processing. Still, the areas that consistently light up when a task assesses number magnitude processing—the left and right intraparietal sulci—fall outside the zone, as can be seen in Fig. 1.6. Bueti and Walsh (2009) reviewed the literature indicating that this region is active not only in number processing but also in time and space understanding. Van Opstal and Verguts (2013), however, pointed to problems with this view of the intraparietal sulcus as a generalized magnitude system.

For the sake of completeness, it is important to keep in mind that the intraparietal sulci do not work in isolation but form part of larger networks. In particular, interactions with the lateral prefrontal cortex are important (Nieder & Dehaene, 2009).

THINGS WE HAVE LEARNED III: THERE IS A DIRECT ARABIC-VERBAL TRANSLATION ROUTE

An element from the triple-code model that elicited some controversy was whether it was possible to directly translate Arabic numbers into spoken (and written) numbers. An alternative model proposed by McCloskey, Caramazza, and Basili (1985) postulated that all numerical processing required mediation by the semantic system. Some evidence pointed in this direction. Fias, Reynvoet, and Brysbaert (2001), for instance, presented a digit and a number word on the same screen and asked the participants to name the word or the digit. The word and the digit either pointed to the same number (e.g., 6—six) or to different numbers (6—five). Fias et al. observed that the digit was named faster when the two stimuli referred to the same number than when they referred to different numbers. No such interference effect was observed for the naming of number words. In contrast, when the participants had to indicate whether the digit or the word was an odd or an even number, there was equivalent interference for both notations. On the basis of this pattern of results, Fias et al. concluded that digits were processed like pictures and could not be named via a nonsemantic, direct translation route.

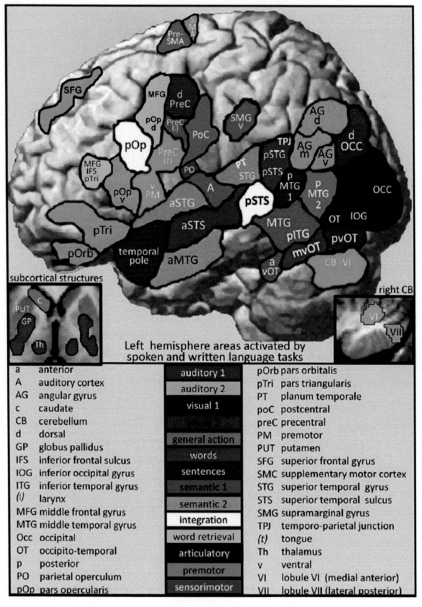

FIGURE 1.5   Brain areas of the left hemisphere active in language processing. *From Price, C. J. (2012). A review and synthesis of the first 20 years of PET and fMRI studies of heard speech, spoken language and reading. Neuroimage, 62(2), 816–847.*

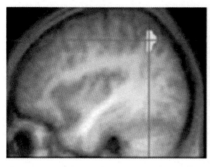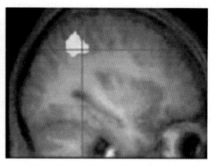

FIGURE 1.6   The intraparietal sulci (left and right) are active whenever number magnitude is addressed in a task. *From Piazza, M., Izard, V., Pinel, P., Le Bihan, D., & Dehaene, S. (2004). Tuning curves for approximate numerosity in the human intraparietal sulcus.* Neuron, 44(3), 547–555.

In recent years, however, several paradigms have shown that nonsemantic naming of Arabic numbers is possible (see also Roelofs, 2006, for evidence based on the interference effect used by Fias et al., 2001). One series of experiments made use of the semantic blocking paradigm (Herrera & Macizo, 2012). In this paradigm, five stimuli are presented over and over again to be named. Two different conditions are distinguished: A blocked condition in which the stimuli come from the same semantic category (e.g., five animals) and a mixed condition in which the stimuli come from different semantic categories (e.g., an animal, a body part, a piece of furniture, a vehicle, and a piece of clothing). The typical finding in this paradigm is that words are named faster in the blocked condition than in the mixed condition but that pictures are named *more slowly* in the blocked condition. The difference in naming cost is explained by assuming that words can be named directly, whereas pictures require semantic mediation to be named. In the blocked picture naming condition, the various concepts and names compete and hinder each other. The semantic blocking paradigm is ideal to test whether digits are named like words or pictures, as the alternative interpretations predict opposite effects. In a series of experiments, Herrera and Macizo (2012) showed that digits are named *faster* in a blocked condition than in a mixed condition, thus resembling words and deviating from pictures.

THINGS WE HAVE LEARNED IV: THERE ARE DIFFERENCES IN PROCESSING ARABIC NUMBERS AND VERBAL NUMBERS

Arabic and verbal numbers are not interchangeable, even not for numbers below 10 (it was traditionally thought that the Arabic notation was particularly efficient for multidigit numbers). An important finding is that calculations are more efficient when the problems are given in Arabic notation than in verbal notation. Therefore, "4 + 2" is solved faster than "four + two," even though naming times of digits and number words are the same (Clark & Campbell, 1991; Noël et al., 1997; see also Megías & Macizo, 2016, for other evidence that digits activate arithmetic information more strongly than words).

The advantage of digits over words is also true when the numbers are presented as part of word problems. Therefore, children are better at solving the visually presented problem "Manuel had 3 marbles and then Pedro gave him 5" than at solving the problem "Manuel had three marbles and then Pedro gave him five" (Orrantia, Múñez, San Romualdo, & Verschaffel, 2015). More in general, magnitude information is activated faster by Arabic numbers than by verbal numbers (Ford & Reynolds, 2016; Kadosh, Henik, & Rubinsten, 2008).

THINGS WE HAVE LEARNED V: INDIVIDUALS WITH DYSLEXIA HAVE POORER ARITHMETIC PERFORMANCE

Despite the differences between Arabic number processing and verbal number processing, people with reading difficulties are likely to experience mathematical deficits as well. For a start, there is a high comorbidity of dyslexia and dyscalculia. In a sample of 2586 primary school children, Landerl and Moll (2010) observed 181 children (7%) with a reading deficit, of whom 23% had an additional arithmetic deficit. Toffalini, Giofrè, and Cornoldi (2017) analyzed the data of 1049 children referred to psychologists for assessment of learning difficulties. Of these children, 308 (29%) had a specific reading difficulty, 147 (14%) a specific spelling problem, 93 (9%) a specific calculation deficit, and 501 (48%) a mixed deficit (not further specified). At present, it is not clear whether the comorbidity of dyslexia and dyscalculia is due to common underlying processes or to divergent processes that have joint risks of malfunctioning (e.g., due to genetic influences; Moll, Goebel, Gooch, Landerl, & Snowling, 2016).

Second, high-performing university students with dyslexia are slower at naming digits and at doing elementary arithmetic (Callens, Tops, & Brysbaert, 2012; De Smedt & Boets, 2010). The effect sizes are large (Cohen's $d \approx 1.0$), although not as large as those seen in word naming speed and spelling accuracy ($d \approx 2.0$; Callens et al., 2012). This again suggests an overlap of the processes involved in verbal and arithmetical skills. One element of overlap could be that the addition and multiplication tables are stored in verbal memory.

THINGS WE ARE STILL TRYING TO DECIDE I: WHAT IS THE NATURE OF THE NUMBER QUANTITY SYSTEM?

Dehaene (1992) put forward a few strong hypotheses about the number quantity system. The first was that it was an analog system, based on a combination of summation and place coding (for computational implementations, see Dehaene & Changeux, 1993; Verguts & Fias, 2004). The activations of the various elements in the input were summed and then translated into activation patterns on an ordered and compressed (e.g., logarithmic) "number line."

The second element was that the number quantity system was not really a magnitude system but an abstract number system (ANS), based on modality-independent, discrete amounts. Dehaene used the term "numerosity" to refer to the number of elements perceived, rather than to the summed magnitude (mass, density, surface,...) of the elements. When all items to be counted are of the same size, magnitude and numerosity are perfectly correlated. However, this is no longer the case if the items differ in magnitude: Two big items can have a bigger mass than three small elements. The ANS was supposed to respond to the discrete number of elements and not to the continuous magnitude correlates (mass, density, surface covered).

The third element proposed by Dehaene was that the number line was oriented along the reading direction. Therefore, for languages read from left to right, the small numbers were located on the left side of the number line and the large numbers on the right side.

It is fair to say that all three hypotheses are still heavily contested. First, the compressed nature of the number magnitude system has been questioned. Other explanations are: more noise for large numbers than for small numbers (Gallistel & Gelman, 1992), differences in frequency of occurrence between numbers (Piantadosi, 2016), and asymmetries because of the task rather than the nature of the number line (Cohen & Quinlan, 2016; Verguts, Fias, & Stevens, 2005).

Second, the idea of the ANS being unresponsive to magnitude differences between the discrete elements has been questioned as well, given that in real life there are virtually no situations in which numerosity and mass are uncorrelated (Cantrell & Smith, 2013; Gebuis, Kadosh, & Gevers, 2016; Reynvoet & Sasanguie, 2016). Some authors have proposed that a discrete number system may have evolved next to a continuous magnitude system (Leibovich, Vogel, Henik, & Ansari, 2015).

Still related to the issue of the true nature of the ANS system, other authors have argued that the system may be order related rather than (or in addition to) numerosity related (Berteletti, Lucangeli, & Zorzi, 2012; Goffin & Ansari, 2016; Merkley, Shimi, & Scerif, 2016; Van Opstal, Gevers,

De Moor, & Verguts, 2008). Just like there is a high correlation between numerosity and magnitude, there is a high correlation between numerosity and order. The main difference is that order applies to more stimuli than to numbers.

Finally, the left–right orientation of the number line has been questioned as well, based on the finding that the orientation is mostly observed when numbers must be kept in working memory, leading to the proposal that the orientation is limited to numbers held in working memory (van Dijck, Abrahamse, Acar, Ketels, & Fias, 2014; but see Huber, Klein, Moeller, & Willmes, 2016). There is also some evidence that the spatial-numerical association of response codes effect may not be reversed in people with a language read from right to left (Zohar-Shai, Tzelgov, Karni, & Rubinsten, 2017).

THINGS WE ARE STILL TRYING TO DECIDE II: HOW DOES KNOWLEDGE OF NUMBER SYMBOLS AFFECT/ SHARPEN THE NUMBER MAGNITUDE SYSTEM?

Given that there are differences because of number notation, a straightforward question is to what extent the use of number symbols alters the meaning of numerosities. To what extent do the semantic representations of educated human adults differ from those of preverbal children and animals? Some authors have suggested that the use of symbols makes the number line linear rather than compressed (Siegler & Opfer, 2003), but others have doubted the empirical evidence for this claim (Huber, Moeller, & Nuerk, 2014). Others have argued that number symbols make the semantic representations sharper so that there are less confusions between numerosities (Verguts & Fias, 2004). Still others have proposed that symbolic numbers form a separate type of representations, as indicated earlier (Leibovich et al., 2015; Sasanguie, De Smedt, & Reynvoet, 2017).

THINGS WE ARE STILL TRYING TO DECIDE III: WHAT IS THE RELATIVE IMPORTANCE OF THE ANS TO MATHEMATICAL PERFORMANCE?

A third topic of discussion is to what extent the approximate number system contributes to mathematical achievement. Dehaene saw the ANS as the core of number knowledge from which all other number-related information emerged. A similar view was defended by Butterworth (2005; see also Landerl, Bevan, & Butterworth, 2004), and some authors found evidence in line with this hypothesis (Schleepen, Van Mier, & De Smedt, 2016; Zhang, Chen, Liu, Cui, & Zhou, 2016). Others, however, failed to find evidence (Cipora & Nuerk, 2013; Geary & Vanmarle, 2016) or found a stronger effect for symbolic comparison rather than nonsymbolic magnitude comparison (Fazio, Bailey, Thompson, & Siegler, 2014;

Honoré & Noël, 2016; Vanbinst, Ansari, Ghesquière, & De Smedt, 2016; Vanbinst, Ghesquière, & De Smedt, 2012).

All in all, it seems unlikely that the ANS is strongly related to mathematical achievements in healthy participants. A remaining possibility is that ANS malfunctioning is rare but with grave consequences so that people with a deficient ANS have severe dyscalculia but are too rare to influence correlations in large-scale population studies.

THINGS WE ARE STILL TRYING TO DECIDE IV: DOES LANGUAGE HAVE AN EFFECT ON HOW MATHEMATICAL OPERATIONS ARE PERFORMED?

A basic question about cognitive performance is to what extent language affects thought (also known as linguistic relativity or the Whorfian hypothesis). Is it possible to think without language and is thinking different in languages that carve reality in dissimilar ways? Importantly, the language differences should point to fundamental differences in processing, not simply to differences in strategies to cope with the language difference. For instance, Brysbaert, Fias, and Noël (1998) reported that Dutch-speaking participants are faster to name the solution of problem 4 + 21 than of the problem 21 + 4, whereas French-speaking participants show the opposite effect, in line with the observation that two-digit numbers in Dutch but not in French are pronounced in the reverse way: five and twenty instead of twenty-five. Crucially, the language difference disappeared when both groups of participants were asked to type the answers. With this task, both Dutch-speaking and French-speaking participants were faster to solve 21 + 4 than 4 + 21, leading Brysbaert et al. (1998) to conclude that the language difference in arithmetic was not a true Whorfian effect.

An argument sometimes used against the idea that language shapes thought is the observation that aphasic people are not obviously deficient in their thinking (e.g., Siegal, Varley, & Want, 2001). This rules out a strong version of the linguistic relativity (thought is impossible without language) but not necessarily a weaker version (language affects thought). Indeed, there is evidence that some nonverbal functions such as picture categorization are hindered in people with aphasia (Lupyan & Mirman, 2013), in line with a weak version of linguistic relativity (Lupyan, 2015; Wolff & Holmes, 2011).

Several researchers have taken issue with the initial negative evidence for Whorfian effects in number processing. For instance, Colomé, Laka, and Sebastián-Gallés (2010) reported that Basque speakers solve problems such as 20 + 15 faster than Italian or Catalan speakers, both when the solution had to be named and typed, in line with the observation that the Basque language combines multiples of 20 in its number naming system (e.g., 35 is said as "twenty and fifteen"). A language effect between Basque and Spanish was also reported by Salillas and Carreiras (2014).

Pixner, Moeller, Hermanova, Nuerk, and Kaufmann (2011) showed that participants find it harder to decide that 47 is smaller than 62 than that 42 is smaller than 57 because in the former case there is an incongruity between the response required (47 < 62) and the response elicited by the units (7 > 2). Critically, the incongruity effect was larger in German, which names the units before the tens (seven and forty), than in Italian or Czech, which put the tens before the units (forty-seven). Moeller, Shaki, Göbel, and Nuerk (2015) successfully replicated the effect.

A related topic is how bilinguals with contradicting number names cope. It is well documented that bilinguals continue to do arithmetic in the language they used in school, leading to confusions when the school language changes or when people later have a different dominant language (Prior, Katz, Mahajna, & Rubinsten, 2015; Van Rinsveld, Brunner, Landerl, Schiltz, & Ugen, 2015).

NEW THINGS LANGUAGE RESEARCHERS HAVE TO OFFER I: SEMANTIC VECTORS

Research on the commonalities and the differences between language and numerical cognition can take inspiration from new developments on the other side. In the remainder of this Chapter 1 two new developments in language research are mentioned, which may be of interest to colleagues working in numerical cognition.

A first interesting development in language research is that the meaning of a target word can be approximated by studying the words surrounding the target word (Landauer & Dumais, 1997; Lazaridou, Marelli, & Baroni, 2017; Mandera, Keuleers, & Brysbaert, 2017; Sadeghi, McClelland, & Hoffman, 2015). A successful way of doing so is to use a three-layer connectionist network (Mikolov, Sutskever, Chen, Corrado, & Dean, 2013, pp. 3111–3119). The input layer consists of some 100 thousand nodes representing the words found in a billion-word corpus. The output layer consists of the same nodes. In between, there is a hidden layer with 200–300 nodes. The network is trained as follows. A corpus is read word by word. For each target word activated in the output layer, a few words before the target word and a few words after the target word in the corpus are activated at the input layer. The weights are changed so that the prediction of the target word improves, given the surrounding words in the input layer. At the end of the training, the activity vector in the hidden layer activated by a target word represents the semantic representation of that word (see Fig. 1.7).

Such semantic vectors are quite good predictors of semantic distance judgments, semantic priming data, and synonym judgment (Mandera et al., 2017). Fig. 1.8 shows the semantic distances between the number

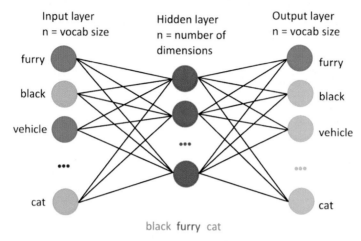

FIGURE 1.7 Network to calculate semantic vectors. A network is trained to predict target words on the basis of the surrounding words. In this example, the weights are changed so that the activity of the word node "furry" in the output layer increases when the words "black" and "cat" are activated in the input layer. The activity patterns in the hidden layer at the end of the training form the words' semantic vectors. *From Mandera, P., Keuleers, E., & Brysbaert, M. (2017). Explaining human performance in psycholinguistic tasks with models of semantic similarity based on prediction and counting: A review and empirical validation.* Journal of Memory and Language, 92, 57–78.

	One	two	three	four	five	six	seven	eight	nine	ten
zero	0.66	0.69	0.70	0.66	0.59	0.65	0.63	0.65	0.59	0.66
one		0.35	0.40	0.42	0.48	0.50	0.51	0.51	0.52	0.54
two			0.13	0.18	0.27	0.28	0.33	0.33	0.39	0.38
three				0.11	0.24	0.23	0.29	0.27	0.35	0.36
four					0.20	0.20	0.22	0.21	0.29	0.31
five						0.23	0.23	0.20	0.25	0.15
six							0.20	0.15	0.22	0.29
seven								0.14	0.17	0.29
eight									0.13	0.24
nine										0.29

FIGURE 1.8 Semantic distances between number words based on semantic vectors (0.00 = no distance, 1.00 = maximal distance). The font size stresses the semantic similarity. This shows a distance-related similarity effect: Numbers close in value have a larger similarity than numbers farther away. The number zero is not much related to any other number. *Mandera, P., Keuleers, E., & Brysbaert, M. (2017). From Explaining human performance in psycholinguistic tasks with models of semantic similarity based on prediction and counting: A review and empirical validation.* Journal of Memory and Language, 92, 57–78.

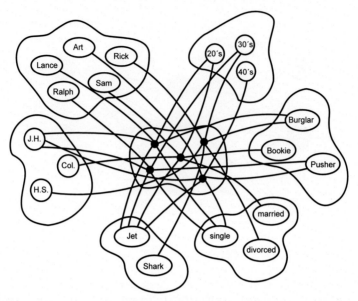

FIGURE 1.9   Hub-and-spoke model to represent information about people in an imaginary world. This model represents information about five people (Lance, Art, Rick, Sam, Ralph) who have different ages (20, 30, 40), have different occupations (burglar, bookie, pusher), have different marital status (married, single, divorced), belong to different groups (jet, shark), and have finished different education levels (junior high school, high school, college). Important in the model is the use of central person nodes (the hub), linking the various aspects belonging to an individual (spokes). *From McClelland, J. L. (1981). Retrieving general and specific information from stored knowledge of specifics. In:* Proceedings of the third annual meeting of the cognitive science society.

words zero to ten (in English). This figure shows that numbers close in value resemble each other more than numbers further apart, simply because they more often co-occur in texts. It will be interesting to examine to what extent the semantic vectors agree with the distance-related effects reported in number comparison and number priming (see also Krajcsi, Lengyel, & Kojouharova, 2016). Another aspect shown in Fig. 1.8 is that the number zero is rather unrelated to the other numbers, in line with the finding that the position of the number zero on the number line is uncertain (Brysbaert, 1995; Pinhas & Tzelgov, 2012).

NEW THINGS LANGUAGE RESEARCHERS HAVE TO OFFER II: THE HUB-AND- SPOKE MODEL OF SEMANTIC REPRESENTATION

Another interesting idea is one originally proposed by McClelland (1981), which has been applied several times in his simulation work (e.g., Rogers et al., 2004). Fig. 1.9 illustrates the idea. Whenever various characteristics of a stimulus must be combined, it makes sense to represent the stimulus as a central, amodal node (hub) in a network connected to modal feature nodes within separate layers (spokes). Such a model successfully predicted the neuroscientific finding that the meaning of stimuli can involve brain areas as diverse as the visual cortex and the motor cortex. It also successfully accounts for the progression of meaning loss in semantic dementia (Rogers et al., 2004). A hypothesis is that the hub of the system is situated in the anterior temporal lobes (Ralph, Jefferies, Patterson, & Rogers, 2017).

It is not difficult to reformulate Dehaene's (1992) triple-code model into a hub-and-spoke model. Rather than having the mutual interactions between the three codes, each code would send activation to a central hub of amodal nodes representing the various numbers. In this way, various types of information can be integrated, including new information (such as that coming from semantic vectors and historical facts). As a matter of fact, one of the first models proposed for number representations— the encoding-complex model—came very close to such a hub-and-spoke model. The encoding-complex model (Campbell & Clark, 1988) stated that numerals become associated with a variety of numerical functions (number reading or transcoding, number comparison, estimation, arithmetic facts,...), which interact with each other and are activated to various extents depending on the task to be performed. One of the factors that have hindered acceptance of the model was that it seemed difficult to implement. The hub-and-spoke model may be a way forward.

CONCLUSIONS

In this chapter, I have reviewed the developments of the past 25 years in our knowledge about number processing and its relationship to language proficiency. First, I showed that arithmetic in intelligence tests initially was seen as part of the verbal scale but later became part of a different (though correlated) factor. Then, I discussed the cognitive models from the 1990s, which focused on the differences between numerical and verbal performance. I argue that research on this topic has led to five established findings and four issues that are still hotly debated. Finally, I presented two new findings from psycholinguistics, which may be of interest to researchers on number processing.

References

Berteletti, I., Lucangeli, D., & Zorzi, M. (2012). Representation of numerical and non-numerical order in children. *Cognition, 124*(3), 304–313.

Binet, A., & Simon, T. (1907). Le développement de l'intelligence chez les enfants [The development of intelligence in children]. *L'Année Psychologique, 14*, 1–94.

Brysbaert, M. (1995). Arabic number reading: On the nature of the numerical scale and the origin of phonological recoding. *Journal of Experimental Psychology: General, 124*, 434–452.

Brysbaert, M., Fias, W., & Noël, M. P. (1998). The Whorfian hypothesis and numerical cognition: Is "twenty-four" processed in the same way as "four-and-twenty"? *Cognition, 66*, 51–77.

Bueti, D., & Walsh, V. (2009). The parietal cortex and the representation of time, space, number and other magnitudes. *Philosophical Transactions of the Royal Society of London B: Biological Sciences, 364*(1525), 1831–1840.

Butterworth, B. (2005). The development of arithmetical abilities. *Journal of Child Psychology and Psychiatry, 46*(1), 3–18.

Calderón-Tena, C. O., & Caterino, L. C. (2016). Mathematics learning development: The role of long-term retrieval. *International Journal of Science and Mathematics Education, 14*(7), 1377–1385.

Callens, M., Tops, W., & Brysbaert, M. (2012). Cognitive profile of students who enter higher education with an indication of dyslexia. *PLoS One, 7*(6), e38081. https://doi.org/10.1371/journal.pone.0038081.

Campbell, J. I. D. (Ed.). (2005). *The handbook of mathematical cognition.* Hove: Psychology Press.

Campbell, J. I. D., & Clark, J. M. (1988). An encoding-complex view of cognitive number processing: Comment on McCloskey, Sokol, and Goodman (1986). *Journal of Experimental Psychology: General, 117*(2), 204–214.

Cantrell, L., & Smith, L. B. (2013). Open questions and a proposal: A critical review of the evidence on infant numerical abilities. *Cognition, 128*(3), 331–352.

Cipora, K., & Nuerk, H. C. (2013). Is the SNARC effect related to the level of mathematics? No systematic relationship observed despite more power, more repetitions, and more direct assessment of arithmetic skill. *The Quarterly Journal of Experimental Psychology, 66*(10), 1974–1991.

Clark, J. M., & Campbell, J. I. D. (1991). Integrated versus modular theories of number skills and acalculia. *Brain and Cognition, 17*(2), 204–239.

Cohen, D. J., & Quinlan, P. T. (2016). How numbers mean: Comparing random walk models of numerical cognition varying both encoding processes and underlying quantity representations. *Cognitive Psychology, 91*, 63–81.

Colomé, À., Laka, I., & Sebastián-Gallés, N. (2010). Language effects in addition: How you say it counts. *The Quarterly Journal of Experimental Psychology, 63*(5), 965–983.

De Smedt, B., & Boets, B. (2010). Phonological processing and arithmetic fact retrieval: Evidence from developmental dyslexia. *Neuropsychologia, 48*(14), 3973–3981.

Dehaene, S. (1992). Varieties of numerical abilities. *Cognition, 44*(1), 1–42.

Dehaene, S., & Changeux, J. P. (1993). Development of elementary numerical abilities—A neuronal model. *Journal of Cognitive Neuroscience, 5*, 390–407.

Dehaene, S., Dehaene-Lambertz, G., & Cohen, L. (1998). Abstract representations of numbers in the animal and human brain. *Trends in Neurosciences, 21*(8), 355–361.

Fazio, L. K., Bailey, D. H., Thompson, C. A., & Siegler, R. S. (2014). Relations of different types of numerical magnitude representations to each other and to mathematics achievement. *Journal of Experimental Child Psychology, 123*, 53–72.

Fias, W., Reynvoet, B., & Brysbaert, M. (2001). Are Arabic numerals processed as pictures in a Stroop interference task? *Psychological Research, 65*, 242–249.

Ford, N., & Reynolds, M. G. (2016). Do Arabic numerals activate magnitude automatically? Evidence from the psychological refractory period paradigm. *Psychonomic Bulletin and Review, 23*(5), 1528–1533.

Gallistel, C. R., & Gelman, R. (1992). Preverbal and verbal counting and computation. *Cognition, 44*(1), 43–74.

Geary, D. C., & Vanmarle, K. (2016). Young Children's core symbolic and nonsymbolic quantitative knowledge in the prediction of later mathematics achievement. *Developmental Psychology, 52*(12), 2130–2144.

Gebuis, T., Kadosh, R. C., & Gevers, W. (2016). Sensory-integration system rather than approximate number system underlies numerosity processing: A critical review. *Acta Psychologica, 171*, 17–35.

Goffin, C., & Ansari, D. (2016). Beyond magnitude: Judging ordinality of symbolic number is unrelated to magnitude comparison and independently relates to individual differences in arithmetic. *Cognition, 150*, 68–76.

Herrera, A., & Macizo, P. (2012). Semantic processing in the production of numerals across notations. *Journal of Experimental Psychology: Learning, Memory, and Cognition, 38*(1), 40–51.

Honoré, N., & Noël, M. P. (2016). Improving preschoolers' arithmetic through number magnitude training: The impact of non-symbolic and symbolic training. *PLoS One, 11*(11), e0166685.

Huber, S., Klein, E., Moeller, K., & Willmes, K. (2016). Spatial–numerical and ordinal positional associations coexist in parallel. *Frontiers in Psychology, 7*, 438. https://doi.org/10.3389/fpsyg.2016.00438.

Huber, S., Moeller, K., & Nuerk, H. C. (2014). Dissociating number line estimations from underlying numerical representations. *The Quarterly Journal of Experimental Psychology, 67*(5), 991–1003.

Jewsbury, P. A., Bowden, S. C., & Strauss, M. E. (2016). Integrating the switching, inhibition, and updating model of executive function with the Cattell—Horn—Carroll model. *Journal of Experimental Psychology: General, 145*(2), 220–245.

Kadosh, C. R., & Dowker, A. (Eds.). (2015). *The Oxford handbook of numerical cognition*. Oxford: Oxford University Press.

Kadosh, R. C., Henik, A., & Rubinsten, O. (2008). Are Arabic and verbal numbers processed in different ways? *Journal of Experimental Psychology: Learning, Memory, and Cognition, 34*(6), 1377–1391.

Kaufman, E. L., Lord, M. W., Reese, T. W., & Volkmann, J. (1949). The discrimination of visual number. *The American Journal of Psychology, 62*(4), 498–525.

Keith, T. Z., Fine, J. G., Taub, G. E., Reynolds, M. R., & Kranzler, J. H. (2006). Higher order, multisample, confirmatory factor analysis of the Wechsler Intelligence Scale for Children – fourth edition: What does it measure? *School Psychology Review, 35*(1), 108–127.

Keith, T. Z., Low, J. A., Reynolds, M. R., Patel, P. G., & Ridley, K. P. (2010). Higher-order factor structure of the differential ability scales–II: Consistency across ages 4 to 17. *Psychology in the Schools, 47*(7), 676–697.

Keith, T. Z., & Witta, E. L. (1997). Hierarchical and cross-age confirmatory factor analysis of the WISC-III: What does it measure? *School Psychology Quarterly, 12*(2), 89–107.

Knight, F. B., & Behrens, M. (1928). *The learning of the 100 addition combinations and the 100 subtraction combinations*. New York: Longmans.

Krajcsi, A., Lengyel, G., & Kojouharova, P. (2016). The source of the symbolic numerical distance and size effects. *Frontiers in Psychology, 7*, 1795.

Landauer, T. K., & Dumais, S. T. (1997). A solution to Plato's problem: The latent semantic analysis theory of acquisition, induction, and representation of knowledge. *Psychological Review, 104*(2), 211–240.

Landerl, K., Bevan, A., & Butterworth, B. (2004). Developmental dyscalculia and basic numerical capacities: A study of 8–9-year-old students. *Cognition, 93*(2), 99–125.

Landerl, K., & Moll, K. (2010). Comorbidity of learning disorders: Prevalence and familial transmission. *Journal of Child Psychology and Psychiatry, 51*(3), 287–294.

Lazaridou, A., Marelli, M., & Baroni, M. (2017). Multimodal word meaning induction from minimal exposure to natural text. *Cognitive Science.* https://doi.org/10.1111/cogs.12481.

Leibovich, T., Vogel, S. E., Henik, A., & Ansari, D. (2015). Asymmetric processing of numerical and nonnumerical magnitudes in the brain: An fMRI study. *Journal of Cognitive Neuroscience, 28*, 166–176.

Lupyan, G. (2015). The centrality of language in human cognition. *Language Learning, 66*(3), 516–553.

Lupyan, G., & Mirman, D. (2013). Linking language and categorization: Evidence from aphasia. *Cortex, 49*(5), 1187–1194.

Mandera, P., Keuleers, E., & Brysbaert, M. (2017). Explaining human performance in psycholinguistic tasks with models of semantic similarity based on prediction and counting: A review and empirical validation. *Journal of Memory and Language, 92*, 57–78.

McClelland, J. L. (1981). Retrieving general and specific information from stored knowledge of specifics. In *Proceedings of the third annual meeting of the cognitive science society.*

McCloskey, M., Caramazza, A., & Basili, A. (1985). Cognitive mechanisms in number processing and calculation: Evidence from dyscalculia. *Brain and Cognition, 4*(2), 171–196.

McGrew, K. S. (2009). CHC theory and the human cognitive abilities project: Standing on the shoulders of the giants of psychometric intelligence research. *Intelligence, 37*(1), 1–10.

Megías, P., & Macizo, P. (2016). Activation and selection of arithmetic facts: The role of numerical format. *Memory and Cognition, 44*(2), 350–364.

Merkley, R., Shimi, A., & Scerif, G. (2016). Electrophysiological markers of newly acquired symbolic numerical representations: The role of magnitude and ordinal information. *ZDM, 48*(3), 279–289.

Mikolov, T., Sutskever, I., Chen, K., Corrado, G. S., & Dean, J. (2013). Distributed representations of words and phrases and their compositionality. In *Advances in neural information processing systems.*

Moeller, K., Shaki, S., Göbel, S. M., & Nuerk, H. C. (2015). Language influences number processing–a quadrilingual study. *Cognition, 136*, 150–155.

Moll, K., Göbel, S. M., Gooch, D., Landerl, K., & Snowling, M. J. (2016). Cognitive risk factors for specific learning disorder: Processing speed, temporal processing, and working memory. *Journal of Learning Disabilities, 49*(3), 272–281.

Moyer, R. S., & Landauer, T. K. (1967). Time required for judgements of numerical inequality. *Nature, 215*(5109), 1519–1520.

Nieder, A., & Dehaene, S. (2009). Representation of number in the brain. *Annual Review of Neuroscience, 32*, 185–208.

Noël, M. P., Fias, W., & Brysbaert, M. (1997). About the influence of the presentation format on arithmetical-fact retrieval processes. *Cognition, 63*, 335–374.

Orrantia, J., Múñez, D., San Romualdo, S., & Verschaffel, L. (2015). Effects of numerical surface form in arithmetic word problems. *Psicológica, 36*(2), 265–281.

Piantadosi, S. T. (2016). A rational analysis of the approximate number system. *Psychonomic Bulletin and Review, 23*(3), 877–886.

Piazza, M., Izard, V., Pinel, P., Le Bihan, D., & Dehaene, S. (2004). Tuning curves for approximate numerosity in the human intraparietal sulcus. *Neuron, 44*(3), 547–555.

Pinhas, M., & Tzelgov, J. (2012). Expanding on the mental number line: Zero is perceived as the "smallest". *Journal of Experimental Psychology: Learning, Memory, and Cognition, 38*(5), 1187–1205.

Pixner, S., Moeller, K., Hermanova, V., Nuerk, H. C., & Kaufmann, L. (2011). Whorf reloaded: language effects on nonverbal number processing in first grade—A trilingual study. *Journal of Experimental Child Psychology, 108*(2), 371–382.

Price, C. J. (2012). A review and synthesis of the first 20 years of PET and fMRI studies of heard speech, spoken language and reading. *Neuroimage, 62*(2), 816–847.

Primi, R., Ferrão, M. E., & Almeida, L. S. (2010). Fluid intelligence as a predictor of learning: A longitudinal multilevel approach applied to math. *Learning and Individual Differences, 20*(5), 446–451.

Prior, A., Katz, M., Mahajna, I., & Rubinsten, O. (2015). Number word structure in first and second language influences arithmetic skills. *Frontiers in Psychology, 6*, 266.

Ralph, M. L., Jefferies, E., Patterson, K., & Rogers, T. T. (2017). The neural and computational bases of semantic cognition. *Nature Reviews Neuroscience, 18*, 42–55.

Reynolds, M. R., Keith, T. Z., Flanagan, D. P., & Alfonso, V. C. (2013). A cross-battery, reference variable, confirmatory factor analytic investigation of the CHC taxonomy. *Journal of School Psychology, 51*(4), 535–555.

Reynvoet, B., & Sasanguie, D. (2016). The symbol grounding problem revisited: A thorough evaluation of the ANS mapping account and the proposal of an alternative account based on symbol–symbol associations. *Frontiers in Psychology, 7*, 1581. https://doi.org/10.3389/fpsyg.2016.01581.

Roelofs, A. (2006). Functional architecture of naming dice, digits, and number words. *Language and Cognitive Processes, 21*(1–3), 78–111.

Rogers, T. T., Lambon Ralph, M. A., Garrard, P., Bozeat, S., McClelland, J. L., Hodges, J. R., et al. (2004). Structure and deterioration of semantic memory: A neuropsychological and computational investigation. *Psychological Review, 111*(1), 205–235.

Sadeghi, Z., McClelland, J. L., & Hoffman, P. (2015). You shall know an object by the company it keeps: An investigation of semantic representations derived from object co-occurrence in visual scenes. *Neuropsychologia, 76*, 52–61.

Salillas, E., & Carreiras, M. (2014). Core number representations are shaped by language. *Cortex, 52*, 1–11.

Sasanguie, D., De Smedt, B., & Reynvoet, B. (2017). Evidence for distinct magnitude systems for symbolic and non-symbolic number. *Psychological Research, 81*(1), 231–242.

Schleepen, T. M., Van Mier, H. I., & De Smedt, B. (2016). The contribution of numerical magnitude comparison and phonological processing to individual differences in fourth Graders' multiplication fact ability. *PLoS One, 11*(6), e0158335.

Seashore, H., Wesman, A., & Doppelt, J. (1950). The standardization of the Wechsler intelligence scale for children. *Journal of Consulting Psychology, 14*(2), 99–110.

Siegal, M., Varley, R., & Want, S. C. (2001). Mind over grammar: Reasoning in aphasia and development. *Trends in Cognitive Sciences, 5*(7), 296–301.

Siegler, R. S., & Opfer, J. (2003). The developmental numerical estimation: Evidence for multiple representation of mental quantity. *Psychological Science, 14*, 237–243.

Taub, G. E., Keith, T. Z., Floyd, R. G., & McGrew, K. S. (2008). Effects of general and broad cognitive abilities on mathematics achievement. *School Psychology Quarterly, 23*(2), 187–198.

Taves, E. H. (1941). Two mechanisms for the perception of visual numerousness. *Archives of Psychology, 265,* 47.

Toffalini, E., Giofrè, D., & Cornoldi, C. (2017). Strengths and weaknesses in the intellectual profile of different subtypes of specific learning disorder: A study on 1,049 diagnosed children. *Clinical Psychological Science.* Electronic preprint https://doi.org/10.1177/2167702616672038.

van Dijck, J. P., Abrahamse, E. L., Acar, F., Ketels, B., & Fias, W. (2014). A working memory account of the interaction between numbers and spatial attention. *The Quarterly Journal of Experimental Psychology, 67*(8), 1500–1513.

Van Opstal, F., Gevers, W., De Moor, W., & Verguts, T. (2008). Dissecting the symbolic distance effect: Comparison and priming effects in numerical and nonnumerical orders. *Psychonomic Bulletin and Review, 15*(2), 419–425.

Van Opstal, F., & Verguts, T. (2013). Is there a generalized magnitude system in the brain? Behavioral, neuroimaging, and computational evidence. *Frontiers in Psychology, 4,* 435.

Van Rinsveld, A., Brunner, M., Landerl, K., Schiltz, C., & Ugen, S. (2015). The relation between language and arithmetic in bilinguals: Insights from different stages of language acquisition. *Frontiers in Psychology, 6,* 265.

Vanbinst, K., Ansari, D., Ghesquière, P., & De Smedt, B. (2016). Symbolic numerical magnitude processing is as important to arithmetic as phonological awareness is to reading. *PLoS One, 11*(3), e0151045.

Vanbinst, K., Ghesquière, P., & De Smedt, B. (2012). Numerical magnitude representations and individual differences in children's arithmetic strategy use. *Mind, Brain, and Education, 6*(3), 129–136.

Verguts, T., & Fias, W. (2004). Representation of number in animals and humans: A neural model. *Journal of Cognitive Neuroscience, 16*(9), 1493–1504.

Verguts, T., Fias, W., & Stevens, M. (2005). A model of exact small-number representation. *Psychonomic Bulletin and Review, 12*(1), 66–80.

Wolff, P., & Holmes, K. J. (2011). Linguistic relativity. *Wiley Interdisciplinary Reviews: Cognitive Science, 2*(3), 253–265.

Zhang, Y., Chen, C., Liu, H., Cui, J., & Zhou, X. (2016). Both non-symbolic and symbolic quantity processing are important for arithmetical computation but not for mathematical reasoning. *Journal of Cognitive Psychology, 28*(7), 807–824.

Zohar-Shai, B., Tzelgov, J., Karni, A., & Rubinsten, O. (2017). It does exist! A left-to-right spatial–numerical association of response codes (SNARC) effect among native Hebrew speakers. *Journal of Experimental Psychology: Human Perception and Performance, 43*(4), 719–728.

The Interplay Between Learning Arithmetic and Learning to Read: Insights From Developmental Cognitive Neuroscience

Jérôme Prado

Institut des Sciences Cognitives Marc Jeannerod – UMR 5304,
Centre National de la Recherche Scientifique (CNRS) &
Université de Lyon, Bron, France

OUTLINE

Heterogeneity of Function in Numerical Cognition
https://doi.org/10.1016/B978-0-12-811529-9.00002-9

27

INTRODUCTION

Over the past two decades or so, studies from both psychology and cognitive neuroscience have converged on the view that mathematical and linguistic abilities are largely separated. For example, behavioral studies have found that preverbal infants (Cordes & Brannon, 2008), as well as individuals with very limited mathematical language (Pica, Lemer, Izard, & Dehaene, 2004), can intuitively process numerical quantities when these are presented in a nonsymbolic format (e.g., dot patterns). Because these intuitions appear to be related to the acquisition of formal mathematical skills later on (Feigenson, Libertus, & Halberda, 2013), it is increasingly believed that these core numerical skills—rather than linguistic skills— provide the foundation for the emergence of abstract mathematical concepts (Dehaene & Cohen, 2007). In line with this idea, neuroimaging studies in adults indicate that many numerical tasks consistently recruit a brain system that is largely dissociated from regions involved in linguistic computations (Nieder & Dehaene, 2009). This system, notably, includes the intraparietal sulcus (IPS), a region that is believed to house a representation of numerical quantities (Nieder & Dehaene, 2009).

Therefore, one would think that the acquisition of mathematical skills in children should be fairly independent from the acquisition of linguistic skills. Yet, this does not appear to be the case, at least as far as two of the most foundational mathematical and linguistic skills are concerned: learning arithmetic and learning to read. Indeed, two relatively independent lines of evidence argue in favor of a relationship between the acquisition of these skills. The first line of evidence comes from correlational studies. These studies show that there is a correlation between arithmetic and reading abilities across children (Durand, Hulme, Larkin, & Snowling, 2005; Hart, Petrill, Thompson, & Plomin, 2009; Hecht, Torgesen, Wagner, & Rashotte, 2001). For example, in a study conducted on 162 children from 7 to 10 years of age, correlations between arithmetic skills and single-word reading abilities ranged from .50 to .60 (Durand et al., 2005). Conversely, studies have also found that mathematical abilities in children can predict later reading outcomes, sometimes even better than early reading skills (Duncan et al., 2007; Lerkkanen, Rasku-Puttonen, Aunola, & Nurmi, 2005).

Therefore, correlational studies point to a relatively clear relationship between the acquisition of arithmetic and reading skills in children.

The second line of evidence that suggests a relationship between learning math and learning to read comes from the study of learning difficulties. Some children can show persistent difficulties acquiring basic arithmetic skills, even though they may have average intelligence and benefit from adequate schooling (Kaufmann & von Aster, 2012). This condition, called developmental dyscalculia or math learning disability, affects from 5% to 10% of children worldwide (prevalence varies depending on diagnostic criteria) (Kaufmann & von Aster, 2012). Although such prevalence rates are similar to those observed with children who show persistent reading difficulties (i.e., developmental dyslexia), dyscalculia is much less researched than dyslexia. Yet, both disabilities may be related because estimates of the prevalence of one disability given the other (i.e., comorbidity rate) are estimated at around 40% (Wilson et al., 2015). Therefore, many children who show persistent struggles with learning math also show persistent struggles with learning to read.

Overall, both (1) the correlation between arithmetic and reading scores and (2) the comorbidity between dyscalculia and dyslexia suggest some overlap between the acquisition of arithmetic and reading skills in children. At first glance, this appears to somewhat conflict with the aforementioned evidence that the neural mechanisms supporting mathematical skills in adults are largely independent from those supporting linguistic skills. Against this background, the goal of this chapter is to review some recent developmental cognitive neuroscience findings that are relevant for understanding the observed relationship between the acquisition of arithmetic and reading skills in children. First, we will evaluate whether developmental neuroimaging studies support what is perhaps the most popular explanation of the link between arithmetic and reading skills: the idea that arithmetic learning in children involves phonological processing mechanisms also involved in reading (Ashkenazi, Black, Abrams, Hoeft, & Menon, 2013). Second, we will review the more recent proposal that the relationship between arithmetic and reading skills may also be explained by the fact that both rely on procedural memory and more specifically on the ability to automatize procedural knowledge.

ARITHMETIC LEARNING AND VERBAL-PHONOLOGICAL PROCESSING

Acquiring the ability to identify and manipulate the sound structure (i.e., the phonology) of a word is a critical component of learning to read. Early phonological processing skills are strong predictors of later reading performance in typically developing children (Melby-Lervåg, Lyster, & Hulme, 2012) and are impaired in dyslexia (Vellutino, Fletcher, Snowling, & Scanlon, 2004). The neural mechanisms supporting phonological processing have been thoroughly researched over the past decades. Taken together, neuroimaging studies indicate that these mechanisms are located in regions of the left temporoparietal cortex, including the left superior and middle temporal gyri (STG and MTG, respectively) and the left angular gyrus (AG) (Vigneau et al., 2006) (see Fig. 2.1). Studies further indicate that reading acquisition is associated with developmental decreases of activity in left temporoparietal regions (Martin, Schurz, Kronbichler, & Richlan, 2015). This suggests that children are more likely than adults to rely on phonology-based reading. Critically, it has often been argued that such verbal-phonological mechanisms in the left temporoparietal cortex might also contribute to arithmetic learning (Prado, Mutreja, & Booth, 2014; Zamarian, Ischebeck, & Delazer, 2009). This idea mainly comes from the triple-code model (Dehaene & Cohen, 1995), a popular neurocognitive model of numerical processing, which is briefly described in the following.

The Triple-Code Model

One of the motivations for the triple-code model proposed by Dehaene and Cohen (1995) was to account for task-specific patterns of

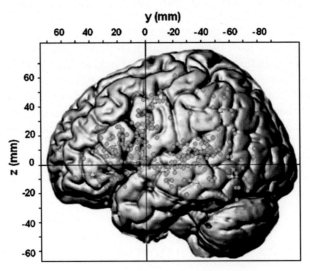

FIGURE 2.1 Location of activation peaks in fMRI studies on phonological processing. Peaks are plotted in blue on a 3D rendering of a left hemisphere. *Reproduced from Vigneau, M., Beaucousin, V., Herve, P. Y., Duffau, H., Crivello, F., Houde, O., et al. (2006). Meta-analyzing left hemisphere language areas: Phonology, semantics, and sentence processing. Neuroimage, 30(4), 1414–1432.*

mathematical impairments (as well as dissociated patterns of impaired performance in mathematical tasks) that are often observed in brain-damaged patients (Cohen & Dehaene, 2000; Lemer, Dehaene, Spelke, & Cohen, 2003; van Harskamp, Rudge, & Cipolotti, 2002, 2005). Dehaene and Cohen hypothesized that numbers may be represented according to three different codes in the adult brain: a visual-Arabic code that would support the identification of visually presented (strings of) digits, a magnitude code that would support semantic knowledge about numerical quantities, and a verbal code that would support the representation of numbers as sequences of written or spoken words. Because one code could be impaired while the others would be spared, the triple-code model constitutes a powerful framework for explaining dissociations in patients (Dehaene, Molko, Cohen, & Wilson, 2004). Of particular interest here is the verbal code, which is posited to be important when retrieving answers of well-known arithmetic facts from memory. For instance, Dehaene and Cohen argue that "arithmetic facts such as $2 \times 3 = 6$ cannot be retrieved unless the problem is coded into a verbal code 'two times three…' which then triggers the retrieval of the result 'six' in the same verbal format" (Dehaene & Cohen, 1995, p. 87). Thus, although there are bidirectional links between codes allowing for indirect routes, the model proposes that arithmetic problems whose answers are well known are not associated with any particular semantic access to underlying quantities. Rather, these problems might be solved by directly retrieving the associated answer from declarative memory. This may of course be case of problems that are explicitly learned by rote in school, such as single-digit multiplication problems. But it should be noted that the verbal-phonological code is also thought to be used for problems that are not necessarily explicitly learned by rote but are particularly well practiced in school (and solved with no apparent difficulty by educated adults). This is, for instance, the case of single-digit addition problems (e.g., $2 + 3 = 5$). Therefore, the triple-code model suggests that some arithmetic skills might rely on the brain mechanisms underlying verbal-phonological processing in the left temporoparietal cortex (as well as the basal ganglia) (Dehaene, Piazza, Pinel, & Cohen, 2003).

Evidence From Studies in Adults

Looking over two decades of neuroimaging research, there seems to be converging support for the triple-code model's assumption that left temporoparietal areas are involved in processing well-learned arithmetic facts in adults. For example, studies that have contrasted the solving of multiplication problems to various control tasks (e.g., number comparison, letter or digit matching, number storage in working memory) have consistently found multiplication-specific activity in regions of the left temporoparietal cortex, including the left AG (Chochon, Cohen, van de Moortele, & Dehaene, 1999; Fulbright, Manson, Skudlarski, Lacadie, & Gore, 2003; Gruber, Indefrey, Steinmetz, & Kleinschmidt, 2001; Jost, Khader, Burke, Bien, & Rosler, 2009). Other studies have found that multiplication is also associated with greater activity than subtraction (i.e., an operation that is thought to rely to a lesser extent on rote memorization than multiplication, Campbell & Xue, 2001) in the left AG (Lee, 2000) and left STG/MTG (Andres, Michaux, & Pesenti, 2012; Andres, Pelgrims, Michaux, Olivier, & Pesenti, 2011; Prado et al., 2011). Activity in the left MTG and the left AG has also been shown to increase as fluency with multiplication problems increases (Ischebeck, Zamarian, Egger, Schocke, & Delazer, 2007). Finally, the left MTG has been found to be more responsive to single-digit multiplication than to single-digit addition problems (Zhou et al., 2007), perhaps because single-digit multiplication problems are even more likely to be directly retrieved from memory than single-digit addition problems (Campbell & Xue, 2001). Finally, the involvement of the left temporoparietal cortex in processing well-known arithmetic facts is also supported by studies that (1) explicitly linked brain activity to self-report of strategies (problems reported to be retrieved vs. calculated) (Grabner et al., 2009) and (2) compared simple with more complex arithmetic problem solving (Grabner et al., 2007; Stanescu-Cosson et al., 2000). Overall, then, there is converging support for the idea that well-known arithmetic facts activate regions of the left temporoparietal cortex in adults.

However, there is at least one important issue with the aforementioned studies: none of these studies have independently localized the brain mechanisms supporting verbal-phonological processing in the left temporoparietal cortex. This issue is particularly problematic with the AG, which has consistently been found to be a component of the default mode network (DMN) (Raichle, 2015; Seghier, 2013). The DMN is a network of regions that contribute to internal modes of cognition and are typically less *deactivated* (with respect to some low-level baseline) when a condition is not attention demanding (and therefore relatively easy) than when it is attention demanding (and therefore more difficult) (Raichle, 2015). It is thus worrisome that most of the studies that have identified the left AG in arithmetic tasks precisely report less *deactivation* in this region for contrasts that systematically involve a comparison between a relatively easy and a relatively difficult condition. This is, for example, almost systematically the case when comparing simple with complex problems (Grabner et al., 2007; Stanescu-Cosson et al., 2000), well-known multiplication problems with lesser-known subtraction problems (Lee, 2000), trained problems with untrained problems (Ischebeck et al., 2007), or problems reported to be retrieved from memory with problems reported to be calculated (Grabner et al., 2009). In other words, differences in levels of deactivation in the left AG during arithmetic problem solving are often correlated with differences in behavioral performance. Therefore, the activation of the left temporoparietal cortex in arithmetic tasks may be an artifact of a difference in difficulty between conditions.[1]

To directly investigate whether phonological processing mechanisms of the left temporoparietal cortex specifically contribute to arithmetic problem solving, Prado et al. (2011) asked the same group of adult participants to perform both a word-rhyming task and an arithmetic task in an MRI scanner. The rhyming task was associated with activity in the left MTG. Brain activity was then measured in this specific cluster when participants were asked to evaluate the validity of either single-digit multiplication or subtraction problems. Not only was the MTG cluster *activated* in the multiplication condition (rather than *deactivated*) but also activity was significantly higher in the multiplication than in the subtraction condition. Thus, processing multiplication facts does involve a brain region involved in verbal-phonological processing in adults and more so than processing subtraction problems. This may be because unlike subtraction problems, multiplication problems are learned by rote in school and more likely to systematically be associated with fact retrieval rather than numerical manipulation (Campbell & Xue, 2001).

It is important to consider, however, that evidence coming from adult studies mainly provides an indirect understanding of the brain systems that underlie simple arithmetic learning. First, simple arithmetic is learned in elementary school and adults can be considered experts in those tasks. Therefore, such studies may not inform on the mechanisms enabling learning per se. Second, although training studies in adults (e.g., Bloechle et al., 2016; Ischebeck et al., 2007) can provide valuable information regarding such learning mechanisms, it is unclear to what extent the neural changes observed in a mature brain mimic the neural changes that occur during initial learning in children. Therefore, it is critical to study in what respect brain activity changes *during* arithmetic learning in children.

---

[1] In a recent study, Bloechle et al. (2016) also speculated that the activation of the left temporoparietal cortex (particularly at the level of the AG) during arithmetic tasks might be explained by the involvement of this region in attention to memory (Cabeza, Ciaramelli, & Moscovitch, 2012). That is, solutions of well-known problems encoded in long-term memory may enter working memory during arithmetic tasks and capture bottom-up attention. Although this proposal differs from the DMN account because it posits that left temporoparietal activity specifically reflects a switch in attentional demands, both accounts are similar in that they assume that this region may not be involved in fact retrieval per se.

Evidence From Studies in Children

Several longitudinal or cross-sectional neuroimaging studies of arithmetic learning in children have been performed over the past few years. Somewhat surprisingly, these studies have provided very little support for age-related (or fluency-related) increases of activity in the left temporoparietal cortex. In a seminal cross-sectional study, Rivera, Reiss, Eckert, and Menon (2005) used a single-digit addition and subtraction task to investigate the neural changes associated with arithmetic learning in children and adults. They found age-related increases of activity in the left anterior IPS but not in the left AG or around the regions of the STG/MTG, which are typically associated with verbal-phonological processing. Rosenberg-Lee, Barth, and Menon (2011) further found greater activity in third than second graders for single-digit addition in the right but not left AG and no change of activity in either the left STG or the left MTG. Kucian, von Aster, Loenneker, Dietrich, and Martin (2008) found greater activity in adults than children in the left IPS but not in the left temporoparietal cortex in a single-digit addition task. In yet another study using a single-digit addition task in children, Cho et al. (2012) found that the only region in which there was an age-related increase of activity was the right superior parietal cortex. Finally, Qin et al. (2015) found longitudinal increases of activity in prefrontal, posterior parietal, and occipital cortex in an addition task but again not in left temporoparietal cortex.

How can one make sense of the discrepancy between adult studies showing activation of left temporoparietal regions during arithmetic tasks on the one hand and developmental studies showing a lack of age-related changes of activity in these regions on the other hand? There might be at least two ways to explain this discrepancy. First, it is clear from the review of adult neuroimaging studies that not all arithmetic tasks recruit left temporoparietal mechanisms. For example, these regions appear to be more activated in single-digit multiplication tasks than in single-digit addition and subtraction tasks (Lee, 2000; Prado et al., 2011; Zhou et al., 2007). This is also consistent with a study by Delazer et al. (2005), showing that only facts that are learned by drill might lead to increased activity in the left AG. It is thus interesting to note that all of the developmental studies described earlier have used tasks that involve arithmetic skills that are (largely) never drilled in school, such as single-digit addition and subtraction problems. Rather, such problems are mastered in adults because they are repeatedly practiced over the years. Therefore, it remains possible that age-related increases of activity would be observed in the left temporoparietal cortex for problems that are drilled, i.e., single-digit multiplication problems. Second, it is also possible that studies might have missed developmental changes of activity in the left temporoparietal cortex because these changes are subtle and require substantial power to be detected with whole-brain analyses of brain imaging data (with larger sample sizes needed to detect relatively small effects).

Prado et al. (2014) attempted to shed light on these two assumptions by investigating the neural bases of single-digit multiplication problem solving in a cross-sectional study of 34 children from second to seventh grade. Critically, brain activity was measured in the whole brain and in a specific region of the left MTG that was localized using a word-rhyming task. Multiplication problems were found to be associated with a developmental increase of activity in that specific MTG cluster. This increase of activity, however, was absent in a single-digit subtraction task. Therefore, at least as far as single-digit multiplication problems are concerned, greater involvement of mechanisms supporting verbal-phonological processing can be observed. This is not to say that learning other types of arithmetic facts cannot engage left temporoparietal regions in children. For example, recent studies in 9- to 12-year-olds have found greater activity in the left AG for (1) small versus large addition and subtraction (De Smedt, Holloway, & Ansari, 2011) and (2) symbolic versus nonsymbolic subtraction across participants (Peters, Polspoel, Op de Beeck, & De Smedt, 2016). Therefore, learning addition facts might also rely to some extent on verbal-phonological mechanisms, albeit perhaps less uniformly than multiplication facts.

Overall, the evidence reviewed earlier suggests that integrity of the phonological processing mechanisms supporting the acquisition of reading might influence at least some aspects of the neural mechanisms that support arithmetic learning. This hypothesis has been recently tested in a study by Evans, Flowers, Napoliello, Olulade, and Eden (2014). The authors investigated the neural bases of arithmetic processing in 14 children with developmental dyslexia (compared with 14 typically developing children). Previous neuroimaging studies have found that individuals with dyslexia have anatomic and functional impairments in brain regions supporting phonological processing in the left temporoparietal cortex (Richlan, Kronbichler, & Wimmer, 2009). Behavioral studies have also identified arithmetic difficulties in children with dyslexia (Simmons & Singleton, 2008). Therefore, it is possible that these difficulties lead to abnormal processing of arithmetic facts in left temporoparietal regions. In line with this hypothesis, less activity was found in the left supramarginal gyrus, a region adjacent to both the left STG and the left AG, in response to single-digit addition and subtraction in dyslexic compared with typically developing children. Thus, it is possible that anatomic and functional impairments in left temporoparietal phonological mechanisms might affect at least some aspects of arithmetic processing in dyslexic children. However, as we have seen here, the involvement of temporoparietal mechanisms in arithmetic learning is very task specific. It is thus unlikely to account in itself for the strong relationship between arithmetic learning and reading acquisition in children. Other explanations might then need to be considered. In the following, we discuss the recent proposal that learning arithmetic and learning to read may both require the skill to automatize procedures until complete fluency.

PROCEDURAL AUTOMATIZATION: A COMMON DENOMINATOR BETWEEN LEARNING TO READ AND LEARNING ARITHMETIC?

Perhaps the most striking and intuitive similarity between learning to read and learning arithmetic is that they arguably share a similar goal: both involve automatizing a skill until complete fluency. By the end of elementary school, children are expected to read and understand common words with no apparent effort in much the same way as they should quickly respond 5 when faced with 2+3. Interestingly, it has long been proposed that a domain-general deficit in procedural automatization may be at the source of dyslexia in children (Lum, Ullman, & Conti-Ramsden, 2013; Nicolson & Fawcett, 2007), and this hypothesis has been recently extended to dyscalculia (Evans & Ullman, 2016). This raises the possibility that at least some aspects of arithmetic learning and reading acquisition both rely on the memory system that supports the automatization of procedures through practice, i.e., procedural memory. This hypothesis is interesting for the current purpose of this chapter because it may explain some of the interactions between learning to read and learning arithmetic. In the following, we describe how theories have explained dyslexia in terms of impaired automatization of procedures and how these theories have been recently extended to dyscalculia. We then examine to what extent this idea is supported by developmental behavioral and neuroimaging research.

The Procedural Learning View of Dyslexia and Dyscalculia

Clearly, the most popular explanation of dyslexia is that it is caused by a deficit in accessing and manipulating phonological information (Vellutino et al., 2004). This phonological deficit view is widely supported by studies showing that children with dyslexia do have impairments in phonological processing, including phonological awareness (sensitivity to the sound structure of oral language) (Vellutino et al., 2004). Yet, studies also show that children with dyslexia often exhibit impairments in other domains, such as attention (Facoetti, Paganoni, Turatto, Marzola, & Mascetti, 2000; Varvara, Varuzza, Sorrentino, Vicari, & Menghini, 2014), motor control (Fawcett & Nicolson, 1995;

Nicolson & Fawcett, 1994), and implicit sequence learning (Gabay, Thiessen, & Holt, 2015; Hedenius et al., 2013; Kelly, Griffiths, & Frith, 2002). Because these impairments are not easily explained by a phonological deficit account, several researchers have argued that dyslexia may also arise from a more general learning disorder (Lum et al., 2013; Nicolson & Fawcett, 2007; Nicolson, Fawcett, & Dean, 2001; Ullman, 2004). Specifically, it has been proposed that dyslexia may stem from a general impairment in the learning and memory system that supports the acquisition of skills and habits through repeated practice, i.e., procedural memory (Ullman, 2016, pp. 953–968). Not only can this hypothesis explain the range of nonreading deficits observed in dyslexic children but also it can account for the reading deficits. That is, impairments in the procedural memory system may disrupt "automatization of skill and knowledge, which may potentially affect grapheme–phoneme conversion, word recognition, verbal working memory, and learning orthographic regularities, thereby contributing to reading impairment" (Gabay et al., 2015, p. 935). The brain system supporting procedural memory has been documented in the literature. It relies on a network of brain regions, which includes the basal ganglia, the premotor cortex, the cerebellum, and parts of the inferior parietal cortex (Ullman, 2016, pp. 953–968). Interestingly, brain imaging studies have found that children with dyslexia do exhibit anatomic and functional impairments in several of these regions, including the basal ganglia (Brunswick, McCrory, Price, Frith, & Frith, 1999; Kita et al., 2013), cerebellum (Eckert et al., 2003; Pernet, Poline, Demonet, & Rousselet, 2009; Rae et al., 2002), frontal cortex (Eckert et al., 2003; Richlan et al., 2009), and parietal cortex (Richlan et al., 2009).

Recently, Evans and Ullman (2016) have proposed an extension of the procedural learning deficit theory to dyscalculia (Evans & Ullman, 2016). This extension is partly motivated by the observed comorbidity between dyslexia and dyscalculia, estimated at around 40% (Wilson et al., 2015). Evans and Ullman (2016) proposed that this comorbidity may be explained by the fact that (at least) some children may have a general learning deficit that would affect both reading and arithmetic learning. This deficit may stem from impaired procedural memory. The proposal relies on the assumption that, much like learning to read, acquiring arithmetic skills involves using increasingly efficient procedures that need to be automatized by the end of the learning process. Impairments in procedural memory may hinder this progressive automatization of procedures, leading to both dyslexia and dyscalculia. Although that specific hypothesis remains to be thoroughly explored, we will see in the following that there is growing support for the idea that automatized procedures may indeed underlie at least some aspects of arithmetic skills in typically developing children and educated adults.

Arithmetic Learning and the Use of Increasingly Efficient Procedures

A defining feature of arithmetic acquisition in children is the use of increasingly efficient procedures over the course of learning. This has been consistently demonstrated in studies in which children are asked to report the strategies they use when solving problems (Baroody & Tiilikainen, 2003; Carpenter & Moser, 1984). Consider, for example, the strategies that young children report relying on in simple addition tasks. Using external aids such as fingers or objects, young children may first use what is often called the *counting-all* procedure: they count out two sets of objects before combining them and counting the newly formed set. Older children may then realize that objects can be replaced by counting words. This realization is often accompanied by the appearance of the more sophisticated *counting-on* procedure, according to which children start to count from one of the two number words. In using this *counting-on* procedure, many children will also realize that starting from the largest number (*minimum strategy*) is more efficient than starting from the smaller number (*maximum strategy*).

In many ways, this use of increasingly efficient procedures over development parallels what is observed when children learn to read. That is, becoming an efficient reader also involves a refinement of procedures underlying word decoding (Farrington-Flint, Coyne, Stiller, & Heath, 2008). For example, young children typically start by identifying words using a *sounding out* strategy, in which graphemes are matched onto phonemes. This *sounding out* strategy can be far from accurate because in many languages (but not all) words may have irregular mappings from orthography to phonology (Farrington-Flint et al., 2008). With reading practice, the *sounding out* strategy will be gradually replaced by more sophisticated (and more efficient) procedures. For instance, a more elaborate procedure might involve making an analogy from the spelling sound pattern of a familiar word to an unfamiliar one. An even more efficient procedure might involve using morphological rules (Farrington-Flint et al., 2008). Therefore, both learning to read and learning arithmetic are characterized by the use of increasingly efficient procedures with age.

Procedural Automatization as a Critical Element of Arithmetic Fluency

Of course, the end product of both learning to read and learning arithmetic ought to be fluency. For instance, educated adults must be able to very quickly recognize a written word. They must also be able to quickly come up with the answer of a single-digit addition problem. Over the past decades, the dominant view has been that procedures can never be efficient enough to attain arithmetic fluency and that there is necessarily a shift from procedural to retrieval-based strategies over development (Barrouillet & Fayol, 1998; Groen & Parkman, 1972). That is, the repetitive use of counting procedures over the course learning is thought to lead to the repetitive co-occurrence of operands and answers in working memory, which would result in the emergence of associations between numbers (Logan, 1988). Eventually, adults and educated children might then directly retrieve these associations from declarative memory when faced with simple arithmetic problems— without relying on procedural knowledge (Siegler & Shipley, 1995, pp. 31–76).

More recently, however, several studies have suggested that counting procedures might be so practiced over the course of learning that they might actually be executed automatically and unconsciously in adults and educated children. These automatized procedures may then be very efficient and also account for basic arithmetic skills of fluent individuals. Evidence for this alternative view comes from behavioral studies showing that adults systematically show a problem size effect when they solve even very simple addition problems (i.e., problems with operands smaller than 4) (Barrouillet & Thevenot, 2013; Thevenot, Barrouillet, Castel, & Uittenhove, 2015; Uittenhove, Thevenot, & Barrouillet, 2015). Specifically, the time it takes to solve these problems increases linearly with the distance between the original operand and the sum (i.e., adults take 20 ms longer to solve $1+3$ than $1+2$ and 20 ms longer to solve $1+4$ than $1+3$) (Barrouillet & Thevenot, 2013; Uittenhove et al., 2015) (see Fig. 2.2). This pattern is difficult to explain with the view that associations are retrieved from a network of facts. Rather, it suggests that, without being aware of it, adults might solve these problems by using an automatized counting procedure that may involve scanning a sequence of numbers oriented from left to right (i.e., the so-called mental number line or MNL) (Barrouillet & Thevenot, 2013; Mathieu, Gourjon, Couderc, Thevenot, & Prado, 2016). Solving time would then depend on the distance between the original operand and the target sum to be reached. In support for this view, Mathieu et al. (2016) have recently found that solving addition and subtraction problems is associated with rightward and leftward shifts of attention (respectively) in adults. Such an automatized procedure may be the result of extensive practice with counting in children, in line with the long-standing idea that the repetitive practice of a procedure can lead to its automatization (Baroody, 1983). Therefore, without denying that retrieval from memory might occur in some instances, it is conceivable that learning arithmetic might also rely on the progressive automatization of procedures and therefore involve to a significant extent procedural memory.

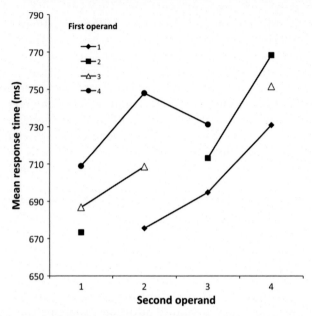

FIGURE 2.2   Mean resolution times of simple addition problems (with operands from 1 to 4 but not including tie problems) as a function of the magnitude of the first and second operands in adults. *Reproduced from Barrouillet, P., & Thevenot, C. (2013). On the problem-size effect in small additions: Can we really discard any counting-based account?* Cognition, 128(1), 35–44.

Developmental Neuroimaging Evidence for Procedural Automatization

The idea that arithmetic learning and reading acquisition gradually involves an automatization of procedural knowledge makes at least three interesting neural predictions. First, if procedures become automatic with expertise, they should require fewer and fewer executive resources. This should translate into decreases of activity in brain regions that support executive control and working memory. This idea is fairly well supported by developmental neuroimaging studies of arithmetic learning. Indeed, studies have consistently demonstrated decreases of activity in regions of the frontal cortex with age in arithmetic tasks. For example, this developmental pattern was first observed in the seminal cross-sectional study by Rivera et al. (2005). There, the development of skills for solving addition problems from ages 8 to 19 was related to decreases of activity in several regions of the inferior, middle, and superior frontal gyri. Rosenberg-Lee, Barth, et al. (2011) similarly found less activity in the ventral medial prefrontal cortex in third graders compared with second graders in an addition task. Recently, longitudinal decreases of activity were found in the bilateral dorsolateral prefrontal cortex as children in elementary school become increasingly proficient in solving addition problems (Qin et al., 2015). Decreases of activity in prefrontal cortex were also noticed in subtraction and multiplication problem solving in a cross-sectional study of children from second to seventh grade (Prado et al., 2014). Finally, differences in prefrontal activity are also observed between adults and children in arithmetic tasks (with less activity in adults than children) (Kucian et al., 2008).

Second, if children increasingly rely on procedures that involve automatized shifts of attention along the MNL (Barrouillet & Thevenot, 2013; Mathieu et al., 2016; Thevenot et al., 2015; Uittenhove et al., 2015), one should observe developmental increases of arithmetic-related activity in the posterior superior parietal lobule, the main region that is thought to support attentional orientation along the MNL (Dehaene et al., 2003). This is exactly what we observed in a recent cross-sectional study in which fMRI activity of children from second to seventh grade was measured while they were solving single-digit problems (Prado et al., 2014). Importantly, these changes were operation specific as they were observed for subtraction problems (a type of operation that is not learned by rote in school) but not for multiplication problems (which are mostly learned by rote). More generally, these results are consistent with studies showing that age-related decreases of prefrontal activity in arithmetic tasks are often accompanied by age-related increases of activity in regions of the parietal cortex (Prado et al., 2014; Rivera et al., 2005; Rosenberg-Lee, Barth, et al., 2011). This has led several researchers to propose that arithmetic learning may be characterized by a frontal-to-parietal developmental shift of activity, with parietal regions becoming progressively specialized for arithmetic tasks (Ansari, 2008). This increase in specialization of the parietal cortex for arithmetic tasks is broadly consistent with the claim that parietal regions might increasingly support automatized numerical-based procedures with development (Prado et al., 2014).

Interestingly, a similar shift from prefrontal to posterior brain regions can be observed during reading acquisition. For example, Martin et al. (2015) recently compared in a metaanalysis the results of 20 adult neuroimaging studies on reading with the results of 20 neuroimaging studies performed with children. They only found two brain regions that were more consistently activated across children studies than across adult studies: the left STG and the medial prefrontal cortex. This is consistent with the idea that, compared with adults, children might use phonology-based strategies that might be more effortful and require executive control mechanisms in the prefrontal cortex (as well as access to phonological representation in the STG). In contrast, compared with children studies, adult studies more consistently activated posterior regions, such as the ventral occipital cortex and the cerebellum. This latter finding points to an anterior-to-posterior shift of activity for reading with development, echoing the frontal-to-parietal shift observed during arithmetic learning.

A third hypothesis, related to the idea that both arithmetic learning and reading acquisition involve an automatization of procedural knowledge, is that arithmetic and reading disabilities should be related to impairments in brain structures that underlie procedural memory. These may include the posterior parietal cortex, the cerebellum, and subcortical structures such as the basal ganglia (Evans & Ullman, 2016). All of these regions are known to be affected in dyslexia (Richlan et al., 2009). Research on dyscalculia also suggests impairments in regions that support procedural memory. Arguably the most consistent locus of anatomic and functional impairments in dyscalculia is the posterior parietal cortex (Ashkenazi, Rosenberg-Lee, Tenison, & Menon, 2012; Iuculano et al., 2015; Molko et al., 2003; Rosenberg-Lee et al., 2015; Rotzer et al., 2008, 2009; Rykhlevskaia, Uddin, Kondos, & Menon, 2009). However, abnormalities have also been observed in the basal ganglia (Molko et al., 2003) and the cerebellum (Rykhlevskaia et al., 2009). Therefore, although more studies are needed, the available evidence suggests that at least some structures supporting procedural memory may be impaired in dyscalculia.

CONCLUSION AND FUTURE DIRECTIONS

Cognitive neuroscience studies largely indicate that human mathematical skills are rooted in nonverbal mechanisms. It is thus somewhat paradoxical that there appears to be a link between the development of arithmetic and reading skills in children (Durand et al., 2005; Hart et al., 2009; Hecht et al., 2001; Wilson et al., 2015). The goal of this chapter was to provide an overview of two possible explanations for this link and confront these explanations to available evidence from developmental cognitive neuroscience.

First, the triple-code model suggests that answers of the most familiar arithmetic facts may be increasingly retrieved from verbal-phonological codes as individuals become fluent (Dehaene et al., 2003). Yet, developmental cognitive neuroscience studies provide limited evidence for this assumption (Cho et al., 2012; Rivera et al., 2005; Rosenberg-Lee, Chang, Young, Wu, & Menon, 2011). One study suggests that increases of activity may be observed in the left temporal cortex as children become increasingly proficient with single-digit multiplication problems (Prado et al., 2014), but this effect appears to be task dependent. The contribution of verbal-phonological mechanisms to arithmetic learning may thus be restricted to facts that are explicitly learned by rote in school.

Second, it has been proposed that learning arithmetic and learning to read may both rely on the automatization of rules and procedures (Barrouillet & Thevenot, 2013; Thevenot et al., 2015; Uittenhove et al., 2015). Procedural memory systems may thus be critical to the acquisition of both skills, and impairments in procedural memory may be at the source of both dyscalculia and dyslexia. Not only does that hypothesis explain why one of the most consistent results obtained in developmental neuroimaging studies is an increase of activity in the parietal cortex (rather than in the left temporoparietal cortex) but also it highlights the importance of procedural memory for arithmetic learning. Because procedural memory has long been hypothesized to also be central to reading acquisition (Ullman, 2016, pp. 953–968), this idea may explain the link between arithmetic and reading acquisition and the comorbidity between dyslexia and dyscalculia (Wilson et al., 2015).

Overall, given the limited neurodevelopmental support for the idea that verbal-phonological processing underlies arithmetic learning in children, the procedural hypothesis is an interesting explanation for the link between arithmetic and reading skills. Of course, this does not mean that other domain-general factors cannot also account for that link in some children. For example, it is clear that both arithmetic and reading tasks involve working memory, attention, or cognitive control (Ashkenazi et al., 2013). It is possible that disruptions in these domain-general mechanisms may also lead to both reading and arithmetic impairments in children (as well as impairments in other skills). This is generally consistent with the idea that arithmetic learning involves a wide range of skills and that dyscalculia may be a heterogeneous disorder (Fias, Menon, & Szucs, 2013). Nevertheless, learning to read and learning arithmetic may both place important demands on procedural memory and automatization of skills, a factor that may explain a large part of the relationship between arithmetic and reading performance in children.

Acknowledgments

I am grateful to Pr. Avishai Henik and Pr. Klaus Willmes for their insightful comments on an earlier version of this chapter. I am supported by grants from the Agence Nationale pour la Recherche (ANR-14-CE30-0002-01) and the European Commission (Marie-Curie Carreer Integration Grant PCIG12-GA-2012-333602).

References

Andres, M., Michaux, N., & Pesenti, M. (2012). Common substrate for mental arithmetic and finger representation in the parietal cortex. *Neuroimage, 62*(3), 1520–1528. https://doi.org/10.1016/j.neuroimage.2012.05.047.

Andres, M., Pelgrims, B., Michaux, N., Olivier, E., & Pesenti, M. (2011). Role of distinct parietal areas in arithmetic: An fMRI-guided TMS study. *Neuroimage, 54*(4), 3048–3056. https://doi.org/10.1016/j.neuroimage.2010.11.009.

Ansari, D. (2008). Effects of development and enculturation on number representation in the brain. *Nature Reviews Neuroscience, 9*(4), 278–291. https://doi.org/10.1038/nrn2334.

Ashkenazi, S., Black, J. M., Abrams, D. A., Hoeft, F., & Menon, V. (2013). Neurobiological underpinnings of math and reading learning disabilities. *Journal of Learning Disabilities, 46*(6), 549–569.

Ashkenazi, S., Rosenberg-Lee, M., Tenison, C., & Menon, V. (2012). Weak task-related modulation and stimulus representations during arithmetic problem solving in children with developmental dyscalculia. *Developmental Cognitive Neuroscience,* 2(Suppl. 1), S152–S166. https://doi.org/10.1016/j.dcn.2011.09.006.

Baroody, A. J. (1983). The development of procedural knowledge: An alternative explanation for chronometric trends of mental arithmetic. *Developmental Review, 3,* 225–230.

Baroody, A. J., & Tiilikainen, S. H. (2003). Two perspectives on addition development. In A. J. Baroody, & A. Dowker (Eds.), *The development of arithmetic concepts and skills: Constructive adaptive expertise.* London, UK: Lawrence Elbaum Associates.

Barrouillet, P., & Fayol, M. (1998). From algorithmic computing to direct retrieval: Evidence from number and alphabetic arithmetic in children and adults. *Memory and Cognition, 26,* 355–368.

Barrouillet, P., & Thevenot, C. (2013). On the problem-size effect in small additions: Can we really discard any counting-based account? *Cognition, 128*(1), 35–44. https://doi.org/10.1016/j.cognition.2013.02.018.

Bloechle, J., Huber, S., Bahnmueller, J., Rennig, J., Willmes, K., Cavdaroglu, S., et al. (2016). Fact learning in complex arithmetic—the role of the angular gyrus revisited. *Human Brain Mapping, 37*(9), 3061–3079.

Brunswick, N., McCrory, E., Price, C. J., Frith, C. D., & Frith, U. (1999). Explicit and implicit processing of words and pseudowords by adult developmental dyslexics: A search for Wernicke's Wortschatz? *Brain, 122*(Pt 10), 1901–1917.

Cabeza, R., Ciaramelli, E., & Moscovitch, M. (2012). Cognitive contributions of the ventral parietal cortex: An integrative theoretical account. *Trends in Cognitive Sciences, 16,* 338–352.

Campbell, J. I., & Xue, Q. (2001). Cognitive arithmetic across cultures. *Journal of Experimental Psychology: General, 130*(2), 299–315.

Carpenter, T., & Moser, J. M. (1984). The acquisition of addition and subtraction concepts in grades one through three. *Journal for Research in Mathematics Education, 15,* 179–202.

Chochon, F., Cohen, L., van de Moortele, P. F., & Dehaene, S. (1999). Differential contributions of the left and right inferior parietal lobules to number processing. *Journal of Cognitive Neuroscience, 11*(6), 617–630.

Cho, S., Metcalfe, A. W., Young, C. B., Ryali, S., Geary, D. C., & Menon, V. (2012). Hippocampal-prefrontal engagement and dynamic causal interactions in the maturation of children's fact retrieval. *Journal of Cognitive Neuroscience, 24*(9), 1849–1866. https://doi.org/10.1162/jocn_a_00246.

Cohen, L., & Dehaene, S. (2000). Calculating without reading: Unsuspected residual abilities in pure alexia. *Cognitive Neuropsychology, 17*(6), 563–583. https://doi.org/10.1080/02643290050110656.

Cordes, S., & Brannon, E. M. (2008). Quantitative competencies in infancy. *Developmental Science, 11*(6), 803–808. https://doi.org/10.1111/j.1467-7687.2008.00770.x.

De Smedt, B., Holloway, I. D., & Ansari, D. (2011). Effects of problem size and arithmetic operation on brain activation during calculation in children with varying levels of arithmetical fluency. *Neuroimage, 57*(3), 771–781. https://doi.org/10.1016/j.neuroimage.2010.12.037.

Dehaene, S., & Cohen, D. (1995). Towards an anatomical and functional model of number processing. *Mathematical Cognition, 1*, 83–120.

Dehaene, S., & Cohen, L. (2007). Cultural recycling of cortical maps. *Neuron, 56*(2), 384–398. https://doi.org/10.1016/j.neuron.2007.10.004.

Dehaene, S., Molko, N., Cohen, L., & Wilson, A. J. (2004). Arithmetic and the brain. *Current Opinion in Neurobiology, 14*(2), 218–224. https://doi.org/10.1016/j.conb.2004.03.008.

Dehaene, S., Piazza, M., Pinel, P., & Cohen, L. (2003). Three parietal circuits for number processing. *Cognitive Neuropsychology, 20*(3), 487–506. https://doi.org/10.1080/02643290244000239.

Delazer, M., Ischebeck, A., Domahs, F., Zamarian, L., Koppelstaetter, F., Siedentopf, C. M., et al. (2005). Learning by strategies and learning by drill–evidence from an fMRI study. *Neuroimage, 25*(3), 838–849. https://doi.org/10.1016/j.neuroimage.2004.12.009.

Duncan, G. J., Dowsett, C. J., Claessens, A., Magnuson, K., Huston, A. C., Pagani, L. S., et al. (2007). School readiness and later achievement. *Developmental Psychology, 43*(6), 1428–1446.

Durand, M., Hulme, C., Larkin, R., & Snowling, M. (2005). The cognitive foundations of reading and arithmetic skills in 7-to 10-year-olds. *Journal of Experimental Child Psychology, 91*(2), 113–136.

Eckert, M. A., Leonard, C. M., Richards, T. L., Aylward, E. H., Thomson, J., & Berninger, V. W. (2003). Anatomical correlates of dyslexia: Frontal and cerebellar findings. *Brain, 126*(Pt 2), 482–494.

Evans, T. M., Flowers, D. L., Napoliello, E. M., Olulade, O. A., & Eden, G. F. (2014). The functional anatomy of single-digit arithmetic in children with developmental dyslexia. *Neuroimage, 101*, 644–652. https://doi.org/10.1016/j.neuroimage.2014.07.028.

Evans, T. M., & Ullman, M. T. (2016). An extension of the procedural deficit hypothesis from developmental language disorders to mathematical disability. *Frontiers in Psychology, 7*, 1318. https://doi.org/10.3389/fpsyg.2016.01318.

Facoetti, A., Paganoni, P., Turatto, M., Marzola, V., & Mascetti, G. G. (2000). Visual-spatial attention in developmental dyslexia. *Cortex, 36*(1), 109–123.

Farrington-Flint, L., Coyne, E., Stiller, J., & Heath, E. (2008). Variability in children's early reading strategies. *Educational Psychology, 28*(6), 643–661.

Fawcett, A. J., & Nicolson, R. I. (1995). Persistent deficits in motor skill of children with dyslexia. *Journal of Motor Behavior, 27*, 235–240.

Feigenson, L., Libertus, M. E., & Halberda, J. (2013). Links between the intuitive sense of number and formal mathematics ability. *Child Development Perspectives, 7*(2), 74–79.

Fias, W., Menon, V., & Szucs, D. (2013). Multiple components of developmental dyscalculia. *Trends in Neuroscience and Education, 2*(2), 43–47.

Fulbright, R., Manson, S., Skudlarski, P., Lacadie, C., & Gore, J. (2003). Quantity determination and the distance effect with letters, numbers, and shapes: A functional MR imaging study of number processing. *American Journal of Neuroradiology, 24*(2), 193–200.

Gabay, Y., Thiessen, E. D., & Holt, L. L. (2015). Impaired statistical learning in developmental dyslexia. *Journal of Speech, Language, and Hearing Research: JSLHR, 58*(3), 934–945. https://doi.org/10.1044/2015_JSLHR-L-14-0324.

Grabner, R. H., Ansari, D., Koschutnig, K., Reishofer, G., Ebner, F., & Neuper, C. (2009). To retrieve or to calculate? Left angular gyrus mediates the retrieval of arithmetic facts during problem solving. *Neuropsychologia, 47*(2), 604–608. https://doi.org/10.1016/j.neuropsychologia.2008.10.013.

Grabner, R. H., Ansari, D., Reishofer, G., Stern, E., Ebner, F., & Neuper, C. (2007). Individual differences in mathematical competence predict parietal brain activation during mental calculation. *Neuroimage, 38*(2), 346–356. https://doi.org/10.1016/j.neuroimage.2007.07.041.

Groen, G. J., & Parkman, J. M. (1972). A chronometric analysis of simple addition. *Psychological Review, 79*, 329–343.

Gruber, O., Indefrey, P., Steinmetz, H., & Kleinschmidt, A. (2001). Dissociating neural correlates of cognitive components in mental calculation. *Cerebral Cortex, 11*(4), 350–359.

Hart, S. A., Petrill, S. A., Thompson, L. A., & Plomin, R. (2009). The ABCs of math: A genetic analysis of mathematics and its links with reading ability and general cognitive ability. *Journal of educational psychology, 101*(2), 388–402.

Hecht, S. A., Torgesen, J. K., Wagner, R. K., & Rashotte, C. A. (2001). The relations between phonological processing abilities and emerging individual differences in mathematical computation skills: A longitudinal study from second to fifth grades. *Journal of Experimental Child Psychology, 79*(2), 192–227.

Hedenius, M., Persson, J., Alm, P. A., Ullman, M. T., Howard, J. H., Howard, D. V., et al. (2013). Impaired implicit sequence learning in children with developmental dyslexia. *Research in Developmental Disabilities, 34*(11), 3924–3935.

Ischebeck, A., Zamarian, L., Egger, K., Schocke, M., & Delazer, M. (2007). Imaging early practice effects in arithmetic. *Neuroimage, 36*(3), 993–1003. https://doi.org/10.1016/j.neuroimage.2007.03.051.

Iuculano, T., Rosenberg-Lee, M., Richardson, J., Tenison, C., Fuchs, L., Supekar, K., et al. (2015). Cognitive tutoring induces widespread neuroplasticity and remediates brain function in children with mathematical learning disabilities. *Nature Communications, 6*, 8453. https://doi.org/10.1038/ncomms9453.

Jost, K., Khader, P., Burke, M., Bien, S., & Rosler, F. (2009). Dissociating the solution processes of small, large, and zero multiplications by means of fMRI. *Neuroimage, 46*(1), 308–318. https://doi.org/10.1016/j.neuroimage.2009.01.044.

Kaufmann, L., & von Aster, M. (2012). The diagnosis and management of dyscalculia. *Deutsches Ärzteblatt International, 109*(45), 767–778.

Kelly, S. W., Griffiths, S., & Frith, U. (2002). Evidence for implicit sequence learning in dyslexia. *Dyslexia, 8*(1), 43–52.

Kita, Y., Yamamoto, H., Oba, K., Terasawa, Y., Moriguchi, Y., Uchiyama, H., et al. (2013). Altered brain activity for phonological manipulation in dyslexic Japanese children. *Brain, 136*(Pt 12), 3696–3708. https://doi.org/10.1093/brain/awt248.

Kucian, K., von Aster, M., Loenneker, T., Dietrich, T., & Martin, E. (2008). Development of neural networks for exact and approximate calculation: A fMRI study. *Developmental Neuropsychology, 33*(4), 447–473.

Lee, K. M. (2000). Cortical areas differentially involved in multiplication and subtraction: A functional magnetic resonance imaging study and correlation with a case of selective acalculia. *Annals of Neurology, 48*(4), 657–661.

Lemer, C., Dehaene, S., Spelke, E., & Cohen, L. (2003). Approximate quantities and exact number words: Dissociable systems. *Neuropsychologia, 41*(14), 1942–1958.

Lerkkanen, M.-K., Rasku-Puttonen, H., Aunola, K., & Nurmi, J.-E. (2005). Mathematical performance predicts progress in reading comprehension among 7-year-olds. *European Journal of Psychology of Education*, *20*(121).

Logan, G. D. (1988). Toward an instance theory of automatization. *Psychological Review*, *95*, 492–527.

Lum, J. A. G., Ullman, M. T., & Conti-Ramsden, G. (2013). Procedural learning is impaired in dyslexia: Evidence from a meta-analysis of serial reaction time studies. *Research in Developmental Disabilities*, *34*, 3460–3476.

Martin, A., Schurz, M., Kronbichler, M., & Richlan, F. (2015). Reading in the brain of children and adults: A meta-analysis of 40 functional magnetic resonance imaging studies. *Human Brain Mapping*, *36*(5), 1963–1981. https://doi.org/10.1002/hbm.22749.

Mathieu, R., Gourjon, A., Couderc, A., Thevenot, C., & Prado, J. (2016). Running the number line: Rapid shifts of attention in single-digit arithmetic. *Cognition*, *146*, 229–239. https://doi.org/10.1016/j.cognition.2015.10.002.

Melby-Lervåg, M., Lyster, S. A. H., & Hulme, C. (2012). Phonological skills and their role in learning to read: A meta-analytic review. *Psychological Bulletin*, *132*(2), 322–352.

Molko, N., Cachia, A., Riviere, D., Mangin, J. F., Bruandet, M., Le Bihan, D., et al. (2003). Functional and structural alterations of the intraparietal sulcus in a developmental dyscalculia of genetic origin. *Neuron*, *40*(4), 847–858.

Nicolson, R. I., & Fawcett, A. J. (1994). Comparison of deficits in cognitive and motor skills among children with dyslexia. *Annals of Dyslexia*, *44*(1), 147–164.

Nicolson, R. I., & Fawcett, A. J. (2007). Procedural learning difficulties: Reuniting the developmental disorders? *Trends in Neurosciences*, *30*(4), 135–141. https://doi.org/10.1016/j.tins.2007.02.003.

Nicolson, R. I., Fawcett, A. J., & Dean, P. (2001). Developmental dyslexia: The cerebellar deficit hypothesis. *Trends in Neurosciences*, *24*(9), 508–511.

Nieder, A., & Dehaene, S. (2009). Representation of number in the brain. *Annual Review of Neuroscience*, *32*, 185–208. https://doi.org/10.1146/annurev.neuro.051508.135550.

Pernet, C. R., Poline, J. B., Demonet, J. F., & Rousselet, G. A. (2009). Brain classification reveals the right cerebellum as the best biomarker of dyslexia. *BMC Neuroscience*, *10*, 67. https://doi.org/10.1186/1471-2202-10-67.

Peters, L., Polspoel, B., Op de Beeck, H., & De Smedt, B. (2016). Brain activity during arithmetic in symbolic and non-symbolic formats in 9-12 year old children. *Neuropsychologia*, *86*, 19–28. https://doi.org/10.1016/j.neuropsychologia.2016.04.001.

Pica, P., Lemer, C., Izard, V., & Dehaene, S. (2004). Exact and approximate arithmetic in an Amazonian indigene group. *Science*, *306*(5695), 499–503. https://doi.org/10.1126/science.1102085.

Prado, J., Mutreja, R., & Booth, J. R. (2014). Developmental dissociation in the neural responses to simple multiplication and subtraction problems. *Developmental Science*, *17*(4), 537–552.

Prado, J., Mutreja, R., Zhang, H., Mehta, R., Desroches, A. S., Minas, J. E., et al. (2011). Distinct representations of subtraction and multiplication in the neural systems for numerosity and language. *Human Brain Mapping*, *32*(11), 1932–1947. https://doi.org/10.1002/hbm.21159.

Qin, S., Cho, S., Chen, T., Rosenberg-Lee, M., Geary, D. C., & Menon, V. (2015). Hippocampal-neocortical functional reorganization underlies children's cognitive development. *Nature Neuroscience*, *17*(9), 1263–1269. https://doi.org/10.1038/nn.3788.

Rae, C., Harasty, J. A., Dzendrowskyj, T. E., Talcott, J. B., Simpson, J. M., Blamire, A. M., et al. (2002). Cerebellar morphology in developmental dyslexia. *Neuropsychologia*, *40*(8), 1285–1292.

Raichle, M. E. (2015). The brain's default mode network. *Annual Review of Neuroscience*, *38*, 433–447. https://doi.org/10.1146/annurev-neuro-071013-014030.

Richlan, F., Kronbichler, M., & Wimmer, H. (2009). Functional abnormalities in the dyslexic brain: A quantitative meta-analysis of neuroimaging studies. *Human Brain Mapping*, *30*(10), 3299–3308. https://doi.org/10.1002/hbm.20752.

Rivera, S. M., Reiss, A. L., Eckert, M. A., & Menon, V. (2005). Developmental changes in mental arithmetic: Evidence for increased functional specialization in the left inferior parietal cortex. *Cerebral Cortex*, *15*(11), 1779–1790. https://doi.org/10.1093/cercor/bhi055.

Rosenberg-Lee, M., Ashkenazi, S., Chen, T., Young, C. B., Geary, D. C., & Menon, V. (2015). Brain hyper-connectivity and operation-specific deficits during arithmetic problem solving in children with developmental dyscalculia. *Developmental Science*, *18*(3), 351–372. https://doi.org/10.1111/desc.12216.

Rosenberg-Lee, M., Barth, M., & Menon, V. (2011). What difference does a year of schooling make? Maturation of brain response and connectivity between 2nd and 3rd grades during arithmetic problem solving. *Neuroimage*, *57*(3), 796–808. https://doi.org/10.1016/j.neuroimage.2011.05.013.

Rosenberg-Lee, M., Chang, T. T., Young, C. B., Wu, S., & Menon, V. (2011). Functional dissociations between four basic arithmetic operations in the human posterior parietal cortex: A cytoarchitectonic mapping study. *Neuropsychologia*, *49*(9), 2592–2608. https://doi.org/10.1016/j.neuropsychologia.2011.04.035.

Rotzer, S., Kucian, K., Martin, E., Von Aster, M., Klaver, P., & Loenneker, T. (2008). Optimized voxel-based morphometry in children with developmental dyscalculia. *Neuroimage*, *39*, 417–422.

Rotzer, S., Loenneker, T., Kucian, K., Martin, E., Klaver, P., & von Aster, M. (2009). Dysfunctional neural network of spatial working memory contributes to developmental dyscalculia. *Neuropsychologia*, *47*(13), 2859–2865. https://doi.org/10.1016/j.neuropsychologia.2009.06.009.

Rykhlevskaia, E., Uddin, L. Q., Kondos, L., & Menon, V. (2009). Neuroanatomical correlates of developmental dyscalculia: Combined evidence from morphometry and tractography. *Frontiers in Human Neuroscience*, *3*, 51. https://doi.org/10.3389/neuro.09.051.2009.

Seghier, M. L. (2013). The angular gyrus: Multiple functions and multiple subdivisions. *The Neuroscientist*, *19*(1), 43–61. https://doi.org/10.1177/1073858412440596.

Siegler, R. S., & Shipley, C. (1995). Variation, selection, and cognitive change. In G. Halford, & T. Simon (Eds.), *Developing cognitive competence: New approaches to process modeling*. Hillsdale, NJ: Erlbaum.

Simmons, F. R., & Singleton, C. (2008). Do weak phonological representations impact on arithmetic development? A review of research into arithmetic and dyslexia. *Dyslexia*, *14*(2), 77–94.

Stanescu-Cosson, R., Pinel, P., van De Moortele, P., Le Bihan, D., Cohen, L., & Dehaene, S. (2000). Understanding dissociations in dyscalculia: A brain imaging study of the impact of number size on the cerebral networks for exact and approximate calculation. *Brain*, *123*(Pt 11), 2240–2255.

Thevenot, C., Barrouillet, P., Castel, C., & Uittenhove, K. (2015). Ten-year-old children strategies in mental addition: A counting model account. *Cognition*, *146*, 48–57. https://doi.org/10.1016/j.cognition.2015.09.003.

Uittenhove, K., Thevenot, C., & Barrouillet, P. (2015). Fast automated counting procedures in addition problem solving: When are they used and why are they mistaken for retrieval? *Cognition*, *146*, 289–303. https://doi.org/10.1016/j.cognition.2015.10.008.

Ullman, M. T. (2004). Contributions of memory circuits to language: The declarative/procedural model. *Cognition*, *92*(1–2), 231–270. https://doi.org/10.1016/j.cognition.2003.10.008.

Ullman, M. T. (2016). The declarative/procedural model: A neurobiological model of language learning, knowledge, and use. In G. Hickok, & S. A. Small (Eds.), *Neurobiology of language*. Amsterdam, Netherlands: Elsevier.

van Harskamp, N. J., Rudge, P., & Cipolotti, L. (2002). Are multiplication facts implemented by the left supramarginal and angular gyri? *Neuropsychologia*, *40*(11), 1786–1793.

van Harskamp, N. J., Rudge, P., & Cipolotti, L. (2005). Does the left inferior parietal lobule contribute to multiplication facts? *Cortex*, *41*(6), 742–752.

Varvara, P., Varuzza, C., Sorrentino, A. C., Vicari, S., & Menghini, D. (2014). Executive functions in developmental dyslexia. *Frontiers in Human Neuroscience*, *8*, 120. https://doi.org/10.3389/fnhum.2014.00120.

Vellutino, F. R., Fletcher, J. M., Snowling, M. J., & Scanlon, D. M. (2004). Specific reading disability (dyslexia): What have we learned in the past four decades? *Journal of Child Psychology and Psychiatry*, *45*(1), 2–40.

Vigneau, M., Beaucousin, V., Herve, P. Y., Duffau, H., Crivello, F., Houde, O., et al. (2006). Meta-analyzing left hemisphere language areas: Phonology, semantics, and sentence processing. *Neuroimage*, *30*(4), 1414–1432. https://doi.org/10.1016/j.neuroimage.2005.11.002.

Wilson, A. J., Andrewes, S. G., Struthers, H., Rowe, V. M., Bogdanovic, R., & Waldie, K. E. (2015). Dyscalculia and dyslexia in adults: Cognitive bases of comorbidity. *Learning and Individual Differences*, *37*, 118–132.

Zamarian, L., Ischebeck, A., & Delazer, M. (2009). Neuroscience of learning arithmetic–evidence from brain imaging studies. *Neuroscience and Biobehavioral Reviews*, *33*(6), 909–925. https://doi.org/10.1016/j.neubiorev.2009.03.005.

Zhou, X., Chen, C., Zang, Y., Dong, Q., Chen, C., Qiao, S., et al. (2007). Dissociated brain organization for single-digit addition and multiplication. *Neuroimage*, *35*(2), 871–880. https://doi.org/10.1016/j.neuroimage.2006.12.017.

CHAPTER

3

Language and Arithmetic: The Potential Role of Phonological Processing

Bert De Smedt

Faculty of Psychology and Educational Sciences, University of Leuven, Leuven, Belgium

OUTLINE

51

INTRODUCTION

The association between language processing and numerical cognition, or the absence thereof, has been subject to a long-standing debate (Amalric & Dehaene, 2016; Carey, 2009; Dehaene, 1992; Gelman & Butterworth, 2005; Hurford, 1987). As mathematics consists of multiple domains, including calculation, geometry, measurement, or algebra, it is likely that only some of them depend on language processing, and this dependence might vary across development. Likewise, language also involves different domains, ranging from phonology to morphology and syntax, and some of these may be more closely related to mathematics than others. In this contribution, I restrict the focus to one aspect of mathematics, i.e., arithmetic, and its connection to one aspect of language, i.e., phonological processing. This was done because one of the most influential models of numerical cognition, the triple-code model (Dehaene, 1992), already postulated a link between arithmetic fact retrieval and phonological processing in the left angular gyrus. On the other hand, brain imaging studies of arithmetic and reading suggest overlapping neural networks in the left temporoparietal cortex, including the angular gyrus (see Fig. 3.1), and this overlap has been often interpreted as reflecting a common reliance on phonological codes in reading and arithmetic fact retrieval (De Smedt, Taylor, Archibald, & Ansari, 2010), although alternative interpretations, such as symbol-referent mapping mechanisms (e.g., Andin, Fransson, Ronnberg, & Rudner, 2015; Price & Ansari, 2011), are gaining increasing interest, as I will describe in the discussion. In this chapter, I will review brain imaging and behavioral data that have revealed associations between arithmetic and

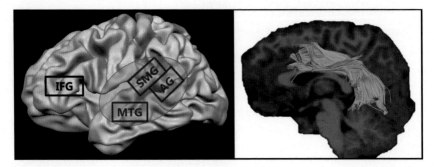

FIGURE 3.1 The left panel shows a lateral view of the left hemisphere. The *blue ellipse* represents the areas in the temporoparietal cortex that have been shown to be overlapping between arithmetic fact retrieval and phonological processing, including the angular gyrus (AG), supramarginal gyrus (SMG), and middle temporal gyrus (MTG). The right panel shows a sagittal slice of the brain on which the left arcuate fasciculus, a white matter tract that connects the inferior frontal gyrus (IFG) with the temporoparietal cortex, is depicted in blue.

phonological processing. I will end this chapter with a critical discussion of the existing evidence and some avenues for further study.

ARITHMETIC FACT RETRIEVAL AND THE TRIPLE-CODE MODEL

Decades of cognitive research have shown that the development of arithmetic consists of a change in the mix of strategies that people use to calculate the answer to a given problem (e.g., Siegler, 1996; for a review). Initially, children use a variety of procedural strategies, such as counting or decomposing a problem into smaller problems, as in $9+3=9+1+2=12$), to calculate the answer to a problem. Through the repeated use of these procedural strategies, they develop associations between problems $(4+3)$ and their answers (7), which are stored in long-term memory in an associative network. These associations are called arithmetic facts. The acquisition of these facts is important as arithmetic fact retrieval is more efficient (i.e., accurate and faster) than the error-prone and more cognitively demanding use of procedures. Over development, children develop an increasing reliance on arithmetic fact retrieval, yet procedural strategies remain available (Siegler, 1996), even in adulthood (Ashcraft, 1992; LeFevre, Sadesky, & Bisanz, 1996). It is important to note that the types of procedural strategies that are being used vary across countries and mathematics curricula. For example, this may vary depending on whether a curriculum stimulates or discourages a particular strategy, such as counting (De Smedt, 2016, pp. 219–243). In any case, across countries and curricula, individuals develop an increasing reliance on arithmetic fact retrieval and this becomes the dominant strategy for solving single-digit arithmetic.

There exist various cognitive models on how these arithmetic facts are represented in long-term memory (see Ashcraft, 1992 and Campbell, 2015, pp. 140–157, for an excellent descriptions and comparisons of these models). Earlier models of arithmetic fact retrieval, such as Ashcraft's *network retrieval model* (Ashcraft, 1992) and Siegler's *distributions of associations model* (Siegler & Shrager, 1984, pp. 229–293), did not explicitly address whether these facts were represented in a linguistic format or not. Subsequent cognitive architectures of arithmetic fact retrieval, including the *abstract code model* (McCloskey, 1992), *triple-code model* (Dehaene, 1992), and *encoding complex model* (Campbell, 1994) explicitly dealt with this issue. Although the *abstract code model*, which is exclusively based on patients with brain injury, posits a language-independent and amodal representation of arithmetic facts (McCloskey, 1992), both the *triple-code model* (Dehaene, 1992) and *encoding complex model* (Campbell, 1994) postulate the existence of language-based representations of arithmetic facts, yet the latter two models differ as to whether these arithmetic facts are exclusively (Dehaene, 1992) or dominantly but not exclusively (Campbell, 1994) represented in a linguistic format. This chapter aligns with these last two models of arithmetic fact retrieval, which both emphasize that linguistic codes have a key role in arithmetic fact retrieval. This chapter further focuses on the *triple-code model* as this model has been the dominant framework in studies on the association between phonological processing and arithmetic fact retrieval and because the *triple-code model* included detailed neuroanatomical predictions that have been tested in subsequent brain imaging studies (see Brain Imaging Data section). The triple-code model (Dehaene, 1992) posits that there are different representational systems to mentally process number, each of which is related to different aspects of calculation. One of these systems is a verbal system. This system is part of the general language network and is not specific to the processing of number. It was originally (tentatively) tied to operations that require the retrieval of arithmetic facts from memory, as is the case in multiplication and (small) addition. In this verbal system, numbers, as well as arithmetic facts, are represented lexically, phonologically, and syntactically, similar to any other type of word. This system was linked to the left-lateralized perisylvian language network (including superior and middle temporal gyri) and more specifically tied to the left angular gyrus (Dehaene, Piazza, Pinel, & Cohen, 2003). Evidence for this model originally accumulated via neuropsychological case studies in adult patients with acalculia (Dehaene & Cohen, 1995). These case studies reported double dissociations between arithmetic operations, with some patients showing deficits in subtraction but not multiplication, whereas other patients showing the opposite pattern. Patients with this latter pattern of deficits typically had lesions in the left perisylvian cortex and they often had associated aphasia (Dehaene et al., 2003). Functional magnetic resonance imaging (fMRI) studies in healthy adults additionally reported an association between activity in these left temporoparietal areas and arithmetic (Arsalidou & Taylor, 2011; Dehaene et al., 2003), particularly in multiplication (Grabner et al., 2007; Prado et al., 2011). It is important to emphasize that such operation effects on brain activity likely reflect differences in the arithmetic strategies that were used during different operations rather than effects of operation per se. Indeed, retrieving arithmetic facts is associated with temporoparietal brain activity, independent of the operation that is performed (Grabner et al., 2009; Polspoel, Peters, Vandermosten, & De Smedt, 2017; Tschentscher & Hauk, 2014).

ARITHMETIC AND READING: THE ROLE OF PHONOLOGICAL CODES

Various studies have reported strong and stable associations between arithmetic and reading, independently of more general cognitive variables (e.g., Fuchs et al., 2005; Hecht, Torgesen, Wagner, & Rashotte, 2001), and disorders in learning to calculate (dyscalculia) and learning to read (dyslexia) show frequent comorbidity (e.g., von Aster & Shalev, 2007; Kovas et al., 2007; Landerl & Moll, 2010; Lewis, Hitch, & Walker, 1994). A close inspection of the available data on the brain networks of reading (e.g., Martin, Schurz, Kronbichler, & Richlan, 2015; Schlaggar & McCandliss, 2007) and arithmetic (e.g., Arsalidou & Taylor, 2011; Dehaene et al., 2003; Peters & De Smedt, 2018) suggests that both abilities have a neural overlap in the temporoparietal cortex (De Smedt et al., 2010; Simmons & Singleton, 2008, for a discussion). More specifically, fMRI studies in reading have revealed that this temporoparietal area is particularly active during the reading of words that require extensive phonological decoding, which is most prominent in the reading of pseudowords and in the early stages of reading (Eden, Olulade, Evans, Krafnick, & Alkire, 2016, pp. 815–826; Martin et al., 2015; Schlaggar & McCandliss, 2007). On the other hand, fMRI studies in arithmetic have shown that brain activity in this temporoparietal cortex is associated with multiplication (Dehaene et al., 2003) and symbolic formats (Peters, Polspoel, Op de Beeck, & De Smedt, 2016) and that this activity increases with training (Zamarian, Ischebeck, & Delazer, 2009) and expertise (Grabner et al., 2007) in adults, conditions which are all reflective of an increasing reliance on arithmetic fact retrieval. Grabner et al. (2009) and Tschentscher and Hauk (2014) even directly correlated brain activity with strategy reports in arithmetic and observed that this temporoparietal activity is correlated with those items of which participants indicated that they solved them via fact retrieval, independently of the operation. Recent evidence by Polspoel et al. (2017), which compared subtractions and multiplications that were either solved by retrieval or by procedures, reported similar findings in fourth-grade children, clearly indicating that it is the strategy and not the operation, which is determining brain activity. These data all suggest a neural overlap between phonological processing and arithmetic fact retrieval.

Van Beek, Ghesquière, Lagae, and De Smedt (2014) used diffusion tensor imaging to examine this hypothesis of a neural overlap between reading and arithmetic at the level of structural brain connectivity. The authors embarked on the well-established association between the integrity of the arcuate fasciculus—a white matter tract that connects frontal and parietal areas, running through the temporoparietal cortex (Fig. 3.1)—and individual differences in language and reading, in particular in phonological processing (Vandermosten et al., 2012; Yeatman et al., 2011). Van Beek et al. (2014) investigated in typically developing 12-year-olds to which extent individual differences in the white matter of the arcuate fasciculus correlated with performance in arithmetic operations that differ in their reliance on arithmetic fact retrieval. Their data revealed that it was specifically the integrity of the anterior portion of the left arcuate fasciculus, which connects the inferior frontal gyrus with the temporoparietal cortex that was uniquely (i.e., independently of intelligence quotient (IQ) and working memory) correlated with arithmetic. Crucially, these associations were observed for addition and multiplication but not for subtraction and division. Follow-up analyses revealed that this association remained when word reading was controlled for, but disappeared when pseudoword reading was used as a covariate. Pseudoword reading draws much more heavily on the processing of phonological codes and consequently the temporoparietal cortex (Pugh et al., 2001; Schlaggar & McCandliss, 2007). These data suggest common processes in addition/multiplication and pseudoword reading, and this might reflect their common reliance on phonological codes.

PHONOLOGICAL PROCESSING AND ARITHMETIC FACT RETRIEVAL

Brain imaging and behavioral studies have more directly investigated this association between arithmetic fact retrieval and phonological processing. Phonological processing involves three interdependent dimensions (Wagner & Torgesen, 1987): (1) *phonological awareness* or the conscious sensitivity to the sound structure of the language; (2) *lexical access* or *rapid automatized naming* of phonological information in long-term memory, which involves the recoding of a visual symbol onto a sound-based representation by retrieving its lexical referent from long-term memory; and (3) *phonological working memory* or the temporary memory storage of phonological information similar to the phonological loop in Baddeley's, (2003) working memory model. Prototypical measures of phonological processing include rhyme judgment tasks (indicate which of two words rhyme) or tasks that require participants to identify or delete particular phonemes, all of which activate phonological representations. These tasks have been used as functional localizer tasks in brain imaging studies to delineate phonological regions of interest in which subsequently the brain activity during an arithmetic task was investigated (see Brain Imaging Data section). On the other hand, these tasks have been widely used in behavioral studies in which their association with measures of arithmetic has been examined (see Behavioral Evidence section).

Brain Imaging Data

Simon, Mangin, Cohen, Le Bihan, and Dehaene (2002) were the first to directly investigate the neural overlap between phonological processing and arithmetic. They investigated brain activity during a phonological task and a subtraction task, yet restricted their analyses to the parietal cortex. In the phonological task, participants had to verify whether a particular phoneme was included in a visually presented word or not. As predicted, this task, compared with a letter-detection control, elicited increased activity in the left inferior frontal gyrus and the left angular gyrus. In the subtraction task, participants had to subtract two numbers, a task that elicited a widespread bilateral frontoparietal network. Studying the overlap in brain activity between these two tasks, Simon et al. observed that a small region at the intersection of the left posterior segment of the intraparietal sulcus and the angular gyrus was coactivated during these two tasks. Even more striking, single-subject analyses revealed that this coactivation was observed in all participants under study.

Extending Simon et al. (2002), Prado et al. (2011) investigated brain activity during different arithmetic operations, i.e., subtraction and multiplication, which were predicted to differentially rely on arithmetic fact retrieval and consequently on phonological processing. Their phonological localizer task, a word rhyme judgment task, activated the left inferior frontal gyrus and the temporoparietal cortex, including the middle temporal gyrus, areas that were subsequently used as regions of interest in which brain activity during subtraction versus multiplication was examined. This analysis revealed that activity in this left temporoparietal area was larger during multiplication than during subtraction, suggestive of a more direct link between multiplication and phonological processing. Such neural overlap does not necessarily mean that neural processes are shared between these two abilities. It could be that the two tasks recruit intertwined yet functionally distinct neural mechanisms. To verify this, one needs to correlate the brain activity patterns of the two tasks, for example, via multivoxel pattern analysis (see Bulthé, De Smedt, & Op de Beeck, 2014, for an example in the field of numerical cognition). Prado et al. (2011), unfortunately, observed no association between the neural activation patterns during multiplication and phonological processing, which suggests that, despite the observed neural overlap, the activation patterns during multiplication and phonological processing seem to be qualitatively distinct. This interpretation, however, should be treated with caution as the two tasks that were correlated differed largely in their surface characteristics (e.g., visual input, type of task), which might explain the absence of associations between neural patterns.

Similar conclusions were reached by Andin et al. (2015), who used the same set of visual stimuli (e.g., 2E 3T 6Ö) in three task conditions in which participants had to indicate whether the product of the two digits equaled the third (multiplication), whether the difference of the two digits equaled the third (subtraction), or whether one letter and digit in a pair rhymed (phonological processing). Although these authors also observed a general neural overlap between the three tasks in the left angular gyrus, a more fine-grained analysis suggested more divergence between phonological processing and arithmetic. Anterior parts of the left angular gyrus (PGa) were more activated during the phonological task, whereas the posterior left angular gyrus (PGp) was more activated during arithmetic. This pattern is not unexpected because the PGa is close to the classic language areas. On the other hand, PGp regions have been implicated in symbol-referent mapping, suggesting that the activity of the angular gyrus during arithmetic may not necessarily be reflective of phonological processing, as I will return to in the discussion of this chapter.

To conclude, the above-reviewed data support the prediction of overlapping networks of phonological processing and arithmetic. On the other hand, the available data are not completely straightforward because more specific analyses suggest regional differences in the way phonological and arithmetic tasks recruit the left temporoparietal cortex. We also do not understand the precise convergence of both abilities because none of the fMRI studies applied conjunction analyses that could determine the coactivation of tasks. As the abovementioned studies almost exclusively employed rhyme judgment tasks, it is also unclear to which extent different dimensions of phonological processing relate to arithmetic. Finally, it is unknown to which extent the arithmetic tasks used in these studies were tapping into arithmetic fact retrieval, as no strategy data were collected to verify if retrieval was used or not. These latter two issues have been investigated at a more fine-grained level in behavioral studies, as evidenced in the next section.

Behavioral Evidence

Various developmental studies have observed a behavioral association between phonological processing and arithmetic, particularly in more advanced stages of arithmetic development (e.g., De Smedt et al., 2009; Fuchs et al., 2005; Hecht et al., 2001; Koponen, Aunola, Ahonen, & Nurmi, 2007; Koponen, Salmi, Eklund, & Aro, 2013; LeFevre et al., 2010; Noël, Seron, & Trovarelli, 2004; Schleepen, Van Mier, & De Smedt, 2016; Vanbinst, Ansari, Ghesquière, & De Smedt, 2016; Vukovic & Lesaux, 2013; however, see; Passolunghi, Mammarella, & Altoe, 2008). These studies investigated the various dimensions of phonological processing and revealed that each of these is related to arithmetic. One of the most comprehensive studies has been provided by Hecht et al. (2001), who studied the association between phonological processing and arithmetic in a longitudinal study, following children from second to fifth grade. They observed that phonological awareness was a powerful unique predictor of growth in calculation from second to fifth grade, whereas the roles of lexical access and phonological memory were much smaller and time-limited.

The association between phonological processing and arithmetic has been interpreted as reflecting a common reliance on phonological codes and phonological representations (e.g., Simmons & Singleton, 2008). Poor phonological representations hamper the ability to manipulate, retrieve, and maintain phonological codes. Likewise, poor phonological representations of arithmetic facts hamper their efficient retrieval. Although measures of phonological awareness, lexical access, and phonological memory all draw on phonological representations, phonological awareness tasks may be taxing the most on such representations (Boada & Pennington, 2006; Elbro, 1996), explaining why most strong and consistent associations have been observed with this measure (e.g., Hecht et al., 2001).

An outstanding issue in the abovementioned studies is whether the association between phonological processing and arithmetic is specific to the retrieval of arithmetic facts, as predicted by the triple-code model (Dehaene, 1992), or whether this association is more general, such that it also extends to procedural strategies, such as counting or multistep procedures that may require subvocal rehearsal. De Smedt et al. (2010) tested this in 9- to 11-year-old typically developing Canadian children, who solved single-digit arithmetic problems that could be categorized as problems that at this age are likely to be solved via fact retrieval (multiplications and additions ≤ 10) and problems that are likely to be solved via procedural strategies (additions and subtractions > 10), which typically involve at this age a decomposition of a problem into smaller problems. Children also completed a phoneme elision task, in which they had to indicate what a word would be if a phoneme were to be deleted. De Smedt et al. observed that phonological processing was significantly related to the retrieval but not to the procedural problems. This association was unique and remained when reading ability was controlled for. This association was specific to phonological awareness and not explained by phonological memory, leading De Smedt et al. (2010) to conclude that it is specifically the quality of phonological representations, which is explaining the association between phonological processing and arithmetic fact retrieval.

Vanbinst, Ceulemans, Ghesquière, and De Smedt (2015) longitudinally investigated children's development of arithmetic facts in a study from third to fifth grade, measuring children's arithmetic in each grade. These authors used a model-based cluster analysis to determine different developmental trajectories of arithmetic fact development. Parameters of children's fact retrieval mastery, which were collected in each grade, comprised retrieval frequency, as measured by verbal self-reports, reaction time, and the variance in reaction time. These data served as a basis for the model-based cluster analysis, which revealed three clusters of arithmetic fact development: slow and variable, average, and efficient. These clusters did not differ in age, sex, or IQ and did not include children with learning difficulties. Crucially in the context of this contribution, children's phonological processing was investigated by tasks that tapped into its three different dimensions, i.e., phonological awareness (phoneme deletion), lexical access (rapid automatized color naming), and phonological memory (nonword repetition). Data revealed that the clusters differed in phonological awareness and lexical access but not in phonological memory. These cluster differences were entirely explained by the poor performance of children in the slow and variable cluster, who performed significantly more poorly than the other two clusters, who in turn did not differ. It is important to note that the arithmetic task in this study only comprised addition and subtraction but did not include multiplication. On the other hand, these data also suggest that the association between phonological processing and arithmetic may be nonlinear. In other words, this association might only appear in the context of poor arithmetic performance and consequently, poor phonological processing might act as a risk factor for atypical arithmetic development.

Atypical Development

Arithmetic fact retrieval deficits are the hallmark of children with dyscalculia or mathematical learning difficulties (American Psychiatric Association, 2013; Geary, 1993, 2004, 2011; Robinson, Menchetti, & Torgesen, 2002). Are these deficits then associated with poor phonological processing? Phonological processing deficits have been put forward as the key feature of dyslexia, a neurodevelopmental learning disorder that is characterized by persistent difficulties in word decoding that are not better accounted for by intellectual disabilities, sensory problems, or inadequate educational opportunities (American Psychiatric Association, 2013; Snowling, 2000; Vellutino, Fletcher, Snowling, & Scanlon, 2004). Do these individuals with dyslexia show weaknesses in arithmetic fact retrieval?

Phonological Processing Deficits in Dyscalculia

In his seminal description of subtypes of dyscalculia, Geary (1993) hypothesized a fact retrieval subtype, which he explained in terms of a core deficit in the retrieval of (phonological) information from semantic memory, although these fact retrieval difficulties could also be due to difficulties in inhibitory control (Geary, 2004). Relatedly, Robinson et al. (2002) suggested poor phonological abilities as one potential source of fact retrieval deficits in dyscalculia, in particular in those conditions of dyscalculia that also cooccur with reading disabilities. As a result, they predicted that poor phonological processing might explain the comorbidity between dyscalculia and dyslexia, which has a high prevalence (Landerl & Moll, 2010).

Various studies revealed poorer performance in phonological processing in children with dyscalculia (e.g., Chong & Siegel, 2008; Murphy, Mazzocco, Hanich, & Early, 2007; Vukovic & Siegel, 2010), although it needs to be noted that in these studies, the presence of reading disabilities was not specifically investigated. It is possible that the observed phonological processing difficulties in samples of dyscalculia are due to the comorbidity of reading disabilities. On the other hand, others have argued that these phonological processing difficulties in dyscalculia are independent of their reading difficulties (Simmons & Singleton, 2008). To verify this, one needs to study phonological processing in children with dyscalculia, who do not show difficulties in reading. Such studies have been carried out, yet the existing evidence continues to be mixed. Vanbinst, Ghesquière, and De Smedt (2014) investigated children with persistent dyscalculia, who also showed persistent difficulties in arithmetic fact retrieval, had average reading abilities and did not differ from matched controls in IQ and working memory. These authors observed that children with dyscalculia performed significantly more poorly on all dimensions of phonological processing, and these differences remained when reading abilities were additionally controlled for. A reanalysis of these data further revealed that the group differences in arithmetic fact retrieval remained significant when differences in phonological processing were controlled for, although the effect size for the group difference dropped considerably, yet remained high in its size. This suggests that phonological processing played a role in the fact retrieval deficits observed in these children with dyscalculia, but it certainly did not fully explain these difficulties. On the other hand, Moll, Göbel, and Snowling (2015) reported that children with dyscalculia without reading difficulties did not differ from controls in their phonological awareness and lexical access; these phonological abilities were not related to their poor arithmetic performance.

It is unlikely that poor phonological processing skills alone cause deficits in arithmetic fact retrieval. Fact retrieval deficits also occur in children with good phonological skills (e.g., Landerl, Bevan, & Butterworth, 2004), and they are particularly prominent in the context of poor symbolic numerical magnitude processing (e.g., Vanbinst & De Smedt, 2016, pp. 105–130, for a review). It remains unclear to which extent phonological processing and symbolic numerical processing uniquely contribute to individual differences in arithmetic fact retrieval. Vanbinst et al. (2016) tested this in typically developing children. Phonological awareness was correlated with arithmetic (cross-sectionally but not longitudinally) and regression analyses, as well as Bayesian analyses, indicated that the evidential strength for an association between phonological processing and arithmetic was far less strong compared with the simultaneously considered and very strong association between symbolic numerical magnitude processing and arithmetic. It remains to be determined whether this also generalizes to atypical development.

On the other hand, poor phonological processing could act as an additional risk factor for the development of fact retrieval difficulties, pointing to the possibility that there are multiple pathways to develop arithmetic fact retrieval difficulties. Such a possibility of a similar behavioral phenotype (fact retrieval difficulties) that emerges from different cognitive factors was already theoretically postulated by Robinson et al. (2002), yet empirical studies in, preferably large, samples of children with dyscalculia are needed to verify this.

Arithmetic Fact Retrieval Deficits in Dyslexia

Deficits in phonological processing have been indicated to be a core deficit in dyslexia (e.g., Snowling, 2000; Vellutino et al., 2004). If a reliable association between phonological processing and arithmetic fact retrieval exists, then one would predict arithmetic fact retrieval difficulties in dyslexia. Temple (1991) described a 19-year-old female patient with dyslexia and also analyzed her calculation abilities. Interestingly, this patient was specifically impaired in multiplication. She did not show other calculation deficits and her number processing skills were in the normal range. Difficulties in arithmetic fact retrieval in dyslexia have been reported for a long time (Geary, Hamson, & Hoard, 2000; Miles, 1983; Simmons & Singleton, 2008; Träff & Passolunghi, 2015), but the presence of dyscalculia in these studies was not controlled for, leaving it unresolved whether the poor arithmetic in dyslexia is explained by comorbid dyscalculia, and consequently impairments number processing, or by poor phonological processing. To test this latter explanation, one should observe arithmetic fact retrieval difficulties in individuals with dyslexia without dyscalculia (and consequently without deficits in number processing).

This prediction was tested in children (Boets & De Smedt, 2010; Moll et al., 2015; Träff, Desoete, & Passolunghi, 2017) and adults (De Smedt & Boets, 2010; Göbel & Snowling, 2010) with dyslexia without difficulties in their general mathematical achievement (and who consequently did not meet criteria for dyscalculia). Boets and De Smedt (2010) showed that third graders with dyslexia performed significantly more poorly than matched controls in single-digit arithmetic, a finding that was later replicated by Moll et al. (2015) and Träff et al. (2017). Boets and De Smedt (2010) showed that these group differences were most prominent in multiplication. More specifically, control children were significantly faster in multiplication than in subtraction, potentially indicating the reaction time advantage of retrieval in multiplication compared with procedure use in subtraction. Children with dyslexia did not show such operation effect, suggesting that they used less retrieval or less efficient retrieval or both during multiplication. Similar findings were observed in adults with dyslexia without dyscalculia (Göbel & Snowling, 2010). Using fMRI, Evans, Flowers, Napoliello, Olulade, and Eden (2014) observed in children with dyslexia less activity in the left supramarginal gyrus during addition and subtraction compared with healthy matched controls, revealing an overlap between less optimal fact retrieval and atypical activity in left perisylvian language regions. Interestingly, hypoactivity in the left supramarginal gyrus has been frequently observed during reading-related tasks in dyslexia (Richlan, Kronbichler, & Wimmer, 2011). A limitation of these data is that none of these studies have provided direct assessments of the strategies participants used. To test the hypothesis of fact retrieval difficulties in dyslexia, it needs to be demonstrated that individuals with dyslexia either report less use of fact retrieval strategies or that they perform more poorly when they retrieve facts from their memory. Even more, to test the hypothesis that phonological processing and arithmetic fact retrieval are linked, one needs to demonstrate that poor arithmetic fact retrieval in individuals with dyslexia is directly related to their poor phonological skills.

De Smedt and Boets (2010) tested these predictions in university students with a clinical diagnosis of dyslexia but without a diagnosis of dyscalculia and no history of remedial teaching for mathematics. This group was compared with a group of healthy university students who were individually matched to the students with dyslexia in terms of their study discipline, age, sex, and IQ. The groups did not differ in their nonsymbolic magnitude processing skills, indicating that their nonsymbolic representations of number were not impaired. All completed an extensive phonological processing battery, tapping into all its different dimensions, and they solved subtraction and multiplication problems during which their strategy use was recorded on a trial-by-trial basis. As predicted, individuals with dyslexia had significantly lower frequencies of fact retrieval than controls and they were significantly slower, particularly in multiplication. Crucially, the frequency of fact retrieval was related to phonological processing in both adults with dyslexia and controls, such that higher frequencies of fact retrieval were associated with better phonological abilities. These associations were most prominent for phonological awareness, again suggesting that it is particularly the quality of phonological representations that explains the association between phonological processing and arithmetic fact retrieval.

Experimental Data

The above-reviewed data are all correlational, and the vast majority is even cross-sectional, which leaves it unresolved whether the association between phonological processing and fact retrieval is causal or not; note that the same critique applies to the above-reviewed brain imaging data. One way to shed light on this issue is via dual-task experiments where arithmetic performance with and without phonological load is compared. In their dual-task experiment, Lee and Kang (2002) investigated the effect of phonological suppression (i.e., performing a phonological secondary (short-term) memory task, e.g., repeating a nonword, while performing a primary task of interest, in this case arithmetic) on arithmetic in healthy adults. They observed that phonological suppression significantly impaired performance during multiplication but not during subtraction.

These experimental data suggest that the association between phonological processing and fact retrieval could be causal, at least in multiplication. Another way to investigate this would be to examine the effect of an intervention that taps into phonological processing on arithmetic (fact retrieval). To the best of my knowledge, such studies have not been conducted so far, yet they might be important to further unravel the association between phonological processing and fact retrieval.

DISCUSSION

The current overview of brain and behavioral studies points to an association between phonological processing and arithmetic. This association is most prominent in the context of the retrieval of arithmetic facts, which might be stored in a phonological format, as originally postulated by the triple-code model (Dehaene, 1992). The association is more frequently observed in multiplication and might be related to the way in which multiplication is taught (i.e., via rote recitation of multiplication tables). Even when reading ability is taken into account, the association holds and this suggests that it is not the mere covariation of arithmetic and reading that is driving it. Against the background of different dimensions of phonological processing (Wagner & Torgesen, 1987), the association is most often found for phonological awareness. The above-reviewed studies also indicate that the association is particularly observed in atypical development, suggesting that poor phonological skills might be one risk factor for arithmetic fact retrieval difficulties.

It is important to point out that the bulk of the available evidence is correlational (except for Lee & Kang, 2002), and even cross-sectional, except for Fuchs et al. (2005) and Hecht et al. (2001). In view of the lack of longitudinal data, one should be very cautious in interpreting these associations. There are various interpretations of these associations: (1) phonological processing could act as a prerequisite for arithmetic fact retrieval, (2) changes in phonological processing could be a consequence of developing arithmetic facts, (3) phonological processing could be a facilitator for arithmetic fact retrieval, such that it is not a necessary skill to develop arithmetic facts but that those who have good phonological skills progress faster in their arithmetic fact development than those who do not, and (4) phonological processing could be functionally independent of arithmetic fact retrieval, but a correlation between phonological processing and arithmetic fact retrieval occurs because of a mediating third variable, such as school experience. In the absence of longitudinal data, it is not possible to distinguish between these four possibilities, and therefore longitudinal data are crucial to make headway in our understanding of this association. On the other hand, if longitudinal data would suggest a causal connection between phonological processing and arithmetic fact retrieval, this needs to be tested in an experimental design, for example, by investigating the effect of phonologically based (reading) instruction on arithmetic fact retrieval.

Longitudinal data at various time points throughout development are also critical to unravel the time course of the association. The current body of studies includes samples that vary considerably in their age ranges. One crucial question is to determine at which point in the development of arithmetic facts phonological skills play their most prominent role. For example, phonological processing might play a role in the transition from procedural strategies to fact retrieval, or it might play a role when arithmetic facts become more consolidated and automatized, or both. The answer to this question will additionally be determined by educational curricula and the transparency of the instructional language, which differ across countries. More specifically, educational curricula will differ in the extent to which instruction emphasizes the reliance on fact retrieval (Campbell & Xue, 2001; De Smedt, 2016, pp. 219–243). It could therefore be predicted that the association between phonological processing and arithmetic will be stronger in contexts that place a larger emphasis on retrieval. On the other hand, the transparency of the language, i.e., the consistency between spelling and sounds, affects children's phonological skills and determines the effects of specific phonological skills on reading development, i.e., languages with more inconsistent spelling-sound correspondence, such as English, show a longer and larger effect of phonological awareness on reading (e.g., Torgesen, Wagner, Rashotte, & Burgess, 1997), whereas in languages with more consistent spelling-sound mappings, lexical access plays a more prominent role (e.g., Landerl & Wimmer, 2008). It is not unlikely that these language effects will play a role in the association between phonological processing and arithmetic too, and this should be tested in future studies.

Another outstanding question deals with the mechanism behind the association between phonological processing and arithmetic fact retrieval. Tasks that measure phonological awareness, which show the most consistent associations with arithmetic, are also tapping to a large extent in working memory because they require individuals to maintain phonological information while doing some additional processing on the input task (e.g., deleting a phoneme). On the other hand, there are consistent associations between individual differences in working memory and arithmetic (Peng, Namkung, Barnes, & Sun, 2016, for a metaanalysis). This might suggest that working memory acts a third variable in explaining the phonological processing–arithmetic association. Two observations render this explanation unlikely. Firstly, the associations remain when additional measures of phonological memory are taken into account (De Smedt & Boets, 2010; De Smedt et al., 2010; Vanbinst et al., 2014). Secondly, if working memory is driving the association between phonological processing and arithmetic, the association should be the most prominent for arithmetic problems that put the highest load on working memory, i.e., larger problems or problems that are solved via procedural strategies. This is not what is observed. Instead, the association is the strongest for those problems that require less working memory, i.e., smaller problems or problems that are solved via retrieval (De Smedt & Boets, 2010; De Smedt et al., 2010). This indicates that working memory does not merely account for the association between phonological processing and arithmetic.

The involvement of phonological representations in both phonological processing and arithmetic fact retrieval has been one prominent explanation for their association, and this has been fueled by the observation that phonological processing and arithmetic fact retrieval coactivate the left temporoparietal cortex (e.g., Simmons & Singleton, 2008). Indeed, the phonological tasks that are the most sensitive to phonological representations, i.e., phonological awareness (Boada & Pennington, 2006; Elbro, 1996), showed the most consistent associations with arithmetic fact retrieval in children (De Smedt et al., 2010; Hecht et al., 2001) and adults (De Smedt & Boets, 2010). This suggest that more distinct phonological representations will add to more efficient arithmetic fact retrieval as more precise phonological representations are easier to retrieve. There is a need for future studies to test this prediction more rigorously with tasks that measure phonological representations more directly, i.e., without metalinguistic processing requirements (see Boada & Pennington, 2006), and examine their associations with arithmetic fact retrieval. Such studies should also verify the locus of this association, i.e., whether it occurs at the level of encoding of facts, storing them in memory, or mapping them to a phonological output.

The above-reviewed studies on the coactivation of arithmetic fact retrieval and phonological processing in the left temporoparietal cortex have not always been very precise on the exact neuroanatomical location of this coactivation. Indeed, the temporoparietal areas that have been referred to comprise a variety of regions (Fig. 3.1), including the angular gyrus and its subdivisions (Andin et al., 2015), the adjacent supramarginal gyrus (Evans et al., 2014), and the more distant middle temporal gyrus (Prado et al., 2011). Although these regions are all in the classic left temporoparietal language areas, each of them might reflect distinct aspects of linguistic processing (see Price, 2012, for a review), and these aspects might play different roles in arithmetic fact retrieval. It is therefore important for future studies to carefully specify the exact anatomical locations of overlap and to flesh out the overlap between arithmetic fact retrieval and specific linguistic processes.

The interpretation of overlapping brain activity between phonological processing and arithmetic fact retrieval in the left temporoparietal cortex, in particular the left angular gyrus, as reflecting a common reliance on phonological processing, has been questioned. Regional differences in angular gyrus activity between phonological processing tasks and arithmetic have been observed (Andin et al., 2015), and multivariate analyses of brain activity patterns indicated that their patterns of brain activity are not correlated (Prado et al., 2011). This indicates that the neuronal processes in the angular gyrus during phonological processing and arithmetic may not be the same. It also might be that this neural overlap is not necessarily linguistic (see also Benn, Zheng, Wilkinson, Siegal, & Varley, 2012).

Another interpretation of the overlapping brain activity in the left angular gyrus might be that it reflects domain-general processing of semantically weighted stimuli and that it may be related to processing the mappings between visual symbols and their semantic referents (Price & Ansari, 2011) or, applied to arithmetic, between arithmetic problems and their solutions stored in memory (Ansari, 2008; Grabner, Ansari, Koschutnig, Reishofer, & Ebner, 2013). When a particular arithmetic problem becomes overlearned, it might be stored as higher-order (chunk of) symbols that automatically activate their solution in memory. This interpretation has been framed as the symbol-referent mapping hypothesis (Ansari, 2008; Grabner et al., 2013). For example, Grabner et al. (2013) examined the associative confusion effect, comparing verification problems where the solution to the problem corresponded with another operation on the same operands or confusion problems (e.g., $9 \times 6 = 15$) with problems where the solution was not related to both operations (e.g., $9 \times 6 = 52$). During all these fact retrieval problems, it were the confusion problems that showed consistently higher activity in the (left) angular gyrus. The confusion effect can only arise if there is a mapping between the symbols (operands) and their semantic referent (solution). In other words, the mapping process is more reflected in the confusion problems (where operands are associated with correct solutions) compared with the nonconfusion problems. These data are not so easy to explain by recourse to the role of phonological representations per se, given that the critical difference between the task conditions is not straightforwardly related to a difference in phonological representations. On the other hand, these differences can be predicted by the symbol-referent mapping account as it is the difference in automatic mapping between chunks of symbols that explains this difference in brain activity. Future imaging studies are needed to directly contrast the phonological representation versus symbol-referent mapping accounts.

Interestingly, Koponen et al. (2013) observed in their longitudinal data that lexical access remained a key factor in predicting calculation fluency and reading fluency, once measures of phonological awareness were controlled for, but it needs to be noted that their measure of calculation included fact retrieval and multistep arithmetic. This suggests that such association might also be explained by arbitrary mapping mechanisms between visual symbols (i.e., numbers or even problems) and their (verbal) referents (i.e., number words/sounds and even solutions) that must be learned and retrieved quickly (see also Bull & Johnston, 1997; Göbel, Watson, Lervag, & Hulme, 2014) and consequently might question the hypothesis of phonological representations as a key factor in explaining these associations. On the other hand, Hecht et al. (2001) and De Smedt and Boets (2010) did not observe a consistent association between lexical access and arithmetic fact retrieval, once phonological awareness was additionally controlled for.

It is unlikely that only one cognitive factor or brain area is critical for the development of number and arithmetic, its individual differences, and its impairments (e.g., Fias, Menon, & Szucs, 2013; LeFevre et al., 2010; Peters & De Smedt, 2018). It is becoming increasingly clear that a simple linear causal pathway from one factor to a behavioral outcome is too simplified and does not provide a satisfactory account of atypical development (Pennington, 2006), as this is now increasingly recognized in many psychiatric conditions (Fried et al., 2016). Instead, various heterogeneous contributions should be considered, resulting in multiple brain and behavioral pathways of typical arithmetic development, as well as multiple risk and protective factors in the context of atypical development. Phonological processing could therefore act as a risk or protective factor for arithmetic development. Such approach, however, requires research to move beyond single-factor studies and to consider multiple factors or brain networks that affect numerical and arithmetical cognition and its difficulties. From a practical point of view, this all suggest that multiple measures, which are not necessarily all domain-specific, should be considered when diagnosing children with difficulties in arithmetic.

References

Amalric, M., & Dehaene, S. (2016). Origins of the brain networks for advanced mathematics in expert mathematicians. *Proceedings of the National Academy of Sciences of the United States of America, 113*(18), 4909–4917. https://doi.org/10.1073/pnas.1603205113.

American Psychiatric Association. (2013). *Diagnostic and statistical manual of mental disorders* (5th ed.). Washington, DC: Author.

Andin, J., Fransson, P., Ronnberg, J., & Rudner, M. (2015). Phonology and arithmetic in the language-calculation network. *Brain and Language, 143*, 97–105. https://doi.org/10.1016/j.bandl.2015.02.004.

Ansari, D. (2008). Effects of development and enculturation on number representation in the brain. *Nature Reviews Neuroscience, 9*(4), 278–291. https://doi.org/10.1038/nrn2334.

Arsalidou, M., & Taylor, M. J. (2011). Is 2+2=4? Meta-analyses of brain areas needed for numbers and calculations. *Neuroimage, 54*(3), 2382–2393. https://doi.org/10.1016/j.neuroimage.2010.10.009.

Ashcraft, M. H. (1992). Cognitive arithmetic: A review of data and theory. *Cognition, 44*, 75–106.

von Aster, M. G., & Shalev, R. S. (2007). Number development and developmental dyscalculia. *Developmental Medicine and Child Neurology, 49*(11), 868–873.

Baddeley, A. (2003). Working memory: Looking back and looking forward. *Nature Reviews Neuroscience, 4*(10), 829–839. https://doi.org/10.1038/nrn1201.

Benn, Y., Zheng, Y., Wilkinson, I. D., Siegal, M., & Varley, R. (2012). Language in calculation: A core mechanism? *Neuropsychologia*, *50*(1), 1–10. https://doi.org/10.1016/j.neuropsychologia.2011.09.045.

Boada, R., & Pennington, B. F. (2006). Deficient implicit phonological representations in children with dyslexia. *Journal of Experimental Child Psychology*, *95*(3), 153–193. https://doi.org/10.1016/j.jecp.2006.04.003.

Boets, B., & De Smedt, B. (2010). Single-digit arithmetic in children with dyslexia. *Dyslexia*, *16*(2), 183–191. https://doi.org/10.1002/dys.403.

Bull, R., & Johnston, R. S. (1997). Children's arithmetical difficulties: Contributions from processing speed, item identification, and short-term memory. *Journal of Experimental Child Psychology*, *65*(1), 1–24. https://doi.org/10.1006/jecp.1996.2358.

Bulthé, J., De Smedt, B., & Op de Beeck, H. P. (2014). Format-dependent representations of symbolic and non-symbolic numbers in the human cortex as revealed by multi-voxel pattern analyses. *Neuroimage*, *87*, 311–322. https://doi.org/10.1016/j.neuroimage.2013.10.049.

Campbell, J. I. D. (1994). Architectures for numerical cognition. *Cognition*, *53*, 1–44.

Campbell, J. I. D. (2015). How abstract is arithmetic? In R. Cohen-Kadosh, & A. Dowker (Eds.), *The Oxford handbook of numerical cognition*. Oxford, UK: Oxford University Press.

Campbell, J. I. D., & Xue, Q. L. (2001). Cognitive arithmetic across cultures. *Journal of Experimental Psychology-General*, *130*(2), 299–315. https://doi.org/10.1037//0096-3445.130.2.299.

Carey, S. (2009). Where our number concepts come from. *Journal of Philosophy*, *106*(4), 220–254.

Chong, S. L., & Siegel, L. S. (2008). Stability of computational deficits in math learning disability from second through fifth grades. *Developmental Neuropsychology*, *33*(3), 300–317. https://doi.org/10.1080/87565640801982387.

De Smedt, B. (2016). Individual differences in arithmetic fact retrieval. In D. Berch, D. Geary, & K. Mann-Koepke (Eds.). D. Berch, D. Geary, & K. Mann-Koepke (Eds.), *Mathematical cognition and learning: (Vol. 2)*. San Diego, CA: Elsevier Academic Press.

De Smedt, B., & Boets, B. (2010). Phonological processing and arithmetic fact retrieval: Evidence from developmental dyslexia. *Neuropsychologia*, *48*(14), 3973–3981. https://doi.org/10.1016/j.neuropsychologia.2010.10.018.

De Smedt, B., Janssen, R., Bouwens, K., Verschaffel, L., Boets, B., & Ghesquière, P. (2009). Working memory and individual differences in mathematics achievement: A longitudinal study from first grade to second grade. *Journal of Experimental Child Psychology*, *103*(2), 186–201. https://doi.org/10.1016/j.jecp.2009.01.004.

De Smedt, B., Taylor, J., Archibald, L., & Ansari, D. (2010). How is phonological processing related to individual differences in children's arithmetic skills? *Developmental Science*, *13*(3), 508–520. https://doi.org/10.1111/j.1467-7687.2009.00897.x.

Dehaene, S. (1992). Varieties of numerical abilities. *Cognition*, *44*(1–2), 1–42. https://doi.org/10.1016/0010-0277(92)90049-n.

Dehaene, S., & Cohen, L. (1995). Towards an anatomical and functional model of number processing. *Mathematical Cognition*, *1*(1), 83–120.

Dehaene, S., Piazza, M., Pinel, P., & Cohen, L. (2003). Three parietal circuits for number processing. *Cognitive Neuropsychology*, *20*(3–6), 487–506. https://doi.org/10.1080/02643290244000239.

Eden, G. F., Olulade, O. A., Evans, T. M., Krafnick, A. J., & Alkire, D. A. (2016). Developmental dyslexia. In G. Hickok, & S. Small (Eds.), *Neurobiology of language*. Oxford, UK: Elsevier.

Elbro, C. (1996). Early linguistic abilities and reading development: A review and a hypothesis. *Reading and Writing, 8*(6), 453–485. https://doi.org/10.1007/bf00577023.

Evans, T. M., Flowers, D. L., Napoliello, E. M., Olulade, O. A., & Eden, G. F. (2014). The functional anatomy of single-digit arithmetic in children with developmental dyslexia. *Neuroimage, 101*, 644–652. https://doi.org/10.1016/j.neuroimage.2014.07.028.

Fias, W., Menon, V., & Szucs, D. (2013). Multiple components of developmental dyscalculia. *Trends in Neuroscience and Education, 2*(2), 43–47.

Fried, E. I., van Borkulo, C. D., Cramer, A. O. J., Lynn, B., Schoevers, R. A., & Borsboom, D. (2016). Mental disorders as networks of problems: A review of recent insights. *Social Psychiatry and Psychiatric Epidemiology*. https://doi.org/10.1007/s00127-016-1319-z.

Fuchs, L. S., Compton, D. L., Fuchs, D., Paulsen, K., Bryant, J. D., & Hamlett, C. L. (2005). The prevention, identification, and cognitive determinants of math difficulty. *Journal of Educational Psychology, 97*(3), 493–513. https://doi.org/10.1037/0022-0663.97.3.493.

Geary, D. C. (1993). Mathematical disabilities – cognitive, neuropsychological, and genetic components. *Psychological Bulletin, 114*(2), 345–362. https://doi.org/10.1037//0033-2909.114.2.345.

Geary, D. C. (2004). Mathematics and learning disabilities. *Journal of Learning Disabilities, 37*(1), 4–15. https://doi.org/10.1177/00222194040370010201.

Geary, D. C. (2011). Consequences, characteristics, and causes of mathematical learning disabilities and persistent low achievement in mathematics. *Journal of Developmental and Behavioral Pediatrics, 32*(3), 250–263. https://doi.org/10.1097/DBP.0b013e318209edef.

Geary, D. C., Hamson, C. O., & Hoard, M. K. (2000). Numerical and arithmetical cognition: A longitudinal study of process and concept deficits in children with learning disability. *Journal of Experimental Child Psychology, 77*(3), 236–263. https://doi.org/10.1006/jecp.2000.2561.

Gelman, R., & Butterworth, B. (2005). Number and language: How are they related? *Trends in Cognitive Sciences, 9*(1), 6–10. https://doi.org/10.1016/j.tics.2004.11.004.

Göbel, S. M., & Snowling, M. J. (2010). Number-processing skills in adults with dyslexia. *Quarterly Journal of Experimental Psychology, 63*(7), 1361–1373. https://doi.org/10.1080/17470210903359206.

Göbel, S. M., Watson, S. E., Lervag, A., & Hulme, C. (2014). Children's arithmetic development it is number knowledge, not the approximate number sense, that counts. *Psychological Science, 25*(3), 789–798. https://doi.org/10.1177/0956797613516471.

Grabner, R. H., Ansari, D., Koschutnig, K., Reishofer, G., & Ebner, F. (2013). The function of the left angular gyrus in mental arithmetic: Evidence from the associative confusion effect. *Human Brain Mapping, 34*(5), 1013–1024. https://doi.org/10.1002/hbm.21489.

Grabner, R. H., Ansari, D., Koschutnig, K., Reishofer, G., Ebner, F., & Neuper, C. (2009). To retrieve or to calculate? Left angular gyrus mediates the retrieval of arithmetic facts during problem solving. *Neuropsychologia, 47*(2), 604–608. https://doi.org/10.1016/j.neuropsychologia.2008.10.013.

Grabner, R. H., Ansari, D., Reishofer, G., Stern, E., Ebner, F., & Neuper, C. (2007). Individual differences in mathematical competence predict parietal brain activation during mental calculation. *Neuroimage, 38*(2), 346–356. https://doi.org/10.1016/j.neuroimage.2007.07.041.

Hecht, S. A., Torgesen, J. K., Wagner, R. K., & Rashotte, C. A. (2001). The relations between phonological processing abilities and emerging individual differences in mathematical computation skills: A longitudinal study from second to fifth grades. *Journal of Experimental Child Psychology, 79*(2), 192–227. https://doi.org/10.1006/jecp.2000.2586.

Hurford, J. R. (1987). *Language and number.* Oxford, UK: Basil Blackwell.

Koponen, T., Aunola, K., Ahonen, T., & Nurmi, J. E. (2007). Cognitive predictors of single-digit and procedural calculation skills and their covariation with reading skill. *Journal of Experimental Child Psychology, 97*(3), 220–241. https://doi.org/10.1016/j.jecp.2007.03.001.

Koponen, T., Salmi, P., Eklund, K., & Aro, T. (2013). Counting and RAN: Predictors of arithmetic calculation and reading fluency. *Journal of Educational Psychology, 105*(1), 162–175. https://doi.org/10.1037/a0029285.

Kovas, Y., Haworth, C. M. A., Harlaar, N., Petrill, S. A., Dale, P. S., & Plomin, R. (2007). Overlap and specificity of genetic and environmental influences on mathematics and reading disability in 10-year-old twins. *Journal of Child Psychology and Psychiatry, 48*(9), 914–922. https://doi.org/10.1111/j.1469-7610.2007.01748.x.

Landerl, K., Bevan, A., & Butterworth, B. (2004). Developmental dyscalculia and basic numerical capacities: A study of 8-9-year-old students. *Cognition, 93*(2), 99–125. https://doi.org/10.1016/j.cognition.2003.11.004.

Landerl, K., & Moll, K. (2010). Comorbidity of learning disorders: Prevalence and familial transmission. *Journal of Child Psychology and Psychiatry, 51*(3), 287–294. https://doi.org/10.1111/j.1469-7610.2009.02164.x.

Landerl, K., & Wimmer, H. (2008). Development of word reading fluency and spelling in a consistent orthography: An 8-year follow-up. *Journal of Educational Psychology, 100*(1), 150–161. https://doi.org/10.1037/0022-0663.100.1.150.

Lee, K. M., & Kang, S. Y. (2002). Arithmetic operation and working memory: Differential suppression in dual tasks. *Cognition, 83*(3), B63–B68. https://doi.org/10.1016/s0010-0277(02)00010-0.

LeFevre, J. A., Fast, L., Skwarchuk, S. L., Smith-Chant, B. L., Bisanz, J., Kamawar, D., et al. (2010). Pathways to mathematics: Longitudinal predictors of performance. *Child Development, 81*(6), 1753–1767. https://doi.org/10.1111/j.1467-8624.2010.01508.x.

LeFevre, J., Sadesky, G. S., & Bisanz, J. (1996). Selection of procedures in mental addition: Reassessing the problem size effect in adults. *Journal of Experimental Psychology: Learning, Memory and Cognition, 22*, 216–230. https://doi.org/10.1037/0278-7393.22.1.216.

Lewis, C., Hitch, G. J., & Walker, P. (1994). The prevalence of specific arithmetic difficulties and specific reading difficulties in 9-year-old to 10-year-old boys and girls. *Journal of Child Psychology and Psychiatry and Allied Disciplines, 35*(2), 283–292. https://doi.org/10.1111/j.1469-7610.1994.tb01162.x.

Martin, A., Schurz, M., Kronbichler, M., & Richlan, F. (2015). Reading in the brain of children and adults: A meta-analysis of 40 functional magnetic resonance imaging studies. *Human Brain Mapping, 36*(5), 1963–1981. https://doi.org/10.1002/hbm.22749.

McCloskey, M. (1992). Cognitive mechanisms in numerical processing: Evidence from acquired dyscalculia. *Cognition, 44*, 107–157.

Miles, T. R. (1983). *Dyslexia: The pattern of difficulties*. Springfield, IL: Charles C Thomas.

Moll, K., Göbel, S. M., & Snowling, M. J. (2015). Basic number processing in children with specific learning disorders: Comorbidity of reading and mathematics disorders. *Child Neuropsychology, 21*(3), 399–417. https://doi.org/10.1080/09297049.2014.899570.

Murphy, M. M., Mazzocco, M. M. M., Hanich, L. B., & Early, M. C. (2007). Cognitive characteristics of children with mathematics learning disability (MLD) vary as a function of the cutoff criterion used to define MLD. *Journal of Learning Disabilities, 40*(5), 458–478. https://doi.org/10.1177/00222194070400050901.

Noël, M. P., Seron, X., & Trovarelli, F. (2004). Working memory as a predictor of addition skills and addition strategies in children. *Cahiers De Psychologie Cognitive-Current Psychology of Cognition, 22*(1), 3–25.

Passolunghi, M. C., Mammarella, I. C., & Altoe, G. (2008). Cognitive abilities as precursors of the early acquisition of mathematical skills during first through second grades. *Developmental Neuropsychology, 33*(3), 229–250. https://doi.org/10.1080/87565640801982320.

Peng, P., Namkung, J., Barnes, M., & Sun, C. Y. (2016). A meta-analysis of mathematics and working memory: Moderating effects of working memory domain, type of mathematics skill, and sample characteristics. *Journal of Educational Psychology, 108*(4), 455–473. https://doi.org/10.1037/edu0000079.

Pennington, B. F. (2006). From single to multiple deficit models of developmental disorders. *Cognition, 101*(2), 385–413. https://doi.org/10.1016/j.cognition.2006.04.008.

Peters, L., & De Smedt, B. (2018). Arithmetic in the developing brain: A review of brain imaging studies. *Developmental Cognitive Neuroscience.* https://doi.org/10.1016/j.dcn.2017.05.002.

Peters, L., Polspoel, B., Op de Beeck, H., & De Smedt, B. (2016). Brain activity during arithmetic in symbolic and non-symbolic formats in 9-12 year old children. *Neuropsychologia, 86*, 19–28. https://doi.org/10.1016/j.neuropsychologia.2016.04.001.

Polspoel, B., Peters, L., Vandermosten, M., & De Smedt, B. (2017). Strategy over operation: Neural activation in subtraction and multiplication during fact retrieval and procedural strategy use in children. *Human Brain Mapping, 38*, 4657–4670. https://doi.org/10.1002/hbm.23691.

Prado, J., Mutreja, R., Zhang, H. C., Mehta, R., Desroches, A. S., Minas, J. E., et al. (2011). Distinct representations of subtraction and multiplication in the neural systems for numerosity and language. *Human Brain Mapping, 32*(11), 1932–1947. https://doi.org/10.1002/hbm.21159.

Price, C. J. (2012). A review and synthesis of the first 20 years of PET and fMRI studies of heard speech, spoken language and reading. *Neuroimage, 62*, 816–847. https://doi.org/10.1016/j.neuroimage.2012.04.062.

Price, G. R., & Ansari, D. (2011). Symbol processing in the left angular gyrus: Evidence from passive perception of digits. *Neuroimage, 57*(3), 1205–1211. https://doi.org/10.1016/j.neuroimage.2011.05.035.

Pugh, K. R., Mencl, W. E., Jenner, A. R., Katz, L., Frost, S. J., Lee, J. R., et al. (2001). Neurobiological studies of reading and reading disability. *Journal of Communication Disorders*, *34*(6), 479–492. https://doi.org/10.1016/s0021-9924(01)00060-0.

Richlan, F., Kronbichler, M., & Wimmer, H. (2011). Meta-analyzing brain dysfunctions in dyslexic children and adults. *Neuroimage*, *56*(3), 1735–1742. https://doi.org/10.1016/j.neuroimage.2011.02.040.

Robinson, C. S., Menchetti, B. M., & Torgesen, J. K. (2002). Toward a two-factor theory of one type of mathematics disabilities. *Learning Disabilities Research and Practice*, *17*, 81–89.

Schlaggar, B. L., & McCandliss, B. D. (2007). Development of neural systems for reading. *Annual Review of Neuroscience*, *30*, 475–503.

Schleepen, T. M. J., Van Mier, H. I., & De Smedt, B. (2016). The contribution of numerical magnitude comparison and phonological processing to individual differences in fourth graders' multiplication fact ability. *PLoS One*, *11*(6). https://doi.org/10.1371/journal.pone.0158335.

Siegler, R. S. (1996). *Emerging minds: The process of change in children's thinking.* Oxford, UK: Oxford University Press.

Siegler, R. S., & Shrager, J. (1984). Strategy choices in addition and subtraction: How do children know what to do? In C. Sophian (Ed.), *The origins of cognitive skills*. Hillsdale, NJ: Erlbaum.

Simmons, F. R., & Singleton, C. (2008). Do weak phonological representations impact on arithmetic development? A review of research into arithmetic and dyslexia. *Dyslexia*, *14*(2), 77–94. https://doi.org/10.1002/dys.341.

Simon, O., Mangin, J. F., Cohen, L., Le Bihan, D., & Dehaene, S. (2002). Topographical layout of hand, eye, calculation, and language-related areas in the human parietal lobe. *Neuron*, *33*(3), 475–487. https://doi.org/10.1016/s0896-6273(02)00575-5.

Snowling, M. J. (2000). *Dyslexia* (2nd ed.). Malden, MA: Blackwell Publishers.

Temple, C. M. (1991). Procedural dyscalculia and number fact dyscalculia – double dissociation in developmental dyscalculia. *Cognitive Neuropsychology*, *8*(2), 155–176. https://doi.org/10.1080/02643299108253370.

Torgesen, J. K., Wagner, R. K., Rashotte, C. A., & Burgess, S. H. S. (1997). Contributions of phonological awareness and rapid automatic naming ability to the growth of word-reading skills in second- to fifth-grade children. *Scientific Studies of Reading*, *1*(2), 161–185.

Träff, U., Desoete, A., & Passolunghi, M. C. (2017). Symbolic and non-symbolic number processing in children with developmental dyslexia. *Learning and Individual Differences*, *56*, 105–111. https://doi.org/10.1016/j.lindif.2016.10.010.

Träff, U., & Passolunghi, M. C. (2015). Mathematical skills in children with dyslexia. *Learning and Individual Differences*, *40*, 108–114. https://doi.org/10.1016/j.lindif.2015.03.024.

Tschentscher, N., & Hauk, O. (2014). How are things adding up? Neural differences between arithmetic operations are due to general problem solving strategies. *Neuroimage*, *92*, 369–380. https://doi.org/10.1016/j.neuroimage.2014.01.061.

Van Beek, L., Ghesquière, P., Lagae, L., & De Smedt, B. (2014). Left fronto-parietal white matter correlates with individual differences in children's ability to solve additions and multiplications: A tractography study. *Neuroimage, 90*, 117–127. https://doi.org/10.1016/j.neuroimage.2013.12.030.

Vanbinst, K., Ansari, D., Ghesquière, P., & De Smedt, B. (2016). Symbolic numerical magnitude processing is as important to arithmetic as phonological awareness is to reading. *PLoS One, 11*(3). https://doi.org/10.1371/journal.pone.0151045.

Vanbinst, K., Ceulemans, E., Ghesquière, P., & De Smedt, B. (2015). Profiles of children's arithmetic fact development: A model-based clustering approach. *Journal of Experimental Child Psychology, 133*, 29–46. https://doi.org/10.1016/j.jecp.2015.01.003.

Vanbinst, K., & De Smedt, B. (2016). Individual differences in children's mathematics achievement: The roles of symbolic numerical magnitude processing and domain-general cognitive functions. In M. Cappelletti, & W. Fias (Eds.)M. Cappelletti, & W. Fias (Eds.), *Mathematical brain across the lifespan: (Vol. 227)*.

Vanbinst, K., Ghesquière, P., & De Smedt, B. (2014). Arithmetic strategy development and its domain-specific and domain-general cognitive correlates: A longitudinal study in children with persistent mathematical learning difficulties. *Research in Developmental Disabilities, 35*(11), 3001–3013. https://doi.org/10.1016/j.ridd.2014.06.023.

Vandermosten, M., Boets, B., Poelmans, H., Sunaert, S., Wouters, J., & Ghesquière, P. (2012). A tractography study in dyslexia: Neuroanatomic correlates of orthographic, phonological and speech processing. *Brain, 135*, 935–948. https://doi.org/10.1093/brain/awr363.

Vellutino, F. R., Fletcher, J. M., Snowling, M. J., & Scanlon, D. M. (2004). Specific reading disability (dyslexia): What have we learned in the past four decades? *Journal of Child Psychology and Psychiatry, 45*(1), 2–40. https://doi.org/10.1046/j.0021-9630.2003.00305.x.

Vukovic, R. K., & Lesaux, N. K. (2013). The language of mathematics: Investigating the ways language counts for children's mathematical development. *Journal of Experimental Child Psychology, 115*(2), 227–244. https://doi.org/10.1016/j.jecp.2013.02.002.

Vukovic, R. K., & Siegel, L. S. (2010). Academic and cognitive characteristics of persistent mathematics difficulty from first through fourth grade. *Learning Disabilities Research and Practice, 25*, 25–38.

Wagner, R. K., & Torgesen, J. K. (1987). The nature of phonological processing and its causal role in the acquisition of reading skills. *Psychological Bulletin, 101*(2), 192–212. https://doi.org/10.1037//0033-2909.101.2.192.

Yeatman, J. D., Dougherty, R. F., Rykhlevskaia, E., Sherbondy, A. J., Deutsch, G. K., Wandell, B. A., et al. (2011). Anatomical properties of the arcuate fasciculus predict phonological and reading skills in children. *Journal of Cognitive Neuroscience, 23*(11), 3304–3317.

Zamarian, L., Ischebeck, A., & Delazer, M. (2009). Neuroscience of learning arithmetic-Evidence from brain imaging studies. *Neuroscience and Biobehavioral Reviews, 33*(6), 909–925. https://doi.org/10.1016/j.neubiorev.2009.03.005.

Discussion: Specific Contributions of Language Functions to Numerical Cognition

Klaus Willmes

RWTH Aachen University, Aachen, Germany

OUTLINE

INTRODUCTION

Historically, a close link and a clear distinction were drawn between language functions and numerical cognition. Right from the beginning of research on impairments of number processing and calculation, coined acalculia by Solomon Henschen (1920), he considered calculation

Heterogeneity of Function in Numerical Cognition
https://doi.org/10.1016/B978-0-12-811529-9.00004-2

mechanisms to be similar in complexity to language- and music mechanisms. Henschen offered a systemic view and considered numerical cognition to form an anatomo-functional system, which he conceived of as a "psychic formation" ("psychischer Verband"). Henschen could base his theoretical thinking on an impressively large collection of postmortem anatomical brain studies of patients carefully examined for impairment, and he viewed such a functional formation to be subserved by distinct cortical centers and their related association fibers, the cortical centers showing a certain degree of independence but also interdependencies mediated by possibly topologically distant association fibers. Around the same time, on the other hand, Berger (1926) introduced the distinction between primary and secondary acalculia, the former one designating impairments in calculation and number processing not reducible to impairments of short- or long-term memory, language, reading, attention, or other cognitive functions.

One privileged approach in neuropsychology to understanding the dependence or independence of cognitive functions as implemented in the brain is to look for performance dissociations as an indication of fractionated language and numerical cognition functions possibly relying on separable brain regions or systems. Like for many other cognitive functions, the neuropsychological literature comprises also a substantial body of cases in which rather specific performance dissociations between (often rather basic) language and numerical tasks were documented as a demonstration of the fractionation of the number system (McCloskey, Caramazza, & Basili, 1985) and for which a very comprehensive and scholarly review was presented by Cappelletti (2015, chap. 44). She starts from a characterization of the number system as including a "number semantic system" for the understanding and manipulation of number concepts, a "verbal system" for numerical processes preferentially mediated by language, and an "executive control/attention/memory system" managing numerical manipulations, which calls for monitoring and attentional resources. These systems can be discerned conceptually, but regular numerical processing as required in a given laboratory or everyday task may require more than one system at a time, in particular when the retrieval of numerical information in a verbal format is implied. A short sketch of a clinical case may exemplify this perspective.

Before doing so I just want to mention that I will not touch on the debate of whether or to what extent high(er)-level mathematics is also rooted in the language faculty, e.g., claimed by Chomsky, or whether there are "multiple languages" of the brain, as argued—and backed up by an fMRI study—e.g., by Amalric and Dehaene (2016), with "high-level mathematical expertise and basic number sense sharing common roots in a nonlinguistic brain circuit." These authors also argue that only rote memory for arithmetic facts, which constitutes jut a small portion of arithmetic knowledge, is encoded linguistically and that most of number comprehension and algebraic manipulations may be preserved even in patients with global aphasia or semantic dementia. One should obviously also list counting (forwards and backwards), reading out/naming numbers, and writing numbers to dictation as activities that require language. As pointed out by Piazza, Pinel, Le Bihan, and Dehaene (2007), there remains an unresolved difficulty that symbolic numerals do not merely refer to approximate numerosities. There must be a (last) step (labeled "crystallization" by these authors) in processing where a numeral comes to acquire an exact meaning (e.g., "42" is exactly "forty-two"), a process which requires language and some degree of education. Related to this problem is the perspective of Wiese (2003). She claims that number words and Arabic numbers are not different labels for the same abstract numbers but elements of different systems of "numerical tools" (a nonverbal, visual system of digits; a verbal system with phonological representations of number words) with generative rules for the production of infinite sequences (progressions) of distinct elements. As numerical tools, Arabic numbers and number words are used in cardinal, ordinal, and nominal assignments; therefore, they should have associative connections (termed "noniconic system-dependent linking" by Wiese) with concepts of numerical quantities, numerical rank orders, and numerical labels: (four—4 pens; fourth—fourth runner; bus line number four/ bus #4). Similarly, symbolic number systems are also viewed by Lefevre, Wells, and Sowinski (2015, p. 908, chap. 48) as cognitive tools developed to circumvent cognitive limitations and not a reflection of a universal linguistic/cognitive capacity.

CLINICAL CASE

This quite typical case seen by me in routine clinical diagnostic at the aphasia ward of the University Clinic of RWTH Aachen University will serve to demonstrate what would have been called secondary acalculia subsequent to aphasic language impairment. The patient had suffered his first-ever stroke at the age of 40 years, 6 months before having been referred to the intensive 7 weeks language therapy program. He was right handed with German as his first language but speaking several other languages working as a bank manager. MRI in the subacute phase revealed a left temporoparietal lesion; the patient had neither hemiplegia nor any visual or auditory problems. Routine language assessment during the first week on the ward with the Aachen Aphasia Test (AAT; Huber, Poeck, & Willmes, 1984) revealed fluent spontaneous speech in the semistructured interview of the AAT with semantic and phonemic paraphasias, strong word-finding problems, and paragrammatic syntax. The AAT performance profile resulted in the psychometrically based diagnosis of Wernicke's aphasia, which also matched the clinical impression.

Routine diagnostics for acalculia were based primarily on the German version of the EC 301 R (Claros Salinas, 1994, original version by Deloche et al., 1993) and some additional in-house tasks complementing the EC 301 R. Writing Arabic numbers to dictation revealed no problems for less complex one- to six-digit numbers (25; 75; 118; 700; 400,000) with a limited number of lexical constituents. Problems emerged for morphosyntactically more complex numbers, e.g., for the inversion property of German number words with an inverse linear order of the tens and unit number word (examiner: *siebenhundertfünfundachtzig* [seven hundred five and eighty, literally]—patient: *758* with subsequent self-correction to *785*) or a combined syntactic and lexical error, possibly also because of an overload of verbal working memory (examiner: *elftausenddreihundertsechzig* [eleven thousand three hundred sixty]—patient: *218,000*). Reading aloud (naming) of numbers was severely compromised, with lexical substitutions already for one-digit numbers (*5*—*sechs* [six]), but always an element from the number word lexicon and no responses for more complex numbers (*1854*, *4712*). The problem of the patient was, however, not at the semantic stage because when presenting structurally equivalent "encyclopedic numbers" (Cappelletti, 2015, chap. 44) such as *1945*, the patient could not produce the complex number word sequence but responded *Hitler weg* (Hitler gone). When being presented 4711, the patient recognized this very familiar German eau-de-cologne brand label and uttered *Oma hat das* (grandmother has it), making an additional pantomime of spraying perfume. A clear demonstration that Arabic digit numbers could be understood at the semantic level can be made by presenting pairs of multidigit numbers, asking the patient for pointing to the larger number (e.g., 354–345; 2057–2065; 13,648–12,648; 29,400–29,399; 10,100–10,010; 675,342–576,243); the patient's performance was fast without any hesitations (29/29 correct). Symbolic numerical magnitude comparison may therefore be flawless even for severely language impaired patients; in my clinical experience even for globally aphasic patients, intact semantic magnitude representation remains accessible through this nonlanguage symbolic notation (see the triple-code model in the following). To name small, mostly single-digit numbers, which regularly cannot be named spontaneously in case of more severe aphasia, patients spontaneously develop (or can be instructed easily to use) the compensatory strategy to start counting upward from one up to the number that cannot be named spontaneously. This highly overlearned sequence seems to be recited automatically and the intact semantic magnitude representation seems to "signal" when to stop counting. Frequently this oral counting strategy is accompanied by a finger tapping or pressing sequence on the table or somewhere at the body. Occasionally patients find this preschool or first-grade counting support embarrassing and try to hide it from the examiner. But patients should be supported by the examiner to apply this strategy if helpful. In addition, if oral counting backward (e.g., down to "one" and starting from "twenty-five") cannot be done, these more strongly affected aphasic patients can always write down the appropriate sequence using Arabic digits.

A caveat seems appropriate here: This traditional (cognitive) neuropsychological approach of acquired lesion studies in single patients or lesion-function studies in large groups of patients looking for dissociations in performance for language versus numerical tasks does provide relevant information but is, e.g., silent with respect to whether the better preserved or even unimpaired task(s) in a given patient indicates the premorbid level of performance or whether there may be compensation going on.

ANATOMO-FUNCTIONAL
SYSTEMS-LEVEL APPROACH

The anatomo-functional "triple-code model" (TCM) by Dehaene and Cohen (1995) and its elaboration of the differential contributions of parietal areas (bilateral horizontal intraparietal sulcus, left angular gyrus, bilateral posterior parietal cortex) (Dehaene, Piazza. Pinel, Cohen, 2003) has broadened the view to account for dissociations and associations between the number domain and other cognitive domains also summarized by Cappelletti (2015, chap. 44). The comprehensive quantitative metaanalysis of fMRI studies about number processing and calculation by Arsalidou and Taylor (2011) also revealed that the TCM framework is still not comprehensive enough to cover the brain areas subserving mathematical performance. Cingulate gyri, the insula, the cerebellum, and a hierarchy of dorsolateral and frontopolar regions of the prefrontal cortices were proposed as additional candidates for an updated TCM.

The latter metaanalysis is still centered on the joint contribution of cortical (and subcortical) brain areas, but brain imaging methods comprising different variants of DTI also helped to pave the way for a modern hodological approach to the study of brain functioning (Catani & ffytche, 2005). According to this latter perspective, intact respectively disconnected or affected interhemispheric commissural fibers, cortico-subcortical projection fibers, and cortico-cortical association fibers were held responsible or at least contributing to behaviorally observable intact respectively disordered higher brain functions such as language and numerical cognition. This disconnection perspective is also compatible with the view of the brain being composed of localized and segregated but connected functionally specialized areas working together both in parallel and in a hierarchically ordered, sequential way (Catani, 2011) via all three types of connecting fiber tracts. Broadly speaking, this view can be characterized as a systems-level perspective on cognitive functions and disorders. Already in 1985 Ivar Reinvang coined the term "systemic localization" to characterize the overarching principle of brain organization: the function of some brain area is not determined by its anatomical structure alone but by "its relationship to other areas within the bounds determined by its anatomical structure" (Reinvang, 1985, p. 18). This view allows also for change and recovery of function after acquired brain damage with white matter pathways mediating such systemic adjustments. Reinvang considered aphasia as his test case: starting from well-known perisylvian language areas of the (left) hemisphere dominant for language (Broca's area, Wernicke's area, supramarginal, and angular gyrus), interconnections (at his time, prominently the arcuate fasciculus) between areas should be considered all potentially contributing not only to language impairments when lesioned but also to include neighboring areas, which—when lesioned in isolation—would mostly not lead to aphasia but contribute to "symptom formation." New candidates for being assigned a status as language areas in the latter sense were—according to Reinvang—several structures of the lenticular zone comprising insular cortex, capsula extrema, claustrum, capsula externa, and basal ganglia nuclei, as well as structures of the limbic system, among them cingulate and parahippocampal gyri, the hippocampal formation, and thalamic nuclei, together with their fiber connections.

A still more "radical" perspective regarding the systems-level approach with respect to language is taken up by Skipper, Devlin, and Lametti (2017), where distributed brain regions involved in producing and comprehending speech play specific, dynamic, and contextually determined roles depending on task requirements. These authors presented a quantitative review that employed region- and network-based neuroimaging metaanalyses, as well as a novel text mining method to describe relative contributions of nodes in distributed brain networks. The authors identified a distributed set of cortical and subcortical speech production regions to be ubiquitously active and to form multiple networks whose composition dynamically changes with listening context. In their view, the findings were inconsistent with motor and acoustic only models of speech perception and also classical and contemporary dual-stream models of the organization of language and the brain, but more consistent with complex network models in which multiple speech production related networks and subnetworks dynamically self-organize. Such a comprehensive analysis has not been undertaken for the numerical cognition domain so far.

Some performance dissociations may even be better accommodated with a hodological perspective (Klein, Moeller, & Willmes, 2013). As an example I want to consider arithmetic fact retrieval for multiplication tables. According to the TCM the retrieval of arithmetic fact knowledge is verbally mediated and subserved by a neural system relying on left-hemispheric perisylvian language areas and the angular gyrus. It is thus assumed that arithmetic facts are retrieved directly from long-term memory without the need of numerical magnitude manipulations. Instead, the verbally mediated representation format is claimed to trigger the retrieval of a word sequence from memory, which is associated with the angular gyrus (e.g., *"three times four equals twelve"*). We did a reassessment of the lesion of our patient WT, whom Zaunmueller et al. (2009) in a training study had reported to be suffering from a severe multiplication fact retrieval deficit compared with addition and subtraction, although his brain lesion due to hemorrhagic stroke did not involve the left angular gyrus or left-hemispheric cortical language areas. Instead there was a perifocal edema affecting his left basal ganglia. We showed recently, again with a hodological perspective, that fact retrieval–related processing was subserved by a frontoparietal network, comprising dorsal and ventral connections (Klein et al., 2013), which is also distinct from the other most important number magnitude processing frontoparietal network, and also dorsal and ventral connections (see below). For arithmetic fact retrieval, these connections between parietal and frontal cortex encompass the extreme and the external capsule system for the ventral pathway(s) and the superior longitudinal fascicule for the dorsal pathway(s). One interpretation of the initial performance pattern before training, which is also in line with the original TCM perspective, would suggest that the basal ganglia affection is responsible for the selective performance deficit. A close inspect of the lesion revealed that there was probably a disconnection of both dorsal and ventral fiber pathway systems, which would also account for the multiplication impairment of WT. Left angular gyrus and left-hemispheric perisylvian language areas were not affected by the lesion but no longer connected to frontal areas such as Broca's area and to the basal ganglia.

WITHIN-DOMAIN AND CROSS-DOMAIN COGNITIVE PROCESSES PERSPECTIVES

From a methodological perspective, it seems reasonable, whenever possible, to study the joint versus distinct working of language-based and semantic numerical magnitude-based processes in a within-task approach, to avoid differences in, e.g., fMRI-detectable activation patterns, just because of accessory differences in stimulus and/or response properties comprising any task. We exemplified this approach (Klein, Moeller, Glauche, et al., 2013; Klein, Moeller, & Willmes, 2013) for mental arithmetic by combining a reanalysis of previously published fMRI activation peak data with probabilistic fiber-tracking data from an independent sample of participants. We could differentiate neural correlates and connectivity for relatively easy (e.g., 14 + 3) and more difficult (e.g., 54 + 38) addition problems in healthy adults and their association with either rather verbally mediated fact retrieval or magnitude manipulations, respectively. The brain activation data suggested that magnitude- and fact retrieval–related processing may be subserved by two largely separate networks, which both comprise dorsal and ventral connections. Both networks not only differed in the localization of activation patterns but also in the white matter connections between the cortical areas involved. Nevertheless, these networks, although seemingly distinct anatomically, operated as a functionally integrated circuit for mental calculation as revealed by a parametric analysis of brain activation to be depending on the two parametric predictors sum of decade digits and sum of units digits. For both predictors, increasing values were associated with an increase of activation in areas subserving magnitude-related processing (e.g., intraparietal sulcus, posterior superior parietal lobule), as well as frontal activation related to working memory (e.g., BA 44), the application of rules (BA 47) and processes supporting operation procedures (supplementary motor area). Decreasing values were associated with increasing activation in areas assumed to be recruited in arithmetic fact retrieval (e.g., left angular gyrus) and left-hemispheric language areas. More simple additions were also related to increased retrosplenial cortical activation, possibly as an indication of familiarity in the recognition process of a small addition problem.

More compelling, a significant conjunction of concordant activations of these predictors indicated areas subserving number magnitude processing, whereas a significant conjunction of discordant activations was associated with activation in areas subserving verbally mediated processes of arithmetic fact retrieval. In summary, magnitude- and fact retrieval–related processing were subserved by two largely separate networks, both comprising dorsal and ventral connections. These networks not only differed in localization of activation but also in the connections between the cortical areas involved.

Using the same type of approach, we could corroborate the distinction between two largely separate dorsal and ventral processing networks with their connecting pathways (Klein et al., 2016) for the so-called number bisection task, in which participants have to decide whether the middle number of three numbers arranged in ascending order from left to right is also the arithmetical middle of the two outer numbers. In the same probabilistic DTI study, a very similar distinction could be established for data from a previous activation study concerning exact versus approximate additions.

The joint consideration of patterns of activated brain areas and their white matter connections led us to proposing an update for the TCM (cf. Fig. 3 in Klein et al., 2016). The number magnitude processing network would not only comprise sections of the intraparietal sulcus but also prefrontal areas were engaged: Broca's Area (precisely: inferior frontal gyrus, pars triangularis, area 45) as well as BA 47 and the supplementary motor area including their ventral and dorsal connecting fiber pathways with the intraparietal areas. These additional areas are not confined to the tasks actually analyzed but considered essential for number processing in general. With respect to the verbally mediated representation of arithmetic facts, we also suggest that the verbal representation needs to be extended to a network subserving fact retrieval, including connections of the angular gyrus to retrosplenial cortex and further on via dorsal and ventral fiber pathways to ventromedial prefrontal cortex, a subcomponent of the mentalizing network, and the hippocampus. Both latter regions may subserve the automatic recognition of familiar or anticipated arithmetic facts, while the hippocampus not only reflects retrieval of arithmetic facts from long-term memory but has also been associated with the recognition of familiarity as well.

The exact way in which the magnitude-related network and the verbally mediated fact retrieval network operate in a closely integrated fashion as well as through which anatomical structures this cooperation is specified, awaits further clarification. We suggest invariant and variable components of both number magnitude processing and fact retrieval to be activated whenever one encounters a numerical problem irrespective of the task or its difficulty. It has been shown repeatedly that the intraparietal sulcus is activated automatically whenever we see or hear numerical stimuli. On the other hand, it was observed that even in a nondemanding number-matching task, the product of the two digits was also activated automatically. We assume that stable components of number magnitude and arithmetic facts processing are complemented by variable ones, whose activation depends on task requirements. With increasing task difficulty, the influence of fact retrieval processes may decrease, whereas the influence of magnitude-related processes may increase complimentarily. However, we assume processes of fact retrieval to also take effect in difficult arithmetic. This proposition is corroborated by participants breaking down complex problems into more tractable subproblems, the solutions of which are retrieved from long-term memory. In this way, our model is not only capable of accounting for fact retrieval deficits often exhibited by brain-damaged patients, by assuming an interruption of the pathways underlying arithmetic fact retrieval (e.g., Zaunmueller et al., 2009). This also offers an explanation for the often severe problems in more difficult arithmetic tasks observed after brain damage. Because a stable component of arithmetic fact retrieval is also involved in more difficult problems, loss of arithmetic fact knowledge will influence difficult arithmetic as well. Furthermore, we do not claim that the brain networks identified are specific to numerical tasks only (cf. also Willmes, Moeller, & Klein, 2014). (Parts of) the fact retrieval network may be involved in nonnumerical tasks, which require retrieval of factual knowledge from long-term memory.

As regard the anatomical correlates of the interaction of the two networks, the present study did not aim to find the exact location of this interaction. Thus, one may only speculate in which structures the two networks might interact. From an anatomical point of view, this might most probably be between the angular gyrus and the intraparietal sulcus, which are not only anatomically close but also closely connected via association fibers and most probably via U-fibers as well.

In a next step, it would be interesting to investigate the interrelation between the two networks in a more dynamic context. So far, we looked for stable structural connectivity. In a training study, it seems reasonable to assume that tasks, which are predominantly solved by number magnitude manipulations in the beginning, may be solved by arithmetic fact retrieval with increasing proficiency. In an extensive training study in healthy participants for multiplication fact problems, we measured brain activation before and after training to evaluate neural correlates of arithmetic fact acquisition. When comparing activation patterns for trained and untrained problems of the posttraining session, higher angular gyrus activation for trained problems was replicated. But when activation for trained problems was compared with activation for the same problems in the pretraining session, no signal change in the angular gyrus was observed (Bloechle et al., 2016). Rather the results indicate a central role of hippocampal, parahippocampal, and retrosplenial structures in arithmetic fact retrieval. The angular gyrus might not be associated with the actual retrieval of arithmetic facts but be involved in attention to memory processes for carrying out the fact retrieval in other brain areas.

In this context, it would also be desirable to examine not only the structural but also the effective connectivity among these networks. For instance, employing multivariate dynamic causal modeling, it would be required to further specify the direction and strength of the connections identified in the present study.

Finally, I want to go a step further and pursue the hypothesis of underlying joint cognitive processes or functions in a cross-domain rather than a domain-specific manner. In line with the importance of white matter connections of cortex sites associated with specific domains, these common underlying cognitive processes or functions should be indicated by the involvement of specific white matter tracts connecting core regions subserving domain-specific aspects of these processes. This argument was studied by us (Willmes et al., 2014) when evaluating white matter connectivity for the specific cognitive process of semantic classification, which is an integral part of tasks commonly employed to investigate neural correlates of language (words vs. nonwords; living vs. nonliving) and number processing (e.g., odd/even digit; number magnitude as smaller/larger than a standard). The notion of a common classification process subserved by specific white matter structures would imply common ventral frontoparietal white matter connections for semantic classification in both the domains of language and number processing involving the angular gyrus, supramarginal gyrus and superior temporal gyrus at the temporo-parietal, and the inferior frontal gyrus (BA 44 and 45) at the frontal end. In line with our expectations, fiber tracking results clearly indicated a common ventral network for semantic classification for the domains of language and number processing including white matter connectivity that should be considered an integral part of the neural underpinnings of human cognition.

A more refined analysis of the impact of language on numerical cognition tasks would also benefit from studying closely matched tasks in behavioral experiments in which either numbers in symbolic notation or linguistic entities (letter, words) had to be processed (Rath et al., 2015). Aphasic patients ($n = 60$) were assessed with a battery of 11 linguistic and numerical tasks comparable with their cognitive processing levels (e.g., perceptual, morpholexical, semantic), and associations and dissociations of function were studied in a within-participant approach for the whole profile of task performances (cf. Table 2 for a description of the asemantic numbers vs. letters, asemantic numbers vs. words, and semantic numbers vs. words tasks in Rath et al., 2015). Our findings revealed that the extent of numerical advantages depends on the cognitive level of the task (asemantic vs. semantic) and the type of material (numbers compared with letters vs. words). Mean performance differences and frequencies of (complementary) dissociations in individual patients revealed the most prominent numerical advantage for asemantic tasks when comparing the processing of numbers versus letters, whereas the least numerical advantage was found for semantic tasks when comparing the processing of numbers versus words. Even if not required by the task, automatic access to the number semantic level may have made processing easier for numbers. Overall the numerical advantage found across tasks and the dissociations between performance in numerical and linguistic tasks is consistent with the notion of a certain degree of independence of numerical processes from language.

SOME METHODOLOGICAL CONSIDERATIONS

The scientific study of numerical cognition has reached a stage, where experiments and the study of individual differences have made fruitful contact (cf. Dowker, 2015, chap. 47; Lefevre et al., 2015, chap. 48), getting over the historical separation, once diagnosed for all of general psychology, to a more timely "crossbreeding" of the "manipulating and the correlating schools of research" (Cronbach, 1975). One such field of research is about the relation between basic representations of number magnitude and arithmetic skills in early school years, which has provided conflicting results (cf. for an overview Donlan, 2015, chap. 46) and often numerically rather low correlations (accounts of small proportions of variance in the criterion measure) with measures of arithmetic performance or when relating behavioral measures with indicators of brain activation. Brunswik's (1955) general perspective on representative research design captured in his lens model takes recourse to Gestalt principles, which imply to consider symmetry principles in the degree of aggregation, level of generality, and correspondence between predictor and criterion constructs. Only when these symmetry principles regarding the degree of granularity of predictor and criterion variables hold, can validity in terms of correlation coefficients or variance accounted for be possibly maximized. Wittmann and Klumb (2006, pp. 185–211) provide instructive schematic drawings of different violations of the symmetry condition. The special case of asymmetry, because of a more narrow/lower level predictor combined with a broader criterion, seems to be quite characteristic of studies on the impact of basic numerical magnitude processing on mathematical skills in early school grades. Similarly relevant would also be the partial asymmetry case of broader constructs both at the predictor and criterion level assessed with several lower-level indicators, which only partly correspond at the lower level (cf. Fig. 3 in Wittmann & Klumb, 2006, pp. 185–211). This only partial overlap would also lead to underestimation of the true degree of relation among both constructs. Therefore, looking for corresponding levels of generality in relating predictors (or treatments) and criteria is mandatory. Otherwise, asymmetry results and unnecessarily low relationships will be observed, followed by misleading conclusions.

Brunswik just used regression/correlation analysis strategies. Yet, the same pitfalls will be present when using more advanced structural equation modeling or path analysis to look for causal relationships in nonexperimental designs. To give some example, Goebel, Watson, Lervag, and Hulme (2014) studied the predictability of arithmetic skills from multiple measures of symbolic and nonsymbolic magnitude comparison tasks as well as a number identification task (as a measure of number knowledge) with structural equation modeling tools in an 11 months longitudinal study comprising a large sample of first graders. There was a strong (symmetric) relation (cf. Fig. 2 in Goebel et al., 2014) between the latent variable for number magnitude processing assessed at both time points about a year apart with no significant contribution by number identification and arithmetic at the first assessment let alone measures of letter comparison, vocabulary, and nonverbal abilities, all of which would constitute aspects of too broad predictor constructs. On the other hand, the inclusion of the measure of number knowledge turned out to be the only substantial predictor of later arithmetic besides the arithmetic skills themselves, suggesting that the children's ability to relate Arabic numerals to their spoken forms, rather than to representations of their magnitude, is critical in constituting the broader criterion construct of arithmetic skills (cf. Fig. 1 in Goebel et al., 2014). Therefore, care has to be taken when looking for corresponding levels of generality in relating predictors or treatments and criteria.

Another related point concerns selection artifacts affecting the relation (ratio $u = SD_{sample}/SD_{population}$) between sample and population (systematic) variance. Wittmann and Klumb (2006, pp. 185–211, Fig. 5) present an instructive nomogram illustrating the modification of the true correlation between predictor and criterion construct depending on the restriction (e.g., undersampling low- and high-performing participants) respectively enhancement (e.g., oversampling low- and high-performing participants) of range in the sample compared with the population. For true correlations of 0.5 (strong), 0.3 (medium), and 0.1 (weak) alike underestimation respectively overestimation of the true correlation may easily be substantial. In case of u ranging between 0.5 and 2.0, the estimated correlations for a strong, medium, and weak relation would also vary substantially between 0.27 and 0.75, 0.16 and 0.54, and 0.03 and 0.20, respectively.

CONCLUSION

The cross-domain multicomponential character of more complex functions such as language and numerical cognition was already acknowledged in the classical era of neuropsychology by, e.g., Carl Wernicke and Salomon Eberhard Henschen pointing out the systemic character of these cognitive functions. More modern functional and structural imaging methods in accord with more fine-grained behavioral tasks also revealed a pattern of both segregated and overlapping brain areas and fiber pathways in more basic and advanced processing of numbers, calculation, and mathematical reasoning. Therefore, an association of symptoms and impairments both in language and numerical cognition could frequently be expected after acquired brain damage, which is "an experiment of nature," either, following the vascular pattern of blood supply in vascular cases, the biomechanics of traumatic affection of the brain or biochemical processes in degenerative diseases. Although (double) dissociations of functions need not be the rule under such conditions, one can nevertheless expect and predict them in accord with current cognitive processing models, the TCM of numerical cognition in particular. Here much more work has still to be done in a comprehensive equally fine-grained evaluation of both language and numerical cognition functions in patients and a within-subject experimental (and functional brain imaging) contrasting of comparable language and numerical cognition tasks. Nevertheless, it may be that at a more fine-grained neural microlevel of synapses and transmitter systems, there may still be structural and/or functional segregation, yet beyond the current methods of structural and functional brain imaging in healthy subjects or patients. With increased refinement of assessment tools, an individual differences perspective also becomes more feasible to probe the universal validity of current processing models.

Acknowledgments

I would like to thank my coauthors, in particular, Elise Klein and Korbinian Möller for their continuous cooperation on experimental and structural and functional imaging aspects of numerical cognition.

References

Amalric, M., & Dehaene, S. (2016). Origins of the brain networks for advanced mathematics in expert mathematicians. *Proceedings of the National Academy of Sciences of the United States of America*, *113*, 4909–4917.

Arsalidou, M., & Taylor, M. J. (2011). Is 2+2=4? Meta-analyses of brain areas needed for numbers and calculations. *Neuroimage*, *54*, 2382–2393.

Berger, H. (1926). Über Rechenstörungen bei Herderkrankungen des Großhirns. *Archiv Psychiatrie Nervenkrankheiten*, *78*, 238–263.

Bloechle, J., Huber, S., Bahnmueller, J., Rennig, J., Willmes, K., Cavdaroglu, S., et al. (2016). Fact learning in complex arithmetic – the role of the angular gyrus revisited. *Human Brain Mapping*, *37*, 3061–3079.

Brunswik, E. (1955). Representative design and probabilistic theory in a functional psychology. *Psychological Review*, *62*, 193–217.

Cappelletti, M. (2015). The neuropsychology of acquired number and calculation disorders. In R. Cohen Kadosh, & A. Dowker (Eds.), *The Oxford handbook of numerical cognition*. Oxford: Oxford University Press.

Catani, M. (2011). John Hughlings Jackson and the clinico-anatomical correlation method. *Cortex*, *47*, 905–907.

Catani, M., & ffytche, D. (2005). The rise and fall of disconnection syndromes. *Brain*, *128*, 2224–2239.

Claros Salinas, D. (1994). *EC 301 R: Untersuchungsmaterial zu Störungen des Rechnens und der Zahlenverarbeitung*. Konstanz: Kliniken Schmieder.

Cronbach, L. J. (1975). Beyond the two disciplines of scientific psychology. *The American Psychologist*, *30*, 116–127.

Dehaene, S., & Cohen, L. (1995). Towards an anatomical and functional model of number processing. *Mathematical Cognition*, *1*, 83–120.

Dehaene, S., Piazza, M., Pinel, P., & Cohen, L. (2003). Three parietal circuits for number processing. *Cognitive Neuropsychology*, *20*, 487–506.

Deloche, G., Seron, X., Metz-Lutz, M. N., Baeta, E., Basso, A., Claros-Salinas, et al. (1993). Calculation and number processing: The EC301 assessment battery for brain-damaged adults. In F. Stachowiak (Ed.), *Developments in the assessment and rehabilitation of brain-damaged patients*. Tübingen: Narr.

Donlan, C. (2015). Individual differences. In R. Cohen Kadosh, & A. Dowker (Eds.), *The Oxford handbook of numerical cognition*. Oxford: Oxford University Press.

Dowker, A. (2015). Individual differences in arithmetical abilities – the componential nature of arithmetic. In R. Cohen Kadosh, & A. Dowker (Eds.), *The Oxford handbook of numerical cognition*. Oxford: Oxford University Press.

Goebel, S. M., Watson, S. E., Lervag, A., & Hulme, C. (2014). Children's arithmetic development: It is number knowledge, not the approximate number sense that counts. *Psychological Science, 25*, 789–798.

Henschen, S. E. (1920). *Klinische und anatomische Beitrage zur Pathologie des Gehirns, pt 5* Nordiska Bukhandeln, Stockholm.

Huber, W., Poeck, K., & Willmes, K. (1984). The Aachen aphasia test. *Advances in Neurology, 42*, 291–303.

Klein, E., Moeller, K., Glauche, V., Weiller, C., & Willmes, K. (2013). Processing pathways in mental arithmetic – evidence from probabilistic fiber tracking. *PLoS One, 8*(1), e55455.

Klein, E., Moeller, K., & Willmes, K. (2013). A neural disconnection hypothesis on impaired numerical processing. *Frontiers in Human Neuroscience, 7*, 663.

Klein, E., Suchan, J., Moeller, K., Karnath, H.-O., Knops, A., Wood, G., et al. (2016). Considering structural connectivity in the triple code model of numerical cognition: Differential connectivity for magnitude processing and arithmetic facts. *Brain Structure and Function, 221*, 979–995.

Lefevre, J.-A., Wells, E., & Sowinski, C. (2015). Individual differences in basic arithmetical processes in children and adults. In R. Cohen Kadosh, & A. Dowker (Eds.), *The Oxford handbook of numerical cognition*. Oxford: Oxford University Press.

McCloskey, M., Caramazza, A., & Basili, A. (1985). Cognitive mechanisms in number processing and calculation: Evidence from dyscalculia. *Brain and Cognition, 4*, 171–196.

Piazza, M., Pinel, P., Le Bihan, D., & Dehaene, S. (2007). A magnitude code common to numerosities and number symbols in human intraparietal cortex. *Neuron, 53*, 293–305.

Rath, D., Domahs, F., Dressel, K., Claros-Salinas, D., Klein, E., Willmes, K., et al. (2015). Patterns of linguistic and numerical performance in aphasia. *Behavioral and Brain Functions, 11*, 2.

Reinvang, I. (1985). *Aphasia and brain organization*. Berlin: Springer.

Skipper, J. I., Devlin, J. T., & Lametti, D. R. (2017). The hearing ear is always found close to the speaking tongue: Review of the role of the motor system in speech perception. *Brain and Language, 164*, 77–105.

Wiese, H. (2003). *Numbers, language, and the human mind*. Cambridge, MA: Cambridge University Press.

Willmes, K., Moeller, K., & Klein, E. (2014). Where numbers meet words – a common ventral network for semantic classification. *Scandinavian Journal of Psychology, 55*, 202–211.

Wittmann, W. W., & Klumb, P. L. (2006). How to fool yourself with experiments in testing theories in psychological research. In R. R. Bootzin, & P. E. McKnight (Eds.), *Strengthening research methodology: Psychological measurement and evaluation*. Washington, DC: American Psychological Association Press.

Zaunmueller, L., Domahs, F., Dressel, K., Lonnemann, J., Klein, E., Ischebeck, A., et al. (2009). Rehabilitation of arithmetic fact retrieval via extensive practice: A combined fMRI and behavioural case-study. *Neuropsychological Rehabilitation, 19*, 422–443.

PERFORMANCE CONTROL AND SELECTIVE ATTENTION

An Introduction to Attention and Its Implication for Numerical Cognition

Mattan S. Ben-Shachar, Andrea Berger

Ben-Gurion University of the Negev, Beer-Sheva, Israel

Numerical cognition is considered to be a core knowledge: an innate and adaptive ability that plays a crucial role in human cognition (Carey, 2009; Spelke & Kinzler, 2007). Yet, numerical information, even if present, is not always available to us, either because it is not salient in our environment or because we ignore it because of its irrelevance to the task at hand. In other words, this numerical information does not always come to our attention. Studies in children have shown individual differences in the degree to which numerical information is attended to and that these differences have implications for the development of mathematical skills (Hannula & Lehtinen, 2005). In this chapter, we will first define the major aspects of attention, the classic behavioral and neuroimaging methods used to study them, and their proposed underlying neural networks. We will then illustrate the various roles attention can play in numerical cognition and specifically discuss our own findings regarding individual differences in spontaneous orienting of attention to the numerosity dimension.

FUNCTIONS OF ATTENTION

Although attention is colloquially regarded as a single function, it can be broken into several functions, all serving a single purpose, which is to enable adaptive behavior by the selection, integration, and prioritization of information among competing demands on our sensory and cognitive systems. According to Posner and Petersen's model, attention encompasses three interrelated functions, identified by three neural networks (Petersen & Posner, 2012; Posner & Petersen, 1990); these functions are alertness, orienting of attention, and executive control.

Alerting

Alerting, or activation, refers to the change and maintenance of an arousal level that optimally prepares the sensory systems to receive incoming information from the environment (Petersen & Posner, 2012). Changes in alertness can be short lived when induced extrinsically (such as the sound of a car horn leading to a momentary heightened sense of focus) or the changes can be maintained over longer periods (such as when taking an exam). When heightened alertness is short lived, it is called *phasic alertness*, whereas maintaining the state of alertness over time is a function often referred to as *sustained attention* (Matthias et al., 2010; Posner, Rueda, & Kanske, 2007; Weinbach & Henik, 2011). In the brain, both types of changes in alertness are modulated by activation in the reticular activating system of the brain stem, specifically the locus coeruleus (LC; Aston-Jones & Cohen, 2005; Rajkowski, Kubiak, & Aston-Jones, 1994). Increased activation in the LC induces greater activity in the cortex and is generally associated with the sense of "arousal" (Moruzzi & Magoun, 1949).

Phasic changes to the state of alertness affect the readiness to respond to incoming sensory information (Weinbach & Henik, 2011). Presenting alerting cues prior to target stimuli have been found to facilitate the detection of target stimuli, even when the warning cues provide no additional information regarding the location or identity of the target (Coull, Nobre, & Frith, 2001; Posner & Boies, 1971). Increased alertness has also been found to increase the probability of consciously perceiving near-threshold visual targets (Kusnir, Chica, Mitsumasu, & Bartolomeo, 2011). At the brain activity level, alerting cues have been found to produce a sustained negative component in event-related potentials (ERPs) called the contingent negative variation (Walter, Cooper, Aldridge, & McCallum, 1964). This negative deflection seems to reflect the suppression of ongoing activities, preparing the system for the appearance of a target (Petersen & Posner, 2012).

While phasic changes in alertness are externally driven, sustained attention is a controlled behavior, the effects of which are often studied using long and boring tasks. For example, in the continuous performance task (CPT), participants are instructed to respond only to an infrequently presented target within a stream of frequent nontargets. Successfully sustaining attention is marked by lower miss rates, with response times (RTs) and error rates normally increasing as the task progresses (e.g., Shalev, Ben-Simon, Mevorach, Cohen, & Tsal, 2011). In patients with lesions involving the right frontal lobe, the ability to voluntarily sustain attention over prolonged periods is impaired, as marked by a faster decay in sustained attention throughout the CPT, compared with controls (Rueckert & Grafman, 1996; Wilkins, Shallice, & McCarthy, 1987).

Executive Control

Executive control is the function by which other cognitive processes are regulated to produce desired behaviors (Petersen & Posner, 2012). It is concerned with the execution of controlled responses by inhibiting dominant, possibly automatic tendencies, as well as monitoring for context-relevant information that might be relevant for the execution of adaptive and desired behaviors (Diamond, 2013). Cognitive control is composed of several mental processes, including conscious detection, inhibition, error detection, and conflict processing and resolution, as well as the top-down regulation of other attention networks, including alerting and orienting of attention (Rueda, Pozuelos, & Cómbita, 2015).

Many tasks can be used in the study of executive control, all of which include the introduction of some form of conflict. For example, in the Flanker task, participants are presented with five side-by-side arrows and instructed to respond only to the middle arrow. RTs and error rates are typically lower when the surrounding arrows are pointing in the same direction as the middle arrow compared with when the surrounding arrows are pointing in the opposite direction, indicating difficulty in ignoring salient, task-irrelevant information (Eriksen & Eriksen, 1974). Another task used is the classic Stroop task (Stroop, 1935). In this task, participants are presented with the name of a color written in a different colored ink (e.g., the word "RED" written in blue ink). Participants are instructed to name the color of the word, while ignoring the word itself. Success on this task relies on the ability to inhibit the dominant response of reading the word to execute the subdominant response of naming the color. The typical finding on this task and its variants is that participants find it more difficult to respond correctly when the name and color are incongruent (e.g., the word "YELLOW" written in red) compared with when they are congruent (e.g., the word "YELLOW" written in yellow). Using such tasks, the anterior cingulate cortex (ACC) and the dorsolateral prefrontal cortex have been consistently found to be involved in tasks that require the use of cognitive control, showing increased activation during tasks that require conflict monitoring or resolution (Bush, Luu, & Posner, 2000; Fan, 2014; Petersen & Posner, 2012; van Veen & Carter, 2002).

Maturation of gray matter in the ACC has also been associated with the development of executive control in children (Casey et al., 1997). Using a modified Flanker task known as the child attention network test, children's ability to monitor and resolve conflicts has been found to increase with age (Buss, Dennis, Brooker, & Sippel, 2011; Rueda et al., 2004). These age-related changes in executive control have been associated with the maturation of gray matter, particularly in the ACC (Casey et al., 1997).

Orienting

Orienting is the mechanism by which incoming information (external sensory input or internal thoughts and emotions) is selectively chosen for further processing (Petersen & Posner, 2012). Orienting can be externally driven, such as when changes to an object or a location "draw" our attention to those objects or locations. For example, our eyes are automatically drawn to the "check engine" light when it turns on. Orientation can also be goal driven, where a stimulus is not prioritized for processing because of any changes but rather because of some knowledge-based reason. For example, when searching for our lost car keys, we will prioritize the scanning of small, metal objects. When orientation is externally driven (bottom-up), it is referred to as *exogenous orientation*, and when it is goal driven (top-down), it is referred to as *endogenous orientation* (Corbetta & Shulman, 2002). Both forms of attention can be studied in the context of spatial attention (attending to an object in space) or in the context of featural selective attention (attending to a certain aspect of an object).

Spatial attention has been studied using variations of the cued-target task developed by Posner (1980). In the basic form of this task, participants are instructed to respond as quickly as possible to the appearance of a target stimulus, which may be preceded by valid or invalid cues. The basic finding in such tasks is that orienting attention—either overtly or covertly (with or without eye movements)—to a location in space where a target is set to appear facilitates the detection of targets at that location.

Using variants of this task, two different brain networks have been identified as being involved in spatial attention (Petersen & Posner, 2012). Exogenous orientation has been related to increases in blood flow and electrical activity in a right-lateralized network of structures, including the fusiform gyrus (Mangun, Buonocore, Girelli, & Jha, 1998), as well as the temporoparietal junction and inferior frontal cortex (Corbetta & Shulman, 2002). On the other hand, endogenous orientation has been associated with activation of a bilateral network that involves the intraparietal sulcus, the superior parietal lobule, and the frontal eye fields (Thompson, Biscoe, & Sato, 2005).

Selective featural attention has been studied using tasks in which stimuli comprise multiple features. One such task is the Navon task (Navon, 1977), in which the stimuli consist of a large letter formed by the configuration of smaller letters (e.g., a large H made of small Ss). In this task, participants are asked to identify either the large letter (the global feature of the stimulus) or the smaller letters (the local feature). When asked to identify the large letter, performance is typically only minimally affected by the identity of the smaller letters, but when identifying the smaller letters, performance is significantly hindered when the local and global dimensions conflict with each other (Beaucousin et al., 2013; Navon, 1977). Global and local processing involve the temporoparietal lobes but in an asymmetric fashion; in other words, global processing involves the right TPJ, whereas local processing involves the left TPJ (Weissman & Woldorff, 2004; Yamaguchi, Yamagata, & Kobayashi, 2000).

Although most studies find an attentional bias for global features (for a review, see Kimchi, 1992), works focusing on individual differences have shown individual differences in preference to global information over local information. For example, a bias toward local features is a common characteristic of individuals on the autistic spectrum (Plaisted, Swettenham, & Rees, 1999), as well as individuals diagnosed with depression (Basso, Schefft, Ris, & Dember, 1996). A reduced global bias has also been found in attention deficit hyperactivity disorder (ADHD; Kalanthroff, Naparstek, & Henik, 2013). Individual differences have also been found in the general population, with some individuals showing a greater bias for global features than others (Peterson & Deary, 2006), indicating that different people might attend to different types of information within the same stimulus.

Regardless of the differences between orienting in space and orienting to specific features, both are the result of modulating the functioning of sensory systems in two complementary ways: on the one hand, relevant information is enhanced, whereas, on the other hand, irrelevant or distracting information is suppressed. The result is an increased signal-to-noise ratio, affecting the saliency of relevant incoming sensory information (Briggs, Mangun, & Usrey, 2013; Verghese, 2001). Functional brain studies have demonstrated these modulating effects of attention: when stimuli are attended to, early ERP components have larger magnitudes, whereas the magnitude in response to unattended stimuli is smaller (Näätänen & Michie, 1979; Valdes-Sosa, Bobes, Rodriguez, & Pinilla, 1998). Attention seems to modulate even the lowest levels of perceptual processing, with an enhanced neural response to attended visual stimuli in the lateral geniculate nucleus of the thalamus (Briggs et al., 2013; O'Connor, Fukui, Pinsk, & Kastner, 2002), suggesting that attention facilitates perceptual processing from very early stages of neural processing. These mechanisms of orienting not only improve the ability to detect stimuli but also to discriminate between similar stimuli by improving the resolution of the perceived stimuli (Murray & Wojciulik, 2004; Yeshurun & Carrasco, 1999). For example, activity in the auditory cortex, as measured by means of magnetoencephalography, was found to be more distinct for tones of different frequencies when the tones were attended to compared with when they were not (Okamoto, Stracke, Wolters, Schmael, & Pantev, 2007).

NUMERICAL COGNITION AND ATTENTION

In this section, we will illustrate some of the effects of the described types of attention on processes of numerical cognition, with an emphasis on orienting of attention toward the numerical information presented in the environment.

Alerting and Numerosity Perception

One example of the effect of attention on numerical cognition can be seen in the effect of alertness on enumeration within the subitizing range. When enumerating a small set of one to four items, enumeration is quick, perfectly accurate, and accomplished with little or no feeling of conscious effort, a phenomenon known as *subitizing* (Kaufman, Lord, Reese, & Volkmann, 1949). If more than four elements are presented, enumeration either becomes slow and serial (through the process of *counting*) or it becomes inaccurate, with the numerosity only being approximately *estimated* (Dehaene, 1992).

The process of subitizing has been argued to require no attention because it is seen as a preattentive process (e.g., Trick & Pylyshyn, 1994); however, more recent work suggests that even the quick and accurate process of subitizing is dependent on attentional resources. For example, Gliksman, Weinbach, and Henik (2016) examined the effect of pretarget warning cues on the subitizing process. They found that cued arrays within the subitizing range were enumerated faster than uncued arrays, indicating that subitizing is an attention-dependent process and can be manipulated through enhanced alertness. This alerting effect, when enumerating arrays within the subitizing range, has also been found in individuals diagnosed with developmental dyscalculia, although such cuing was unable to expand their smaller-than-normal subitizing range (Gliksman & Henik, 2018).

Inhibiting Numerical Processing

The role of executive attention on numerical cognition has been repeatedly demonstrated using a numerical Stroop task (Tzelgov, Meyer, & Henik, 1992). In this task, participants are required to indicate which of two presented stimuli is larger. Both stimuli are digits (1–9) of varying physical sizes. Thus, determining which stimulus is larger can be accomplished based on the physical size of the digit or the numerical magnitude (e.g., 8 is larger than 1). Typically, participants are instructed to ignore numerical magnitude and to instead respond according to physical size only. The general finding in this task is similar to those in classic Stroop tasks: compared with trials on which the stimuli differ only in terms of physical size (neutral condition), reaction times and error rates are lower when the larger digit is also greater in numerical magnitude (congruent condition) and greater when the larger digit is smaller in numerical magnitude (incongruent condition; e.g., Szűcs, Soltész, Jármi, & Csépe, 2007). This compatibility effect, known as the size compatibility effect (SiCE), indicates that subjects cannot ignore numerical values of numerals; in other words, the numerical dimension is processed in a nonintentional and automatic manner (Henik & Tzelgov, 1982; Tzelgov et al., 1992). Interestingly, this effect is usually not fully observable before the second grade. Specifically, younger children either fail to present a significant SiCE or only show interference of the numerical dimension (i.e., incongruent trials are slower than neutral ones), without showing facilitation (i.e., neutral trials are slower than congruent ones; Ben-Shalom, Berger, & Henik, 2012, pp. 195–208; Girelli, Lucangeli, & Butterworth, 2000; Mussolin & Noël, 2007; Rubinsten, Henik, Berger, & Shahar-Shalev, 2002). Furthermore, such lack of facilitation has also been found in both acquired acalculia (Ashkenazi, Henik, Ifergane, & Shelef, 2008) and developmental dyscalculia (Rubinsten & Henik, 2006).

The lack of SiCE in young children and those suffering from dyscalculia has been explained as resulting from an immature association between numerical symbols and their corresponding magnitudes, which, in typical mature development, become automatic (Girelli et al., 2000; Rubinsten et al., 2002). Later works by Soltész, Goswami, White, and Szűcs (2011) and Soltész, Szűcs, Dékány, Márkus, and Csépe (2007) utilizing ERP have concluded that this effect is not because of difficulty in access to numerical representations. They suggest, instead, that early interference of irrelevant information is the cause of a weaker interference effect. In adults, a diminished interference effect has been found to accrue under a high cognitive load, which has also been related to lower activation in the motor area for task-irrelevant information (Cohen Kadosh et al., 2007). These findings support the notion that a reduced interference effect stems from reduced executive control and not from a difficulty in access to numerical representations. Thus, deficits in children's mathematical development could stem from deficits in executive control, a notion that is also supported by deficits in numerical processing found in individuals diagnosed with ADHD (Benedetto-Nasho & Tannock, 1999). During childhood, individual differences in executive attention seem to be related to, and even predict, later mathematical achievements. This literature is reviewed by Merkley, Matusz, and Scerif (Chapter 6).

The importance of executive attention and, specifically, the ability to inhibit irrelevant information have also been demonstrated for more complex aspects of numerical cognition. For example, in the area of arithmetic, attention effects have been shown for mental calculations (e.g., Cowell, Egan, Code, Harasty, & Watson, 2000). Moreover, executive attention effects have been shown within logical and probabilistic reasoning (Babai, Brecher, Stavy, & Tirosh, 2006) and even in geometry (Stavy & Babai, 2008, 2010). For example, shapes' areas have been found to affect the comparison of two shapes' perimeters: RTs and error rates are lower when the shape with the longer perimeter also has a larger area (congruent condition) compared with when the shape with the longer perimeter has a smaller area (incongruent condition; Stavy & Babai, 2008). Accuracy in the incongruent condition has been associated with activation in prefrontal areas, indicating the involvement of executive control in inhibiting responses to the irrelevant area dimension (Stavy & Babai, 2010).

Orienting Toward Numerosity

When observing an object or situation, many different aspects may be either attended or unattended to, for example, color, size, shape, distance, and numerosity. Attention to any specific feature may be driven by the task at hand (goal directed) or because of its perceived saliency. In the context of enumeration, individual differences have been found in the extent to which attention is given to numerical information. Specifically, children seem to differ in their tendency to *spontaneously focus on numerosity* (SFON), which is the tendency to spontaneously attend to and use numerical information present in the environment without being instructed to do so. This tendency has been defined and explored mostly in young children and has been found to be a domain-specific marker for later mathematical abilities in school.

Hannula and Lehtinen (2005) developed a number of imitation tasks that enabled evaluation of this spontaneous tendency among young children. All the tasks have some numerical information presented by the experimenter, but at no point is this information explicitly marked as important, and special care is taken to avoid any wording that could suggest that the tasks are mathematical or quantitative in nature. For example, in one of the tasks, the child is only instructed to do what the experimenter does ("Watch carefully what I do, then you do what I just did"); the experimenter then "feeds" a puppet a number of "sweets," one at a time. When the experimenter is finished, the child is again instructed only to imitate what the experimenter has done ("Now you do exactly what I did"). By repeating these procedures several times with different quantities, children are assigned a SFON score based on the number of times the child incorporated the numerical information into their behavior. Using these tasks, individual differences in children's tendencies to spontaneously direct attention to numerosity were found: some children will imitate the actions of the experimenter (i.e., "feeding" a puppet "sweets"; see Fig. 5.1) without regard to numerosity, whereas others will imitate the action while also attending to the numerical information (i.e., "feeding" the puppet the exact number of "sweets" that the experimenter did).

Longitudinal studies have found that preschool children who showed higher SFON tendencies later developed higher mathematical skills (Hannula & Lehtinen, 2005), yet it is unclear if this predictive relationship is because of SFON tendencies reflecting individual differences in sensitivity to numerical information or greater resolution of quantitative processing. To test this, we measured children's ratios of discrimination between numerosities using the Panamath task. These discrimination ratios have been found to serve as domain-specific markers for numerical

FIGURE 5.1 Child imitating an experimenter in one of the *spontaneously focus on numerosity* imitation tasks. Here, the experimenter feeds the parrot puppet a fixed number of "sweets." The child is asked to imitate the experimenter's action; however, the number of sweets that the parrot is fed is never mentioned. Children differ in their spontaneous attention toward this dimension. See a full description of the task in Hannula and Lehtinen (2005).

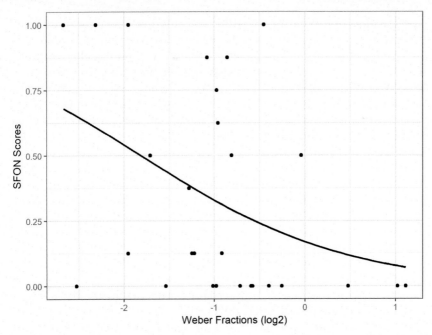

FIGURE 5.2   Children's *spontaneously focus on numerosity* (SFON) scores as a function of their Weber fractions (log2). The *black line* is the estimated logistic regression line.

abilities (Halberda & Feigenson, 2008). During the task, participants select which of two dot arrays is more numerous, with the task becoming more difficult when the ratio between both dot arrays is close (as a result of Weber's Law[1]). Using this task, we determined each child's individual threshold of discrimination, resulting in an individual Weber fraction.

We conducted a logistic regression analysis to predict SFON scores, measured with Hannula and Lehtinen's imitation tasks, from discrimination ratios and age (because the ability to discriminate between ratios typically improves with age). As expected, we found that preschool children who had smaller (better) discrimination ratios were more likely to spontaneously attend to numerical information in the imitation tasks (see Fig. 5.2; Ben-Shachar, Shotts-Peretz, Hannula-Sormunen, & Berger, 2018). The directionality of the relationship between SFON tendencies and numerical discrimination is not yet clear. It might suggest that higher numerical processing abilities make numerical information more salient and thus

---

[1] There is an open debate in the literature regarding the difficulty to separate changes in numerosity from simultaneous changes in continuous dimensions of the stimuli, such as extended area, item size, total surface, density, and circumference (e.g., Leibovich, Kallai, & Itamar, 2016, pp. 355–374). However, this debate is beyond the scope of the present chapter.

children are more likely to integrate numerical information into their behaviors. However, it might also suggest that SFON tendencies improve numerical processing in some manner.

As mentioned, research on SFON has focused on this tendency in children. In adults, it is more difficult to covertly measure a spontaneous focus on a specific type of information, as the imitation tasks developed for children are not suitable for adults. Therefore, we designed a computerized task that would fulfill the demands created by the children's SFON tasks but that would be suitable for adults: (1) a task that would have no explicit request to regard numerosity as a relevant factor but would (2) include numerical information that could be processed if attended to, as well as other aspects that would compete with the numerical dimension.

In each block of the task, participants were presented with two stimuli. The selection of one would award them points, but selecting the other would result in a deduction of points. The explicit goal of the task was to learn which stimulus was the beneficial one ("Your goal is to learn in each [block] which stimulus earns you the most points"). The two stimuli always differed from one another on two dimensions: the number of objects comprising the stimulus (1, 2, 3, or 4) and the object's color (red, green, blue, or yellow). For example, in a given block, the two stimuli could be three red dots and two blue dots.

In all but the last trial of the block, the pairing between color and quantity was kept constant, such that, for example, the three dots were always red and the two dots were always blue. This allowed participants to learn fairly quickly, through trial and error, which stimulus earned them points (e.g., 3-red; see Fig. 5.3A). Because the two stimuli differed on two dimensions, the manner in which participants identified the beneficial stimulus could be based on either the color of the stimulus ("choosing red awards me points") or on the number of items comprising the stimulus—its numerosity—("choosing three dots awards me points"). Both strategies would lead to the same performance because the color and numerosity were paired throughout these trials in each block.

The last trial in each block was designed to test whether identification was based on color or quantity. This was done by reverse pairing the two dimensions; for example, if the stimuli in the leading trials were **3-red** versus 2-*blue*, the stimuli in the final trial would then be 2-**red** versus 3-*blue* (see Fig. 5.3B). If the identification was color based, the participant would, in this final trial, select the stimulus comprising the color that previously awarded points (e.g., 2-**red**), but if learning was number based, the participant would select the stimulus comprising the numerosity that previously awarded points (e.g., 3-*blue*). This allowed us to measure which dimension (numerosity vs. color) was more salient to the participant and thus assess the participant's SFON. No feedback was given for the last trial.

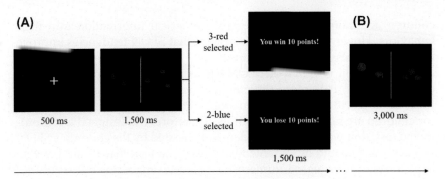

FIGURE 5.3   Example number versus color block. (A) Example learning trial, with stimuli *3-red dots* and *2-blue dots*. Feedback was given according to selection, allowing participants to learn to select *3-red dots*. (B) In the test trial, color and numerosity were reverse paired, giving the stimuli *2-red dots* versus *3-blue dots*. No feedback was given for this trial.

To statistically control for our measure reflecting a bias against attending to color, we also measured participants' spontaneous focus on color. This was measured via an additional block in which the stimuli were also comprising two dimensions: color and shape (e.g., blue-triangle vs. red-square), with the number of objects kept constant at 1.

Our results indicate that, as in children, individual differences exist in the tendency to spontaneously regard numerical information. Individual SFON scores, based on the last trial of each of the blocks, ranged from 0 (responses were never based on the quantitative dimension of the stimuli) to 1 (all responses were based on the quantitative dimension of the stimuli), with a median score of 0.13. When measuring the reliability of these scores using the split-test method (Webb, Shavelson, & Haertel, 2006)—comparing the scores calculated based on the odd blocks with those based on the even blocks—we found a reliability of 0.96, indicating that participants were consistent in their selection throughout the task and that individual differences do exist in the tendency to spontaneously regard numerical information.

We were also interested in testing whether individual differences in SFON among adults were related to numerical discrimination, such as we have found in children. To that end, we measured participants' ability to discriminate between numerosities, their intelligence using the Raven's matrices test, and their quantitative reasoning (scores ranging from 50 to 150, taken from participants' university psychometric entrance test). We expected SFON to correlate with discrimination ratios and quantitative reasoning scores but not to Raven scores because in children SFON has been shown to be domain specific.

Results from logistic regression support this hypothesis: as with the children, adults with smaller (better) discrimination ratios had a higher

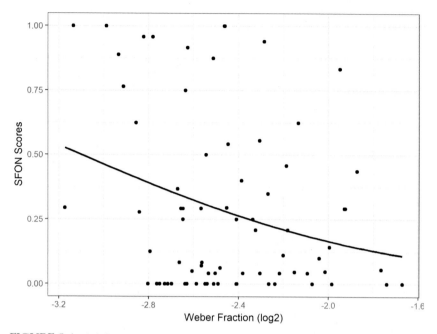

FIGURE 5.4 Adults' *spontaneously focus on numerosity* (SFON) scores as a function of their Weber fractions (log2). The *black line* is the estimated logistic regression line.

likelihood of responding according to the number of objects comprising the stimulus (and not to the stimulus' color; see Fig. 5.4). A separate logistic regression analysis found that adults with higher-quantitative reasoning scores also had a higher likelihood of attending to the stimuli's numerosities. Raven scores were not found to have any predictive power, as expected (Ben-Shachar et al., 2018).

Whether children and adults diagnosed with dyscalculia have atypical SFON tendencies has not been studied in the extant literature. Therefore, it is still an open question whether spontaneous tendency to attend numerical information in the environment has direct implications for mathematical abilities in the atypical developmental range of mathematical abilities.

It should be mentioned that orienting of attention is highly relevant for numerical cognition also when referring to *spatial* orienting. This relation has been studied and demonstrated mainly with the spatial-numerical association of response codes effect (Dehaene, Bossini, & Giraux, 1993), in which responses to large numbers are faster with the right hand, whereas responses to small numbers are faster with the left hand. Readers can see examples for the effects of orienting of attention within this context in the works of Fischer, Castel, Dodd, and Pratt (2003), Hubbard, Piazza, Pinel, and Dehaene (2005) and van Dijck, Abrahamse, Acar, Ketels, and Fias (2014).

To summarize, the study of the behavioral and neuropsychological effects of attention in the context of numerical cognition can shed light on the processes and mechanisms behind numerical cognition, as well as on atypical numerical cognition, such as that found in dyscalculia. For example, studying the effects of cuing and changes in alertness on the enumeration of small sets demonstrated the dependence of subitizing on attention allocation (Gliksman et al., 2016), and studying the brain activity of children during a numerical Stroop task led to a better understanding of the trajectories of numerical skill development (Rubinsten et al., 2002; Soltész et al., 2007). We have also demonstrated how differences in attentional biases toward numerical information are related to the resolution of numerical perception and mathematical skill. Using a newly developed task, we found an association between SFON and mathematical skill in adulthood. This finding is on par with previous work on SFON among children, which found that the tendency to spontaneously attend to numerical information predicted later mathematical skill (Hannula & Lehtinen, 2005). Furthermore, we found, in both children and adults, that the tendency to attend to numerical information is related to individuals' numerical abilities. This might suggest that better processing of numerical information makes numerical information more salient, increasing the likelihood that information will be attended to and used or that the tendency to attend to numerical information improves numerical processing in some manner.

Acknowledgments

This research was partially supported by the Israeli Science Foundation grant no. 1799/12 awarded to the Center for the Study of the Neurocognitive Basis of Numerical Cognition.

We would like to thank Prof. Minna M. Hannula for her collaboration with the children's SFON tasks, and Mrs. Dalit Shotts-Peretz for her contribution to the experiments described in this chapter.

References

Ashkenazi, S., Henik, A., Ifergane, G., & Shelef, I. (2008). Basic numerical processing in left intraparietal sulcus (IPS) acalculia. *Cortex, 44*(4), 439–448.

Aston-Jones, G., & Cohen, J. D. (2005). An integrative theory of locus coeruleus-norepinephrine function: Adaptive gain and optimal performance. *Annual Review of Neuroscience, 28*, 403–450.

Babai, R., Brecher, T., Stavy, R., & Tirosh, D. (2006). Intuitive interference in probabilistic reasoning. *International Journal of Science and Mathematics Education, 4*(4), 627–639.

Basso, M. R., Schefft, B. K., Ris, M. D., & Dember, W. N. (1996). Mood and global-local visual processing. *Journal of the International Neuropsychological Society, 2*(03), 249–255.

Beaucousin, V., Simon, G., Cassotti, M., Pineau, A., Houdé, O., & Poirel, N. (2013). Global interference during early visual processing: ERP evidence from a rapid global/local selective task. *Frontiers in Psychology, 4.*

Ben-Shachar, M. S., Shotts-Peretz, D., Hannula-Sormunen, M., & Berger, A. (2018). *Spontaneous focusing on numerosity among adults and children* (in preparation).

Ben-Shalom, T., Berger, A., & Henik, A. (2012). The beginning of the road: Learning mathematics for the first time. In *Reading, writing, mathematics and the developing brain: Listening to many voices.* Dordrecht: Springer.

Benedetto-Nasho, E., & Tannock, R. (1999). Math computation, error patterns and stimulant effects in children with attention deficit hyperactivity disorder. *Journal of Attention Disorders, 3*(3), 121–134.

Briggs, F., Mangun, G. R., & Usrey, W. M. (2013). Attention enhances synaptic efficacy and the signal-to-noise ratio in neural circuits. *Nature, 499*(7459), 476–480.

Bush, G., Luu, P., & Posner, M. I. (2000). Cognitive and emotional influences in anterior cingulate cortex. *Trends in Cognitive Sciences, 4*(6), 215–222.

Buss, K. A., Dennis, T. A., Brooker, R. J., & Sippel, L. M. (2011). An ERP study of conflict monitoring in 4-8-year old children: Associations with temperament. *Developmental Cognitive Neuroscience, 1*(2), 131–140.

Carey, S. (2009). *The origin of concepts.* New York, NY: Oxford University Press.

Casey, B. J., Trainor, R., Giedd, J., Vauss, Y., Vaituzis, C. K., Hamburger, S., et al. (1997). The role of the anterior cingulate in automatic and controlled processes: A developmental neuroanatomical study. *Developmental Psychobiology, 30*(1), 61–69.

Cohen Kadosh, R., Cohen Kadosh, K., Linden, D. E., Gevers, W., Berger, A., & Henik, A. (2007). The brain locus of interaction between number and size: A combined functional magnetic resonance imaging and event-related potential study. *Journal of Cognitive Neuroscience, 19*(6), 957–970.

Corbetta, M., & Shulman, G. L. (2002). Control of goal-directed and stimulus-driven attention in the brain. *Nature Reviews Neuroscience, 3*(3), 201–215.

Coull, J. T., Nobre, A. C., & Frith, C. D. (2001). The noradrenergic α2 agonist clonidine modulates behavioural and neuroanatomical correlates of human attentional orienting and alerting. *Cerebral Cortex, 11*(1), 73–84.

Cowell, S. F., Egan, G. F., Code, C., Harasty, J., & Watson, J. D. G. (2000). The functional neuroanatomy of simple calculation and number repetition: A parametric PET activation study. *Neuroimage, 12*(5), 565–573.

Dehaene, S. (1992). Varieties of numerical abilities. *Cognition, 44*(1), 1–42.

Dehaene, S., Bossini, S., & Giraux, P. (1993). The mental representation of parity and number magnitude. *Journal of Experimental Psychology: General, 122*(3), 371.

Diamond, A. (2013). Executive functions. *Annual Review of Psychology, 64*, 135–168.

Eriksen, B. A., & Eriksen, C. W. (1974). Effects of noise letters upon the identification of a target letter in a nonsearch task. *Attention, Perception, and Psychophysics, 16*(1), 143–149.

Fan, J. (2014). An information theory account of cognitive control. *Frontiers in Human Neuroscience, 8*, 680.

Fischer, M. H., Castel, A. D., Dodd, M. D., & Pratt, J. (2003). Perceiving numbers causes spatial shifts of attention. *Nature Neuroscience, 6*(6), 555.

Girelli, L., Lucangeli, D., & Butterworth, B. (2000). The development of automaticity in accessing number magnitude. *Journal of Experimental Child Psychology, 76*(2), 104–122.

Gliksman, Y., & Henik, A. (2018). *Enumerating and alertness in developmental dyscalculia* (Submitted for publication).

Gliksman, Y. Weinbach, N., & Henik, A. (2016). Alerting cues enhance the subitizing process. *Acta Psychologica, 170*, 139 145

Halberda, J., & Feigenson, L. (2008). Developmental change in the acuity of the "Number Sense": The approximate number system in 3-, 4-, 5-, and 6-year-olds and adults. *Developmental Psychology, 44*(5), 1457–1465.

Hannula, M. M., & Lehtinen, E. (2005). Spontaneous focusing on numerosity and mathematical skills of young children. *Learning and Instruction, 15*(3), 237–256.

Henik, A., & Tzelgov, J. (1982). Is three greater than five: The relation between physical and semantic size in comparison tasks. *Memory and Cognition, 10*(4), 389–395.

Hubbard, E. M., Piazza, M., Pinel, P., & Dehaene, S. (2005). Interactions between number and space in parietal cortex. *Nature Reviews Neuroscience, 6*(6), 435–448.

Kalanthroff, E., Naparstek, S., & Henik, A. (2013). Spatial processing in adults with attention deficit hyperactivity disorder. *Neuropsychology, 27*(5), 546.

Kaufman, E. L., Lord, M. W., Reese, T. W., & Volkmann, J. (1949). The discrimination of visual number. *The American Journal of Psychology*, 498–525.

Kimchi, R. (1992). Primacy of wholistic processing and global/local paradigm: A critical review. *Psychological Bulletin, 112*(1), 24.

Kusnir, F., Chica, A. B., Mitsumasu, M. A., & Bartolomeo, P. (2011). Phasic auditory alerting improves visual conscious perception. *Consciousness and Cognition, 20*(4), 1201–1210.

Leibovich, T., Kallai, A., & Itamar, S. (2016). What do we measure when we measure magnitudes. In A. Henik (Ed.), *Continuous issues in numerical cognition: How many or how much*. New York, NY: Elsevier.

Mangun, G. R., Buonocore, M. H., Girelli, M., & Jha, A. P. (1998). ERP and fMRI measures of visual spatial selective attention. *Human Brain Mapping, 6*(5–6), 383–389.

Matthias, E., Bublak, P., Müller, H. J., Schneider, W. X., Krummenacher, J., & Finke, K. (2010). The influence of alertness on spatial and nonspatial components of visual attention. *Journal of Experimental Psychology: Human Perception and Performance, 36*(1), 38.

Moruzzi, G., & Magoun, H. W. (1949). Brain stem reticular formation and activation of the EEG. *Electroencephalography and Clinical Neurophysiology, 1*(1), 455–473.

Murray, S. O., & Wojciulik, E. (2004). Attention increases neural selectivity in the human lateral occipital complex. *Nature Neuroscience, 7*(1), 70–74.

Mussolin, C., & Noël, M.-P. (2007). The nonintentional processing of Arabic numbers in children. *Journal of Clinical and Experimental Neuropsychology, 29*(3), 225–234.

Näätänen, R., & Michie, P. T. (1979). Early selective-attention effects on the evoked potential: A critical review and reinterpretation. *Biological Psychology, 8*(2), 81–136.

Navon, D. (1977). Forest before trees: The precedence of global features in visual perception. *Cognitive Psychology, 9*(3), 353–383.

O'Connor, D. H., Fukui, M. M., Pinsk, M. A., & Kastner, S. (2002). Attention modulates responses in the human lateral geniculate nucleus. *Nature Neuroscience, 5*(11), 1203–1209.

Okamoto, H., Stracke, H., Wolters, C. H., Schmael, F., & Pantev, C. (2007). Attention improves population-level frequency tuning in human auditory cortex. *Journal of Neuroscience, 27*(39), 10383–10390.

Petersen, S. E., & Posner, M. I. (2012). The attention system of the human brain: 20 years after. *Annual Review of Neuroscience, 35*, 73.

Peterson, E. R., & Deary, I. J. (2006). Examining wholistic–analytic style using preferences in early information processing. *Personality and Individual Differences, 41*(1), 3–14.

Plaisted, K., Swettenham, J., & Rees, L. (1999). Children with autism show local precedence in a divided attention task and global precedence in a selective attention task. *Journal of Child Psychology and Psychiatry, 40*(5), 733–742.

Posner, M. I. (1980). Orienting of attention. *Quarterly Journal of Experimental Psychology, 32*(1), 3–25.

Posner, M. I., & Boies, S. J. (1971). Components of attention. *Psychological Review, 78*(5), 391.

Posner, M. I., & Petersen, S. E. (1990). The attention system of the human brain. *Annual review of neuroscience, 13*(1), 25–42.

Posner, M. I., Rueda, M. R., & Kanske, P. (2007). Probing the mechanisms of attention. In *Handbook of psychophysiology* (Vol. 410).

Rajkowski, J., Kubiak, P., & Aston-Jones, G. (1994). Locus coeruleus activity in monkey: Phasic and tonic changes are associated with altered vigilance. *Brain Research Bulletin, 35*(5), 607–616.

Rubinsten, O., & Henik, A. (2006). Double dissociation of functions in developmental dyslexia and dyscalculia. *Journal of Educational Psychology, 98*(4), 854.

Rubinsten, O., Henik, A., Berger, A., & Shahar-Shalev, S. (2002). The development of internal representations of magnitude and their association with Arabic numerals. *Journal of Experimental Child Psychology, 81*(1), 74–92.

Rueckert, L., & Grafman, J. (1996). Sustained attention deficits in patients with right frontal lesions. *Neuropsychologia, 34*(10), 953–963.

Rueda, M. R., Fan, J., McCandliss, B. D., Halparin, J. D., Gruber, D. B., Lercari, L. P., et al. (2004). Development of attentional networks in childhood. *Neuropsychologia, 42*(8), 1029–1040.

Rueda, M. R., Pozuelos, J. P., & Cómbita, L. M. (2015). Cognitive neuroscience of attention. *AIMS Neuroscience, 2*(4), 183–202.

Shalev, L., Ben-Simon, A., Mevorach, C., Cohen, Y., & Tsal, Y. (2011). Conjunctive continuous performance task (CCPT)—a pure measure of sustained attention. *Neuropsychologia, 49*(9), 2584–2591.

Soltész, F., Goswami, U., White, S., & Szűcs, D. (2011). Executive function effects and numerical development in children: Behavioural and ERP evidence from a numerical Stroop paradigm. *Learning and Individual Differences, 21*(6), 662–671.

Soltész, F., Szűcs, D., Dékány, J., Márkus, A., & Csépe, V. (2007). A combined event-related potential and neuropsychological investigation of developmental dyscalculia. *Neuroscience Letters, 417*(2), 181–186.

Spelke, E. S., & Kinzler, K. D. (2007). Core knowledge. *Developmental Science, 10*(1), 89–96.

Stavy, R., & Babai, R. (2008). Complexity of shapes and quantitative reasoning in geometry. *Mind, Brain, and Education, 2*(4), 170–176.

Stavy, R., & Babai, R. (2010). Overcoming intuitive interference in mathematics: Insights from behavioral, brain imaging and intervention studies. *ZDM, 42*(6), 621 633

Stroop, J. R. (1935). Studies of interference in serial verbal reactions. *Journal of Experimental Psychology, 18*(6), 643.

Szűcs, D., Soltész, F., Jármi, É., & Csépe, V. (2007). The speed of magnitude processing and executive functions in controlled and automatic number comparison in children: An electro-encephalography study. *Behavioral and Brain Functions, 3*(1), 23.

Thompson, K. G., Biscoe, K. L., & Sato, T. R. (2005). Neuronal basis of covert spatial attention in the frontal eye field. *Journal of Neuroscience, 25*(41), 9479–9487.

Trick, L. M., & Pylyshyn, Z. W. (1994). Why are small and large numbers enumerated differently? A limited-capacity preattentive stage in vision. *Psychological Review, 101*(1), 80–102.

Tzelgov, J., Meyer, J., & Henik, A. (1992). Automatic and intentional processing of numerical information. *Journal of Experimental Psychology: Learning, Memory, and Cognition, 18*(1), 166.

Valdes-Sosa, M., Bobes, M. A., Rodriguez, V., & Pinilla, T. (1998). Switching attention without shifting the spotlight: Object-based attentional modulation of brain potentials. *Journal of Cognitive Neuroscience, 10*(1), 137–151.

van Dijck, J.-P., Abrahamse, E. L., Acar, F., Ketels, B., & Fias, W. (2014). A working memory account of the interaction between numbers and spatial attention. *The Quarterly Journal of Experimental Psychology, 67*(8), 1500–1513.

van Veen, V., & Carter, C. S. (2002). The anterior cingulate as a conflict monitor: fMRI and ERP studies. *Physiology and Behavior, 77*(4), 477–482.

Verghese, P. (2001). Visual search and attention: A signal detection theory approach. *Neuron, 31*(4), 523–535.

Walter, W. G., Cooper, R., Aldridge, W. J., & McCallum, W. C. (1964). Contingent negative variation: An electrophysiological sign of sensoriomotor association and expectancy in the human brain. *Nature, 203*, 308–384.

Webb, N. M., Shavelson, R. J., & Haertel, E. H. (2006). Reliability coefficients and generalizability theory. In *Handbook of statistics* (Vol. 26) (pp. 81–124).

Weinbach, N., & Henik, A. (2011). Phasic alertness can modulate executive control by enhancing global processing of visual stimuli. *Cognition, 121*(3), 454–458.

Weissman, D. H., & Woldorff, M. G. (2004). Hemispheric asymmetries for different components of global/local attention occur in distinct temporo-parietal loci. *Cerebral Cortex, 15*(6), 870–876.

Wilkins, A. J., Shallice, T., & McCarthy, R. (1987). Frontal lesions and sustained attention. *Neuropsychologia, 25*(2), 359–365.

Yamaguchi, S., Yamagata, S., & Kobayashi, S. (2000). Cerebral asymmetry of the top-down allocation of attention to global and local features. *Journal of Neuroscience, 20*(9), RC72.

Yeshurun, Y., & Carrasco, M. (1999). Spatial attention improves performance in spatial resolution tasks. *Vision Research, 39*(2), 293–306.

The Control of Selective Attention and Emerging Mathematical Cognition: Beyond Unidirectional Influences

Rebecca Merkley[1], Pawel J. Matusz[2], Gaia Scerif[3]

[1]University of Western Ontario, London, ON, Canada; [2]University of Hospital Centre – University of Lausanne, Lausanne, Switzerland; [3]University of Oxford, Oxford, United Kingdom

It has been well established that the control of attention is correlated with mathematics ability, but the causal direction of the relationship remains unclear. In this chapter, we explore how the control of attention plays a role in emerging numerical cognition. First, we provide a brief operationalization of the constructs encompassed by the term "selective attention" and their overlap with "executive functions (EFs)" and "cognitive control," as they are studied in the context of mathematical development. Following from this, we summarize the correlational evidence for relationships between the control of attention and the development of mathematical abilities. We then critically review research exploring causal mechanisms underlying this relationship and highlight directions for future research. We argue that executive control of attention operates in conjunction with relevant mathematics-specific knowledge (e.g., Amso & Scerif, 2015; Johnson, 2011) to support the acquisition of mathematical skills.

DEFINING THE CONTROL OF ATTENTION: OPERATIONALIZING CONSTRUCT OVERLAP AND DIFFERENCES

Different terms are used throughout the literature to describe the control of attention. The idea of "attentional control" or a "central executive" is a crucial component of both prominent models of attention (e.g., Desimone & Duncan, 1995; Petersen & Posner, 2012; Posner & Petersen, 1990) and working memory (WM) (e.g., Baddeley, 2000; Baddeley & Hitch, 1974). One of the most popular models of attention in adults (Posner & Petersen, 1990) and of its development (Rueda et al., 2004) operationalizes the control of attention (termed "executive attention") as the ability to resolve the conflict between competing stimuli and responses (as classically measured by the flanker task, Eriksen & Eriksen, 1974). This is one of the three key processes encompassing attention, alongside orienting of attention in space, and alerting of attention to particular moments in time (see Petersen & Posner, 2012 for an updated overview of this model). In this context, the control of attention encompasses the bias imposed on incoming information to prioritize task-relevant stimuli or responses and to suppress or ignore what is not task relevant (Desimone & Duncan, 1995).

The supramodal "central executive" within the WM model (Baddeley & Hitch, 1974) and EFs have been likened to a chief executive officer, as they are responsible for directing and monitoring all other cognitive processes, especially important in situations involving processing unfamiliar stimuli (Goldberg, 2002). Crucially, performance on tasks used to measure WM and tasks used to measure executive functioning are extremely highly correlated, and the two share a common attentional control component (McCabe, Roediger, McDaniel, Balota, & Hambrick, 2010).

There is some consensus that EF consists of several domains, including complex reasoning and problem-solving, WM, attentional control, cognitive flexibility, self-monitoring, and regulation of cognition, emotion, and behavior (Miyake et al., 2000). However, there have been many challenges for defining and measuring cognitive control and EFs. As such there currently is no universally accepted model. One popular model is Miyake et al. (2000) model of EFs, which proposes three major EF processes: updating, shifting, and inhibition. *Shifting* refers to flexibility of attention and ability to switch between tasks, *updating* is short for monitoring and updating information in WM, and *inhibition* is defined as the ability to inhibit automatic responses. However, this model has not been replicated in children younger than 6 years old, and EFs are best captured by a single factor in the preschool age group (3–5-year-olds) (Garon, Bryson, & Smith, 2008; Wiebe et al., 2011). The control of attention is required for any EF task, and Garon et al. (2008) argued that the development of children's voluntary ability to control attention underlies the improvements and dissociation in EFs observed over the preschool period.

Throughout this chapter, we focus on discussing the evidence for the role of these executive controls of attention functions on emerging mathematical cognition. We note that we will use the term "control of attention" to refer to the broader construct that encompasses attentional skills such as executive control, executive attention, selective, and sustained attention (e.g., Posner & Petersen, 1990). We note that attention control and EFs differentiate (see the section on cognitive control within this volume), but they also overlap (McCabe et al., 2010). For example, models of the mechanisms of selective attention have explicitly emphasized the role of representations held in WM as the source of attentional bias on visual selection (Desimone & Duncan, 1995). Therefore, we use the term control of attention to capture the interrelated cognitive processes involved in selecting the information important for the task at hand.

CORRELATIONAL EVIDENCE FOR RELATIONSHIPS BETWEEN THE CONTROL OF ATTENTION AND EARLY NUMERACY

Correlational relationships between the control of attention and mathematics have been found both concurrently and longitudinally in school-aged children (e.g., Hassinger-Das, Jordan, Glutting, Irwin, & Dyson, 2014; LeFevre et al., 2010, 2013) and in preschoolers (e.g., Steele, Karmiloff-Smith, Cornish, & Scerif, 2012) (for reviews see Bull & Lee, 2014; Cragg & Gilmore, 2014, pp. 1–16). For example, in one study modeling relationships between multiple domain-general and domain-specific contributors to the development of numeracy in young children, 4- and 5-year-old children, completed behavioral assessments of linguistic skills, numeracy skills, and spatial attention and were subsequently assessed on mathematics achievement 2 years later (LeFevre et al., 2010). The spatial span task used to measure spatial attention captures overlapping skills, including spatial WM, making it impossible to isolate attentional capacity from visual spatial skills. Results supported the hypothesized relationships between linguistic skills and symbolic mathematics, as well as between early quantitative skills and numerical magnitude processing. Crucially, spatial selective attention was related to both symbolic and nonsymbolic number skills, both concurrently and longitudinally.

Similarly, teacher's classroom observations support the importance of cognitive control of attention processes for mathematics education. A qualitative study of teachers' perceptions of the role of EFs in mathematics found that most teachers believed that skills that could be classified as EFs were important for learning math, despite the fact that only 20% of the participating teachers were familiar with the term "EF" (Gilmore & Cragg, 2014). Teachers rated the EF skills almost as highly as the math-specific skills in terms of importance for successful development of mathematics skills. Notably, both child-based assessments and teacher reports of children's EFs (Fuhs, Farran, & Nesbitt, 2015) were significantly predictive of gains in mathematic skills over 8 months in a preschool program.

As mathematics and control of attention and EFs are multicomponential, some studies have aimed to tease apart relationships between specific processes across these domains. For example, one study found that inhibition was related to all components of early numeracy in 3–5-year-old children, whereas shifting was related specifically to digit identification and cardinal knowledge of number symbols (Purpura, Schmitt, & Ganley, 2017). Similarly, in older children, inhibition has been repeatedly found to be related to multiple components of mathematics ability (Gilmore et al., 2013; Gilmore, Keeble, Richardson, & Cragg, 2015; Robinson & Dubé, 2013). Notably, inhibition has been specifically associated with performance on nonsymbolic magnitude comparison tasks (e.g., Clayton & Gilmore, 2014), with this in turn often considered a key marker task of emerging numerical cognition (Halberda, Mazzocco, & Feigenson, 2008). These tasks require participants to choose the more numerous of two simultaneously presented arrays. The number of objects in an array is sometimes in conflict with visual cues from the continuous perceptual dimensions of the array (such as the size of objects, the total area they subtend), and inhibitory control plays a role in resolving this conflict (Clayton & Gilmore, 2014). Resolving this conflict also requires having a good understanding of what discrete number is, and young children seem to struggle with this (Merkley, Thompson, & Scerif, 2016; Rousselle & Noël, 2008). In fact, a recent review challenged the notion that the sense of number is innate and proposed instead that separation of discrete numerosity and continuous quantity is something that children have to consciously learn (Leibovich, Katzin, Harel, & Henik, 2016). However, it remains unclear exactly how children learn to select number as the relevant dimension on tasks such as nonsymbolic magnitude comparison. In other words, it is unclear how selective attention bolsters and is bolstered by an increasing awareness of number in the context (and in conflict with) other dimensions representing quantity (Merkley, Scerif, & Ansari, 2017).

Thus, there is substantial evidence that the control of attention is important for effectively deploying knowledge when doing mathematics (especially as indexed by the role of individual differences in EFs), but it could additionally be that it plays an essential role in the acquisition of basic numerical knowledge. Although correlational evidence supports the importance of the control of attention for performing numerical operations, it cannot verify whether and, if so, how exactly selective attention matters for the *learning* of numerical skills. One study found that executive attention predicted growth in arithmetic fluency in 8–10-year-old children (Lefevre et al., 2013), which suggests that the control of attention played a role in the acquisition of these skills. Similarly, Clark, Pritchard, and Woodward (2010) found that preschool executive control at 3 years of age predicted performance on mathematics assessments at 5 years of age, and an earlier study demonstrated the role of EF prior to school entry as a predictor of growth in both mathematics and reading over the first 2 years of primary school (Bull, Espy, & Wiebe, 2008). These findings support the importance of cognitive control also for the learning of mathematical skills, rather than just their mere implementation. At the same time, they are insufficient as support for the causality of this influence. Equally, these studies provide no insights as into the potential mechanisms whereby cognitive, goal-based attentional control could shape mathematics skills acquisition. Cognitive control facilitates the preferential allocation of attention to information that is relevant to learning, relative to other learning-irrelevant information, and maintenance of such focus for prolonged periods so that the selected information can be encoded into long-term memory (Posner & Rothbart, 2007). As such, theoretically, cognitive control should be particularly relevant to educational achievement. However, currently the existing causal evidence supporting this hypothesis encompasses longitudinal data, rather than explicit experimental manipulations. However suggestive, the former are always liable to alternative variables driving significant correlations between early attentional control skills and growth in math (for example, correlated differences in the environment children experience and are able to learn more from). We therefore move onto discussing causal evidence emerging from direct manipulations of attention control skills.

II. PERFORMANCE CONTROL AND SELECTIVE ATTENTION

DOES ATTENTION PLAY A CAUSAL ROLE IN LEARNING MATHEMATICS?

Investigating whether the control of attention plays a role in the acquisition of basic numerical competencies requires experimental intervention, such as, for example, design and systematic evaluation of preschool curricula (Diamond, Barnett, Thomas, & Munro, 2008). One type of intervention that has gained a substantial amount of attention during the last decade is training of domain-general functions, such as attention. It usually takes the form of a computerized regime designed to improve a specific cognitive function, such as updating in WM. These interventions are based on the idea that the adaptive nature of the training task, i.e., increasing difficulty with increasing on-task performance, will lead not only to, first, long-term improvements in the trained functions (tested with untrained tasks, i.e., near transfer) but also to improvements in untrained functions, skills, or behaviors (i.e., far transfer). The latter assumption is based on the discussed correlational evidence reviewed in the previous section. That is, if the control of attention and EFs are malleable and causally related to mathematics achievement, then training-related improvements should, in principle, lead to higher math performance or growth in math skills. In this section, we will first discuss critically the early studies on far transfer of attention or EF training and then present the existing evidence for transfer specifically to mathematical abilities. We will then place these findings within the broader context of current evidence for far transfer of domain-general training and delineate the prerequisite conditions for a sound experimental study design to test for causal links between control of attention and mathematical abilities.

Early training studies suggested strongly that control of attention (e.g., Rueda, Posner, & Rothbart, 2005) and EFs (e.g., Klingberg, Forssberg, & Westerberg, 2002) can be improved by computerized interventions training the construct of interest (see Diamond, 2012 for a review), and some have even shown that training WM (updating) can lead to improvements in mathematics (e.g., Kroesbergen, van't Noordende, & Kolkman, 2014). However, verifying causal links between cognitive control and math skills through this direct intervention route remains difficult for several reasons. First, the degree to which attention or executive training leads to far transfer effects to mathematics skills varies greatly across studies, which is likely linked to the methodological limitations that characterize most existing, published studies on computerized training of domain-general functions such as WM (Melby-Lervåg, Redick, & Hulme, 2016). Second, many existing studies in the area have been motivated by application rather than theory. By this we mean, more specifically, that training could be used with the goal of improving outcomes in populations with learning difficulties or used to investigate causal relationships between cognitive processes.

It can be challenging or even unacceptable for a study to try accomplishing both aims (Jolles & Crone, 2012; Wass, Scerif, & Johnson, 2012).

The majority of existing domain-general training studies were aimed at improving educational outcomes in children who struggle with both attention and learning, such as children with attention deficit hyperactivity disorder (ADHD) (e.g., Green et al. 2012; Klingberg et al., 2002). The initial, highly promising studies demonstrated that training on a mixed regime of tasks targeting attention and EFs led to improvements on the trained tasks, as well as on untrained tasks of visual spatial updating and nonverbal reasoning (Klingberg et al., 2002) or conflict resolution (Klingberg et al. 2005). Crucially, when the same training protocol was tested on a larger, more heterogeneous sample of ADHD children and using more rigorous control of confounding factors across intervention and control groups (e.g., contingent reinforcement, time-on-task with computer training, parent–child interactions, supportive coaching), no difference was found in parental and teacher ratings of behavior across groups (Chacko et al., 2014). Thus, the efficacy of domain-general training in generating transfer to classroom-relevant behavior might have been overestimated (Melby-Lervåg et al., 2016). Given the controversy surrounding far transfer of attention and EF training, it is necessary to critically evaluate the evidence for it for numeracy.

Fewer studies have investigated transfer from attention control training to academic achievement, including mathematics outcomes. At the same time, many of the existing studies in this area share both the theoretical and methodological limitations characterizing the early ADHD training studies. In a study evaluating a computerized WM training program, 8 weeks of training did not transfer to standardized mathematics assessments in a sample of typically developing 5–8-year-old children (St Clair-Thompson, Stevens, Hunt, & Bolder, 2010). Adults with dyscalculia who completed attentional control training did show improvements in their control of attention but failed to show far transfer to mathematics measures (Ashkenazi & Henik, 2012). Witt (2011) found that 9–10-year-old children who completed a WM training intervention did show greater improvements in mental arithmetic. However, the training program consisted of a variety of games, including practicing a backward digit recall task, inhibiting distractors, and verbal rehearsal, thus making it difficult to determine which of the trained tasks led to the observed improvements. This is especially pertinent as the study included only a passive, but no active, control comparison group. In contrast, a study in 6-year-olds from low (socioeconomic status) SES areas showed that children who played adaptive games targeting updating, planning, and inhibition improved at school measures of language and math compared with an active control group who played less cognitively demanding computer games (Goldin et al., 2014). However, the study lacked a passive control group, and therefore it is not possible to rule out the possibility that improvements were because of simple practice effects or change over time on tests administered before and after training, rather than to training benefits exclusively.

It has been postulated that cognitive training in general may be more effective in younger children, possibly because of heightened general plasticity of their brains and/or lack of differentiation of their brain networks, including those involved in attentional control (Jolles & Crone, 2012; Wass et al., 2012). However, to date, both domain-general and domain-specific computerized interventions aiming to improve numeracy skills in preschool have revealed inconsistent results. For example, Kroesbergen et al. (2014) compared training effects across a group of low-performing 5–6-year-old children, where some completed domain-general WM training, some a WM training with number-specific stimuli, and others no training at all. Children in both training groups showed improvements in posttraining measures of WM and on a standardized early numeracy skills assessment. Notably, only the domain-specific WM training group showed significant improvements on a nonsymbolic magnitude comparison task, suggesting that domain-specific training more strongly influences mathematical abilities. Another study that used very similar interventions in typically developing preschoolers found that domain-specific counting training was more effective at improving mathematics performance than a combination of WM and counting training (Kyttälä, Kanerva, & Kroesbergen, 2015). This contrasts with the results of Passolunghi and Costa (2014), who also compared WM and numeracy training interventions in a group of preschoolers. They demonstrated that training numeracy led to improvements in math skills but not in WM, whereas training WM led to transfer in both WM and numeracy. Another EF intervention study found that the training group performed better than the control group on an early numeracy assessment following training but did not measure numeracy at baseline (Blakey & Carroll, 2015). Given the inconsistent results across training studies, evidence for a causal role of control functions in early mathematics learning is at best mixed. However, it is also possible that domain-specific training may act by improving inhibitory functions. Specifically, in one training study aimed at improving nonsymbolic magnitude comparison in preschool-aged children from low-income homes, such improvements were driven by performance on trials on which number was in conflict with continuous quantity (Fuhs & McNeil, 2015). This suggests that inhibitory control may have been influenced by the intervention because improvements were not seen on nonsymbolic magnitude comparison more broadly.

To summarize, currently it remains unclear whether training the control of attention reliably leads to improvements in educationally relevant mathematics outcomes. First, studies on low-SES or low-achieving populations (Fuhs & McNeil, 2015; Goldin et al., 2014; Kroesbergen, van't Noordende, & Kolkman, 2012) have focused on applied goals and therefore are, in terms of experimental design, less appropriate for addressing theoretical questions pertaining to mechanisms underlying transfer. Specifically, the lack of double-blind experimental procedures that frequently characterize the applied studies may contribute to the overestimation of true effects of training cognitive control on far transfer (Sonuga-Barke et al. 2013). This notwithstanding, the current evidence for transfer from training attention to math is mixed. Notably, it shares many of the limitations characterizing other studies on cognitive training and far transfer. A lack of appropriate control group(s) renders impossible to distinguish effects driven by training compared with placebo effects (no active control) or regular development (no passive control). In turn, small sample sizes may lead to overestimations of wider transfer effects. This was demonstrated by a recent metaanalysis of 87 publications with 145 experimental comparisons by Melby-Lervåg et al. (2016) who concluded that wider transfer of the effects of training of cognitive control, such as WM, to other cognitive processes, including mathematics skills, is not convincingly supported by existing data. Notably, the authors suggest that even in the more rigorous studies, which used rigorous control groups and have sufficient statistical power, the effects of wide transfer are only statistically small (Cohen's d ~0.25; e.g., Sonuga-Barke et al. 2013). Crucially, this effect size estimate is likely exaggerated, as many researchers finding null results on the far transfer of EFs, on behavioral outcomes, and/or educational achievements fail to publish their findings (i.e., publication bias). This problem may be alleviated by the greater emphasis on preregistration. More research is necessary to clarify whether training of attentional control reliably leads to improvements in educationally relevant mathematics outcomes. In sum, an ideal training attention/EF intervention study should (1) randomly assign participants to groups and have assessors blind to their assignment (double-blindness), (2) have a sample size that is sufficiently large to afford high statistical power, (3) include a group that does not receive any training and an active control group that does some form of intervention that is not hypothesized to lead to the same improvements as the experimental training regime, and also (4) use multiple measures of the construct being trained and constructs to which far transfer is hypothesized (Gathercole, Dunning, & Holmes, 2012; Rabipour & Raz, 2012; Shipstead, Redick, & Engle, 2012).

BIDIRECTIONAL RELATIONSHIPS BETWEEN ATTENTION AND MATH: EXPERTISE INFLUENCES THE DEPLOYMENT OF ATTENTION

Most previous studies exploring relationships between the control of attention and math have aimed to test the unidirectional causal hypothesis that attention is important for selecting information relevant to learning.

However, bidirectional relationships between EFs and mathematics were found in one longitudinal study (Van der Ven, Kroesbergen, Boom, & Leseman, 2012) and another showed that math achievement predicted EFs, but that EFs did not predict mathematics over and above early math measures (Watts et al., 2015). Therefore, improvements in mathematics could in turn contribute to improvements in attentional control and EFs. Further research is needed to determine the direction of causality between the two cognitive capacities (Clements, Sarama, & Germeroth, 2016). We propose that the control of attention and domain-specific mathematics knowledge dynamically interact over the course of the development of numerical cognition. Specifically, attentional control processes enable children to select information relevant to math to be encoded into short- and then long-term memory. At the same time, previously learned knowledge guides how strongly attentional resources are allocated to task-relevant stimuli in the environment, which in turn improves as knowledge accrues (e.g., Amso & Scerif, 2015; Johnson, 2011).

One learning challenge in the domain of mathematics is acquiring the meaning of numerical symbols (words and Arabic digits). Children learn the meaning of the number words one to four sequentially and there is some evidence that their prior knowledge influences subsequent learning (Huang, Spelke, & Snedeker, 2010). Specifically, a group of 3-year-old children were taught the meaning of a number word they had not yet learned (Huang et al., 2010). Children who knew "two," as demonstrated by the fact that they could reliably give an experimenter two objects when asked, were trained on the word "three." Children who knew "three" were trained on the word "four." The training involved an experimenter showing children cards depicting different numbers of objects and verbally indicating whether the card did or did not have the target number of objects. Results showed that children who learned "four" were more likely to generalize the newly acquired knowledge and apply the word in novel contexts compared with children who learned "three." These results could be interpreted as evidence suggesting that children who have more prior number word knowledge are more likely to select abstract numerosity as the relevant referent when learning a new number word.

Higher proficiency with numerical symbols is also associated with better learning of a novel abstract symbol set (Merkley, 2015). In a series of studies, adults, older children (10-year-olds), and young children (6-year-olds) were taught to associate a set of abstract symbols with numerical meaning. Following the learning phase, they performed a magnitude comparison task with the learned symbols. During the learning phase, the symbols were paired with nonsymbolic arrays of dots, and participants learned to associate the symbols with the approximate magnitude of the corresponding arrays. Half of the participants were also taught the order of the symbols from smallest to largest. Adults and older children performed equally well on the comparison task regardless of whether they had been explicitly cued to the order of the symbols, which suggests that they could infer the order of the symbols based on magnitude information alone. However, the younger children who were not cued to order failed to perform above chance on the comparison task. Thus, it appears that greater experience with real numbers is associated with more efficient formation of novel symbolic representations. In particular, it seems that having more experience with number exerts a top-down guidance on the ability to extract numerosity from nonsymbolic arrays, a process that in turn is likely supported by goal-based attentional control.

In addition to learning the meaning of numerical symbols, young children must also learn to selectively attend to discrete number when faced with competing cues associated with continuous magnitude (Leibovich et al., 2016). In one study, children who did not know the cardinal meaning of number words failed to perform above chance on a nonsymbolic magnitude comparison task when numerosity conflicted with continuous quantity (Negen & Sarnecka, 2014). This suggests that having a better understanding of the cardinal meaning of number symbols may facilitate attention to numerosity, but the direction of this relationship is still debated. Uncovering the causal mechanisms underlying the relationship between number knowledge and attention to number requires testing bidirectional hypotheses and moving beyond correlational, cross-sectional experimental designs (Merkley et al., 2018).

CONCLUSION

Relationships between the control of attention and mathematics abilities have been demonstrated in many correlational studies. However, the causal nature of this relationship remains unclear. Some EF training interventions have been associated with far transfer to mathematics, yet others have failed to find similar results. Furthermore, more rigorous investigations are necessary to determine whether cognitive control interventions are indeed effective at improving mathematical outcomes and which populations stand to benefit most. Moreover, bidirectional relationships between selective attention and the development of numerical cognition should be more systematically investigated. Not only does selective attention influence which information is filtered into memory but also prior knowledge guides the deployment of attention. This interplay between the control of selective attention and emerging numeracy likely plays a role in learning early mathematics.

References

Amso, D., & Scerif, G. (2015). The attentive brain: Insights from developmental cognitive neuroscience. *Nature Reviews Neuroscience, 16*(10), 606–619.

Ashkenazi, S., & Henik, A. (2012). Does attentional training improve numerical processing in developmental dyscalculia? *Neuropsychology, 26*(1), 45.

Baddeley, A. (2000). The episodic buffer: A new component of working memory? *Trends in Cognitive Sciences, 4*(11), 417–423. https://doi.org/10.1016/S1364-6613(00)01538-2.

Baddeley, A. D., & Hitch, G. (1974). Working memory. *Psychology of Learning and Motivation, 8,* 47–89.

Blakey, E., & Carroll, D. J. (November 2015). A short executive function training program improves preschoolers' working memory. *Frontiers in Psychology, 6,* 1–8. https://doi.org/10.3389/fpsyg.2015.01827.

Bull, R., Espy, K. A., & Wiebe, S. A. (2008). Short-term memory, working memory, and executive functioning in preschoolers: Longitudinal predictors of mathematical achievement at age 7 years. *Developmental Neuropsychology, 33*(3), 205–228.

Bull, R., & Lee, K. (2014). Executive functioning and mathematics achievement. *Child Development Perspectives, 8*(1), 36–41. https://doi.org/10.1111/cdep.12059.

Chacko, A., Bedard, A. C., Marks, D. J., Feirsen, N., Uderman, J. Z., Chimiklis, A., et al. (2014). A randomized clinical trial of Cogmed working memory training in school-age children with ADHD: A replication in a diverse sample using a control condition. *Journal of Child Psychology and Psychiatry, 55*(3), 247–255.

Clark, C. A., Pritchard, V. E., & Woodward, L. J. (2010). Preschool executive functioning abilities predict early mathematics achievement. *Developmental Psychology, 46*(5), 1176–1191. https://doi.org/10.1037/a0019672.

Clayton, S., & Gilmore, C. (2014). Inhibition in dot comparison tasks. *Zdm.* https://doi.org/10.1007/s11858-014-0655-2.

Clements, D. H., Sarama, J., & Germeroth, C. (2016). Learning executive function and early mathematics: Directions of causal relations. *Early Childhood Research Quarterly, 36,* 79–90.

Cragg, L., & Gilmore, C. (2014). *Skills underlying mathematics: The role of executive function in the development of mathematics proficiency.* Trends in Neuroscience and Education, 3, 63–68.

Desimone, R., & Duncan, J. (1995). Neural mechanisms of selective visual attention. *Annual Review of Neuroscience, 18*(1), 193–222.

Diamond, A. (2012). Activities and programs that improve children's executive functions. *Current Directions in Psychological Science, 21*(5), 335–341.

Diamond, A., Barnett, W. S., Thomas, J., & Munro, S. (2008). *NIH Public Access, 318*(5855), 1387–1388.

Eriksen, B. A., & Eriksen, C. W. (1974). Effects of noise letters upon the identification of a target letter in a nonsearch task. *Perception and Psychophysics, 16,* 143–149.

Fuhs, M. W., Farran, D. C., & Nesbitt, K. T. (2015). Prekindergarten children's executive functioning skills and achievement gains: The utility of direct assessments and teacher ratings. *Journal of Educational Psychology, 107*(1), 207–221. https://doi.org/10.1037/a0037366.

Fuhs, M. W., & McNeil, N. M. (2015). The role of non-numerical stimulus features in approximate number system training in preschoolers from low-income homes mary. *Statewide Agricultural Land Use Baseline, 1*, 1–58. https://doi.org/10.1017/CBO9781107415324.004.

Garon, N., Bryson, S. E., & Smith, I. M. (2008). Executive function in preschoolers: A review using an integrative framework. *Psychological Bulletin, 134*(1), 31.

Gathercole, S. E., Dunning, D. L., & Holmes, J. (2012). Cogmed training: Let's be realistic about intervention research. *Journal of Applied Research in Memory and Cognition, 1*(3), 201–203. https://doi.org/10.1016/j.jarmac.2012.07.007.

Gilmore, C., Attridge, N., Clayton, S., Cragg, L., Johnson, S., Marlow, N., et al. (2013). Individual differences in inhibitory control, not non-verbal number acuity, correlate with mathematics achievement. *PLoS One, 8*(6), e67374. https://doi.org/10.1371/journal.pone.0067374.

Gilmore, C., & Cragg, L. (2014). *Teachers' understanding of the role of executive functions in maths* 37474.

Gilmore, C., Keeble, S., Richardson, S., & Cragg, L. (2015). The role of cognitive inhibition in different components of arithmetic. *Zdm*. https://doi.org/10.1007/s11858-014-0659-y.

Goldberg, E. (2002). *The executive brain: Frontal lobes and the civilized mind*. USA: Oxford University Press.

Goldin, A. P., Hermida, M. J., Shalom, D. E., Elias Costa, M., Lopez-Rosenfeld, M., Segretin, M. S., et al. (2014). Far transfer to language and math of a short software-based gaming intervention. *Proceedings of the National Academy of Sciences of the United States of America*, 1–6. https://doi.org/10.1073/pnas.1320217111.

Green, C. T., Long, D. L., Green, D., Iosif, A. M., Dixon, J. F., Miller, M. R., et al. (2012). Will working memory training generalize to improve off-task behavior in children with attention-deficit/hyperactivity disorder? *Neurotherapeutics, 9*(3), 639–648.

Halberda, J., Mazzocco, M. M. M., & Feigenson, L. (2008). Individual differences in non-verbal number acuity correlate with maths achievement. *Nature, 455*(7213), 665–668. https://doi.org/10.1038/nature07246.

Hassinger-Das, B., Jordan, N. C., Glutting, J., Irwin, C., & Dyson, N. (2014). Domain-general mediators of the relation between kindergarten number sense and first-grade mathematics achievement. *Journal of Experimental Child Psychology, 118*, 78–92. https://doi.org/10.1016/j.jecp.2013.09.008.

Huang, Y. T., Spelke, E., & Snedeker, J. (2010). When is four far more than three? Children's generalization of newly acquired number words. *Psychological Science, 21*(4), 600–606. https://doi.org/10.1177/0956797610363552.

Johnson, M. H. (2011). Interactive specialization: A domain-general framework for human functional brain development? *Developmental Cognitive Neuroscience, 1*(1), 7–21.

Jolles, D. D., & Crone, E. A. (2012). Training the developing brain: A neurocognitive perspective. *Frontiers in Human Neuroscience, 6*.

Klingberg, T., Fernell, E., Olesen, P. J., Johnson, M., Gustafsson, P., Dahlström, K., et al. (2005). Computerized training of working memory in children with ADHD-a randomized, controlled trial. *Journal of the American Academy of Child and Adolescent Psychiatry, 44*(2), 177–186.

Klingberg, T., Forssberg, H., & Westerberg, H. (2002). Training of working memory in children with ADHD. *Journal of Clinical and Experimental Neuropsychology, 24*(6), 781–791. https://doi.org/10.1076/jcen.24.6.781.8395.

Kroesbergen, E. H., van't Noordende, J. E., & Kolkman, M. E. (January 2012). Training working memory in kindergarten children: Effects on working memory and early numeracy. *Child Neuropsychology: A Journal on Normal and Abnormal Development in Childhood and Adolescence, 37*–41. https://doi.org/10.1080/09297049.2012.736483.

Kroesbergen, E. H., van 't Noordende, J. E., & Kolkman, M. E. (2014). Training working memory in kindergarten children: effects on working memory and early numeracy. *Child Neuropsychology, 20*(1), 23–37. https://doi.org/10.1080/09297049.2012.736483. Epub 2012 Oct 25.

Kyttälä, M., Kanerva, K., & Kroesbergen, E. (2015). Training counting skills and working memory in preschool. *Scandinavian Journal of Psychology.* n/a-n/a https://doi.org/10.1111/sjop.12221.

Lefevre, J.-A., Berrigan, L., Vendetti, C., Kamawar, D., Bisanz, J., Skwarchuk, S.-L., et al. (2013). The role of executive attention in the acquisition of mathematical skills for children in Grades 2 through 4. *Journal of Experimental Child Psychology, 114*(2), 243–261. https://doi.org/10.1016/j.jecp.2012.10.005.

LeFevre, J.-A., Fast, L., Skwarchuk, S.-L., Smith-Chant, B. L., Bisanz, J., Kamawar, D., et al. (2010). Pathways to mathematics: Longitudinal predictors of performance. *Child Development, 81*(6), 1753–1767. https://doi.org/10.1111/j.1467-8624.2010.01508.x.

Leibovich, T., Katzin, N., Harel, M., & Henik, A. (2016). From "sense of number" to "sense of magnitude" – the role of continuous magnitudes in numerical cognition. *Behavioral and Brain Sciences, 1,* 1–62. https://doi.org/10.1017/S0140525X16000960.

McCabe, D. P., Roediger, H. L., III, McDaniel, M. A., Balota, D. A., & Hambrick, D. Z. (2010). The relationship between working memory capacity and executive functioning: Evidence for a common executive attention construct. *Neuropsychology, 24*(2), 222.

Melby-Lervåg, M., Redick, T. S., & Hulme, C. (2016). Working memory training does not improve performance on measures of intelligence or other measures of "far transfer" evidence from a meta-analytic review. *Perspectives on Psychological Science, 11*(4), 512–534.

Merkley, R. (2015). *Beyond number sense: Contributions of domain general processes to the development of numeracy in early childhood* (Doctor of Philosophy thesis). Oxford, UK: University of Oxford.

Merkley, R., Scerif, G., & Ansari, D. (Jan 2017). What is the precise role of cognitive control in the development of a sense of number? Commentary on Leibovich et al. *Behavioral and Brain Sciences, 40,* e179. https://doi.org/10.1017/S0140525X1600217X.

Merkley, R., Thompson, J., & Scerif, G. (January 2016). Of huge mice and tiny elephants: Exploring the relationship between inhibitory processes and preschool math skills. *Frontiers in Psychology, 6.* https://doi.org/10.3389/fpsyg.2015.01903.

Miyake, A., Friedman, N. P., Emerson, M. J., Witzki, A. H., Howerter, A., & Wager, T. D. (2000). The unity and diversity of executive functions and their contributions to complex "Frontal Lobe" tasks: A latent variable analysis. *Cognitive Psychology, 41*(1), 49–100. https://doi.org/10.1006/cogp.1999.0734.

Negen, J., & Sarnecka, B. W. (2014). Is there really a link between exact-number knowledge and approximate number system acuity in young children? *The British Journal of Developmental Psychology*, 1–14. https://doi.org/10.1111/bjdp.12071.

Passolunghi, M. C., & Costa, H. M. (November 2014). Working memory and early numeracy training in preschool children. *Child Neuropsychology: A Journal on Normal and Abnormal Development in Childhood and Adolescence*, 1–18. https://doi.org/10.1080/09297049.2014.971726.

Petersen, S. E., & Posner, M. I. (2012). The attention system of the human brain: 20 years after. *Annual Review of Neuroscience, 35*, 73–89.

Posner, M. I., & Petersen, S. E. (1990). The attention system of the human brain. *Annual Review of Neuroscience, 13*, 25–42. https://doi.org/10.1146/annurev.ne.13.030190.000325.

Posner, M. I., & Rothbart, M. K. (2007). Research on attention networks as a model for the integration of psychological science. *Annual Review of Psychology, 58*, 1–23.

Purpura, D. J., Schmitt, S. A., & Ganley, C. M. (2017). Foundations of mathematics and literacy: The role of executive functioning components. *Journal of Experimental Child Psychology, 153*, 15–34. https://doi.org/10.1016/j.jecp.2016.08.010.

Rabipour, S., & Raz, A. (2012). Training the brain: Fact and fad in cognitive and behavioral remediation. *Brain and Cognition, 79*(2), 159–179.

Robinson, K. M., & Dubé, A. K. (2013). Children's additive concepts: Promoting understanding and the role of inhibition. *Learning and Individual Differences, 23*, 101–107. https://doi.org/10.1016/j.lindif.2012.07.016.

Rousselle, L., & Noël, M.-P. (2008). The development of automatic numerosity processing in preschoolers: Evidence for numerosity-perceptual interference. *Developmental Psychology, 44*(2), 544–560. https://doi.org/10.1037/0012-1649.44.2.544.

Rueda, M. R., Fan, J., McCandliss, B. D., Halparin, J. D., Gruber, D. B., Lercari, L. P., & Posner, M. I. (2004). Development of attentional networks in childhood. *Neuropsychologia, 42*(8), 1029–1040.

Rueda, M. R., Posner, M. I., & Rothbart, M. K. (2005). The development of executive attention: Contributions to the emergence of self-regulation. *Developmental Neuropsychology, 28*(2), 573–594.

Shipstead, Z., Redick, T. S., & Engle, R. W. (2012). Is working memory training effective? *Psychological Bulletin, 138*(4), 628–654. https://doi.org/10.1037/a0027473.

Sonuga-Barke, E. J., Brandeis, D., Cortese, S., Daley, D., Ferrin, M., Holtmann, M., et al. (2013). Nonpharmacological interventions for ADHD: Systematic review and meta-analyses of randomized controlled trials of dietary and psychological treatments. *American Journal of Psychiatry, 170*(3), 275–289.

St Clair-Thompson, H., Stevens, R., Hunt, A., & Bolder, E. (2010). Improving children's working memory and classroom performance. *Educational Psychology, 30*(2), 203–219. https://doi.org/10.1080/01443410903509259.

Steele, A., Karmiloff-Smith, A., Cornish, K., & Scerif, G. (2012). The multiple subfunctions of attention: Differential developmental gateways to literacy and numeracy. *Child Development, 83*(6), 2028–2041. https://doi.org/10.1111/j.1467-8624.2012.01809.x.

Van der Ven, S. H. G., Kroesbergen, E. H., Boom, J., & Leseman, P. P. M. (2012). The development of executive functions and early mathematics A dynamic relationship. *The British Journal of Educational Psychology*, *82*(Pt 1), 100–119. https://doi.org/10.1111/j.2044-8279.2011.02035.x.

Wass, S. V., Scerif, G., & Johnson, M. H. (2012). Training attentional control and working memory–Is younger, better? *Developmental Review*, *32*(4), 360–387.

Watts, T. W., Duncan, G. J., Chen, M., Claessens, A., Davis-Kean, P. E., Duckworth, K., et al. (2015). The role of mediators in the development of longitudinal mathematics achievement associations. *Child Development*, *86*(6), 1892–1907.

Wiebe, S. A., Sheffield, T., Nelson, J. M., Clark, C. A., Chevalier, N., & Espy, K. A. (2011). The structure of executive function in 3-year-olds. *Journal of Experimental Child Psychology*, *108*(3), 436–452.

Witt, M. (2011). School based working memory training : Preliminary finding of improvement in children's mathematical performance. *Advances in Cognitive Psychology*, *7*(0), 7–15.

Further Reading

Alloway, T. P., & Alloway, R. G. (2010). Investigating the predictive roles of working memory and IQ in academic attainment. *Journal of Experimental Child Psychology*, *106*(1), 20–29. https://doi.org/10.1016/j.jecp.2009.11.003.

Best, J. R., Miller, P. H., & Naglieri, J. A. (2011). Relations between executive function and academic achievement from ages 5 to 17 in a large, representative national sample. *Learning and Individual Differences*, *21*(4), 327–336. https://doi.org/10.1016/j.lindif.2011.01.007.

Bull, R., & Scerif, G. (2001). Executive functioning as a predictor of children's mathematics ability: Inhibition, switching, and working memory. *Developmental Neuropsychology*, *19*(3), 273–293. https://doi.org/10.1207/S15326942DN1903_3.

St Clair-Thompson, H. L., & Gathercole, S. E. (2006). Executive functions and achievements in school: Shifting, updating, inhibition, and working memory. *Quarterly Journal of Experimental Psychology*, *59*(4), 745–759. https://doi.org/10.1080/17470210500162854.

Performance Control in Numerical Cognition: Insights From Strategic Variations in Arithmetic During the Life Span

Kim Uittenhove[1], Patrick Lemaire[2]

[1]University of Geneva, Geneva, Switzerland; [2]Aix-Marseille University & CNRS, Marseille, France

Numerical cognition provides an interesting window on human cognition because it involves the combination of different types of processes. Many of the processes are specific and can be automatized. For example, reciting the number chain is an activity that speeds up during childhood through repeated practice. It requires little conscious effort, as do many other crucial numerical operations (e.g., subitizing, number comparison, magnitude, and parity). Besides these specific processes, numerical cognition involves more general processing resources and mechanisms. Examples are executive control, decision-making, and strategy choices. For example, when solving a difficult arithmetic problem such as 389+257, we have to decide which strategy to use among several available strategies. Following such decisions, executing the selected strategy requires high levels of performance control.

In this chapter, we discuss how the bidirectional link between strategies and executive control influences numerical performance. Moreover, we look at this issue from a life span perspective, taking into account the developmental trajectory of children and the effects of cognitive aging. Indeed, performance in tasks assessing numerical cognition, like in many other domains, changes during development and aging. Research has documented age-related changes in several components of numerical cognition, including approximate and exact number systems (e.g., see Halberda, Ly, Wilmer, Naiman, & Germine, 2012; Wood, Ischebeck, Koppelstaetter, Gotwald, & Kaufmann, 2009, for a life span perspective), quantification (i.e., subitizing and counting, see Gallistel & Gelman, 1992; Gelman & Gallistel, 1986, for the development of these capacities in children and see, for example, Sliwinski, 1997, for a cognitive aging perspective), and arithmetic (see Groen & Parkman, 1972, for a seminal study on the development of arithmetic in childhood and Allen, Ashcraft, & Weber, 1992; Duverne & Lemaire, 2005, pp. 397–412; Salthouse & Coon, 1994, for an overview of changes related to cognitive aging). We argue that these effects can be better understood when first adopting a strategy perspective and second when taking into account one crucial determiner of strategic variations during aging and development, which is executive control.

A CONCEPTUAL FRAMEWORK FOR UNDERSTANDING STRATEGIC VARIATIONS

An obvious, yet crucial component of the approach we advocate is to take into account the strategies used by participants in numerical cognition. Strategies have been defined by Lemaire and Reder (1999, p. 365) as a *"procedure or set of procedures to accomplish a high-level goal."* Strategies can be distinguished by the type and number of processes they include. For example, *strategy A* may have a larger number of processes than *strategy B*, which could make it more precise, yet more difficult for execution. Thus, to understand performance, we need to have a thorough understanding of which strategies are used. Siegler (1987) used this strategy approach while investigating how children solve addition problems. He found that their performance was better understood when taking into account the strategies they used (see also Groen & Parkman, 1972; Siegler & Shrager, 1984; Thevenot, Barrouillet, Castel, & Uittenhove, 2016). However, there is more to the story than merely the types of strategies that are being used. Important questions are how efficiently individuals select and execute available strategies.

To further our understanding of strategic variations during the course of children's cognitive development, Lemaire and Siegler (1995) provided a conceptual framework. They distinguished four dimensions of strategic changes, namely strategy repertoire (i.e., which strategies are used), strategy distribution (i.e., relative frequency of use of strategies), strategy execution (i.e., the speed and precision obtained with each strategy), and strategy selection (i.e., how we choose a strategy among available strategies). In this chapter, we use this framework to identify on what levels of strategy use cognitive control plays a role and how this can explain developmental and age-related differences.

EXECUTIVE RESOURCES AND STRATEGIC VARIATIONS

Executive resources are involved in a variety of functions such as planning, maintaining, manipulating, and monitoring more basic cognitive processes. Several taxonomies propose a number of distinct executive functions (EFs). For example, the widely used Miyake et al. (2000) taxonomy of EFs includes updating, inhibition, and flexibility (see also Miyake & Friedman, 2012). These functions are necessary for controlled processing and working memory. Dimensions of strategy use rely on executive resources. First, keeping track of the steps within a strategy that has to be executed relies on working memory. Second, executing this strategy necessitates EFs for controlling the sequence of steps and working memory for maintaining intermediate results. Third, when selecting a strategy, we need to disengage from the previously executed strategy, inhibit competing strategies, and flexibly shift to a new appropriate strategy, thus relying on the EFs of inhibition and flexibility. We will now review the role of executive control in each strategy dimension in more detail.

Strategy Execution

The different steps of a strategy need to be coordinated and often partial results need to be maintained as we are executing a strategy, with a crucial role for executive control mechanisms. Most evidence for the involvement of executive control in strategy execution comes from experimental and correlational studies, Several experimental approaches have been adopted, ranging from dual-task methodology to manipulations of factors, the effects of which are known to result from variations in demands of executive resources. For example, in dual-task studies, performance on simple numerical tasks is compared when numerical tasks are accomplished alone (single-task condition) and when numerical tasks are accomplished together with a secondary task (dual-task condition). These secondary tasks involve executive resources, and as a consequence execution of strategies becomes less efficient. For example, Imbo, Duverne, and Lemaire (2007) had participants execute computational estimation strategies (e.g., doing 40×70 to estimate 43×72) and found that when they presented a simultaneous choice reaction task taxing executive resources (i.e., deciding whether randomly presented tones were high or low), participants executed numerical strategies less efficiently (see also Duverne, Lemaire, & Vandierendonck, 2008; Imbo & Vandierendonck, 2007). In correlational studies, individual differences in working memory capacities or EFs are correlated to efficiency of strategy execution. For example, Andersson (2008) administered tests of EFs (e.g., verbal fluency, trail making test (TMT), Stroop) and working memory (counting span, digit span, Corsi span) to children. He found that performance on these correlated with the number of two-digit written addition and subtraction problems solved in 10 min and the number of single-digit addition problems (out of 14) correctly solved within a reasonable delay (3 s per problem). Thus, these findings link executive control to strategy efficiency measured by fast and correct resolution of arithmetic problems (see also Agostino, Johnson, & Pascual-Leone, 2010; Bull & Scerif, 2001; Rasmussen & Bisanz, 2005).

Strategy Selection

Choosing an adequate strategy relies on decision processes and on executive control. To choose a strategy on a given problem, participants need to analyze problem characteristics and choose strategies accordingly, as problem characteristics are often correlated with relative efficiency of strategies (e.g., counting is more efficient on small problems like 3×4 than on larger problems like 7×8). However, in a sequence of problems, the question becomes even more interesting. To switch to a new strategy, participants need to disengage from the previously executed strategy and shift to another, more appropriate strategy. Research suggests that this ability requires inhibitory control to suppress just-executed and competing strategies and to flexibly alternate between strategies (Ardiale, Hodzik, & Lemaire, 2012; Lemaire & Lecacheur, 2010; Lemaire & Leclère, 2014a,b; Luwel, Schillemans, Onghena, & Verschaffel, 2009).

Strategy Repertoire and Distribution

The impact of executive control mechanisms on strategy execution and selection has implications for strategy repertoire and distribution. Generally, individuals will aim to lessen executive control demands when using strategies to accomplish a task. For example, those strategies that require a high level of executive control will gradually give way to easier strategies and may eventually fall into disuse, and this will lessen overall executive control demands. Another way to reduce the need for executive control is to switch between fewer strategies. Alterations of this type are especially apparent in populations who suffer from less-efficient executive control such as Alzheimer patients (Arnaud, Lemaire, Allen, & Michel, 2008; Gandini, Lemaire, & Michel, 2009). In summary, data are consistent with the hypothesis that cognitive control plays a role in each strategy dimension, thus rendering strategy use sensitive to variable cognitive control capacities in different types of individuals.

STRATEGIC VARIATIONS IN THE LIFE SPAN

A number of studies have documented how development and aging change strategic variations in a wide variety of cognitive domains including numerical cognition (see Siegler & Jenkins, 2014 for an overview of strategic variations in numerical cognition throughout development, and see Lemaire, 2016, for an overview of age-related changes during adulthood).

The developmental trajectory of children's strategy use can be characterized by improvements in strategy execution efficiency, especially on harder strategies that involve more cognitive steps (Imbo & Vandierendonck, 2007). Moreover, throughout development, children select strategies more adaptively (Imbo & Vandierendonck, 2007), use better strategies (Luwel, Verschaffel, Onghena, & De Corte, 2001), and increasingly switch between a larger numbers of strategies in accomplishing the same task (Lemaire & Brun, 2018; Luwel, Verschaffel, Onghena, & De Corte, 2000). One driving factor in this development is captured in models of strategy use like strategy choice and discovery simulation (SCADS, Siegler & Araya, 2005). This model assumes that the execution of specific strategies on specific problems will increase expertise underlying future adaptive strategy choices.

However, an accumulation of evidence also suggests an important role for the development of cognitive control (Andersson, 2008; Bull, Espy, & Wiebe, 2008; Bull & Scerif, 2001; Geary & Brown, 1991) and working memory capacity in arithmetic skills (Bull, Johnston, & Roy, 1999; McKenzie, Bull, & Gray, 2003). Younger children have less-efficient executive control than older children, and this is an important determiner of strategic variations in numerical cognition. On the other end of the life span, older adults also have less-efficient executive control than young adults, following degradation of frontal brain regions (e.g., Reuter-Lorenz & Park, 2010). Interestingly, in older adults, research points to a reduced strategy repertoire, increased use of easier strategies, less-efficient strategy execution, and less-adaptive strategy selection. This pattern of strategic variations in older adults is reminiscent of what is found in children. It seems that strategic variations are very sensitive to how executive control changes throughout the life span. Let us now look at the evidence in more detail.

Strategy Execution and Distribution

Lemaire and Lecacheur (2002) found that 9-year-old children were less efficient in a computational estimation task with three-digit additions (e.g., 458 + 326) than older children (11-year-olds) and adults (see also Lemaire & Callies, 2009, for an overview of developmental improvements in the execution of addition and subtraction strategies and Luwel, Lemaire, & Verschaffel, 2005, for examples in numerosity estimation), especially when executing the more difficult rounding-up strategy. The authors suggested that younger children execute the sequence of cognitive operations included in computational estimation strategies more slowly, leading to a more pronounced effect on more difficult strategies. Similarly, many studies found that older adults execute arithmetic and numerosity estimation strategies less efficiently than younger adults (e.g., Gandini, Lemaire, Anton, & Nazarian, 2008; Hodzik & Lemaire, 2011; Lemaire & Lecacheur, 2001), and this is especially the case for more complex arithmetic strategies (e.g., Salthouse & Coon, 1994). For example, Lemaire, Arnaud, and Lecacheur (2004) found that older adults made more errors and were slower with computational estimation strategies (i.e., estimating the answer to multiple-digit multiplication problems like 43 × 57 by rounding the operands to a nearby decade) and especially when using the harder rounding-up strategies (i.e., rounding both operands up to the closest larger decades, like doing 50 × 60, relative to rounding both operands down to the closest smaller decades like doing 40 × 50) or when solving difficult problems that rely more on cognitive control.

Consequently, children (Lemaire & Lecacheur, 2011) and older adults stick to simple arithmetic strategies more often than younger adults (e.g., Gandini et al., 2008; Siegler & Lemaire, 1997). Indeed, they use strategies that put the smallest burden on their executive resources. For example, Lemaire and Lecacheur (2011) found that children use the easier rounding-down strategy more often than the more difficult rounding-up strategy, even when the latter strategy is the most appropriate choice for the problem.

Strategy Selection and Repertoire

Lemaire and Lecacheur (2002) found that 9-year-old children less often chose the most appropriate strategy for each estimation problem, reflecting immature strategy selection. In another study, Lemaire and Callies (2009) tested 7–10-year-old children and found that while all age groups took into account relative strategy efficiency when selecting strategies on addition problems, 7-year-old children were not able to take into account relative strategy efficiency when solving subtraction problems (see also Luwel et al., 2005). Concerning older adults, they were also found to use the best strategy on each problem less often than young adults (Ardiale & Lemaire, 2012; Barulli, Rakitin, Lemaire, & Stern, 2013; Hodzik & Lemaire, 2011; Lemaire et al., 2004; Lemaire & Leclère, 2014a,b). For example, when young and older adults were asked to choose rounding-down or rounding-up strategies to estimate two-digit products, older adults used the rounding-down strategy on rounding-down problems (e.g., 41×36) and the rounding-up strategy on rounding-up problems (e.g., 34×68) less often than young adults.

Such poorer-strategy choices in children and older adults may have several reasons. Children and older adults could have difficulties correctly assessing problem characteristics (see, for example, Ardiale & Lemaire, 2013), or decision processes could be immature in children and altered with aging. Moreover, switching strategies from one problem to the next when appropriate is resource demanding in terms of switching and flexibility, and children and older adults have fewer of those resources available (e.g., Hodzik & Lemaire, 2011; Lemaire & Lecacheur, 2011). Consistent with this, Peters, Smedt, Torbeyns, Ghesquière, and Verschaffel (2012) tested 8–11-year-old children and found that, unlike young adults, children did not flexibly switch between a direct subtraction and a subtraction-by-addition strategy as a function of problem characteristics when solving subtraction problems. Young children and older adults compensate for their deficits in switching strategies by reducing their strategy repertoire (i.e., switching between fewer strategies), thus reducing the number of switches made during the task. Sometimes, they even reduce their repertoire so that no switching occurs, leading to relatively larger numbers of monostrategic individuals in young children and older adults and especially older adults with AD (e.g., Duverne et al., 2008; Hodzik & Lemaire, 2011; Lemaire & Arnaud, 2008; Lemaire & Brun, 2018).

Some recent evidence regarding a new phenomenon named "strategy combination" (Hinault, Dufau, & Lemaire, 2015) illustrates how older adults use strategy combination less often than young adults. The authors used a multiplication verification task where participants had to verify the veracity of problems such as $4 \times 18 = 72$, by pressing one response key if the proposed solution is correct and another response key if this solution is incorrect. Previous studies have found evidence for the use of multiple strategies in this type of task (e.g., Ashcraft & Battaglia, 1978; Lemaire & Fayol, 1995; Lemaire & Reder, 1999) other than the exhaustive verification strategy where the participant completes the full calculation and compares the result to the proposed solution before making a true/false decision and providing his/her response. These other strategies are called plausibility-checking strategies and are apparent in solution latencies. For example, when the proposed solution violates the five rule (i.e., the solution to a problem where one of the units equals 5 will end in 5 or 0), responses come faster than when the proposed solution respects the five rule (e.g., $7 \times 5 = 43$ vs. $7 \times 5 = 45$). Another example is when the proposed solution violates the parity rule (i.e., when both multiplicands are even, the solution should be even as well; otherwise, it must be odd), responses are also faster relative to no-parity rule violation problems (e.g., $4 \times 6 = 23$ vs. $4 \times 6 = 26$). This suggests that instead of making the full calculation, participants can apply a much faster strategy where they quickly check whether the answer is plausible or not.

Flexible combination of multiple plausibility-checking strategies leads to even faster solution latencies. Indeed, Hinault et al. (2015) found that when a problem violated both the five and the parity rules (e.g., $5 \times 31 = 158$), participants were faster relative to when the problem violated either the parity rule (e.g., $5 \times 31 = 150$) or the five rule (e.g., $5 \times 31 = 153$). Strategy combination leads to best performance on some problems, but it is also costly in terms of executive control. Hinault et al. (2015) compared strategy combination in young and older adults and found that while both groups had similar benefits when implementing the five-rule or the parity-rule strategy in isolation, combining both violations led to comparatively larger benefits in younger adults than in older adults. Absolute benefits associated with two-rule violation (relative to one-rule violation) were significantly larger in younger adults compared with older adults. The reduced benefits for two-rule violation problems in older adults compared with young adults could be because of a less systematic use in older adults than in young adults of strategy combinations, given the executive resources that such a combination incurs and given lower-available executive resources in older adults.

It can be noted that although the evidence just discussed favors the role of executive control in changes to strategic variations during the life span, this evidence is mostly indirect. Two studies tested the relations between executive control and strategic variations more directly in development (Lemaire & Lecacheur, 2011) and in aging (Hodzik & Lemaire, 2011). Lemaire and Lecacheur (2011) tested 8-, 10-, and 12-year-old children, asking them to choose between rounding down and rounding up to estimate sums to two-digit additions. They collected measures of EFs in the same children (i.e., Stroop, TMT, and excluded letter fluency). Consistent with previous results, they found that from 8 to 12 years of age, children's strategy execution efficiency gradually improved, and they chose the best strategy on each problem more and more often. While 8-year-olds selected the best strategy on 65% of problems, 10-year-olds selected the best strategy on 75% of problems and 12-year-olds on 80% of problems. Most importantly, these developmental differences in strategy selection correlated with all three measures of executive control. A hierarchical regression analysis revealed that the proportion of age-related variance on mean percent use of the better strategy decreased by 67% when taking into account EFs, thus proving that EFs are an important sources of development-related changes in strategy choices.

Hodzik and Lemaire (2011) conducted a similar study in young and older adults. In their study, participants were given a series of two-digit addition problems (e.g., 12 + 46) to solve, and verbal protocols were collected on the strategy each participant used on each problem. The researchers were able to distinguish nine different strategies, for example, rounding the first operand down (i.e., 10 + 46 + 2) or borrowing units (i.e., 18 + 40). As it turned out, the employed strategies were the same in young and older adults. However, a different picture arose when looking at young and older individuals. The data revealed that while a young individual would on average use 5.5 strategies, an older individual would alternate between "only" 3.2 strategies on average. Thus, although the same strategies are known and used by young and older adults when analyzed at the group level, individuals' strategy repertoires tend to shrink with age.

In a second experiment, the authors discovered that this reduction of strategy repertoire was linked to measures of executive control. They assessed EFs (i.e., inhibition and flexibility) and tested whether the age-related variance in strategy repertoire was explained by inhibition capacity and flexibility. They pursued this goal in hierarchical regression analyses and found that EFs explained 91% of age-related variance in the number of strategies used by individuals. This suggests that older adults use fewer strategies as a result of decreased efficiency of EFs. However, this is correlational evidence and as such is one step away from proving causality of relationships between EFs, aging, and arithmetic strategies. Fortunately, it has recently been possible to devise experimental settings that can address the issue, which we discuss in the next section.

INVESTIGATING EXECUTIVE CONTROL THROUGH SEQUENTIAL EFFECTS

The role of executive control resources on strategic variations during aging has also been investigated by looking at how the magnitude of experimental effects (known to result from variations in demands in executive resources) changes with age. The particular category of effects we focus on in this section is sequential effects, as they have been recently investigated in great detail. Sequential effects arise from selecting and executing strategies in a sequential context. Current strategy selection and execution are influenced by strategies used on immediately preceding trials. Participants have a tendency to repeat the previous strategy, and strategy execution may be faster or slower depending on the strategy use on the previous trial. These phenomena result from the need for executive control, for example, inhibition and flexibility, to switch between strategies. Here, we discuss the following sequential effects: strategy switch costs, sequence poorer-strategy effects, strategy sequential difficulty effects (SSDE), and strategy perseveration effects.

In numerous studies, we documented strategy switch costs in arithmetic using the computational estimation tasks. In these tasks, participants are presented with two-digit addition or multiplication problems (e.g., 46×52), and they have to estimate, but not calculate, the answer to these problems by using rounding strategies. During computational estimation, we decompose the initial problem (e.g., 43×67) in a number of subsequent retrievals. These retrievals involve (1) the procedure of the strategy (e.g., for rounding up: discard unit digits, increment 10 digits, multiply incremented 10 digits), (2) the rounded operands (e.g., for rounding up: 50 and 70), and (3) the solution to their multiplication (e.g., for rounding up: $50 \times 70 = 3500$). Different rounding strategies will put different demands on executive resources. Retrieval of the procedure, rounded operands, and multiplication should be easiest for rounding down and most difficult for rounding up and mixed rounding. This is because the latter two require the additional step of incrementing at least one operand.

The computational estimation task provides a couple of noteworthy advantages for studying strategic variations. One advantage is that it is easy to determine which strategy is the more difficult strategy to execute and will require more executive resources. Another advantage is that it is easy for us to know what strategy would have been the most appropriate on any problem. For example, 52×43 is solved more correctly by rounding down, whereas 57×48 is solved better by rounding both operands up. It is relatively easy for the participant to determine the best strategy on this type of problems. However, when confronted with problems like 52×48, it is more difficult to know what strategy to use. This offers the opportunity to investigate the strategy selection processes, given that by varying problem features (e.g., size of unit digits), we can assess participants' difficulty to find the best strategy on each problem.

Lemaire and Lecacheur (2010) used this task and imposed the strategies that had to be used by participants on each problem to assess strategy switch costs. The authors cued the rounding-down strategy on rounding-down problems (e.g., 41×76) and the rounding-up strategy on rounding-up problems (e.g., 38×54). On half the trials, participants had to repeat the same strategy across two successive problems, and on the other trials they had to use a different strategy on two successive problems. The authors found that participants are faster when they use the same strategy on two consecutive problems than when they use different strategies, and this is only when they switch from the harder, rounding-up strategy to the easier, rounding-down strategy. No strategy switch costs were observed when participants switch from the easier to the harder strategy (the same asymmetry has often been observed in the task-switching literature; Bryck & Mayr, 2008). These switch costs are consistent with the hypothesis that executive control resources are involved while switching between strategies, especially when switching from the harder rounding-up strategy to the easier rounding-down strategy.

Lemaire and Brun (2014, 2016) tested 8- and 10-year-old children with a computational estimation task and found strategy switch costs when these children had to switch between two rounding strategies on consecutive trials. These switch costs were much larger in the younger (1148 ms) than in the older children (160 ms), suggesting a developmental trend toward more efficient strategy switching. Interestingly, these differences were attenuated when response-stimulus interval (RSI) increased from 900 to 1900 ms, thereby providing more time to complete switching. This reduced the relatively large switch costs in younger children from 1447 to 848 ms. Similarly, Ardiale and Lemaire (2013) found a developmental trend toward smaller strategy switch costs within the same trial in children aged 8, 10, and 12 years. They also used a computational estimation task and found that when children switched for a more appropriate strategy, switch costs were of 2161 ms in 8-year-olds, 996 ms in 10-year-olds, and 659 ms in 12-year-olds.

Ardiale et al. (2012) tested strategy switch costs in young and older adults. They compared a condition in which participants could switch between two strategies versus three strategies, with a two-digit addition problem solving task in which participants had to calculate the exact answer to problems such as 47 + 84. In the two-strategy condition, participants could choose between a strategy where they would first add the units and then the decades (i.e., unit strategy) or a strategy where they would first add the decades and then the units (i.e., decade strategy). In the three-strategy condition, a third strategy was possible, which was to borrow units (i.e., 40 + 84 + 7). Thus, switch costs in young and older adults were tested under two- and three-strategy conditions. Data showed age-related increases in switch costs only in the three-strategy condition. Although younger adults had similar switch costs when switching between three- and two strategies (301 vs. 266 ms), older adults had much larger switch costs in the former than in the latter condition (849 vs. 405 ms). It seems that older adults' reduced switching capacities become increasingly apparent as the number of strategies to switch between becomes larger.

These larger switch costs in younger children and older adults may be the precise reason why, as aforementioned, when allowed to choose any strategies, young children and older adults will choose to switch between fewer strategies (see Arnaud et al., 2008; Hodzik & Lemaire, 2011; Lemaire & Lecacheur, 2011), even though they have access to the same strategies as younger adults. In more extreme cases, only one strategy will be used, something relatively more common in children and older adults than in younger adults (e.g., Duverne et al., 2008; Hodzik & Lemaire, 2011; Lemaire & Arnaud, 2008).

The sequential modulation of poorer-strategy effects is another strategy sequential effect that evidences the role of executive control during strategy execution. This effect is a modulation of the so-called poorer-strategy effect. Poorer-strategy effect is the finding that performance is better when participants are asked to execute a more appropriate strategy on a given problem than when asked to execute a poorer strategy (e.g., that yields an answer further away from the correct answer than that of a better strategy). For example, the best rounding strategy for solving 42×73 is rounding both operands down. Performance will be poorer (i.e., estimates will be further from the exact product, and latencies will be longer) with the rounding-up strategy. Lemaire and Hinault (2014) found that poorer-strategy effects were larger, or only existed, when the previous problem was solved with a better strategy relative to when solved with another poorer strategy. The authors accounted for such sequential modulations by assuming that on previous poorer-strategy problems, participants used executive control processes (i.e., they inhibited the best, most readily activated strategy to activate the required poorer strategy). Such use would have generated expectancies on the next problem (i.e., it is likely that the next problem will also have to be solved with the poorer strategy). These expectancies enabled participants to prepare themselves to process conflicting information (between the required and the best strategy) more quickly. In contrast, when previous problems were solved by the best strategy, such preparation was not triggered, thus increasing time to process conflict on subsequent poorer-strategy items.

Lemaire and Hinault (2014) compared these sequential modulations of poorer-strategy effects between young and older adults in a computational estimation task. They found that older adults modulated poorer-strategy effects to a much smaller extent than young adults (using the poorer strategy on a trial increased latencies by 136 and 750 ms in young adults following a poorer and better strategy, respectively). In older adults, poorer-strategy effects even increased following a previous poorer strategy compared with a previous better strategy (using the poorer strategy on a trial increased latencies by 685 and 590 ms in older adults following a poorer and better strategy, respectively). Most interesting, Lemaire and Hinault found important individual differences among older adults. They divided older adults into two categories based on their performance in a Simon conflict task. High-control older adults showed efficient executive control in a Simon task; they were also able to respond adequately to incongruent trials and sequentially modulated poorer-strategy effects like young adults. When they used the poorer strategy on a trial, latencies increased by 301 ms on current items, following previous poorer-strategy use and 679 ms after previous better-strategy use. In contrast, low-control older adults showed no sequential modulations of poorer-strategy effects. In fact, when low-control older adults used a poorer strategy, latencies were increased by 1177 ms after previous poorer-strategy problems compared with 490 ms following previous better-strategy problems. These aging effects on sequential modulations of poorer-strategy effects and individual differences therein among older adults emphasize the role of executive control processes during strategy execution.

SSDE are further evidence for the role of executive control processes during strategy execution. SSDE is the finding that solution latencies are larger on current problems following execution of a difficult strategy on the previous problems than after execution of an easier strategy. Uittenhove and Lemaire (2012) found this effect while participants accomplished a computational estimation task. Uittenhove and Lemaire found that execution of a mixed-rounding strategy on current problems was slower after execution of a more difficult, rounding-up strategy on immediately preceding problems compared with after executing the easier rounding-down strategy. Moreover, the size of these effects was correlated to working memory capacities (Uittenhove & Lemaire, 2013b) and EFs (Uittenhove, Burger, Taconnat, & Lemaire, 2015), suggesting that these effects on strategy execution crucially involve executive control. Giving participants more time in between problems, so more time for executive control processes to run to completion, yielded smaller SSDE (Uittenhove & Lemaire, 2013b). Interestingly, populations that suffer from severely reduced executive control, such as Alzheimer patients, showed much larger sequential difficulty effects compared with age-matched controls (1591 ms vs. 180, Uittenhove & Lemaire, 2013a). Uittenhove and Lemaire accounted for these effects by assuming that executing a harder strategy on previous problems consumes much available processing resources. As a consequence, on current problems, processing resources are depleted or alternatively suffer from high levels of interference by execution of a previous hard strategy. With time, these processing resources can be restored, or interference between processing previous hard strategies and current strategies can be overcome. Participants need time to replenish processing resources and overcome interference between processing current and previous problems.

The final effects that evidence the crucial role of executive control processes during strategy use and execution are strategy perseveration effects. Lemaire and Brun (2014) studied strategy perseveration effects in children through manipulation of RSI (short: 900 ms and long: 1900 ms). When given little time in the short RSI condition, 8- and 10-year-olds had an enlarged tendency to repeat strategies on consecutive trials (in 61% of cases). When RSI was longer, repetition probabilities returned to chance levels (52% of cases), benefiting adaptive strategy selection. The authors proposed that longer RSI permitted more complete inhibition and switching processes in children, resulting in a reduced tendency to repeat strategies and a more adaptive strategy selection on the next trial.

Lemaire and Leclère (2014a,b) tested strategy perseveration effects in older adults. They studied strategy perseveration through priming participants with a rounding-down or a rounding-up strategy in a computational estimation paradigm. They manipulated the strength of the prime by varying the number of successive problems that were cued with either strategy. In the one-prime condition, the rounding-down or rounding-up strategy was cued on one problem, whereas in the two-prime condition, each strategy was cued on two successive problems. They then measured the repetition rate of the primed strategy on the next problem in young and older adults. They found that older adults repeated the primed strategy more often than younger adults in the one-prime condition (59% compared with 52% strategy repetitions) and that this effect increased in the two-prime condition (85% compared with 68% strategy repetitions). This suggests that young and older adults were both sensitive to the strength of priming and had more difficulties disengaging from a strategy that had been primed twice. However, as a consequence of reduced efficiency of executive control, older adults suffered more from this repetition, leading them to change strategies on smaller percentages of trials.

CONCLUSION

In this chapter, we argued that performance control during numerical cognition tasks occurs via two general nonindependent resources, namely executive control and strategic variations. We discussed the relations among them during development and aging. Indeed, examining the strategies people use, how they use them, and how they select them permits to pinpoint precisely when and how cognitive control will play a role in numerical cognition. First, during strategy execution, cognitive control is employed to coordinate the different steps of the strategy and to maintain interim results. We can divide strategies into their elementary processes to better understand which strategies require more executive control and thus take longer to develop in children and are more likely to show age-related declines. Interestingly, executive control also plays a role in successive strategy execution. Results regarding SSDE show how the availability of executive control fluctuates dynamically over successive trials as a function of previously executed strategies.

Then, during strategy selection, we need executive control to inhibit competing strategies. This need is higher in sequential contexts, where we need to flexibly alternate between strategies, requiring disengaging from previously executed strategies. Results pertaining to strategy switch costs evidence the costs associated to this type of switching. In children, switch costs decrease gradually with increasing age. Interestingly, switching deficits in older adults only become visible when the number of strategies to alternate between is larger. Executive control has also shown to be important in the matching between strategies and problem characteristics. A poor match between the type of problem and the executed strategy leads to longer solution latencies. Anticipatory cognitive control can attenuate this conflict in a situation where participants are required to execute poorer strategies in a sequential context (i.e., sequential modulation of poorer-strategy effects). However, in older adults with reduced cognitive control capacities, such anticipation is no longer possible or much more difficult, and they suffer from an accumulation of conflict when required to execute inappropriate strategies in a sequential context.

In summary, cognitive or executive control is of paramount importance in numerical cognition. We adopted an approach where we looked at the strategies involved and the involvement of executive control in strategy selection, strategy execution, strategy repertoire, and strategy distribution. The role of executive control was evidenced in both correlational and experimental studies (e.g., sequential effects during strategy use and execution). This role was mostly illustrated here in arithmetic problem solving. What was found in arithmetic tasks regarding this role may also be found in any other numerical tasks, an issue that future studies may investigate.

References

Agostino, A., Johnson, J., & Pascual-Leone, J. (2010). Executive functions underlying multiplicative reasoning: Problem type matters. *Journal of Experimental Child Psychology*, 105(4), 286–305.

Allen, P. A., Ashcraft, M. H., & Weber, T. A. (1992). On mental multiplication and age. *Psychology and Aging*, 7(4), 536.

Andersson, U. (2008). Working memory as a predictor of written arithmetical skills in children: The importance of central executive functions. *British Journal of Educational Psychology*, 78(2), 181–203.

Ardiale, E., Hodzik, S., & Lemaire, P. (2012). Aging and strategy switch costs: A study in arithmetic problem solving. *L'Année Psychologique*, 112, 345–360.

Ardiale, E., & Lemaire, P. (2012). Within-item strategy switching: An age comparative study in adults. *Psychology and Aging*, 27(4), 1138.

Ardiale, E., & Lemaire, P. (2013). Within-item strategy switching in arithmetic: A comparative study in children. *Frontiers in Psychology*, 4, 924.

Arnaud, L., Lemaire, P., Allen, P., & Michel, B. F. (2008). Strategic aspects of young, healthy older adults', and Alzheimer patients' arithmetic performance. *Cortex*, *44*(2), 119–130.

Ashcraft, M. H., & Battaglia, J. (1978). Cognitive arithmetic: Evidence for retrieval and decision processes in mental addition. *Journal of Experimental Psychology: Human Learning and Memory*, 4(5), 527.

Barulli, D. J., Rakitin, B. C., Lemaire, P., & Stern, Y. (2013). The influence of cognitive reserve on strategy selection in normal aging. *Journal of the International Neuropsychological Society*, *19*(07), 841–844.

Bryck, R. L., & Mayr, U. (2008). Task selection cost asymmetry without task switching. *Psychonomic Bulletin and Review*, *1*, 128–134.

Bull, R., Espy, K. A., & Wiebe, S. A. (2008). Short-term memory, working memory, and executive functioning in preschoolers: Longitudinal predictors of mathematical achievement at age 7 years. *Developmental Neuropsychology*, *33*(3), 205–228.

Bull, R., Johnston, R. S., & Roy, J. A. (1999). Exploring the roles of the visual-spatial sketch pad and central executive in children's arithmetical skills: Views from cognition and developmental neuropsychology. *Developmental Neuropsychology*, *15*(3), 421–442.

Bull, R., & Scerif, G. (2001). Executive functioning as a predictor of children's mathematics ability: Inhibition, switching, and working memory. *Developmental Neuropsychology*, *19*(3), 273–293.

Duverne, S., & Lemaire, P. (2005). Aging and arithmetic. In J. I. D. Campbell (Ed.), *The handbook of mathematical cognition*. New York: Psychology Press.

Duverne, S., Lemaire, P., & Vandierendonck, A. (2008). Do working-memory executive components mediate the effects of age on strategy selection or on strategy execution? Insights from arithmetic problem solving. *Psychological Research*, *72*(1), 27–38.

Gallistel, C. R., & Gelman, R. (1992). Preverbal and verbal counting and computation. *Cognition*, *44*(1), 43–74.

Gandini, D., Lemaire, P., Anton, J. L., & Nazarian, B. (2008). Neural correlates of approximate quantification strategies in young and older adults: An fMRI study. *Brain Research*, *1246*, 144–157.

Gandini, D., Lemaire, P., & Michel, B. F. (2009). Approximate quantification in young, healthy older adults', and Alzheimer patients. *Brain and Cognition*, *70*(1), 53–61.

Geary, D. C., & Brown, S. C. (1991). Cognitive addition: Strategy choice and speed-of-processing differences in gifted, normal, and mathematically disabled children. *Developmental Psychology*, *27*(3), 398.

Gelman, R., & Gallistel, C. R. (1986). The child's understanding of number. Harvard University Press.

Groen, G. J., & Parkman, J. M. (1972). A chronometric analysis of simple addition. *Psychological Review*, *79*(4), 329.

Halberda, J., Ly, R., Wilmer, J. B., Naiman, D. Q., & Germine, L. (2012). Number sense across the lifespan as revealed by a massive Internet-based sample. *Proceedings of the National Academy of Sciences*, *109*(28), 11116–11120.

Hinault, T., Dufau, S., & Lemaire, P. (2015). Strategy combination in human cognition: A behavioral and ERP study in arithmetic. *Psychonomic Bulletin and Review*, 22(1), 190–199.

Hodzik, S., & Lemaire, P. (2011). Inhibition and shifting capacities mediate adults' age-related differences in strategy selection and repertoire. *Acta Psychologica*, 137, 335–344.

Imbo, I., Duverne, S., & Lemaire, P. (2007). Working memory, strategy execution, and strategy selection in mental arithmetic. *The Quarterly Journal of Experimental Psychology*, 60(9), 1246–1264.

Imbo, I., & Vandierendonck, A. (2007). The development of strategy use in elementary school children: Working memory and individual differences. *Journal of Experimental Child Psychology*, 96(4), 284–309.

Lemaire, P. (2016). *Cognitive aging: The role of strategies*. Psychology Press.

Lemaire, P., & Arnaud, L. (2008). Young and older adults' strategies in complex arithmetic. *The American Journal of Psychology*, 1–16.

Lemaire, P., Arnaud, L., & Lecacheur, M. (2004). Adults' age-related differences in adaptivity of strategy choices: Evidence from computational estimation. *Psychology and Aging*, 19(3), 467.

Lemaire, P., & Brun, F. (2014). Effects of strategy sequences and response–stimulus intervals on children's strategy selection and strategy execution: A study in computational estimation. *Psychological Research*, 78(4), 506–519.

Lemaire, P., & Brun, F. (2016). Effects of problem presentation durations on arithmetic strategies: A study in young and older adults. *Journal of Cognitive Psychology*, 28, 909–922.

Lemaire, P., & Brun, F. (2018). Age-related changes in children's strategies for solving two-digit addition problems. *Journal of Numerical Cognition*, 3(3), 582–597.

Lemaire, P., & Callies, S. (2009). Children's strategies in complex arithmetic. *Journal of Experimental Child Psychology*, 103(1), 49–65.

Lemaire, P., & Fayol, M. (1995). When plausibility judgments supersede fact retrieval: The example of the odd-even effect on product verification. *Memory and Cognition*, 23(1), 34–48.

Lemaire, P., & Hinault, T. (2014). Age-related differences in sequential modulations of poorer-strategy effects, A study in arithmetic problem solving. *Experimental Psychology*, 61(4), 253.

Lemaire, P., & Lecacheur, M. (2001). Older and younger adults' strategy use and execution in currency conversion tasks: Insights from French franc to euro and euro to French franc conversions. *Journal of Experimental Psychology: Applied*, 7(3), 195.

Lemaire, P., & Lecacheur, M. (2002). Children's strategies in computational estimation. *Journal of Experimental Child Psychology*, 82(4), 281–304.

Lemaire, P., & Lecacheur, M. (2010). Strategy switch costs in arithmetic problem solving. *Memory and Cognition*, 38(3), 322–332.

Lemaire, P., & Lecacheur, M. (2011). Age-related changes in children's executive functions and strategy selection: A study in computational estimation. *Cognitive Development*, 26(3), 282–294.

Lemaire, P., & Leclère, M. (2014a). Strategy repetition in young and older adults: A study in arithmetic. *Developmental Psychology*, 50(2), 460.

Lemaire, P., & Leclère, M. (2014b). Strategy selection in Alzheimer patients: A study in arithmetic. *Journal of Clinical and Experimental Neuropsychology, 36*(5), 507–518.

Lemaire, P., & Reder, L. (1999). What affects strategy selection in arithmetic? The example of parity and five effects on product verification. *Memory and Cognition, 27*(2), 364–382.

Lemaire, P., & Siegler, R. S. (1995). Four aspects of strategic change: Contributions to children's learning of multiplication. *Journal of Experimental Psychology: General, 124*(1), 83.

Luwel, K., Lemaire, P., & Verschaffel, L. (2005). Children's strategies in numerosity judgment. *Cognitive Development, 20*(3), 448–471.

Luwel, K., Schillemans, V., Onghena, P., & Verschaffel, L. (2009). Does switching between strategies within the same task involve a cost? *British Journal of Psychology, 100*(4), 753–771.

Luwel, K., Verschaffel, L., Onghena, P., & De Corte, E. (2000). Children's strategies for numerosity judgement in square grids of different sizes. *Psychologica Belgica, 40*, 183–209.

Luwel, K., Verschaffel, L., Onghena, P., & De Corte, E. (2001). Strategic aspects of children's numerosity judgement. *European Journal of Psychology of Education, 16*(2), 233–255.

McKenzie, B., Bull, R., & Gray, C. (2003). The effects of phonological and visual-spatial interference on children's arithmetical performance. *Educational and Child Psychology, 20*(3), 93–108.

Miyake, A., & Friedman, N. P. (2012). The nature and organization of individual differences in executive functions four general conclusions. *Current Directions in Psychological Science, 21*(1), 8–14.

Miyake, A., Friedman, N. P., Emerson, M. J., Witzki, A. H., Howerter, A., & Wager, T. D. (2000). The unity and diversity of executive functions and their contributions to complex "frontal lobe" tasks: A latent variable analysis. *Cognitive Psychology, 41*(1), 49–100.

Peters, G., Smedt, B., Torbeyns, J., Ghesquière, P., & Verschaffel, L. (2012). Children's use of addition to solve two-digit subtraction problems. *British Journal of Psychology, 104*(4), 495–511.

Rasmussen, C., & Bisanz, J. (2005). Representation and working memory in early arithmetic. *Journal of Experimental Child Psychology, 91*(2), 137–157.

Reuter-Lorenz, P. A., & Park, D. C. (2010). Human neuroscience and the aging mind: A new look at old problems. *The Journals of Gerontology Series B: Psychological Sciences and Social Sciences, 65*(4), 405–415.

Salthouse, T. A., & Coon, V. E. (1994). Interpretation of differential deficits: The case of aging and mental arithmetic. *Journal of Experimental Psychology: Learning, Memory, and Cognition, 20*(5), 1172.

Siegler, R. S. (1987). The perils of averaging data over strategies: An example from children's addition. *Journal of Experimental Psychology: General, 116*(3), 250.

Siegler, R., & Araya, R. (2005). A computational model of conscious and unconscious strategy discovery. *Advances in Child Development and Behavior, 33*, 1–42.

Siegler, R., & Jenkins, E. A. (2014). *How children discover new strategies.* Psychology Press.

Siegler, R. S., & Lemaire, P. (1997). Older and younger adults' strategy choices in multiplication: Testing predictions of ASCM using the choice/no-choice method. *Journal of Experimental Psychology: General, 126*(1), 71.

Siegler, R. S., & Shrager, J. (1984). Strategy choices in addition and subtraction: How do children know what to do. *Origins of Cognitive Skills,* 229–293.

Sliwinski, M. (1997). Aging and counting speed: Evidence for process-specific slowing. *Psychology and Aging, 12*(1), 38.

Thevenot, C., Barrouillet, P., Castel, C., & Uittenhove, K. (2016). Ten-year-old children strategies in mental addition: A counting model account. *Cognition, 146,* 48–57.

Uittenhove, K., Burger, L., Taconnat, L., & Lemaire, P. (2015). Sequential difficulty effects during execution of memory strategies in young and older adults. *Memory, 23*(6), 806–816.

Uittenhove, K., & Lemaire, P. (2012). Sequential difficulty effects during strategy execution. *Experimental Psychology, 59*(5), 295–301.

Uittenhove, K., & Lemaire, P. (2013a). Strategy sequential difficulty effects in Alzheimer patients: A study in arithmetic. *Journal of Clinical and Experimental Neuropsychology, 35*(1), 83–89.

Uittenhove, K., & Lemaire, P. (2013b). Strategy sequential difficulty effects vary with working-memory and response–stimulus-intervals: A study in arithmetic. *Acta Psychologica, 143*(1), 113–118.

Wood, G., Ischebeck, A., Koppelstaetter, F., Gotwald, T., & Kaufmann, L. (2009). Developmental trajectories of magnitude processing and interference control: An fMRI study. *Cerebral Cortex, 19*(11), 2755–2765.

The Interplay Between Proficiency and Executive Control

Avishai Henik, Naama Katzin, Shachar Hochman

Ben-Gurion University of the Negev, Beer-Sheva, Israel

O U T L I N E

Paraphrasing William James' well-known line ("Everybody knows what attention is"), it could be suggested that everybody knows that attention affects performance in general and numerical cognition in particular. However, the three chapters included in this book section suggest that this is not so simple. It is true that research has advanced our knowledge of the mental operations involved, but there are still many open questions and issues to be studied. Moreover, it is quite possible that the influence goes in the other direction also. Namely, performance or proficiency in numerical cognition affects attention.

Merkley, Matusz, and Scerif (Chapter 6, this section) seem to focus mainly on the selective aspect of attention and suggest that the effects of attention are not necessarily unidirectional (attention modulates performance) but may work in the other direction (performance proficiency may modulate attention) as well. Ben-Shachar and Berger (Chapter 5, this section) wrote that attention is composed of three separate systems—alerting, orienting, and selection—which modulate different aspects of performance. Uittenhove and Lemaire (Chapter 7, this section) discuss the interaction between executive control and strategic variation. We summarize three contributions and discuss general implications.

PROFICIENCY AND ATTENTION CONTROL

Merkley et al. (Chapter 6, this section) noted that control is used in the literature in many circumstances and in many ways. Some discussions of control note overlap or even identify control with EFs. They mention work by Miyake et al. (2000) who suggested that EFs are composed of three major factors: shifting, updating, and inhibition. Inhibition is the aspect of control mostly discussed in tasks that require selective attention (e.g., Stroop task, flanker task) and is a major component of "control of attention" or cognitive control in general. Goal-directed behavior requires focusing on relevant information and actions and at the same time, ignoring or inhibiting irrelevant and interfering information and actions. Accordingly, as Merkley et al. suggest, many tasks that probe the building blocks of numerical cognition and its development require inhibitory control. For example, in comparative judgments of dot arrays, participants are presented with two arrays of dots that differ in numerosity and are asked to decide which array is more numerous. Because various continuous variables (e.g., density) are correlated with numerosity and among themselves, it is necessary to be able to focus on numerosity and ignore these irrelevant variables to perform the task appropriately. This is where control and selective attention come in. Merkley et al. also note that to develop and acquire the number concept, infants and children have to learn what are the features that should attract our attention, and at the same time, they have to learn what features are irrelevant and interfering and have to be ignored and inhibited. Otherwise, these irrelevant features may blur their numerosity judgments. This is a long process that is crucial for the development of the number concept. Merkley et al. present this line of reasoning and cite quite a few studies that report associations between math ability and development and control of attention. The importance of focusing on numerosity is discussed in the contribution by Ben-Shachar and Berger (Chapter 5, this section) also. We turn to this later.

To examine causal relationship between control and math ability, Merkley et al. (Chapter 6, this section) survey training studies. Researchers suggest that training with various attention or control tasks should improve performance in mathematical tasks. However, it seems that the picture here is not as clear as desired. At best, results of various studies are inconsistent, with some suggesting positive results, and others having no effects of training or no transfer. Here, one could just be discouraged and conclude that this line of work would lead nowhere or is of very limited use, both from the theoretical and the practical points of view. However, there is another possibility. It is possible that to design the "appropriate" training, one needs to be able to tailor train specific mental operations such that they would improve when carrying out the tested tasks. In spite of the fact that some tasks seem very simple in terms of the mental operations involved, years of studies with the Stroop task, a very simple task indeed (see below), suggest that things are not so simple, and additional efforts need to be invested in this line of work before we unravel the exact mental operations involved, and are able to produce positive and consistent training results.

In the last part of their chapter, Merkley et al. (Chapter 6, this section) point out the possible existence of a bidirectional relationship between attention and numerical cognition. In fact, this potential bidirectionality may not be unique to the relation between attention and numerical cognition but may affect other areas of cognition also. This potential bidirectional effect may be related to an old issue in cognition: What is the end point of learning or of gaining proficiency, automatic performance, or controlled performance? It can even be phrased as lack of control or control. Lack of control is a case of extreme automaticity, when the system takes its course of action and cannot change whatever happens during this course.

Automaticity has been debated for many years. In spite of the fact that automaticity has been defined and described in different ways throughout the years (Hasher & Zacks, 1979; Logan, 1988; Tzelgov, 1997), a general description could be as follows: automaticity refers to the ability to carry out a task or an operation smoothly, without the need to monitor it closely, and without the need to invest attentional resources. One task heavily involved in discussions of selective attention and automaticity is the Stroop (1935) task. In Stroop's original work, he presented color words in color (e.g., BLUE in red) and asked participants to respond to the color and ignore the meaning of the word. He compared this incongruent condition (example above) with a neutral condition of color patches, where no interference was supposed to exist. Stroop reported that participants were not able to ignore the meaning of the word and their responses to the color were slower in the incongruent condition than in the neutral condition. In today's experiments, researchers commonly use a congruent condition (e.g., RED in red) in addition to the incongruent and neutral conditions and find that responding to incongruent stimuli is slowest and responding to congruent stimuli is usually faster or similar to that of neutral stimuli.

Quite a few years ago, Tzelgov, Henik, and Leiser (1990) tackled the question posed earlier: What is the end point of learning or gaining proficiency? They used a Stroop task to measure the ability of selection or the failure of this ability, they recruited Hebrew–Arabic bilingual participants to have different levels of language proficiency, and they manipulated control by changing the expectancy for a given language. The Stroop task included only congruent and incongruent trials. All the participants were Hebrew–Arabic bilinguals, but they differed in language proficiency. For example, the group that was more proficient in Arabic was composed of students for whom Arabic was their native language, but they began to study Hebrew in elementary school and thus studied Hebrew for at least 12 years. "They were quite proficient in Hebrew, but knew Arabic better. They spoke only Arabic at home and both Hebrew and Arabic with friends. Outside of the home, they used mainly Hebrew for everyday interaction." (p. 762). The other group was more proficient in Hebrew than in Arabic. Language expectancy was manipulated in separate blocks of trials so that a given block included 80% Hebrew color words and 20% Arabic color words, and in another block, the opposite proportions. Please note that the words were completely irrelevant to the task and were supposed to be ignored. In general, participants showed the Stroop effect in both languages, but they were able to control or to inhibit reading the irrelevant word in their more proficient language. In the Arabic-proficient participants, the Stroop effect was similar for Hebrew stimuli, regardless of the expectancy level for these stimuli. That is, even though in certain sessions the participants knew that the next stimulus would probably be written in Hebrew, they could not suppress reading the irrelevant color name. In contrast, expectation influenced the size of the Stroop effect of color names written in Arabic. The knowledge that in certain sessions the next stimulus would probably be written in Arabic enabled the participants to reduce the amount of interference. A similar pattern was observed in the Hebrew-proficient participants. Namely, they were able to use expectancy to reduce the Stroop effect in Hebrew but not in Arabic. In conclusion, the Stroop effect was always present, but participants were able to control (reduce) it in their native language but not in their second language. Accordingly, automaticity and control are both characteristics of skilled performance. Gaining proficiency means advancing automaticity and letting proficient aspects interfere with our performance, and at the same time, enable control—the ability to stop or reduce interference of irrelevant information or habit. In the terms used by Merkley, Matusz, and Scerif, selective attention influences the selection of information, but prior knowledge or proficiency influences the ability to manage attention.

COMPONENTS OF ATTENTION

Ben-Shachar and Berger (Chapter 5, this section) first describe the components of the attention system according to Posner and Petersen (Petersen & Posner, 2012; Posner & Petersen, 1990). Three attentional systems are described: alerting, executive control, and orienting of attention. They next survey the literature on the effects of each of these three attentional systems on numerical cognition. The discussion of executive control revolves around the contribution of inhibition. One of the studies mentioned in that part was conducted by Rubinsten, Henik, Berger, and Shahar-Shalev (2002). Interestingly, the discussion of this study fits in with the discussion of Merkley et al. (Chapter 6, this section) of bidirectionality between control of attention and numerical cognition, and our comments above regarding the interplay between control and bilingual proficiency. The work of Rubinsten et al. (2002) was a developmental cross-sectional study that examined the size congruity effect in children at the beginning of first grade, end of first grade, third grade, fifth grade, and in college students. Imagine you are presented with the two numbers 3 and 5 that vary both in physical size and in numerical value so that they could be congruent (e.g., 3 5) or incongruent (e.g., 3 5). You may be asked to pay attention to the numerical values and ignore the physical sizes (numerical task) or pay attention to the physical sizes and ignore the numerical values (physical task). It is not surprising that all children and adults show a congruity effect in the numerical task. This is because the physical sizes are right there in front of us on the screen and they are hard to ignore. On the other hand, finding a congruity effect in the physical task, when the numerical values are irrelevant, is surprising because (1) for such an effect to appear, there is a need to derive the irrelevant numerical values that are attached to the symbols on the screen, and (2) this effect appears, even though in many cases the physical comparisons are much faster than the numerical comparisons. In general, the congruity effect in the physical task (numerical values irrelevant) is small at the beginning of first grade, it grows with age up to fifth grade, and then declines so that college students show a smaller congruity effect. Two years earlier than the study of Rubinsten et al., Girelli, Lucangeli, and Butterworth (2000) published a similar one. They examined size congruity in physical and numerical tasks in schoolchildren and college students. However, there were several differences in stimuli, and the children and adults were tested in two separate studies. In general, the pattern of results was similar, and eyeballing the data shows a reduction of the congruity effect from fifth grade children to adults. It seems that the increase in the effect in schoolchildren is related both to improvement in the number concept because of schooling (and age) and to a not fully matured inhibitory system. The smaller size congruity effect presented by college students compared with schoolchildren is probably related to improved inhibitory control and possibly also improved proficiency in using numbers. However, as suggested above, the increase in efficiency of control is not only related to the advance in the control systems but also to the progress in numerical cognition and specifically, in the number concept that is contributed by schooling and probably also by age. As suggested by Merkley et al., the current and other similar studies cannot provide an unequivocal causal explanation regarding the relationship between control and numerical proficiency. Importantly, it is possible that there exists a bidirectional influence, and such influence is probably not restricted to the area of numerical cognition.

The last part of Ben-Shachar and Berger's (Chapter 5, this section) contribution is devoted to the third attentional system: orienting. Here they discuss the spontaneous focus on numerosity (SFON) task (Hannula & Lehtinen, 2005) that was mainly designed for children. It has been shown that children who can be characterized (by SFON) as being spontaneously oriented to numerosity have higher mathematical achievements than children who do not have such an orientation. Similar results were found with adults who underwent a test (similar to the SFON) designed for adults. These results are in line with the suggestion of Merkley et al. (Chapter 6, this section) that during the development of the number concept, infants and children have to learn what are the features that should be focused on and at the same time, what features are irrelevant. However, it is still an open question whether training in orientation to numerosity might help those who have a lower tendency to spontaneously orient to numerosity.

STRATEGIES AND CONTROL

Uittenhove and Lemaire's (Chapter 7, this section) contribution does not stay at the building block level of numerical cognition but rather discusses higher processes. Their contribution revolves around interactions between strategies and control. Strategies are defined as a "procedure or set of procedures to accomplish a high-level goal." Following Lemaire and Siegler (1995), they suggest that the use of strategies differs in strategy repertoire, strategy distribution, strategy execution, and strategy selection. Uittenhove and Lemaire focus on works by Miyake et al. (2000) and Miyake and Friedman (2012) in the discussion of EF. Hence, EF is not only inhibition or attention but also working memory and shifting.

According to Uittenhove and Lemaire (Chapter 7, this section), various aspects of EF (e.g., inhibition, working memory) are heavily involved in the different dimensions of strategy use. For example, strategy execution (i.e., the speed and precision obtained with each strategy) involves working memory and also attention resources.

Accordingly, tying up attention resources by use of a secondary task impairs efficiency of execution of strategies. Similarly, the choice of strategy (i.e., strategy selection) is affected by inhibitory control needed for suppression of competing strategies and shifting (between alternative strategies).

The use of strategies changes with age. Both children and old adults use simpler strategies than young adults do. Children and old adults use strategies that rely less on EF capacity, be it memory or attention, and their strategy choice is described as poor. Other dimensions of strategies are also affected by age. For example, compared to young adults, switch costs between strategies reduce with increase in age of children and increase in older adults. Interestingly, the reasons for these patterns may be different in children and old adults. Poorer strategies in children and higher switch costs may be the result of lack of or low numerical proficiency. In contrast, old adults may present poorer strategies and higher switch costs because of reduced EF resources (e.g., reduced memory capacity). Hence, there seems to be an interplay between EF and proficiency; high numerical proficiency may enable efficient use of EF in selecting strategies and flexibly switching between them, and reduced EF capacity (e.g., reduced switching ability) may be presented as less flexible strategy use, which looks like lower numerical proficiency. This is of course reminiscent of the idea of bidirectional influences between control and numerical cognition suggested by Merkley et al. (Chapter 6, this section) and the interaction between control and automaticity suggested by Tzelgov et al. (1990).

The developmental trajectory of children is characterized by more efficient use of strategies, better selection of strategies, and improved ability of switching between strategies. This developmental trajectory could be the result of improved control and enhanced working memory, but it could also be the result of improved proficiency in numerical cognition or both.

CONCLUSIONS

All contributors to this section—performance control and selective attention—report on association between EFs or specific components of EFs and numerical cognition. They also seem to agree that EF components are necessary for the development of numerical cognition and its proficient use. However, it is not so clear whether the direction of causality is unidirectional (EFs modulate numerical cognition) or bidirectional (EFs modulate numerical cognition and numerical cognition may modulate EFs). Throughout this discussion, it was suggested that proficiency may modulate, in various ways, EFs or use of EFs. In our opinion, this is a viable possibility and should be further studied.

References

Girelli, L., Lucangeli, D., & Butterworth, B. (2000). The development of automaticity in accessing number magnitude. *Journal of Experimental Child Psychology, 76,* 104–122.

Hannula, M. M., & Lehtinen, E. (2005). Spontaneous focusing on numerosity and mathematical skills in young children. *Learning and Instruction, 15,* 237–256.

Hasher, L., & Zacks, R. T. (1979). Automatic and effortful processes in memory. *Journal of Experimental Psychology: General, 108,* 356–388.

Lemaire, P., & Siegler, R. S. (1995). Four aspects of strategic change: Contributions to children's learning of multiplication. *Journal of Experimental Psychology: General, 124,* 83–97.

Logan, G. D. (1988). Toward an instance theory of automatization. *Psychological Review, 95,* 492–527.

Miyake, A., & Friedman, N. P. (2012). The nature and organization of individual differences in executive functions: Four general conclusions. *Current Directions in Psychological Science, 21,* 8–14.

Miyake, A., Friedman, N. P., Emerson, M. J., Witzki, A. H., Howerter, A., & Wager, T. D. (2000). The unity and diversity of executive functions and their contributions to complex "frontal lobe" tasks: A latent variable analysis. *Cognitive Psychology, 41,* 49–100.

Petersen, S. E., & Posner, M. I. (2012). The attention system of the human brain: 20 years after. *Annual Review of Neuroscience, 35,* 73–89.

Posner, M. I., & Petersen, S. E. (1990). The attention system of the human brain. *Annual Review of Neuroscience, 13,* 25–42.

Rubinsten, O., Henik, A., Berger, A., & Shahar-Shalev, S. (2002). The development of internal representations of magnitude and their association with Arabic numerals. *Journal of Experimental Child Psychology, 81,* 74–92.

Stroop, J. R. (1935). Studies of interference in serial verbal reactions. *Journal of Experimental Psychology, 18,* 643–662.

Tzelgov, J. (1997). Specifying the relations between automaticity and consciousness: A theoretical note. *Consciousness and Cognition, 6,* 441–451.

Tzelgov, J., Henik, A., & Leiser, D. (1990). Controlling Stroop interference: Evidence from a bilingual task. *Journal of Experimental Psychology: Learning, Memory and Cognition, 16,* 760–771.

SPATIAL PROCESSING AND MENTAL IMAGERY

How Big Is Many? Development of Spatial and Numerical Magnitude Understanding

Nora S. Newcombe[1,a], Wenke Möhring[2],
Andrea Frick[3,b]

[1]Temple University, Philadelphia, PA, United States; [2]Universität Basel,
Basel, Switzerland; [3]University of Fribourg, Fribourg, Switzerland

One big vase is just one vase, regardless of its height, width, weight, or the volume of water that it can hold. Four tiny bud vases constitute a larger number of vases than that one enormous urn, even if they do not weigh as much taken all together, do not hold as much total water, and so forth. However, this situation is rare. In most cases, a greater number of things

[a]National Science Foundation, SBE 1041707
[b]Swiss National Science Foundation, # PP00P1_150486

will have greater values on other quantitative dimensions than a smaller number of things—for instance, four apples will almost always weigh more than one apple. Put more abstractly, children encounter a physical world in which the number of objects they see in a collection is highly but not perfectly correlated with the continuous physical dimensions of the collection, such as overall height, width, volume, weight, or area. For babies entering the world, this situation represents a potential challenge. Which kind of quantity is more important? If various quantities are highly correlated, are they worth disentangling? How could they best be distinguished?

Furthermore, this difficult problem is not the only challenge that babies face with respect to quantification. They also encounter a world in which they must master two different kinds of quantification systems and discriminate when each is applicable. Fig. 9.1 shows how the same perceptual entities can be conceptualized quite differently in the two systems. On the one hand, quantities can be encoded *extensively*, which refers to an absolute encoding of amounts. Extensive coding can apply both to continuous and to discrete quantities. For example, 1 liter of water is less than 2 liters of water, or we can say that 2 apples are fewer than 4 apples. Thus, when encoded extensively, these quantities would be considered unequal. On the other hand, quantities can be encoded *intensively*, which means that they are encoded in a relational or proportional fashion regardless of the total amount. For example, the number of green apples can be encoded in relation to the number of red apples, and this ratio is the same regardless of the total number of apples (a ratio of 1 green and 1 red apple is equal to the ratio of 2 green and 2 red apples). In this case, the ratio of two extensive properties describes an intensive property. As with extensive coding,

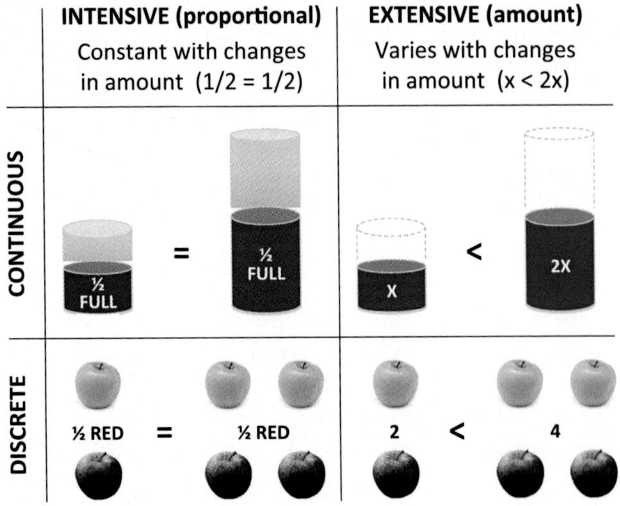

FIGURE 9.1   Different systems for quantification.

intensive coding can apply not only to discrete objects, as in the apple example explained earlier, but also to continuous quantities, e.g., half a glass of water is half a glass regardless of whether the glass itself is small or large. Hence, when encoded intensively, the quantities would be considered equal.

How do children approach these complex learning problems, eventually mastering quantification and acquiring mathematical proficiency? Psychology has long ago moved beyond the initial approach of Piaget (1952), who argued that babies begin with little to no understanding. Contemporary theorists uniformly postulate strong starting points in infancy, although this general approach can be divided into two contrasting theories of quantitative development. One view finesses both problems delineated earlier with postulations of innateness. Nativists claim that infants begin life with an approximate number system (ANS) that is already differentiated from other quantitative dimensions and that specifically functions to enumerate sets of discrete objects (Dehaene & Brannon, 2011; Gallistel & Gelman, 1992; Spelke & Kinzler, 2007). In this view, initial magnitude processing involves more than one system. Infants have a number system applied to discrete and extensive situations (i.e., an integer system, as shown at the bottom right of Fig. 9.1) and additional systems applied to continuous (and often intensive) dimensions such as length, density, or area. Development consists of increasing precision in the ANS and the acquisition of culturally transmitted symbols, such as the count words and numeral representations.

The alternative view postulates that infants begin with a generalized magnitude system extending across various dimensions of quantity, without a separate system for counting discrete objects (Lourenco & Longo, 2011; Mix, Huttenlocher, & Levine, 2002; Walsh, 2003). In this approach, initial magnitude is not only undifferentiated but also continuous and intensive (i.e., the top left of Fig. 9.1). Development consists of differentiation of the various correlated dimensions, formation of a separate integer system for counting sets of discrete objects, and eventual rediscovery of the idea of numerical quantities "between the integers." Overviews of the second view are available elsewhere (Leibovich, Katzin, Harel, & Henik, 2016; Newcombe, Levine, & Mix, 2015). Here, we explore a major implication of this way of thinking, namely that when children learn the count words and the cardinality principle (i.e., the principle that the last count word in a one-to-one series enumerates the set), they are gradually—but not inevitably—less likely to rely on the continuous intensive system with which they began. This access to the intensive system needs to be either maintained or regained to tackle other mathematical topics, notably proportional reasoning and fractions, and use of each system has to be refined so as to target its appropriate range of application.

More specifically, we connect five domains of developmental research which have traditionally been treated as separate. We begin with two spatial tasks, where research has mostly focused on continuous intensive quantity: spatial estimation and spatial scaling. We then consider three mathematical tasks: the use of a number line to represent integers, proportional reasoning, and fraction understanding. The number line task involves mapping the integer system used to count discrete objects onto a continuous (and often bounded) line. In our conceptualization, number line tasks reflect more than simply knowledge of the integer system; they also reflect the skills required to map a system adapted to treat discrete objects that are quantified extensively onto distances on a continuous line generally treated intensively. That is, successful performance requires interrelating the discrete extensive and continuous intensive systems. Proportional reasoning is similar to spatial scaling in that it involves transformation of an intensively defined quantity, as shown in the upper left of Fig. 9.1, but it is generally embedded in a mathematical problem context. Fractions involve moving from a discrete and extensive conceptualization of number to a fuller number system in which all of the four cells of Fig. 9.1 have a place. For these three mathematical topics, we develop the argument that whether tasks are defined in continuous or discrete terms has major implications for children's success because it may influence whether quantities are encoded intensively or extensively. In the last section of the chapter, we consider how this theoretical framework sheds light on the processes underlying the overall relations that have been found between spatial and mathematical thinking and how it may suggest new avenues for future research.

SPATIAL TASKS

Spatial Estimation

Initial work on spatial estimation was aimed at probing the limits of Piaget's assertion that early spatial representation is merely topological, without metric specification. Experiments in which an object was hidden in a long narrow sandbox and young children searched for it after a short delay showed that spatial location is encoded metrically, at least in these simple tasks, by the ages of 18 and 24 months (Huttenlocher, Newcombe, & Sandberg, 1994). Even infants showed sensitivity to metric location in studies using the dependent variable of looking time (Gao, Levine, & Huttenlocher, 2000; Newcombe, Huttenlocher, & Learmonth, 1999; Newcombe, Sluzenski, & Huttenlocher, 2005).

These experiments also suggested that metric coding is not the only kind of coding but is supplemented by categorical coding of location. Such categorical coding is a kind of insurance plan because categories are easy to remember, e.g., someone may know that *my glasses are somewhere on the desk* even after forgetting exactly how far from which edge of the desk. In a model of adult spatial estimation, Huttenlocher, Hedges, and Duncan (1991) proposed that fine-grained estimations are combined with memories for the spatial category in which a location appeared, according to a Bayesian combination rule. The model shed light on the fact that toddlers' searches for a toy hidden in a sandbox were biased toward the center of the box, by suggesting that they used the sandbox as a category (Huttenlocher et al., 1994). However, the spatial categories used become smaller with age and hence more informative (e.g., specifying which *half* of the sandbox contains the hidden toy). Adjustment by a smaller category draws estimates to a prototype value closer to the actual location. Subdivision of a space into more than one category appears between 4 and 8 years, depending on the size of the space, and results in a distinctive bias pattern in children's responses best described as a polynomial function (Huttenlocher et al., 1994). Fig. 9.2 shows spatial memory in older children in the sandbox task—the smaller errors around 15, 30, and 45 inches suggest a subdivision of the space.

Importantly, subsequent studies revealed a vital fact about early metric coding, showing that it is successful only when there is an enclosing frame, such as a sandbox or a container (Duffy, Huttenlocher, & Levine, 2005; Huttenlocher, Duffy, & Levine, 2002). This suggested that spatial estimations are intensive early on. In fact, children rely on intensive (or proportional) coding using a perceptually available standard through the preschool years and into early elementary school (Duffy et al., 2005;

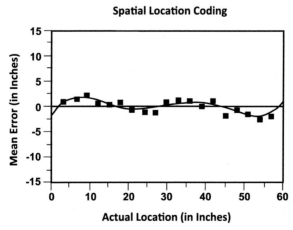

Spatial Location Coding

FIGURE 9.2 Mean signed errors of older children's spatial location coding in the sandbox tasks, with a polynomial curve fit. Positive errors indicate responses too far to the right, and negative errors indicate responses too far to the left of the actual location. *From Huttenlocher, J., Newcombe, N.S., & Sandberg, E. (1994). The coding of spatial location in young children.* Cognitive Psychology, 27, 115–147.

Huttenlocher, Levine, & Ratliff, 2011; Huttenlocher et al , 2002; Vasilyeva, Duffy, & Huttenlocher, 2007). The ability to code extent in the absence of a salient perceptual standard, or *extensive* coding of amount, only begins to emerge after the age of 5 years and is more fully developed by around age 8 years (Duffy et al., 2005; Vasilyeva et al., 2007). Interestingly, it is about at this age that the school curriculum in the United States, where the spatial estimation studies were conducted, generally asks children to learn to differentiate intensive and extensive coding, when they begin to tackle fractions.

Spatial Scaling

If spatial estimation is supported by continuous intensive coding, there is no reason in principle that the kinds of spatial memory just discussed could not be scaled up or down. The top half of a column is the top half no matter the column's height, and a distance defined with respect to such a system then falls into place as the system expands or contracts. If so, toddlers should be able to perform spatial scaling tasks. But scaling also depends on processes other than spatial memory. Success requires appreciating that one stimulus can stand for another. This basic idea of representational correspondence itself takes some time to develop. Research has shown that it is often not until 3 years that children are able to use the information from a smaller map or model to find a toy in a larger room (e.g., DeLoache, 1987, 1991, 2010), showing that they comprehend that objects in a model can "stand for" objects in the referent space. In line with this finding, by the age of 3 years, children can use a small map or model to find a toy hidden in a larger space, at least when the task is constrained to hiding locations along a single dimension (Huttenlocher, Newcombe, & Vasilyeva, 1999; Huttenlocher, Vasilyeva, Newcombe, & Duffy, 2008). However, even after age 3, children still have some difficulties with mapping tasks. When the task required locating a single object along two dimensions, approximately 60% of the 4-year-olds and 90% of the 5-year-olds were able to translate distances from a map to a larger space (Vasilyeva & Huttenlocher, 2004). When there were many objects to place in a scaled representation, 3–5-year-old preschoolers showed considerable difficulties understanding geometric (and sometimes even representational) correspondences between their classrooms and maps or models (Liben & Yekel, 1996; Liben, Moore, & Golbeck, 1982). In fact, even simple scaling tasks that do not require a dimensional translation are somewhat demanding for preschoolers, and the ability undergoes considerable development between 3 and 5 years of age (Frick & Newcombe, 2012).

This descriptive evidence about the developmental trajectory of spatial scaling abilities suggests the need to further specify the cognitive processes and strategies involved. There are at least two possible ways to scale. First, we could form a mental representation of a space or layout and holistically transform it in a way that preserves metric relations, analogous to using a magnifying glass or to zooming in or out of an online map. In zooming, all internal distances are transformed equally and simultaneously (cf. Vasilyeva & Huttenlocher, 2004). Möhring, Newcombe, and Frick (2014) investigated this notion and asked adults, 4-, and 5-year-olds to locate target objects in a constant referent space, based on maps that varied in scaling factor. Results showed a linear increase in response times and localization errors with increasing scaling factor. This indicated that participants applied a mental transformation strategy in analogy to imagistic processes in mental rotation or scanning (Kosslyn, 1975; Shepard & Metzler, 1971), where increasing response time patterns have been typically taken as an index for the use of analog mental transformations. A follow-up study with another group of 4–5-year-olds (Möhring, Newcombe, & Frick, 2015) showed similar linear response patterns, and these were even confirmed using a different paradigm in a study with adults (Möhring, Newcombe, & Frick, 2016). In that latter study, adults were asked to decide whether targets in two pictures presented side by side were in an identical position or not. Participants showed linear response time functions irrespective of whether the task was to scale up or down, supporting the notion that they used mental transformation strategies. Spatial scaling may thus engage similar spatial transformation processes to mental rotation, which leave the internal structure and configuration of an object or layout intact, while changing its spatial relation to the observer (apparent distance or orientation, respectively).

There is also a second way to succeed in a spatial scaling task. As we discussed, according to adaptive combination theory, spatial information from metric and categorical coding is integrated to optimize overall precision (Holden, Newcombe, & Shipley, 2013; Huttenlocher et al., 1991). In the case of scaling, children may encode relative distances, thus preserving the relation between distances regardless of their absolute size. For example, they may think of a distance as one-third of the way across a bounded entity, relating a small entity to a larger one, or vice versa. This intensive coding strategy would not predict the scaling factor effects just discussed, but children may use one or the other strategy, or a combination of both, depending on the characteristics of the situation and their cognitive capacities. Indeed, previous research on children's location coding showed that 4–5-year-olds coded both the relation between a hidden toy and two landmarks and the absolute distance to one of the landmarks and adaptively used this information contingent on which landmarks were present during retrieval (Uttal, Sandstrom, & Newcombe, 2006). The idea that relational coding also guides spatial scaling is supported by participants' accuracy patterns in a scaling assessment (Frick & Newcombe, 2012). Younger children's horizontal errors for locating objects hidden in a narrow horizontal space are smaller toward the midpoint, suggesting that they used only one category with a prototypical center point to guide their responses. For older children, there was increasing accuracy around the quarter points (4 years) and near the edges (5+ years), suggesting that they mentally subdivided the space, similar to the pattern in Fig. 9.2.

MATHEMATICAL TASKS

Number Lines

As we have seen earlier, children are typically able to encode intensive and extensive continuous spatial information (such as relative or absolute length) by the time they enter elementary school. By that time, they have also acquired an increasingly robust and differentiated system of discrete number. They can typically count fairly well and have acquired the cardinality principle. Furthermore, they can interpret and order symbolic numerals, such as 5 < 6. Early math instruction stresses whole numbers and counting, reinforcing these discrete conceptualizations (Mix, Levine, & Huttenlocher, 1999). However, when faced with a number line task, i.e., placing numbers on a number line, children must coordinate discrete and continuous information. Bounded number line tasks present children and adults with various numbers and ask them to indicate the position of these numbers on a physical line that represents a number range, such as from 0 to 10 or from 0 to 100 (e.g., Barth & Paladino, 2011; Booth & Siegler, 2006; Landy, Silbert, & Goldin, 2013; Rips, 2013; Siegler & Opfer, 2003; Slusser, Santiago, & Barth, 2013). In a perfect solution, participants' responses form a linear function of the presented numbers. However, young children often spread the small numbers apart and squeeze the large numbers at the end of the physical line, which results in a response pattern that is best captured by a logarithmic function (Opfer & Siegler, 2007; Siegler & Opfer, 2003; Thompson & Opfer, 2010). Eventually, children's response pattern shifts from this logarithmic to a linear function and this shift occurs several times over individual development, with each number range in turn changing from logarithmic to linear response patterns.

There are several explanations for these response patterns and developmental changes (Barth & Paladino, 2011; Ebersbach, Luwel, Frick, Onghena, & Verschaffel, 2008; Moeller, Pixner, Kaufmann, & Nuerk, 2009; Siegler & Opfer, 2003; Slusser et al., 2013). Our interpretation is that children have difficulties in connecting the discrete extensive information provided by counting (i.e., the number), which is prioritized at the beginning of school, with a continuous intensive conceptualization of quantity (i.e., the space between both ends of the number line). This explanation draws attention to the spatial aspect of the number line task. The number line is a cultural invention that arguably leverages the deep associations between space and number (for an overview of the large literature on spatial–numerical associations, see McCrink & Opfer, 2014). Indeed, it was found that preschool spatial skills longitudinally predicted number line accuracy (Gunderson, Ramirez, Beilock, & Levine, 2012).

Further support for the idea that children may have difficulties connecting discrete extensive and continuous intensive information comes from studies in which children's strategies were analyzed or unbounded lines without a rightmost boundary were presented. For example, Petitto (1990) observed first- to third-graders' estimation strategies on bounded number lines. Results indicated that a majority of the children (around 80%) used a counting strategy and that children progressively shifted from counting to proportional strategies across the first three elementary school years. These results were corroborated in a later study using eye tracking data (Schneider et al., 2008). Such a counting strategy may explain the nonlinear patterns often observed in bounded number line tasks. For instance, children may start off with counting too large units in the smaller and more familiar number range and then have to squeeze the larger numbers in the space that is left. While in the unbounded number line task, such an iterative counting strategy is appropriate, the bounded number line task requires intensive thinking, as the available space has to be subdivided proportionally to determine the appropriate unit size. However, performance on bounded and unbounded number line tasks is similar until about grade 2 (Ebersbach, Luwel, & Verschaffel, 2015; Link, Huber, Nuerk, & Moeller, 2014), suggesting that up to that age, children may erroneously choose an extensive strategy on bounded number lines and treat them essentially like unbounded tasks. This may especially be the case if the upper anchors are large or unfamiliar and thus uninformative to young children. Interestingly, also adults respond logarithmically with unfamiliar symbolic number formats (Chesney & Matthews, 2013) and with large or even fictitious numbers defining the right end of the line (Rips, 2013), indicating that even they may fall back to an extensive strategy when proportional subdivision of the number line is too difficult (see also Hurst, Monahan, Heller, & Cordes, 2014).

Proportional Reasoning

Proportional thinking is required in many activities of our daily lives, such as adjusting a cake recipe after realizing that one has only 200g of flour instead of the required 300g or splitting the cake among 12 people. Additionally, proportional reasoning is important for success in many Science, Technology, Engineering, and Mathematics (STEM) fields, such as geosciences, physics, and chemistry. Unfortunately, children up to 11 years of age often struggle to reason about proportions (e.g., Fujimura, 2001; Moore, Dixon, & Haines, 1991; Noelting, 1980; Piaget & Inhelder, 1975). Proportional reasoning tasks typically require thinking about the ratios of elements in relation to a whole, e.g., judging how intense a drink would taste with a certain amount of juice in it. On the face of it, a problem like this one is very different from spatial scaling. But, in fact, both tasks involve ratios, and both tasks involve thinking about a part in relation to a whole, regardless of absolute extent or quantity. Indeed, proportional reasoning has been shown to correlate to spatial scaling (Möhring et al., 2015). In this study, children of 4–5 years were presented with maps of varying size showing a target position and were asked to point to the same position in a constant referent space. In the proportional reasoning task, children saw different mixtures of cherry juice and water and were asked to indicate on a rating scale how much these mixtures tasted of cherry. Children's accuracy in locating the targets and estimating the cherry taste was highly related, even after controlling for age and verbal intelligence. This shared variance may reflect the fact that relational coding is equally required by both tasks.

Proportional reasoning problems require children to encode the spatial *extent* of the presented components (i.e., amount of juice and water) but to focus on their *intensive* aspects (i.e., their ratio). Interestingly, when proportional reasoning problems are presented in countable units, children from 6- to 10 years focus on counting the discrete units, i.e., on extensive coding (Boyer & Levine, 2012; Boyer, Levine, & Huttenlocher, 2008). For example, in these studies by Boyer et al., children were asked to find the equivalent to a target proportion (e.g., 3 units of juice vs. 6 units of water). Children had to choose their answer from two alternative proportions showing either the correct, smaller-scaled proportion (e.g., 1 unit of juice vs. 3 units of water) or a false proportion showing the same absolute number of units in the more salient juice component (e.g., 3 units of juice vs. 3 units of water). These studies revealed that children often chose the latter alternative and thus relied on an erroneous count-and-match strategy. However, when a counting strategy was prevented by presenting the same proportions continuously instead of discretely, children's performance improved significantly (see also Boyer & Levine, 2015, for initial evidence of a successful intervention).

Similar evidence for an erroneous counting strategy comes from studies investigating probabilistic reasoning (Jeong, Levine, & Huttenlocher, 2007). Here, children were presented with two donut-shaped spinning figures that were divided into red and blue regions. The two figures differed in overall size and in the proportion of red versus blue regions. Children were asked to select the spinning figure that was associated with a higher probability that a central arrow pointed to the red region instead of the blue one if it was spun. Importantly, the different-colored regions were displayed as discrete units in one condition and as continuous areas in another. While children up to 10 years struggled with proportions involving discrete units, even 6-year-olds showed success when presented with continuous proportions. Similar to previous findings (Boyer & Levine, 2012, 2015; Boyer et al., 2008), children may have focused on counting the absolute discrete units instead of encoding the intensive quantities.

Although discrete presentation seems to promote a counting strategy, a bias toward thinking extensively has even been observed with completely nonspatial and nonnumerical continuous quantities. In a study on how children combine intensive quantities, Jäger and Wilkening (2001) presented two glasses of colored liquids, one lighter and one darker, to children at the age of 6-, 8-, 10-, and 12 years. Children were asked to imagine the two liquids of equal volume being poured together and to judge the resulting color intensity on a rating scale. An adding rule was predominant in 8- and 10-year-olds, suggesting that children at this age treated color as an extensive quantity and thought that the resulting liquid would be darker than any of its components. Interestingly, 6- and 12-year-olds did not use additive rules nearly as often. This inverse U-shaped developmental trajectory for the adding rule supports the idea that when extensive thinking is emphasized at the beginning of elementary school, access to the continuous intensive system decreases but is slowly regained later, as evidenced by 12-year-olds' usage of a correct averaging rule. In a second experiment, the authors, additionally, varied the volume of the to-be-combined liquids to investigate children's and adults' understanding of how extensive and intensive quantities are interrelated. Children and adults showed interesting misconceptions. For instance, many of them thought the resulting liquid would become darker if more of the same two liquids were mixed together at the same ratio. These findings suggested a persisting confusion when having to simultaneously deal with extensive and intensive quantities, even into adulthood.

Fractions

The integer system that children have learned during preschool and early elementary school is extensive and discrete. But when they encounter fractions, they encounter numerical quantities that require an understanding of the number system as continuous rather than simply applying to discrete objects and as involving intensive and extensive quantity. Fractions are defined in terms of equal-sized units of a whole. The number of divisions of a whole, indicated by the denominator in the fraction notation, defines the number of discrete entities. Next to the number of divisions of the whole, the fraction notation also entails information about the counts (e.g., 2/8 comprises a count of 2 units with each unit having the size of one part in a whole divided by 8). Thus, reasoning about fractions requires a combination of being able to encode the extensive quantities of the units and the whole, as well as to think about their ratio—which in contrast is an intensive property (cf. Mix, Levine, & Newcombe, 2016). To complicate things, there is an inverse relation between unit size (i.e., continuous quantity) and denominator (i.e., discrete number of parts), as the sizes of the units become larger with decreasing denominator (e.g., as fewer people want a piece of a cake, each person gets a larger piece).

Therefore, reasoning about formal fractions requires children to flexibly and simultaneously integrate intensive and extensive information, similar to what we have seen for proportional reasoning above. Indeed, 8–10-year-old's formal fraction knowledge is correlated with estimations of nonnumerical, spatial proportions (Möhring, Newcombe, Levine, & Frick, 2016). Thus, a propensity to treat quantities as extensive, which has previously been shown in nonnumerical proportional tasks when these were presented in discrete format (cf. Boyer et al., 2008; Boyer & Levine, 2012), may also be responsible for problems in the numerical domain of fractions. This may at least partially explain children's difficulties with formal fractions, which poses substantial challenges to many students (Hecht & Vagi, 2010; Schneider & Siegler, 2010; Stafylidou & Vosniadou, 2004). These profound difficulties are unfortunate, as an understanding of fractions is related to general mathematical proficiency in cross-sectional (Siegler, Thompson, & Schneider, 2011) and longitudinal samples (Bailey, Hoard, Nugent, & Geary, 2012; Siegler et al., 2012). That is, students' fraction knowledge predicts their overall mathematics achievement in high school 6 years later, even after controlling for their knowledge of whole number addition, subtraction, and multiplication, domain-general abilities such as working memory, or socioeconomic status (Siegler et al., 2012). It seems that a better understanding of fractions particularly supports an understanding of higher-order mathematics such as algebra (e.g., Booth, Newton, & Twiss-Garrity, 2014, Siegler et al., 2012).

A typical mistake in fraction calculation is to treat the nominator and denominator as separate whole numbers, rather than related whole numbers, resulting in typical errors such as ¼>½ or 2/3+1/8=3/11 (see Resnick & Ford, 1981). In previous attempts to explain this whole numbers bias, one hypothesis proposed that children are endowed with specialized learning mechanisms that facilitate learning about whole numbers as opposed to other types of numbers and this innate constraint interferes with learning fractions (cf. Gelman, 1991; Wynn, 1995, 1997). This hypothesis was challenged in a study from Mix et al. (1999), which showed some competence on nonsymbolic fraction problems even in preschoolers and parallels in the developmental trajectories of whole number reasoning and fraction calculation. These findings suggested that preschoolers can attend to either whole numbers or fractional amounts and that early quantitative reasoning is not restricted to whole numbers. Instead, Mix and colleagues (Mix et al., 1999; Paik & Mix, 2003) proposed that children's difficulties with fractions are due to their counting and calculation experience, as instruction in elementary school mainly focuses on discrete information, i.e., teaching whole numbers and counting.

Based on this idea and on our review of different domains above showing similar biases in numerical and nonnumerical contexts, we propose an alternative account to the whole number bias: young children might be subject to an "extensitivity bias." That is, they may focus on extensive information, such as the size and number of parts, but have difficulties with putting these in relation to other parts or the whole. This focus on extensive information may be emphasized by discrete presentation and peak at the beginning of formal schooling when integers are frequently trained. This extensive experience could restrict access to the intensive system, even though both systems are developed and available.

USING SPATIAL THINKING IN MATHEMATICS EDUCATION

Spatial and numerical thinking seem to be closely related based on cross-sectional studies of preschoolers, school children, and adolescents (e.g., Casey, Nuttall, Pezaris, & Benbow, 1995; Laski et al., 2013; Reuhkala, 2001; Verdine et al., 2014) and on longitudinal studies that have demonstrated that preschoolers' and kindergarteners' early spatial abilities predict their later performance on mathematical tests (e.g., Carr et al., 2018; Gunderson et al., 2012; Lauer & Lourenco, 2016; LeFevre et al., 2013; Verdine, Golinkoff, Hirsh-Pasek, & Newcombe, 2017). But why do we see this relation and which spatial skills are most relevant for mathematical thinking? With respect to the last question, a cross-sectional study with children from 5- to 13 years of age found that mental rotation and visual-spatial working memory explained the most variance in mathematical performance in kindergarten and sixth grade, respectively (Mix et al., 2016). Recent longitudinal results (Frick, 2018) further suggested that kindergarteners' mental rotation and spatial scaling skills were particularly important predictors of performance on arithmetic operations subtest of a standardized math test 2 years later.

In the context of the overview presented in this chapter, we can better understand why mental rotation, spatial scaling, and spatial working memory may each be important for mathematical achievement. Mental rotation is likely linked to spatial scaling as both involve mental transformation processes. In addition, spatial scaling is a measure that also taps a kind of proportional reasoning in which continuous quantities are subdivided into categories, a process that may also be important for number line estimation. Spatial working memory provides the mental workspace for these operations. In fact, tests of mental rotation and spatial scaling may themselves partially also assess aspects of spatial working memory not picked up by traditional spatial working memory tasks. Children with better mental transformation and spatial working memory skills may be better at translating mathematical magnitudes, such as fractions or proportions, into spatial analogues and at mentally subdividing or transforming these spatial representations. This may be helpful to decide quickly between plausible and implausible answers and to focus on the intensive information in the fraction, instead of erroneously focusing only on the number of counts in one component.

In light of a recent metaanalysis indicating that spatial abilities are malleable (Uttal et al., 2013), researchers have begun to assess whether spatial training, including mental transformation or spatial visualization, can improve mathematics achievement. Some positive evidence is emerging (Cheng & Mix, 2014; Lowrie, Logan, & Ramful, 2017), although there are also some negative findings (Hawes, Moss, Caswell, & Poliszczuk, 2015; Xu & LeFevre, 2016). Other studies investigated the effects of curriculum modifications that intensified spatial activities with mostly positive findings (Cunnington, Kantrowitz, Harnett, & Hill-Ries, 2014; Grissmer et al., 2013; Hawes, Moss, Caswell, Naqvi, & MacKinnon, 2018). The success of the curricular approach may derive from the fact that a broad array of spatial skills was developed, rather than there being a focus on a single spatial task. For example, the visual arts program developed by Cunnington et al. entailed considerable use of spatially challenging exercises and visual illustration of mathematical concepts, such as fractions, and the children in the study by Hawes et al. worked with a range of activities derived from a workbook called *Taking Shape* (Moss, Bruce, Caswell, Flynn, & Hawes, 2016).

However, from the point of view of theory development, these broad curricular studies are less informative than targeted training studies. More targeted studies could identify the specific factors that are critical for positive effects on mathematical development. But first, such targeting requires a more theoretical analysis of spatial and mathematical thinking and their commonalities. The present theoretical analysis suggests such directions. It also emphasizes the need for further research investigating the development of the ability to reason about intensive quantities and to integrate intensive and extensive information. Such research may shed light on children's difficulties with certain mathematical concepts, which may help to find ways for children (and adults) to overcome their inclination to focus on extensive information once they know how to count.

References

Bailey, D. H., Hoard, M. K., Nugent, L., & Geary, D. C. (2012). Competence with fractions predicts gains in mathematics achievement. *Journal of Experimental Child Psychology, 113*, 447–455.

Barth, H. C., & Paladino, A. M. (2011). The development of numerical estimation: Evidence against a representational shift. *Developmental Science, 14*, 125–135. https://doi.org/10.1111/j.1467-7687.2010.00962.x.

Booth, J. L., Newton, K. J., & Twiss-Garrity, L. K. (2014). The impact of fraction magnitude knowledge on algebra performance and learning. *Journal of Experimental Child Psychology, 118*, 110–118.

Booth, J. L., & Siegler, R. S. (2006). Developmental and individual differences in pure numerical estimation. *Developmental Psychology*, *41*, 189–201. https://doi.org/10.1037/0012-1649.41.6.189.

Boyer, T. W., & Levine, S. C. (2012). Child proportional scaling: Is 1/3 = 2/6 = 3/9 = 4/12? *Journal of Experimental Child Psychology*, *111*, 516–533.

Boyer, T. W., & Levine, S. C. (2015). Prompting children to reason proportionally: Processing discrete units and continuous amounts. *Developmental Psychology*, *51*, 615–620.

Boyer, T. W., Levine, S. C., & Huttenlocher, J. (2008). Development of proportional reasoning: Where young children go wrong. *Developmental Psychology*, *44*, 1478–1490.

Carr, M., Alexeev, N., Wang, L., Barned, N., Horan, E., & Reed, A. (2018). The development of spatial skills in elementary school students. *Child Development*, *89*, 446–460.

Casey, M. B., Nuttall, R. L., Pezaris, E., & Benbow, C. P. (1995). The influence of spatial ability on gender differences in math college entrance test scores across diverse samples. *Developmental Psychology*, *31*, 697–705.

Cheng, Y. L., & Mix, K. S. (2014). Spatial training improves children's mathematics ability. *Journal of Cognition and Development*, *15*, 2–11.

Chesney, D. L., & Matthews, P. G. (2013). Knowledge on the line: Manipulating beliefs about the magnitudes of symbolic numbers affects the linearity of line estimation tasks. *Psychonomic Bulletin and Review*, *20*, 1146–1153.

Cunnington, M., Kantrowitz, A., Harnett, S., & Hill-Ries, A. (2014). Cultivating common ground: Integrating standards-based visual arts, math and literacy in high-poverty urban classrooms. *Journal for Learning Through the Arts: A Research Journal on Arts Integration in Schools and Communities*, *10*(1).

Dehaene, S., & Brannon, E. (Eds.). (2011). *Space, time and number in the brain: Searching for the foundations of mathematical thought*. Academic Press.

DeLoache, J. S. (1987). Rapid change in the symbolic functioning of very young children. *Science*, *238*, 1556–1557.

DeLoache, J. S. (1991). Symbolic functioning in very young children: Understanding of pictures and models. *Child Development*, *62*, 736–752.

DeLoache, J. S. (2010). Early development of the understanding and use of symbolic artifacts. In U. Goswami (Ed.), *The Wiley-Blackwell handbook of childhood cognitive development* (2nd ed. pp. 312–336). Hoboken, NJ: Wiley-Blackwell.

Duffy, S., Huttenlocher, J., & Levine, S. (2005). It is all relative: How young children encode extent. *Journal of Cognition and Development*, *6*, 51–63. https://doi.org/10.1207/s15327647jcd0601_4.

Ebersbach, M., Luwel, K., Frick, A., Onghena, P., & Verschaffel, L. (2008). The relationship between the shape of the mental number line and familiarity with numbers in 5- to 9-year old children: Evidence for a segmented linear model. *Journal of Experimental Child Psychology*, *99*, 1–17. https://doi.org/10.1016/j.jecp.2007.08.006.

Ebersbach, M., Luwel, K., & Verschaffel, L. (2015). The relationship between children's familiarity with numbers and their performance in bounded and unbounded number line estimations. *Mathematical Thinking and Learning*, *17*, 136–154. https://doi.org/10.1080/10986065.2015.1016813.

Frick, A. (2018). *Mental transformation abilities and their relation to later academic achievement* (Submitted for publication).

Frick, A., & Newcombe, N. S. (2012). Getting the big picture: Development of spatial scaling abilities. *Cognitive Development, 27*, 270–282.

Fujimura, N. (2001). Facilitating children's proportional reasoning: A model of reasoning processes and effects of intervention on strategy change. *Journal of Educational Psychology, 93*, 589–603.

Gallistel, C. R., & Gelman, R. (1992). Preverbal and verbal counting and computation. *Cognition, 44*, 43–74.

Gao, F., Levine, S. C., & Huttenlocher, J. (2000). What do infants know about continuous quantity? *Journal of Experimental Child Psychology, 77*, 20–29. https://doi.org/10.1006/jecp.1999.2556.

Gelman, R. (1991). Epigenetic foundations of knowledge structures: Initial and transcendent constructions. In S. Carey, & R. Gelman (Eds.), *Epigenesis of mind: Essays on biology and cognition.* Hillsdale, NJ: Erlbaum.

Grissmer, D., Mashburn, A., Cottone, E., Brock, L., Murrah, W., Blodgett, J., et al. (September 26–28, 2013). The efficacy of minds in motion on children's development of executive function, visuo-spatial and math skills. In *Paper presented at the society for research in educational effectiveness conference, Washington, DC.*

Gunderson, E. A., Ramirez, G., Beilock, S. L., & Levine, S. C. (2012). The relation between spatial skill and early number knowledge: The role of the linear number line. *Developmental Psychology, 48*, 1229–1241. https://doi.org/10.10 37/a0027433.

Hawes, Z., Moss, J., Caswell, B., Naqvi, S., & MacKinnon, S. (2018). Enhancing children's spatial and numerical skills through a dynamic spatial approach to early geometry instruction: Effects of a seven-month intervention. *Cognition and Instruction, 35*, 236–264.

Hawes, Z., Moss, J., Caswell, B., & Poliszczuk, D. (2015). Effects of mental rotation training on children's spatial and mathematics performance: A randomized controlled study. *Trends in Neuroscience and Education, 4*, 60–68.

Hecht, S. A., & Vagi, K. J. (2010). Sources of group and individual differences in emerging fraction skills. *Journal of Educational Psychology, 102*, 843–858.

Holden, M. P., Newcombe, N. S., & Shipley, T. F. (2013). Location memory in the real world: Category adjustment effects in 3-dimensional space. *Cognition, 128*, 45–55.

Hurst, M., Monahan, K. L., Heller, E., & Cordes, S. (2014). 123s and ABCs: Developmental shifts in logarithmic-to-linear responding reflect fluency with sequence values. *Developmental Science, 17*, 892–904. https://doi.org/10.1111/desc.12165.

Huttenlocher, J., Duffy, S., & Levine, S. (2002). Infants and toddlers discriminate amount: Are they measuring? *Psychological Science, 13*, 244–249. https://doi.org/10.1111/1467-9280.00445.

Huttenlocher, J., Hedges, L. V., & Duncan, S. (1991). Categories and particulars: Prototype effects in estimating spatial location. *Psychological Review, 98*, 352–376.

Huttenlocher, J., Levine, S. C., & Ratliff, K. R. (2011). The development of measurement: From holistic perceptual comparison to unit understanding. In N. L. Stein, & S. Raudenbush (Eds.), *Developmental science goes to school: Implications for education and public policy research.* New York: Taylor and Francis.

Huttenlocher, J., Newcombe, N. S., & Sandberg, E. (1994). The coding of spatial location in young children. *Cognitive Psychology, 27*, 115–147.

Huttenlocher, J., Newcombe, N. S., & Vasilyeva, M. (1999). Spatial scaling in young children. *Psychological Science, 10*, 393–398.

Huttenlocher, J., Vasilyeva, M., Newcombe, N. S., & Duffy, S. (2008). Developing symbolic capacity one step at a time. *Cognition, 106,* 1–12.

Jäger, S., & Wilkening, F. (2001). Development of cognitive averaging: When light and light make dark. *Journal of Experimental Child Psychology, 79,* 323–345.

Jeong, Y., Levine, S. C., & Huttenlocher, J. (2007). The development of proportional reasoning: Effect of continuous vs. discrete quantities. *Journal of Cognition and Development, 8,* 237–256.

Kosslyn, S. M. (1975). Information representation in visual images. *Cognitive Psychology, 7,* 341–370.

Landy, D., Silbert, N., & Goldin, A. (2013). Estimating large numbers. *Cognitive Science, 37,* 775–799.

Laski, E. V., Casey, B. M., Yu, Q., Dulaney, A., Heyman, M., & Dearing, E. (2013). Spatial skills as a predictor of first grade girls' use of higher level arithmetic strategies. *Learning and Individual Differences, 23,* 123–130.

Lauer, J. E., & Lourenco, S. F. (2016). Spatial processing in infancy predicts both spatial and mathematical aptitude in childhood. *Psychological Science, 27,* 1291–1298.

LeFevre, J. A., Lira, C. J., Sowinski, C., Cankaya, O., Kamawar, D., & Skwarchuk, S. L. (2013). Charting the role of the number line in mathematical development. *Frontiers in Psychology, 4,* 641.

Leibovich, T., Katzin, N., Harel, M., & Henik, A. (2016). From 'sense of number' to 'sense of magnitude' – The role of continuous magnitudes in numerical cognition. *Behavioral and Brain Sciences, 17,* 1–62.

Liben, L. S., Moore, M. L., & Golbeck, S. L. (1982). Preschoolers' knowledge of their classroom environment: Evidence from small-scale and life-size spatial tasks. *Child Development, 53,* 1275–1284.

Liben, L. S., & Yekel, C. A. (1996). Preschoolers' understanding of plan and oblique maps: The role of geometric and representational correspondence. *Child Development, 67,* 2780–2796.

Link, T., Huber, S., Nuerk, H.-C., & Moeller, K. (2014). Unbounding the mental number line – New evidence on children's spatial representation of numbers. *Frontiers in Psychology, 4,* 1021. https://doi.org/10.3389/fpsyg.2013.01021.

Lourenco, S. F., & Longo, M. R. (2011). Origins and the development of generalized magnitude representation. In S. Dehaene, & E. Brannon (Eds.), *Space, time, and number in the brain: Searching for the foundations of mathematical thought* (pp. 225–244). Amsterdam, NL: Elsevier.

Lowrie, T., Logan, T., & Ramful, A. (2017). Visuospatial training improves elementary students' mathematics performance. *British Journal of Educational Psychology, 87,* 170–186.

McCrink, K., & Opfer, J. E. (2014). Development of spatial-numerical associations. *Current Directions in Psychological Science, 23,* 439–445.

Mix, K. S., Huttenlocher, J., & Levine, S. C. (2002). Multiple cues for quantification in infancy: Is number one of them? *Psychological Bulletin, 128,* 278–294.

Mix, K. S., Levine, S. C., & Huttenlocher, J. (1999). Early fraction calculation ability. *Developmental Psychology, 35,* 164–174.

Mix, K. S., Levine, S. C., & Newcombe, N. S. (2016). Development of quantitative thinking across correlated dimensions. In A. Henik (Ed.), *Continuous issues in numerical cognition: How many or how much* (pp. 1–35). Elsevier.

Moeller, K., Pixner, S., Kaufmann, L., & Nuerk, H.-C. (2009). Children's early mental number line: Logarithmic or decomposed linear? *Journal of Experimental Child Psychology, 103,* 300–516.

Möhring, W., Newcombe, N. S., & Frick, A. (2014). Zooming in on spatial scaling: Preschool children and adults use mental transformations to scale spaces. *Developmental Psychology, 50,* 1614–1619.

Möhring, W., Newcombe, N. S., & Frick, A. (2015). The relation between spatial thinking and proportional reasoning in preschoolers. *Journal of Experimental Child Psychology, 132,* 213–220.

Möhring, W., Newcombe, N. S., & Frick, A. (2016). Adults use mental transformation strategies for spatial scaling: Evidence from a discrimination task. *Journal of Experimental Psychology: Learning, Memory, and Cognition, 42,* 1473–1479.

Möhring, W., Newcombe, N. S., Levine, S. C., & Frick, A. (2016). Spatial proportional reasoning is associated with formal knowledge about fractions. *Journal of Cognition and Development, 17,* 67–84.

Moore, C. F., Dixon, J. A., & Haines, B. A. (1991). Components of understanding in proportional reasoning: A fuzzy set representation of developmental progressions. *Child Development, 62,* 441–459.

Moss, J., Bruce, C. D., Caswell, B., Flynn, T., & Hawes, Z. (2016). *Taking shape: Activities to develop geometric and spatial thinking.* Toronto: Pearson School Canada.

Newcombe, N., Huttenlocher, J., & Learmonth, A. (1999). Infants' coding of location in continuous space. *Infant Behavior and Development, 22,* 483–510.

Newcombe, N. S., Levine, S. C., & Mix, K. S. (2015). Thinking about quantity: The intertwined development of spatial and numerical cognition. *WIREs Cognitive Science, 6,* 491–505.

Newcombe, N. S., Sluzenski, J., & Huttenlocher, J. (2005). Preexisting knowledge versus on-line learning: What do young infants really know about spatial location? *Psychological Science, 16,* 222–227.

Noelting, G. (1980). The development of proportional reasoning and the ratio concept, Part I. Differentiation of stages. *Educational Studies in Mathematics, 11,* 217–254.

Opfer, J. E., & Siegler, R. S. (2007). Representational change and children's numerical estimation. *Cognitive Psychology, 55,* 169–195.

Paik, J. H., & Mix, K. S. (2003). U.S. and Korean children's comprehension of fraction names: A re-examination of cross-national differences. *Child Development, 74,* 144–154.

Petitto, A. (1990). Development of numberline and measurement concepts. *Cognition and Instruction, 7,* 55–78. https://doi.org/10.1207/s1532690xci0701_3.

Piaget, J. (1952). *The origins of intelligence in children* (M. Cook, Trans.). New York: International Universities Press (Original work published in 1936).

Piaget, J., & Inhelder, B. (1975). *The origin of the idea of chance in children.* New York: Norton (Original work published in 1951).

Resnick, L. B., & Ford, W. W. (1981). *The psychology of mathematics for instruction.* Hillsdale, NJ: Erlbaum.

Reuhkala, M. (2001). Mathematical skills in ninth-graders: Relationship with visuo-spatial abilities and working memory. *Educational Psychology, 21,* 387–399.

Rips, L. J. (2013). How many is a zillion? Sources of number distortion. *Journal of Experimental Psychology: Learning, Memory, and Cognition, 39*, 1257–1264.

Schneider, M., Heine, A., Thaler, V., Torbeyns, J., de Smedt, B., Verschaffel, L., et al. (2008). A validation of eye movements as a measure of elementary school children's developing number sense. *Cognitive Development, 23*(3), 409–422. https://doi.org/10.1016/j.cogdev.2008.07.002.

Schneider, M., & Siegler, R. S. (2010). Representations of the magnitudes of fractions. *Journal of Experimental Psychology: Human Perception and Performance, 36*, 1227–1238.

Shepard, R. N., & Metzler, J. (1971). Mental rotation of three-dimensional objects. *Science, 171*, 701–703.

Siegler, R. S., Duncan, G. J., Davis-Kean, P. E., Duckworth, K., Claessens, A., Engel, M., et al. (2012). Early predictors of high school mathematics achievement. *Psychological Science, 23*, 691–697.

Siegler, R. S., & Opfer, J. E. (2003). The development of numerical estimation: Evidence for multiple representations of numerical quantity. *Psychological Science, 14*, 237–243. https://doi.org/10.1111/1467-9280.02438.

Siegler, R. S., Thompson, C. A., & Schneider, M. (2011). An integrated theory of whole number and fractions development. *Cognitive Psychology, 62*, 273–296.

Slusser, E. B., Santiago, R. T., & Barth, H. C. (2013). Developmental change in numerical estimation. *Journal of Experimental Psychology: General, 142*, 193–208. https://doi.org/10.1037/a0028560.

Spelke, E. S., & Kinzler, K. D. (2007). Core knowledge. *Developmental Science, 10*, 89–96.

Stafylidou, S., & Vosniadou, S. (2004). The development of students' understanding of the numerical value of fractions. *Learning and Instruction, 14*, 503–518.

Thompson, C. A., & Opfer, J. E. (2010). How 15 hundred is like 15 cherries: Effect of progressive alignment on representational changes in numerical cognition. *Child Development, 81*, 1768–1786.

Uttal, D. H., Meadow, N. G., Tipton, E., Hand, L. L., Alden, A. R., Warren, C., et al. (2013). The malleability of spatial skills: A meta-analysis of training studies. *Psychological Bulletin, 139*(2), 352–402.

Uttal, D. H., Sandstrom, L. B., & Newcombe, N. S. (2006). One hidden object, two spatial codes: Young children's use of relational and vector coding. *Journal of Cognition and Development, 7*, 503–525.

Vasilyeva, M., Duffy, S., & Huttenlocher, J. (2007). Developmental changes in the use of absolute and relative information: The case of spatial extent. *Journal of Cognition and Development, 8*, 455–471. https://doi.org/10.1080/15248370701612985.

Vasilyeva, M., & Huttenlocher, J. (2004). Early development of scaling ability. *Developmental Psychology, 40*(5), 682–690.

Verdine, B. N., Golinkoff, R. M., Hirsh-Pasek, K., & Newcombe, N. S. (2017). Links between spatial and mathematical skills across the preschool years. *Monographs of the Society for Research in Child Development, 82*, 1–150. https://doi.org/10.1111/mono.12280.

Verdine, B. N., Golinkoff, R. M., Hirsh-Pasek, K., Newcombe, N. S., Filipowicz, A. T., & Chang, A. (2014). Deconstructing building blocks: Preschoolers' spatial assembly performance relates to early mathematical skills. *Child Development, 85*, 1062–1076.

Walsh, V. (2003). A theory of magnitude: Common cortical metrics of time, space, and quantity. *Trends in Cognitive Sciences*, 7, 483–488.

Wynn, K. (1995). Origins of numerical knowledge. *Mathematical Cognition*, 1, 35–60.

Wynn, K. (1997). Competence models of numerical development. *Cognitive Development*, 12, 333–339.

Xu, C., & LeFevre, J.-A. (2016). Training young children on sequential relations among numbers and spatial decomposition: Differential transfer to number line and mental transformation tasks. *Developmental Psychology*, 52, 854–866.

Is Visuospatial Reasoning Related to Early Mathematical Development? A Critical Review

Stella F. Lourenco, Chi-Ngai Cheung, Lauren S. Aulet

Emory University, Atlanta, GA, United States

Heterogeneity of Function in Numerical Cognition
https://doi.org/10.1016/B978-0-12-811529-9.00010-8

INTRODUCTION

Visuospatial reasoning—broadly defined as the ability to generate, retain, and transform visual images—is relevant to many of our daily activities, such as assembling furniture, packing the dishwasher or trunks of our cars, interpreting charts and graphs, and projecting the shortest route to a destination (if our GPS is down, of course!). Visuospatial reasoning not only ensures success in everyday functioning but it is also crucial to specialized and more formal computations typical in professions such as architecture, engineering, geoscience, and medicine (Ackerman & Cianciolo, 2002; Hegarty & Waller, 2005; Kozhevnikov, Motes, & Hegarty, 2007; Lohman, 1996). Indeed, the memoirs and introspections of scientists document their reliance on visuospatial reasoning for many of their breakthroughs. For example, Albert Einstein claimed to achieve insights by means of thought experiments into visualized systems of waves and physical bodies in states of relative motion (Lohman, 1996). Well-known discoveries in the history of science benefitting from visuospatial reasoning include Einstein's theory of relativity, Copernicus' heliocentric model of the solar system, and the discovery of the double helix structure of DNA by Watson and Crick (National Research Council (US), Committee on Support for Thinking Spatially, & Downs, 2006). On the flip slide, there are grave errors of visuospatial reasoning, which also emphasize its importance in scientific thinking and pragmatic implications. Among the best known of these examples occurred in the late 1950s when the enantiomer of the drug thalidomide was prescribed to pregnant women for morning sickness with the result being thousands of children born with limb deformities (Fabro, Smith, & Williams, 1967).

Accumulating evidence from longitudinal studies suggests that individual differences in visuospatial reasoning predict school success and professional outcomes related to science, technology, engineering, and mathematics (STEM). For example, high-school students who enter and succeed in STEM careers have appreciably better visuospatial skills than those who go on to non-STEM careers, even when accounting for their verbal and mathematical skills (Wai, Lubinski, & Benbow, 2009). It has also been found that visuospatial reasoning predicts the number of patents and refereed publications over a 30-year period (Kell, Lubinski, Benbow, & Steiger, 2013). Recent research discussed later in this chapter suggests that the predictive value of visuospatial reasoning begins early in development (e.g., Verdine, Golinkoff, Hirsh-Pasek, & Newcombe, 2017), perhaps even in infancy (Lauer & Lourenco, 2016). Findings of such early predictive power dovetail with claims that visuospatial reasoning may serve as a "gatekeeper" to success in STEM. Uttal and Cohen (2012) suggested that visuospatial reasoning in the STEM fields may be especially critical early in learning when the material is particularly challenging and when content knowledge would not allow one to circumvent poor visuo-spatial reasoning.

The idea that mathematical knowledge encompasses spatial understanding is uncontroversial in some circles (e.g., Bishop, 2008; Lohman, 1996). For example, educators who teach geometry likely find statements of the importance of spatial understanding in mathematics trivial, as geometry is considered inherently spatial. In geometry, the composition and decomposition of two-dimensional (2-D) and three-dimensional (3-D) figures involve visuospatial strategies (Clements & Sarama, 2011). More controversial among mathematicians, educators, and psychologists is whether other mathematical domains, including early numerical knowledge, benefit from spatial reasoning such as visualization. More specifically, we can ask whether visualizations such as the mental rotation of objects or mentally representing numbers along a line (i.e., the mental number line) afford any advantages in the acquisition of quantitative concepts and arithmetic calculations.

CHAPTER OVERVIEW

There has been increasing interest in the relation between spatial and mathematical domains, particularly the role that visuospatial reasoning might play in shaping the acquisition and development of mathematical concepts and problem solving. As discussed by Mix and Cheng (2011), there is a long history of research suggesting that spatial and mathematical domains are linked in the mind and brain. In particular, numerous correlational studies indicate that being better in one type of reasoning means being better in the other. There have also been studies employing factor analytic techniques to, again, point to links between the two domains of reasoning (for a recent example, see Mix et al., 2016). Fewer studies have actively manipulated competence in one domain to determine the impact on the other domain, except for recent exceptions, which we discuss in subsequent sections in this chapter. A growing consensus among researchers and educators is that visuospatial reasoning plays a foundational role in math achievement. Thus far, however, the evidence is far from definitive. There are many open questions about *why* the two domains might be related and *how* visuospatial reasoning might facilitate numerical development and math reasoning—questions we take up in this chapter.

Spatial and mathematical domains are not monolithic. Thus, a question that naturally follows is whether relations between spatial and mathematical domains depend on the *types* of reasoning within these domains. Behavioral and neural data suggest that neither the spatial nor mathematical domain is a unitary construct. For example, there is evidence that tests of mental rotation, in which individuals distinguish objects presented from different orientations, do not correlate with performance on navigation tasks, in which individuals are required to plan routes either in the real world or with maps (e.g., Hegarty, Montello, Richardson, Ishikawa, & Lovelace, 2006), despite widespread agreement that mental rotation and navigation both fall within the spatial domain (for other examples, see Miyake, Friedman, Rettinger, Shah, & Hegarty, 2001; Pellegrino, Hunt, Abate, & Farr, 1987). Likewise, within the mathematical domain, there is evidence that different types of math problems, such as geometry and arithmetic, recruit at least partially dissociable systems (e.g., Qin et al., 2014; Zago et al., 2001). For these reasons, questions about the relations between spatial and mathematical reasoning will be affected by the types of spatial and mathematical reasoning. In the present chapter, we take on this issue by focusing specifically on spatial visualization and transformation, particularly as it involves the ability to mentally manipulate (e.g., rotate) objects or to mentally organize numbers. We focus on these visuospatial processes because the majority of the work on the links to mathematics has tested these specific processes. Far less research has been conducted on whether navigation ability, for example, is related to mathematics. Because of our interest in the ontogenetic origins and early development of the links between spatial and mathematical domains, there are necessarily limits on the types of mathematical abilities that can be examined. This research is limited to early quantitative concepts and mathematical computations such as basic arithmetic. Thus, in the present chapter, we review evidence of whether visuospatial reasoning is related to children's early math achievement.

As part of our discussion, we also address questions concerning the *specificity* of the relation between spatial and mathematical domains. There are open questions about whether the association between spatial and mathematical reasoning reflects a unique relation between the two domains or whether it reflects general competencies, such as working memory, executive attention, or inhibitory control, shared across domains. Understanding the extent to which other systems may be implicated in the association between spatial and mathematical domains will be crucial to isolating the mechanisms that support the potential causal links between domains, as well as the development of these mechanisms.

Indeed, there are open questions about whether visuospatial processing is *causally* related to mathematical reasoning. More specifically, we can ask whether better visuospatial reasoning leads to better mathematical reasoning. As noted earlier, accumulating evidence suggests associations between these two domains of reasoning, but such associations are insufficient for making claims about causality. One possibility is that visuospatial reasoning impacts mathematical development, but another possibility is that the causal direction goes in the reverse direction such that greater exposure to mathematics is what impacts visuospatial reasoning.

Although recent intervention studies have begun to test the extent to which trained visuospatial skills affect mathematical performance, there are no studies (to our knowledge) that test the opposite outcome. Moreover, there is a growing need for studies with cross-lagged designs that also address the causal link by comparing whether visuospatial reasoning is more likely to predict later mathematical performance than vice versa. When available, we review these studies in the upcoming sections.

In many cases, the links between spatial and mathematical domains begin early in life. In many cases, these links are also observed in adulthood. Do the links across the two domains change over development? Are they strengthened or weakened? Or is there general continuity across time? There are important questions about *development* in this literature, and in this chapter, we discuss work conducted in children of different ages in an effort to shed light on the nature of the links between visuospatial and mathematical reasoning, particularly the mechanisms underlying these links and the potential causal relations.

MENTAL ROTATION AND VISUALIZATION IN 2-D AND 3-D ACTIVITIES

As noted earlier, there is much evidence to suggest that visuospatial ability, such as the mental rotation of objects, relates to mathematical competence. Numerous studies, many beginning with the work on intelligence and individual differences, show that people who perform better on tests of mental rotation also do better in math classes and standardized math tests (for review, see Mix & Cheng, 2011). In this section, we specifically ask about the developmental origins of these relations by discussing studies conducted with children to test the extent to which their mental rotation abilities influence their later mathematical development. We begin by discussing longitudinal studies that show that visuospatial reasoning, such as mental rotation measured using classic mental rotation tests and block building and puzzle manipulations, each of which may include active mental transformations, are especially good predictors of mathematical competence years later. For example, earlier work by Wolfgang, Stannard, and Jones (2001) found that the quality of 5-year-olds' block play predicted the number of math courses taken and their math performance 10 years later.

Longitudinal Studies

Three studies by Verdine et al. (2014), Verdine, Irwin, Golinkoff & Hirsh-Pasek, 2014, and Verdine et al. (2017) assessed 3-year-olds' visuospatial reasoning using the test of spatial assembly (TOSA), which requires reproducing target models of either 2-D shapes or 3-D block constructions (see Fig. 10.1). Children were given additional measures of mental rotation 1–2 years after the initial assessment (i.e., the Children's Mental Transformation Task or "CMTT" Levine, Huttenlocher, Taylor, & Langrock, 1999, see Fig. 10.1; Spatial Relations subtest of the Woodcock Johnson Tests of Cognitive Abilities Woodcock, Mather, McCrew, & Schrank, 2001), and their math knowledge was tested with age appropriate measures (i.e., Number Sense test Jordan, Glutting, Ramineni, & Watkins, 2010; Math Problem Solving subtest of the Wechsler Individual Achievement Test—Third Ed. Wechsler, 2009). Verdine et al. found that not only did children's visuospatial reasoning correlate with their math knowledge when these abilities were measured concurrently (Verdine, Golinkoff et al., 2014, 2017) but they also found that 3-year-olds' performance on the TOSA was a significant predictor of their math knowledge at age 5. Perhaps not surprisingly, it was also found that visuospatial skills at ages 3 and 5 were correlated, which, importantly, suggests that the relation between children's visuospatial and math reasoning may reflect their developing visuospatial skills. More surprising was that 3-year-olds' visuospatial ability was a better predictor of math performance at age 5 than was math performance at age 3, suggesting a more foundational role for visuospatial reasoning than early math competence in predicting children's mathematical development.

Other work by Gunderson, Ramirez, Beilock, and Levine (2012) suggests a role for a linear mental number line (see Mental Number Line section) in the longitudinal relation between children's mental rotation abilities and their arithmetic understanding. Gunderson et al. (2012) assessed mental rotation ability at age 5 using the CMTT (Levine et al., 1999; see Fig. 10.1). They also assessed the extent to which children's number line estimates were linear by having children mark the position of numbers on physical number lines, labeled 0 at the left and 100 at the right. Arithmetic in this study was assessed using approximate symbolic and nonsymbolic calculation tasks (adapted from Barth et al., 2006; Gilmore, McCarthy, & Spelke, 2007). Gunderson et al. (2012) found that mental rotation ability at age 5 predicted arithmetic competence at age 8, when assessed with symbolic calculation. Moreover, they found that there was a mediating role for a linear mental number line at 6 years of age. The authors proposed that better visuospatial ability early in development may lead to more linear spatial representations of number, which, in turn, exert a positive influence on math development, particularly computations with symbolic values (because of the greater precision afforded by linear representations). These findings are intriguing in that they raise the possibility of intervening variables in the relation between mental rotation and mathematics (but see LeFevre et al., 2013). However, more work will be needed to address the mechanistic question of how a linear mental number line mediates the

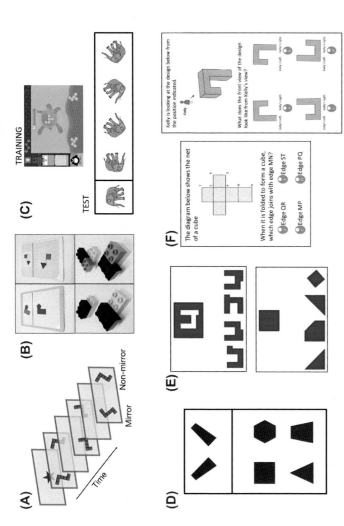

FIGURE 10.1 A sample of the different visuospatial tasks utilized in longitudinal and training studies with children. (A) An illustration of the looking time paradigm in the longitudinal study of Lauer and Lourenco (2016) developed to assess mental rotation in infants. (B) An illustration of the TOSA (Test of Spatial Assembly) developed by Verdine et al. (2017) to assess preschoolers' understanding of spatial relations and visualization with 2-D displays and 3-D constructions. (C) One of the training and test measures utilized in the intervention study of Hawes, Moss, Caswell, and Poliszczuk (2015). The mental rotation training task depicted here required the matching of 2-D shapes from different orientations. The test measure has been used previously to assess mental rotation with children (see Neuburger, Jansen, Heil, & Quaiser-Pohl, 2011). (D) An item from the CMTT (Children's Mental Transformation Task, Levine et al., 1999) has been utilized in both longitudinal (e.g., Gunderson et al., 2012; Lauer & Lourenco, 2016; Verdine et al., 2017) and training (e.g., Cheng & Mix, 2014; Hawes et al., 2015) studies. In this example, the two pieces on top combine to form one of the shapes below. (E) Items from the geometry assessment utilized in the training study of Hawes, Moss, Caswell, Naqvi, and MacKinnon (2017) (see also Hawes et al., 2015). In the top example, children were required to identify the item that fits perfectly within the white center space. In the bottom example, children were required to identify the two shapes that could be put together to make the target shape. Both examples may require mental rotation. (F) Items from the Spatial Reasoning Instrument (Ramful, Lowrie, & Logan, 2017) utilized in the training study of Lowrie, Logan, and Ramful (2017). The left item assesses visualization in the context of folding and the right item assesses perspective taking (referred to as "spatial orientation" on this test).

relation between mental rotation and math development. We discuss candidate mechanisms in Why Would Visuospatial Reasoning Benefit Math Development? Candidate Mechanisms section.

More recent research suggests that the predictive value of visuospatial reasoning on math competence may begin in the infant years. More specifically, Lauer and Lourenco (2016) found that individual differences in infancy predicted the performance of 4-year-olds on a test of mathematical ability (Applied Problems subtest of WJ Test of Achievement Woodcock, McGrew, & Mather, 2001). In this study, children were first tested between 6 and 13 months of age on a measure of visuospatial reasoning. Infants saw two simultaneous streams of images on the left and right sides of the screen. Both streams depicted the same Tetris-like figures in different orientations, but only one of the streams presented mirror images of the figure (see Fig. 10.1). Infants' preferential looking toward the stream with mirror images was used as the measure of their mental rotation ability. Later, at 4 years of age, children were given tests of spatial and mathematical reasoning, as well as controls of general cognitive functioning (e.g., verbal working memory and verbal competence). It was found that the infant measure predicted later mental rotation and math competence, and importantly, these effects could not be fully explained by general cognitive functioning. However, this work leaves open questions about potential variables that might mediate the relation between visuospatial processing in infancy and math competence at 4 years of age. This work also leaves open the question of whether early mental rotation ability is a better predictor of later math achievement than early numerical abilities, which researchers have also found predict later math achievement (Starr, Libertus, & Brannon, 2013).

Taken together, these studies demonstrate that the ability to engage in spatial visualization, which includes mental transformation of objects, and which may be implicated when manipulating 3-D block constructions and solving puzzles, is related to mathematical reasoning, both early and later in development. Longitudinal work has been crucial in establishing the developmental origins and continuity of this link. There is evidence that visuospatial reasoning predicts later math using assessments at ages 3 (Verdine, Golinkoff, et al., 2014; Verdine, Irwin, et al., 2014; Verdine et al., 2017) and 5 years (Gunderson et al., 2012; Wolfgang et al., 2001), as well as in infancy (Lauer & Lourenco, 2016). Although work with infants suggests that both visuospatial (Lauer & Lourenco, 2016) and numerical (Starr et al., 2013) abilities predict later math competence, it is unclear whether both visuospatial and numerical abilities uniquely contribute to math competence later in life. Work by Verdine et al. (2017) suggests that visuospatial skills may be an especially powerful predictor of mathematical development, raising important questions about exactly why this might be and whether this is the case across development.

The extant longitudinal relations are consistent with visuospatial reasoning as an important precursor to mathematical development, but crucial cross-lagged studies are not always available to test the reverse possibility, which is whether early math reasoning is predictive of individual differences in later visuospatial reasoning. More direct tests of the relation between spatial and math understanding, particularly the causal role of visuospatial reasoning in math competence, come from intervention studies in which researchers provide training to children either in the lab or classroom on a specific or range of spatial skills and then test for generalization to math performance. Although such studies have not always been common, the number of experiments taking this question on has increased in recent years, perhaps not surprisingly given the theoretical and practical importance. Several studies have now tested whether training visuospatial reasoning leads to better math performance following training. We discuss these studies next.

Training Studies

To our knowledge, the first training study to demonstrate the effects of spatial training on mathematics in children was performed by Cheng and Mix (2014). In this study, 58 children aged 6–8 years were assigned to either a spatial (mental rotation) training group or a nonspatial control group, in which they received experience with crossword puzzles. Children in the spatial training group were trained on the CMTT (see Fig. 10.1 Levine et al., 1999), which involved asking them to first visualize the solution to each problem and then confirming the accuracy of their response by putting together cardboard pieces of the shapes. In both groups, the interventions were single sessions that lasted 40 min. Immediately before and after the interventions, children completed two visuospatial tasks: the CMTT and the Spatial Relations subtest of Primary Mental Abilities test (Thurstone, 1974). The math test given to children at these two time points included two- and three-digit calculation problems (e.g., $56 + 6 = \_$), as well as missing term problems (e.g., $4 + \_ = 12$). Cheng and Mix (2014) found that children in the spatial training group, but not the control group, improved in their ability to solve both types of math problems, but with greatest improvement on the missing term problems. The authors posited that the spatial intervention may have encouraged reorganization of the problems. More specifically, children may have mentally transformed the missing term problems into the canonical format (e.g., $5 + \_ = 9$ becomes $9 - 5 = \_$). However, a surprising finding in this study, given the authors' interpretation, was that spatial training did not lead to improvement on the Spatial Relations test, the other measure of mental rotation. Thus, although a promising first study on the potential causal relation between mental rotation ability and math performance, more work was needed to address why the spatial training effect occurred for math but not another mental rotation test. Moreover, although Cheng and Mix (2014) suggested that mental rotation training might have facilitated children's performance on the missing terms problems by promoting mental transformations of these problems, there are other more general possibilities that remain to be tested. For example, perhaps mental rotation training improves working memory, which is known to be related to math performance (e.g., Friso-van den Bos, vand de Ven, Kroesbergen & van Luit, 2013; Kyttälä, Aunio, Lehto, Van Luit, & Hautamäki, 2003; Li & Geary, 2017). An important design consideration for future research will be to include control measures, such as tests of working memory, to better understand the nature of the training effects.

Another study by Hawes et al. (2015) failed to replicate the training effect reported by Cheng and Mix (2014). Hawes et al. (2015) similarly tested 6–8-year-old children. Children were randomly assigned to either a mental rotation ($n=32$) or literacy training ($n=29$) group. Training was computerized in both cases and took place over a 6-week period as part of children's regular classroom activities. The mental rotation training consisted of three different computer games—two of which involved matching rotated images and avoiding mirror images (see Fig. 10.1) and the third of which was a puzzle task that involved completing a target shape by putting together component pieces. The literacy training included a collection of language tasks, such as spelling, object naming, sentence construction, and crossword puzzles. Results revealed that children who received spatial training showed significant gains on two measures of 2-D mental rotation (see Fig. 10.1; Neuburger et al., 2011) and marginally significant improvement on another (CMTT, see Fig. 10.1; Levine et al., 1999). However, there was no improvement on any of the math tests including the missing terms problems used by Cheng and Mix (2014).

What might account for the differential findings across these two studies? One possibility is that Cheng and Mix (2014) used a "hands-on" intervention, not a computerized visuospatial training. Perhaps a hands-on manipulation is better for learning. There is certainly evidence suggesting that manipulatives enhance learning (McNeil & Uttal, 2009). However, this explanation likely does not fully account for the differential findings because the computerized spatial training in Hawes et al. (2015) led to improved visuospatial reasoning on multiple measures. A second possibility may be related to the length of the interventions. Hawes et al. (2015) provided children with 6 weeks of training rather than a single session. However, it is unclear why shorter training, as in Cheng and Mix (2014), would result in better math performance. A third possibility concerns the timing of the posttests. Hawes et al. (2015) gave children these tests 3–6 days after the last training session, whereas Cheng and Mix (2014) tested children during the same session as training. Thus, a possibility is that space-to-math transfer effects do exist, but they dissipate over time, as do other training effects (Protzko, 2015). If this turns out to be true, then it will be up to future studies to determine the nature and impact of these short-lived training and generalization effects. Future research would do well to compare the space-to-math effects with other training effects that generalize across domain. An interesting possibility is that additional practice (i.e., "booster") sessions are needed to ensure maintenance over the long term.

We describe three final studies—all of which were published in the last year. Like Hawes et al. (2015) these studies included multiple sessions of intervention and they incorporated varied training approaches. Although we agree with the authors of these studies that such approaches may be more ecologically valid than those that focus on a specific visuospatial measure, we note that there are challenges with interpreting the findings, given the broad training provided to children.

One such study was conducted by Hawes et al. (2017) and, unlike their previous study, they found some evidence of transfer from spatial training to math competence. In this quasi-experimental study, children from three schools serving lower SES environments were assigned to experimental or control groups. The two lowest performing schools were assigned to the experimental group, whereas the remaining school was assigned to the control group. Both groups of students (kindergarten to second grade: ages 4–7 years) received 32 weeks of teacher-led classroom interventions (Moss, Hawes, Naqvi, & Caswell, 2015). The experimental group ($n=39$) received five geometry lessons (approx. 1 h each) and participated in a series of quick challenge activities (10–15 min each), targeting geometry and spatial skills such as mental transformation (e.g., rotation), spatial memory, proportional reasoning, and spatial language. The control group ($n=28$) received an intervention in environmental science. Pre- and post-test assessments were administered immediately before and after the intervention. The spatial intervention was successful in improving students' spatial reasoning—the experimental group showed greater gains than the control group on all three spatial measures in this study (i.e., CMTT, Spatial Language test, and Geometry test; see Fig. 10.1) following the intervention. There was also evidence of transfer to mathematical knowledge—the experimental group performed better on a number comparison task. However, not all of the math measures, including a test of number knowledge, showed improvement following spatial training. Thus, we suggest some caution in interpreting the space-to-math transfer effect. First, although this would seem to suggest a causal relation between visuospatial and mathematical reasoning, the effect could be due to improvements in other abilities such as working memory, executive attention, or effortful control. As noted earlier, most training studies have not included measures that assess such abilities, leaving such a possibility untested.

Second, children's improvement on the number comparison task could have been due to preexisting differences between experimental and control groups, which the authors acknowledged. Compared with the control group, more children in the experimental group failed the number comparison task at pretest, such that improvement could be explained by regression toward the mean. Third, also acknowledged by the authors was that the spatial training program included reference to numerical concepts. Thus, improvement on the number comparison task could have been due to experience with numbers rather than pure visuospatial training.

Another study by Lowrie et al. (2017) also found transfer to a math measure. In their study, children were given a 10-week intervention program that involved 2h of training per week, which was administered by school teachers in the classroom. Children in this study were sixth graders from Australia (10–12 years): 120 in the spatial training group and 66 in the control group. Spatial training focused on three constructs proposed by Linn and Petersen (1985) and Kozhevniko, Hegarty, and Mayer (1999) to represent the spatial domain—namely, mental rotation (2-D and 3-D rotation), spatial visualization (e.g., paper folding), and orientation (e.g., drawing and navigating maps). The control group received their typical math instruction. Children's spatial abilities were assessed using the Spatial Reasoning Instrument (see Fig. 10.1; Ramful et al., 2017), which measures the three spatial constructs described in training. Children's math ability was assessed with the MathT test (consisting of items from Australia's National Assessment Program), in which half of the questions tested number concepts and the other half tested geometry. Pre- and posttests were administered within 2 weeks of the intervention. It was found that children in the spatial training group performed better on the Spatial Reasoning Instrument than the control group, though improvement was not observed on all spatial measures—more specifically, children showed improvement on tests of mental rotation and spatial visualization but not orientation. Crucially, the spatial intervention group showed significantly better performance on the math measure, unlike the control group. However, it is unclear whether spatial training was equally effective for the different types of questions on the MathT test, which contained both items that assessed numerical concepts and geometry. This is an important question because the geometry items may have overlapped with items from the spatial tests. It is also worth noting that other research, namely the study conducted by Hawes et al. (2017), has referred to geometry as a spatial measure, so it is unclear what type of mathematical reasoning the spatial training may have generalized to.

The final recent study was conducted by Cornu, Schiltz, Pazouki, and Martin (2017). They conducted a 10-week intervention with kindergarteners from Luxembourg but found no transfer of spatial training to math performance. In this study, children were again given training on a variety of visuospatial tasks, including copying shapes and drawings, line and figure bisection and rotation, tangrams, and identifying shapes from complex objects. The control classrooms carried on with their normal classroom activities. Pre- and posttests were administered within 3 weeks of the intervention. Results revealed that children who received spatial training ($n = 68$) performed better than the control group ($n = 57$) on two visuospatial measures, one that involved discriminating rotated figures and the other that involved copying figures (i.e., Spatial Orientation and Visual-Motor Integration subtests of the Developmental Test of Visual Perception; Frostig, 1973). Nevertheless, the training effect did not generalize to all spatial abilities (i.e., CMTT; Levine et al., 1999). Moreover, and crucially, there were no differences between intervention and control groups on seven different math measures used in the study (e.g., counting, number naming, number comparison, and arithmetic).

To summarize, extant studies that ask about the causal link between visuospatial reasoning and mathematical understanding have not provided definitive answers to this question. Moreover, there is a lack of consensus as to why the findings are mixed, perhaps largely because of the variability across studies. Cheng and Mix (2014) found a treatment effect (i.e., spatial training improved math performance) but no quality check in terms of the treatment (i.e., spatial training did not improve another spatial measure). Moreover, a recent study with the same training and testing procedure did not find robust transfer from training with the CMTT to arithmetic performance with first to third graders (Sala, Bolognese, & Gobet, 2017). Others have used broader visuospatial training and the findings were, again, mixed. Hawes et al. (2017) found transfer to math and so did Lowrie et al. (2017). In both studies, however, it was unclear whether these effects truly reflected transfer from visuospatial training to math performance. Hawes et al. (2015) and Cornu et al. (2017) did not find a visuospatial training effect on math performance. Notably, though, both studies did show improvement in spatial reasoning, suggesting that the training itself was not problematic. Both of these studies had posttests that were administered several days after training, which was different from the study of Cheng and Mix (2014). Thus, one possibility is that a transfer effect existed, but it was relatively short lived. As noted earlier, if such effects are short lived, it will be important for future studies to include booster sessions to assess durability.

An important question that follows from an analysis of the extant training studies is why the effects in these studies seem less consistent than those in the correlational work. Studies examining both concurrent and longitudinal correlations between visuospatial and mathematical reasoning seem to present a much more consistent picture. One possibility is that the intervention studies that have focused on training visuospatial reasoning and testing the effect on mathematical competence have specified the wrong causal direction. Perhaps the correlations are better accounted for by an impact of mathematical processing on visuospatial reasoning. Although this is certainly a statistical possibility, we would argue that it seems unlikely. Indeed, it is not immediately clear what mechanism would account for a specific benefit of mathematical reasoning on the spatial domain, rather than the reverse (see Why Would Visuospatial Reasoning Benefit Math Development? Candidate Mechanisms section). An exception, however, might be geometry, where there could be a benefit to learning number and quantitative concepts such as area and volume from a spatial perspective. And, consequently, training geometry could also lead to improvement in visuospatial reasoning. Another consideration is that the associations, including a potential causal link between visuospatial and mathematical reasoning, may depend on the shared influence of other systems such as working memory. This is a challenging question to address, and although some studies have made efforts to address the specificity of visuospatial and math reasoning by controlling for individual differences in other abilities, these controls are rarely exhaustive and it is rarely the case that full models are conducted to determine the potential interactions of different systems. Thus, there are unanswered questions about the extent to which other variables may affect the correlations between visuospatial and mathematical reasoning and the extent to which other variables may have to be trained for a transfer effect to be observed.

MENTAL NUMBER LINE

Dating back to the late 1800s, Sir Francis Galton provided evidence that numbers were mentally organized in visuospatial format, occupying distinct locations along a so-called mental number line (Galton, 1880). Subsequent research in cognitive science and neuroscience has provided supporting evidence for the human tendency to represent numbers spatially (for review, see Hubbard, Piazza, Pinel, & Dehaene, 2005). Importantly, this grounding of number in spatial coordinates encompasses different properties of the spatial representation—namely, directionality and scaling. We draw attention to both properties of the mental number line not only to emphasize its multidimensional character but also because the different properties may exhibit unique relations with mathematical development.

Let's begin with *directionality*. We can describe directionality as the horizontal orientation of the mental number line. For example, are numbers oriented from left to right or right to left? Directionality has been shown to vary across culture, though there is also evidence of a default orientation that is present early in human life (de Hevia, Addabbo, Girelli, & Macchi-Cassia, 2014; Opfer, Thompson, & Furlong, 2010) and across species

(Drucker & Brannon, 2014; Rugani, Vallortigara, Priftis, & Regolin, 2015). In Western culture where individuals read from left to right, the mental number line is rightward oriented, with smaller numbers on the left side of space and larger numbers on the right (Dehaene, Bossini, & Giraux, 1993; see also Cheung, Ayzenberg, Diamond, Yousif, & Lourenco, 2015; Wood, Willmes, Nuerk, & Fischer, 2008).

We can also describe the *scaling* of numbers along the mental number line (Siegler & Opfer, 2003). Models of the spatial scaling of the number line are typically characterized as either linear, with equal spatial intervals between numbers, or compressive, such that spatial intervals decrease between pairs of numbers as numerical values increase. A common finding in the literature is the developmental shift from compressive to linear number lines (Siegler & Opfer, 2003; Siegler, Thompson, & Opfer, 2009). However, there is also research suggesting that both types of mental number lines coexist in adulthood, with access to different scales depending on contextual demands (Lourenco & Longo, 2009; Viarouge, Hubbard, Dehaene, & Sackur, 2010). Nevertheless, developmental studies suggest that access to a linear mental number line has relevance for mathematical development, which we discuss further in the following.

Directionality of the Number Line

In adults, the research has focused almost entirely on testing whether a mental number line's direction (e.g., left to right) is related to math achievement. Much of this research has revealed no relation between the extent of directionality and individuals' mathematical competence (e.g., Cipora & Nuerk, 2013; Cipora, Patro, & Nuerk, 2015; Gibson & Maurer, 2016). In cases where significant findings have been obtained, they are suggestive of a negative relation, though these effects appear to be driven by individuals at the extreme ends of the distributions. For example, Hoffmann, Mussolin, Martin, and Schiltz (2014) examined whether a rightward mental number line varied with levels of mathematical expertise. Directionality was measured by performance on the classic "SNARC" (spatial-numerical association of response codes) task (Dehaene et al., 1993), in which participants judged the parity (odd/even) of Arabic numerals (1–9) using left and right response keys. Typically, participants show speeded responses to small numerals when responding with the left key and speeded responses to large numerals when responding with the right key, consistent with left-to-right orientation. In this study, participants who reported math difficulties showed stronger SNARC effects than a math expert group (e.g., mathematicians and engineers). Similarly, Cipora et al. (2016) found that professional mathematicians were less likely than control participants to display evidence of a rightward mental number line on the SNARC task, again suggesting a negative relation between the mental number line's direction and math ability.

The developmental prediction, however, has often been that a mental number line would be beneficial for early math success (e.g., Fischer & Shaki, 2014; Opfer et al., 2010). In other words, the spatial representation of number should be especially helpful when children are first learning mathematics. Yet it remains unclear whether this prediction is borne out by the evidence. Bachot, Gevers, Fias, and Roeyers (2005) found that children (7–12 years) with visuospatial deficits and developmental dyscalculia did not exhibit evidence of a mental number line on a symbolic magnitude comparison task. More specifically, when required to judge whether an Arabic numeral was smaller or larger than 5, these children showed no evidence of space–number congruity, indexing a left-to-right mental number line. By contrast, a control group of children did show space–number congruity consistent with a rightward number line. However, children with visuospatial and math deficits were not analyzed separately, making it unclear whether the lack of a directional mental number line in these children was specifically related to their math deficits. In other work, though, Schneider, Grabner, and Paetsch (2009) conducted a large-scale study ($n = 429$) and found no relation between a SNARC effect and school math achievement in children aged 9–15 years.

Yet it remains possible that the hypothesized positive relation between the mental number line and math ability would be found in younger children. For example, a recent investigation by Georges, Hoffmann, and Schiltz (2017) found that the relation between a directional mental number line and math ability was indeed modulated by age. Specifically, although there was an overall association between the SNARC effect and arithmetic ability, this effect was not observed in children older than 10.25 years of age (1 SD above the mean age of the sample), suggesting that younger children would be more likely to show a positive relation between directionality of the mental number line and math achievement. However, Hoffman, Hornung, Martin, and Schiltz (2013) found that younger children (5-year-olds) showed no relation between a directional mental number line (assessed by an implicit SNARC task) and any measure of numerical competence employed in their study (e.g., verbal counting and digit writing). Likewise, work by Gibson and Mauer (2016) also suggests no relation between the mental number line and math ability in young children (6–8 years). More specifically, they found that children's SNARC scores did not correlate with performance on a standardized measure of general math ability (Ginsberg & Baroody, 2003).

Work from our own lab has recently examined the relation between directionality of the mental number line and math proficiency early in development (5–7-year-olds) with multiple measures of the mental number line to ensure convergent validity and multiple tests of math ability to determine whether the relations vary by the type of mathematics (Aulet & Lourenco, 2017). The two paradigms included to assess directionality of the mental number line were a magnitude comparison task and a novel paradigm known as the "place-the-number" task (Aulet, Yousif, & Lourenco, 2017). Children's math ability was assessed with three measures: cross-modal arithmetic (Barth, Beckmann, & Spelke, 2008), approximate symbolic arithmetic (Gilmore et al., 2007), and exact symbolic arithmetic (Calculation subtest, WJ Achievement; Woodcock, McGrew, et al., 2001). Despite evidence for a rightward mental number line on both paradigms, there was no relation with math performance. Children's individual performance on the place-the-number task was not significantly correlated with any measure of math ability, and although directionality of the mental number line, as assessed by the magnitude comparison task, was correlated with cross-modal arithmetic, this correlation was negative, suggesting a stronger rightward mental number line was associated with worse arithmetic performance.

In summary, the field currently lacks a consensus on whether a directional mental number line is beneficial (or harmful) for math achievement. Studies are, admittedly, limited in this area, but those that do exist, especially with children, do not provide a consistent picture. Some studies report a relation between a directional mental number line and arithmetic skills in children, whereas others do not.

Linear Versus Compressive Number Lines

In contrast to the work described earlier on the relation between directionality and math achievement, studies testing the extent to which linearity predicts mathematical competence have been more definitive. Many researchers now conclude that the more linear one's mental number line, the more competent one appears on a variety of numerical tasks and math measures.

Using a number line estimation task (e.g., Siegler & Opfer, 2003), children are typically asked to mark a number along a physical line bounded by two numbers (e.g., 0 and 100). Performance on this task has been found to predict concurrent and later mathematics performance (Booth & Siegler, 2006; Sasanguie, Göbel, Moll, Smets, & Reynvoet, 2013; Siegler & Booth, 2004). For example, linearity, as indexed by the proportion of variance in a child's estimation explained by numerical magnitude, correlates with standardized math scores ($r \approx .50$; Booth & Siegler, 2006). It has also been found that the relation between number line estimation and arithmetic skill is more pronounced in cases of addition and subtraction than multiplication (Link, Nuerk, & Moeller, 2014), which suggests that the number line is relevant for arithmetic computation because multiplication tasks are typically solved by direct fact retrieval as opposed to online calculation (Dehaene, Piazza, Pinel, & Cohen, 2003; Delazer, Girelli, Granà, & Domahs, 2003).

Despite the consistency in this literature, there is again the question of causality. In particular, we can ask whether linearity leads to better math competence rather than the reverse. Fortunately, studies in which children are trained to develop more linear number lines provide insight into this question. For example, training studies that teach children to play with board games, in which numbers are evenly spaced apart, find that not only do number line estimations improve (becoming more linear) but also children perform better at judging and comparing numerical magnitudes (Siegler & Ramani, 2008, 2009). More recently, Ramani, Siegler, and Hitti (2012) also showed that a board game based on linear number lines improved number line estimation among lower-income preschoolers (3–5.5-year-olds) and, crucially, there was improvement on other numerical tests such as numeral identification and counting, suggesting a role for a linear mental number line in the development of math skills. Moreover, Kucian et al. (2011) found that both children with and without dyscalculia who completed computerized training on number line estimation showed improvements in arithmetic ability.

Yet others have suggested that number line estimation tasks and "linear" training may, instead, elicit proportional reasoning strategies, which are themselves an important part of mathematical development. For example, on number line estimation tasks, older children often make virtual reference points when responding to where the numbers should be located along the number line. This may start with the middle of the line and then dividing virtual halves. With age, there are generally more of these references points and greater accuracy (e.g., Barth & Paladino, 2011; Slusser, Santiago, & Barth, 2013). These proportional strategies could explain the pattern of estimation without a linear mental number line. Separate studies have corroborated this explanation; taken together, they provide support for the alternative explanation that children's and adults' estimations may be better explained by a proportional reasoning strategy, and furthermore, that individual differences in proportional reasoning can account for the relation to math ability (Barth & Paladino, 2011; Cohen & Sarnecka, 2014; Slusser et al., 2013).

WHY WOULD VISUOSPATIAL REASONING BENEFIT MATH DEVELOPMENT? CANDIDATE MECHANISMS

In this section, we discuss candidate mechanisms for why spatial reasoning, particularly visualization as in the case of mental rotation and the mental number line, could prove beneficial for mathematics. Although more research is needed to determine whether spatial reasoning is causally implicated in math development, here we consider the mechanisms that justify research on this question. It is our hope that by delineating potential mechanisms, future research aimed at testing the causal relations between visuospatial and mathematical reasoning will target manipulations that shed light on how visuospatial reasoning precisely affects mathematical thought. It should be noted that although the candidate mechanisms described in the following are discussed separately, they may not be mutually exclusive. Moreover, we do not make claims about the developmental continuity of these mechanisms. It is possible that the relative importance of these mechanisms may change over development and with experience in spatial and mathematical domains.

Visuospatial Reasoning as Mental Model

One possible reason for why visuospatial reasoning might benefit our understanding of quantitative concepts and our ability to solve math problems is that visuospatial reasoning functions as a mental model. More specifically, the idea is that when solving math problems, the visualization of these problems in one's mind, like with real models or diagrams, allows one to "see" not only the individual elements but also the relations of those elements, within the problem space. A well-known example of how real spatial depictions aid in problem solving occurred in 1854 when the English physician John Snow traced the source of a cholera outbreak in London to a public water pump on Broad Street by mapping the clusters of cholera cases in the relevant neighborhoods. Similarly, mental visualizations may facilitate insight by helping one relate the relevant elements of a problem to one another (Huttenlocher, 1968; Johnson-Laird, 1983). In mathematics education, the use of visualizations and encouragement of mental models have become increasingly common. Some educators have argued that all successful calculus courses emphasize the visual elements of the domain (Zimmermann & Cunningham, 1991).

A benefit of visualizing transformations of arrangements of mathematical elements is that one can keep track of which terms should be grouped together and the order of operations to be performed (Barnhardt, Borsting, Deland, Pham, & Vu, 2005; Huttenlocher, Jordan, & Levine, 1994; Laski et al., 2013; Thompson, Nuerk, Moeller, & Kadosh, 2013). To be beneficial, however, such visualizations should provide spatial information not presented in the written description of the problems. They are also generally quite sparse rather than overly pictorial, so as not to overwhelm working memory capacity (e.g., Knauff & Johnson-Laird, 2002). A benefit of spatial visualizations was found in children in the work of Hegarty and Kozhevnikov (1999), in which it was reported that elementary school boys (girls were not tested) who made schematic spatial representations like those in diagrams, rather than pictorial illustrations, were better at solving math problems. It has been suggested that such spatial strategies, whether in real sketches or mental models, may promote mathematical understanding by aiding in the interpretation of word problems and supporting flexibility in solving math equations (Kintsch & Greeno, 1985).

The utility of mental models in problem solving is not specific to mathematics, however. There is much research in other domains, such as seriation and transitive inference, in which participants have been shown to benefit from visualizing the problem space (Huttenlocher, 1968; Johnson-Laird, 1983). Thus, the prediction that follows from this perspective is that if visuospatial reasoning is beneficial for math learning because it involves a mental model strategy, then it should likewise benefit other domains. Indeed, studies interested in shedding light on the mechanisms underlying the causal relation between visuospatial reasoning and mathematical development should examine other domains of reasoning. For example, any benefits of mental rotation training to math should be similarly observed in other domains in which mental models are applicable.

Visuospatial Reasoning as an Act of Grounding

Not unrelated to the mental model strategy just discussed is the possibility that visuospatial reasoning serves to ground abstract concepts. The mental number line is an example of how numbers are grounded in space. FIVE apples are perceptibly quite distinct from FIVE cats and even more so than FIVE ringtones. What unites these examples is their FIVE-ness. To adults, the abstract concept of a small numerical value such as FIVE is trivial, but for children, this explicit understanding takes time to develop (Mix, Huttenlocher, & Levine, 1996). Learning that number is abstract in this way could be facilitated early in development by spatial representations. For example, envisioning the placement of FIVE along a mental number line such that different instances of FIVE all occupy the same location could serve to support the development of such numerical concepts. Although the mental models described earlier could similarly involve grounding information in a spatial layout (cf. Barsalou, 2008; Glenberg & Robertson, 2000), here we consider the possibility that numerical concepts are grounded in space and how they may impact mathematical problem solving.

Some researchers have even emphasized the importance of sensorimotor experience in understanding mathematical concepts, perhaps because of the reference to both location and action from an "embodied" perspective (Barsalou, 2008; Fischer, 2012; Lakoff & Núñez, 2000; Mowat & Davis, 2010). The potential benefit of location and action is evident when numbers are grounded along a spatially ordered line. Indeed, numerical concepts such as cardinality, equivalence, and successor function may become more concrete and, hence, understandable to children because of the discrete positions along the number line (Cheung & Lourenco, 2017; Cipora et al., 2015). The fact that linearity of the number line with distinct positions for number predicts math achievement is consistent with this possibility (Ramani & Siegler, 2008; Siegler & Booth, 2004). Ordinality is also evident when numbers can be straightforwardly related to one another along a fixed spatial structure. Another example is the case of measurement units, which are best understood when applied directly to their spatial coordinates. Principles such as midpoint and equidistance become more apparent when numerical differences can be compared spatially (Cheung & Lourenco, 2015; Cipora et al., 2015). There are also advanced concepts such as ZERO and the negative numbers, which researchers have suggested may be supported by the mental number line and the visualization of symmetry (Blair, Tsang, & Schwartz, 2013). From a grounded perspective, arithmetic can be conceived concretely as increases versus decreases or leftward versus rightward movement in space, with arithmetic operations performed in spatial coordinates. Consistent with this possibility are "operational momentum" effects in which addition and subtraction trigger rightward and leftward biases, respectively (e.g., McCrink, Dehaene, & Dehaene-Lambertz, 2007; Pinhas & Fischer, 2008).

At early ages, spatial grounding may serve to scaffold the acquisition and the understanding of basic mathematical concepts. In adults, congruity effects between physical distance and arithmetic operations reveal the remnants of spatial grounding (e.g., Landy & Goldstone, 2007; Lourenco & Levine, 2008). It has been shown that manipulating the spatial arrangement of mathematical symbols in adults affects performance, consistent with operations as "grounded" in space. For example, in evaluating arithmetic expressions, one must consider the order of operations (e.g., multiplication comes before addition), which researchers have shown to have spatial analogs. Landy and Goldstone (2007) found that participants engaged in more efficient calculation under congruent (e.g., $2 \times 2 + 2 = ?$) than incongruent (e.g., $2 \times 2 + 2 = ?$) spatial arrangements. In these problems, multiplication is computed before addition and, so, participants expected closer physical proximity for multiplication than addition. Goldstone, Weitnauer, Ottmar, Marghetis, and Landy (2016) also found that background motion affected how one solved algebra problems, especially among people who had taken advanced calculus courses. Although these studies do not demonstrate that visuospatial processes are causally implicated in math development, they suggest that spatial processing is automatically triggered during mathematical problem solving.

Cheng and Mix (2014) argued that the mental rotation training in their study led to improved arithmetic performance, particularly in missing term problems because the training may have primed the visualization of the math problem. In particular, children in the spatial training group may have mentally aligned the math problem so as to be more consistent with the canonical format (e.g., $3 + \_\_ = 4$ would become $4 - 3 = \_\_$; see Fig. 10.2). The challenge for this explanation is that it assumes children understand how the operations (+ and −) relate to one another when numerical values cross the equal sign, which mental rotation training does not directly address. Changing the problem into a more canonical format requires knowledge of how the operations must be reconceptualized across contexts (e.g., 3+__=4 becomes 4−3=__ not 4+3=__). We thus suggest an alternative possibility in which mental rotation training could still prove beneficial. In school, children may be instructed to consider 3 objects (e.g., blocks) and then to determine how many they would need to add to get to the answer (4) by counting up (see Fig. 10.2). Without the use of such objects, children could similarly visualize the operation by using their fingers (as they often do) to count up from the addend in the original problem (e.g., 3 and then 4). Such visualizations could also be accomplished by way of a mental number line, with the arithmetic operation simulated as movement along it (see Fig. 10.2).

Importantly, as is the case with mental models described earlier, the specific benefit of grounding abstract concepts makes the directional prediction that training mathematics would not result in improved spatial reasoning because the abstract notions in mathematics are not similarly beneficial for spatial reasoning. Moreover, there is also the prediction of generality. The benefit that grounding affords is not specific to mathematics and, indeed, there are numerous researchers who advocate for the use

FIGURE 10.2 A caricature of how children might solve a missing term problem (e.g., "3+__=4"). The top example depicts visualization and mental rotation of the numbers themselves. The problem "3+__=4" might become "4−3=__" following the mental movement of the number 3. The middle example depicts the visualization of models with common objects (left) or fingers (right), which involve a "counting up" strategy. The bottom example depicts addition in the form of movement along the mental number line. *Images in this figure were obtained from Shutterstock (standard license).*

of spatial "metaphors" to instantiate a variety of other abstract notions such as time, metaphysical concepts, and so on (e.g., Lakoff & Núñez, 2000; Vosniadou & Brewer, 1994).

Shared Representational Format and Neural Mechanisms?

There are commonalities in the representational codes for visuospatial and mathematical processing. For example, mental rotation and representations of numerical magnitude are similarly analog in nature. Consider, first, mental rotation. Rotating an object in one's mind is analog because it tracks closely with the real-world movements of an actual object. More specifically, individuals engaged in mental rotation show classic angular disparity effects such that comparing two objects that deviate by 150 degrees takes longer than two objects that deviate by 30 degrees (e.g., Shepard & Metzler, 1971). Consider, also, numerical magnitude, for which our representations are analog in that their precision decreases as values become larger; moreover, comparing numerosities that are more disparate (e.g., 5 vs. 10 apples) is easier than those that are closer (e.g., 8 vs. 10 apples). In addition to the similar representational format, there is evidence that visuospatial processes recruit brain regions that are also implicated in numerical reasoning and mathematical problem solving (Hubbard et al., 2005; Zacks, 2008). Thus, it is possible that training mental rotation would be beneficial to mathematics because analog processes are "transferable" across domains.

This perspective makes a clear prediction in that there should be no asymmetry in terms of transfer across domains, unlike the mechanisms discussed earlier. For example, training mental rotation ability should transfer to math performance and, similarly, training mathematical understanding (specifically those processes thought to recruit analog representations) should transfer to mental rotation performance.

Alternative Systems? Working Memory as a Possibility

Another type of explanation suggests that visuospatial reasoning is not specifically beneficial for mathematical development, but, instead, that separate systems, which are associated with visuospatial and mathematical reasoning, account for the relation between the two. Here we focus on working memory in particular because it is well known that visuospatial reasoning such as mental rotation and mathematical competence rely on the shared resource of working memory (Carr et al., 2018). This claim is supported by research showing correlations between working memory and both visuospatial and mathematical reasoning (Alloway & Passolunghi, 2011; DeStefano & LeFevre, 2004; Kaufman, 2007; Kyllonen & Christal, 1990; Kyttälä et al., 2003; Li & Geary, 2017; Mix et al., 2016).

These associations are not surprising—after all, the processes that allow individuals to maintain and manipulate information online are also needed for visualization such as mental rotation and mental arithmetic. Future research would do well to test the possibility that working memory may mediate the relation between visuospatial and mathematical reasoning. Such tests will mean ensuring that measures of working memory are included in these studies and that researchers examine whether the effects involving spatial and mathematical reasoning hold when controlling for working memory. Although some correlational studies have made such attempts, there have been fewer such controls in the extant training studies, perhaps in part because of the many practical challenges in implementing such studies. Nevertheless, future training studies should consider testing whether trained visuospatial reasoning and subsequent improvements on math performance are because of working memory. From this perspective, it is possible that training visuospatial reasoning such as mental rotation serves to enhance working memory. After all, when one visualizes and manipulates objects in one's mind, this requires working memory, and engaging in this type of task could serve to enhance working memory. Such studies will need to determine whether improvements in mathematics are accompanied by improvements in working memory and, more specifically, whether they can be accounted for by improvements in working memory over and above those in mental rotation.

Interim Summary

As described at the beginning of this section, there are different, albeit not mutually exclusive, mechanisms, which may account for the relation between visuospatial and mathematical reasoning. We have suggested that some of these mechanisms are not specific to spatial and math domains and, thus, should generalize to other domains of reasoning. We have also suggested that there may be differences in the direction of influence. For example, while the mental model account predicts that visualization supports math development and not vice versa, this is not the case for a mechanism based on shared analog format in which there should be mutually beneficial interactions between visuospatial and mathematical reasoning. We would also posit that the relative importance of these mechanisms may change over development and with experience in spatial and mathematical domains. For example, a mental model strategy may be less likely to be implemented (or less accurately) earlier in development (Mix & Cheng, 2011), whereas transfer based on shared analog format may be present early and throughout development (e.g., Lourenco, 2015; Lourenco & Longo, 2010). In a similar way, Li and Geary (2013) found that visuospatial memory becomes increasingly important for predicting individual differences in math achievement between first and fifth grades.

CONCLUDING THOUGHTS

It has been suggested that visuospatial reasoning offers a different entry point into the math learning environment, with researchers and educators looking to the spatial domain as potential support to math development. In general, there has been increased interest in leveraging visuospatial reasoning in our efforts to both understand and improve mathematical competence in young children. Moreover, although visuospatial reasoning is generally not formally studied in school, young children are already equipped with a variety of geometric "intuitions" (Izard & Spelke, 2009; see also Bonny & Lourenco, 2015), which may serve them well when learning about numerical concepts and solving mathematical problems.

Despite the encouraging views in the literature, we wish to offer some words of caution in conclusion. Although there are certainly good reasons to appeal to spatial reasoning when considering how best to support math learning and development, the extant work is far from definitive. Much of the existing studies are correlational, leaving open questions about causality. Even in the case of longitudinal designs, there is more work to be done to determine definitively whether visuospatial reasoning is a unique precursor to math development and how visuospatial reasoning may interact with other systems such as working memory to support math development. Nevertheless, the extant data suggest that links between visuospatial and math reasoning are present throughout development and that they emerge early. There are also increasing efforts to test the causal links between visuospatial reasoning and mathematical development.

As noted earlier, training studies designed to test whether visuospatial reasoning is causally implicated in mathematical development are not only crucial for constraining our theoretical models of the relations across the domains but also for practical reasons so that appropriate educational practices may be implemented. Although there is preliminary support from the transfer of visuospatial training to mathematical performance, the mechanisms remain unknown. We have discussed potential candidates, but work is needed to test such possibilities directly. There may also be questions about whether such transfer is more likely for the most spatial mathematical domains such as geometry. Future work will need to determine how long such effects last when they do. Other work should also consider assessing the effect sizes of such improvements in relation to math training specifically as opposed to an unrelated control group.

If we find that training spatial reasoning improves math performance, then we may still ask whether it is better to do this than to teach the relevant mathematical concepts. The training studies that exist often compare a spatial training group with a control group that involves training a nonspatial skill (e.g., literacy task). This type of control is critical in any training study, as it ensures that improvements from pre- to posttest are not simply because of practice or the passage of time. However, it would also be useful to test whether spatial training is better than the typical (or enhanced) math lesson. If so, then the answer going forward is a no-brainer. We will need to adopt more spatial training because it benefits both spatial and mathematical reasoning. We should leverage visuospatial reasoning in an effort to support mathematical development (Newcombe, 2013). However, another possibility is that there will be no difference or that spatial training will be worse than a "math as usual" or an enhanced math lesson. In this case, we can ask whether it is beneficial to invest in spatial training. One reason to invest nonetheless is that spatial training may be more fun and engaging than other classroom lessons (Hawes et al., 2017), and students who are engaged are likely to remain more motivated throughout a variety of lessons, such that improvements in learning may be more general and durable as a result. These are some of the issues that we will need to consider going forward so that we are in a better position to understand whether there is a robust relation between visuospatial reasoning and math development, and so that we can make practical recommendations about interventions and educational strategies for early math instruction.

References

Ackerman, P. L., & Cianciolo, A. T. (2002). Ability and task constraint determinants of complex task performance. *Journal of Experimental Psychology: Applied, 8*(3), 194.

Alloway, T. P., & Passolunghi, M. C. (2011). The relationship between working memory, IQ, and mathematical skills in children. *Learning and Individual Differences, 21*(1), 133–137.

Aulet, L. S., & Lourenco, S. F. (October 2017). *Examining the development and functional role of spatial-numerical representations. Poster Presented at the Biennial Meeting of the Cognitive Development Society (CDS).* Portland: OR.

Aulet, L. S., Yousif, S. R., & Lourenco, S. F. (2017). Numbers uniquely bias spatial attention: A novel paradigm for understanding spatial-numerical associations. In G. Gunzelmann, A. Howes, T. Tenbrink, & E. J. Davelaar (Eds.), *Proceedings of the 39th annual conference of the cognitive science society.* Austin, TX: Cognitive Science Society. pp. 75–80.

Bachot, J., Gevers, W., Fias, W., & Roeyers, H. (2005). Number sense in children with visuospatial disabilities: Orientation of the mental number line. *Psychology Science, 47*, 172–183.

Barnhardt, C., Borsting, E., Deland, P., Pham, N., & Vu, T. (2005). Relationship between visual-motor integration and spatial organization of written language and math. *Optometry and Vision Science, 82*(2), 138–143.

Barsalou, L. W. (2008). Grounded cognition. *Annual Review of Psychology, 59*, 617–645.

Barth, H., Beckmann, L., & Spelke, E. (2008). Nonsymbolic, approximate arithmetic in children: Evidence for abstract addition prior to instruction. *Developmental Psychology, 44*, 1466–1477.

Barth, H., La Mont, K., Lipton, J., Dehaene, S., Kanwisher, N., & Spelke, E. S. (2006). Non-symbolic arithmetic in adults and young children. *Cognition, 98*, 199–222. https://doi.org/10.1016/j.cognition.2004.09.011.

Barth, H., & Paladino, A. M. (2011). The development of numerical estimation: Evidence against a representational shift. *Developmental Science, 14*, 125–135.

Bishop, A. J. (2008). Mathematics teaching and values education – an intersection in need of research. *Critical Issues in Mathematics Education*, 231–238.

Blair, K. P., Tsang, I. M., & Schwartz, D. L. (2013). How perception and culture give rise to abstract mathematical concepts in individuals. In *International handbook of research on conceptual change* (Vol. 322).

Bonny, J. W., & Lourenco, S. F. (2015). Individual differences in children's approximations of area correlate with competence in basic geometry. *Learning and Individual Differences, 44*, 16–24.

Booth, J. L., & Siegler, R. S. (2006). Developmental and individual differences in pure numerical estimation. *Developmental Psychology, 42*, 189–201.

Carr, M., Alexeev, N., Wang, L., Barned, N., Horan, E., & Reed, A. (2018). The development of spatial skills in elementary school students. *Child Development, 89*, 446–460.

Cheng, Y.-L., & Mix, K. S. (2014). Spatial training improves children's mathematics ability. *Journal of Cognition and Development, 15*(1), 2–11.

Cheung, C.-N., Ayzenberg, V., Diamond, R. F. L., Yousif, S. R., & Lourenco, S. F. (2015). Probing the mental number line: A between-task analysis of spatial-numerical associations. In D. C. Noelle, R. Dale, A. S. Warlaumont, J. Yoshimi, T. Matlock, C. D. Jennings, et al. (Eds.), *Proceedings of the 37th annual meeting of the cognitive science society*. Austin, TX: Cognitive Science Society. pp. 357–362.

Cheung, C.-N., & Lourenco, S. F. (2015). Representations of numerical sequences and the concept of middle in preschoolers. *Cognitive Processing, 16*(3), 255–268.

Cheung, C.-N., & Lourenco, S. F. (April 2017). The first number is one: The relation between ordinal concepts and children's understanding of cardinality. In *Poster presented at the society for research in child development (SRCD), Austin, TX*.

Cipora, K., Hohol, M., Nuerk, H.-C., Willmes, K., Brożek, B., Kucharzyk, B., et al. (2016). Professional mathematicians differ from controls in their spatial-numerical associations. *Psychological Research, 80*, 710–726.

Cipora, K., & Nuerk, H. C. (2013). Is the SNARC effect related to the level of mathematics? No systematic relationship observed despite more power, more repetitions, and more direct assessment of arithmetic skill. *Quarterly Journal of Experimental Psychology, 66*, 1974–1991.

Cipora, K., Patro, K., & Nuerk, H.-C. (2015). Are spatial-numerical associations a cornerstone for arithmetic learning? The lack of genuine correlations suggests no. *Mind Brain Education, 9*, 190–206.

Clements, D. H., & Sarama, J. (2011). Early childhood mathematics intervention. *Science, 333*(6045), 968–970.

Cohen, D. J., & Sarnecka, B. (2014). Children's number-line estimation shows development of measurement skills (not number representations). *Developmental Psychology, 50*, 1640–1652.

Cornu, V., Schiltz, C., Pazouki, T., & Martin, R. (2017). Training early visuo-spatial abilities: A controlled classroom-based intervention study. *Applied Developmental Science*, 1–21.

de Hevia, M. D., Addabbo, M., Girelli, L., & Macchi-Cassia, V. (2014). Human infants' preference for left-to-right oriented increasing numerical sequences. *PLoS One, 9*, e96412.

Dehaene, S., Bossini, S., & Giraux, P. (1993). The mental representation of parity and number magnitude. *Journal of Experimental Psychology: General, 122*, 371–396.

Dehaene, S., Piazza, M., Pinel, P., & Cohen, L. (2003). Three parietal circuits for number processing. *Cognitive Neuropsychology, 20*, 487–506.

Delazer, M., Girelli, L., Granà, A., & Domahs, F. (2003). Number processing and calculation—Normative data from healthy adults. *The Clinical Neuropsychologist, 17*, 331–350.

DeStefano, D., & LeFevre, J. A. (2004). The role of working memory in mental arithmetic. *European Journal of Cognitive Psychology, 16*(3), 353–386.

Drucker, C., & Brannon, E. M. (2014). Rhesus monkeys (Macaca mulatta) map number onto space. *Cognition, 132*, 57–67.

Fabro, S., Smith, R. L., & Williams, R. T. (1967). Toxicity and teratogenicity of optical isomers of thalidomide. *Nature, 215*(5098), 296.

Fischer, M. H. (2012). A hierarchical view of grounded, embodied, and situated numerical cognition. *Cognitive Processing, 13*(1), 161–164. https://doi.org/10.1007/s10339-012-0477-5.

Fischer, M. H., & Shaki, S. (2014). Spatial associations in numerical cognition – From single digits to arithmetic. *Quarterly Journal of Experimental Psychology, 67*, 1461–1483.

Friso-van den Bos, I., van der Ven, S. H., Kroesbergen, E. H., & van Luit, J. E. (2013). Working memory and mathematics in primary school children: A meta-analysis. *Educational Research Review, 10*, 29–44.

Frostig, M. (1973). *Test de développement de la perception visuelle* [*Developmental test of visual perception*]. Paris, France: Les Editions du Centre de Psychologie Appliquée.

Galton, F. (1880). Visualised numerals. *Nature, 21*, 252–256.

Georges, C., Hoffmann, D., & Schiltz, C. (2017). Mathematical abilities in elementary school: Do they relate to number-space associations? *Journal of Experimental Child Psychology, 161*, 126–147.

Gibson, L. C., & Maurer, D. (2016). Development of SNARC and distance effects and their relation to mathematical and visuospatial abilities. *Journal of Experimental Child Psychology, 150*, 301–313.

Gilmore, C. K., McCarthy, S. E., & Spelke, E. S. (2007). Symbolic arithmetic knowledge without instruction. *Nature, 447*, 589–591.

Ginsberg, H. P., & Baroody, A. J. (2003). *Test of early mathematics ability* (3rd ed.). Austin, TX: Pro-Ed.

Glenberg, A. M., & Robertson, D. A. (2000). Symbol grounding and meaning: A comparison of high-dimensional and embodied theories of meaning. *Journal of Memory and Language, 43*, 379–401.

Goldstone, R. L., Weitnauer, E., Ottmar, E. R., Marghetis, T., & Landy, D. H. (2016). Modeling mathematical reasoning as trained perception-action procedures. *Design Recommendations for Intelligent Tutoring Systems, 213*.

Gunderson, E. A., Ramirez, G., Beilock, S. L., & Levine, S. C. (2012). The relation between spatial skill and early number knowledge: The role of the linear number line. *Developmental Psychology, 48*(5), 1229–1241. https://doi.org/10.1037/a0027433.

Hawes, Z., Moss, J., Caswell, B., Naqvi, S., & MacKinnon, S. (2017). Enhancing children's spatial and numerical skills through a dynamic spatial approach to early geometry instruction: Effects of a 32-week intervention. *Cognition and Instruction*, 1–29. https://doi.org/10.1080/07370008.2017.1323902.

Hawes, Z., Moss, J., Caswell, B., & Poliszczuk, D. (2015). Effects of mental rotation training on children's spatial and mathematics performance: A randomized controlled study. *Trends in Neuroscience and Education*, 4(3), 60–68. https://doi.org/10.1016/j.tine.2015.05.001.

Hegarty, M., & Kozhevnikov, M. (1999). Types of visual–spatial representations and mathematical problem solving. *Journal of Educational Psychology*, 91(4), 684–689.

Hegarty, M., Montello, D. R., Richardson, A. E., Ishikawa, T., & Lovelace, K. (2006). Spatial abilities at different scales: Individual differences in aptitude-test performance and spatial-layout learning. *Intelligence*, 34(2), 151–176.

Hegarty, M., & Waller, D. (2005). Individual differences in spatial abilities. *The Cambridge Handbook of Visuospatial Thinking*, 121–169.

Hoffmann, D., Hornung, C., Martin, R., & Schiltz, C. (2013). Developing number–space associations: SNARC effects using a color discrimination task in 5-year-olds. *Journal of Experimental Child Psychology*, 116, 775–791.

Hoffmann, D., Mussolin, C., Martin, R., & Schiltz, C. (2014). The impact of mathematical proficiency on the number-space association. *PLoS One*, 9, e85048.

Hubbard, E. M., Piazza, M., Pinel, P., & Dehaene, S. (2005). Interactions between number and space in parietal cortex. *Nature Reviews Neuroscience*, 6(6), 435–448.

Huttenlocher, J. (1968). Constructing spatial images: A strategy in reasoning. *Psychological Review*, 75, 550–560.

Huttenlocher, J., Jordan, N. C., & Levine, S. C. (1994). A mental model for early arithmetic. *Journal of Experimental Psychology: General*, 123(3), 284.

Izard, V., & Spelke, E. S. (2009). Development of sensitivity to geometry in visual forms. *Human Evolution*, 23(3), 213.

Johnson-Laird, P. N. (1983). *Mental models: Towards a cognitive science of language, inference, and consciousness (No. 6)*. Harvard University Press.

Jordan, N. C., Glutting, J., Ramineni, C., & Watkins, M. W. (2010). Validating a number sense screening tool for use in kindergarten and first grade: Prediction of mathematics proficiency in third grade. *School Psychology Review*, 39(2), 181–185.

Kaufman, S. B. (2007). Sex differences in mental rotation and spatial visualization ability: Can they be accounted for by differences in working memory capacity? *Intelligence*, 35, 211–223. https://doi.org/10.1016/j.intell.2006.07.00.

Kell, H. J., Lubinski, D., Benbow, C. P., & Steiger, J. H. (2013). Creativity and technical innovation spatial ability's unique role. *Psychological Science*, 24(9), 1831–1836.

Kintsch, W., & Greeno, J. G. (1985). Understanding and solving word arithmetic problems. *Psychological Review*, 92(1), 109.

Knauff, M., & Johnson-Laird, P. N. (2002). Visual imagery can impede reasoning. *Memory and Cognition*, 30(3), 363–371.

Kozhevnikov, M., Hegarty, M., & Mayer, R. (1999). Students' use of imagery in solving qualitative problems in kinematics. (Report No. 143). US. (ERIC Document Reproduction Service No. ED 433 239). Retrieved from ERIC Website http://eric.ed.gov/?id=ED433239.

Kozhevnikov, M., Motes, M. A., & Hegarty, M. (2007). Spatial visualization in physics problem solving. *Cognitive Science, 31*(4), 549–579.

Kucian, K., Grond, U., Kotzer, S., Henzi, B., Schonmann, C., Plangger, F, et al (2011). Mental number line training in children with developmental dyscalculia. *Neuroimage, 57,* 782–795.

Kyllonen, P. C., & Christal, R. E. (1990). Reasoning ability is (little more than) working-memory capacity?! *Intelligence, 14*(4), 389–433.

Kyttälä, M., Aunio, P., Lehto, J. E., Van Luit, J., & Hautamäki, J. (2003). Visuospatial working memory and early numeracy. *Educational and Child Psychology, 20*(3), 65–76.

Lakoff, G., & Núñez, R. E. (2000). *Where mathematics comes from: How the embodied mind brings mathematics into being* (Basic books).

Landy, D., & Goldstone, R. L. (2007). How abstract is symbolic thought? *Journal of Experimental Psychology: Learning, Memory, and Cognition, 33*(4), 720–733.

Laski, E. V., Casey, B. M., Yu, Q., Dulaney, A., Heyman, M., & Dearing, E. (2013). Spatial skills as a predictor of first grade girls' use of higher level arithmetic strategies. *Learning and Individual Differences, 23,* 123–130.

Lauer, J. E., & Lourenco, S. F. (2016). Spatial processing in infancy predicts both spatial and mathematical aptitude in childhood. *Psychological Science, 27*(10), 1291–1298.

LeFevre, J. A., Lira, C. J., Sowinski, C., Cankaya, O., Kamawar, D., & Skwarchuk, S. L. (2013). Charting the role of the number line in mathematical development. *Frontiers in Psychology, 4.*

Levine, S. C., Huttenlocher, J., Taylor, A., & Langrock, A. (1999). Early sex differences in spatial skill. *Developmental Psychology, 35*(4), 940.

Li, Y., & Geary, D. C. (2013). Developmental gains in visuospatial memory predict gains in mathematics achievement. *PloS one, 8*(7), e70160.

Li, Y., & Geary, D. C. (2017). Children's visuospatial memory predicts mathematics achievement through early adolescence. *PLoS One, 12*(2), e0172046. https://doi.org/10.1371/journal.pone.0172046.

Link, T., Nuerk, H. C., & Moeller, K. (2014). On the relation between the mental number line and arithmetic competencies. *The Quarterly Journal of Experimental Psychology, 67,* 1597–1613.

Linn, M. C., & Petersen, A. C. (1985). Emergence and characterization of sex differences in spatial ability: A meta-analysis. *Child Development, 56,* 138–151. https://doi.org/10.2307/1130467.

Lohman, D. F. (1996). Spatial ability and g. In I. Dennis, & P. Tapsfield (Eds.), *Human abilities: Their nature and measurement* (pp. 97–116). Hillsdale, NJ: Erlbaum.

Lourenco, S. F. (2015). On the relation between numerical and non-numerical magnitudes: Evidence for a general magnitude system. In D. C. Geary, D. B. Berch, & K. Mann-Koepke (Eds.). D. C. Geary, D. B. Berch, & K. Mann-Koepke (Eds.), *Mathematical cognition and learning: Evolutionary origins and early development of basic number processing: (Vol. 1).* (pp. 145–174). London: Elsevier Academic Press. https://doi.org/10.1016/B978-0-12-420133-0.00007-7.

Lourenco, S. F., & Levine, S. C. (2008). Early numerical representations and the natural numbers: Is there really a complete disconnect? *Behavioral and Brain Sciences, 31*(06), 660.

Lourenco, S. F., & Longo, M. R. (2009). Multiple spatial representations of number: Evidence for co-existing compressive and linear scales. *Experimental Brain Research, 193*(1), 151–156.

Lourenco, S. F., & Longo, M. R. (2010). General magnitude representation in human infants. *Psychological Science, 21*(6), 873–881.

Lowrie, T., Logan, T., & Ramful, A. (2017). Visuospatial training improves elementary students' mathematics performance. *British Journal of Educational Psychology, 87*(2), 170–186.

McCrink, K., Dehaene, S., & Dehaene-Lambertz, G. (2007). Moving along the number line: Operational momentum in nonsymbolic arithmetic. *Attention, Perception, and Psychophysics, 69*(8), 1324–1333.

McNeil, N. M., & Uttal, D. H. (2009). Rethinking the use of concrete materials in learning: Perspectives from development and education. *Child Development Perspectives, 3*(3), 137–139.

Mix, K. S., & Cheng, Y. L. (2011). The relation between space and math: Developmental and educational implications. *Advances in Child Development and Behavior, 42*, 197–243.

Mix, K. S., Huttenlocher, J., & Levine, S. C. (1996). Do preschool children recognize auditory-visual numerical correspondences? *Child Development*, 1592–1608.

Mix, K. S., Levine, S. C., Cheng, Y. L., Young, C., Hambrick, D. Z., Ping, R., et al. (2016). Separate but correlated: The latent structure of space and mathematics across development. *Journal of Experimental Psychology: General, 145*(9), 1206.

Miyake, A., Friedman, N. P., Rettinger, D. A., Shah, P., & Hegarty, M. (2001). How are visuospatial working memory, executive functioning, and spatial abilities related? A latent-variable analysis. *Journal of Experimental Psychology: General, 130*, 621–640.

Moss, J., Hawes, Z., Naqvi, S., & Caswell, B. (2015). Adapting Japanese lesson study to enhance the teaching and learning of geometry and spatial reasoning in early years classrooms: A case study. *Zdm, 47*(3), 377–390. https://doi.org/10.1007/s11858-015-0679-2.

Mowat, E., & Davis, B. (2010). Interpreting embodied mathematics using network theory: Implications for mathematics education. *Complicity: An International Journal of Complexity and Education, 7*(1).

National Research Council (US), Committee on Support for Thinking Spatially, & Downs, R. M. (2006). *Learning to think spatially*. National Academies Press.

Neuburger, S., Jansen, P., Heil, M., & Quaiser-Pohl, C. (2011). Gender differences in pre-adolescents' mental-rotation performance: Do they depend on grade and stimulus type? *Personality and Individual Differences, 50*(8), 1238–1242. https://doi.org/10.1016/j.paid.2011.02.017.

Newcombe, N. S. (2013). Seeing relationships: Using spatial thinking to teach science, mathematics, and social studies. *American Educator, 37*, 26–31.

Opfer, J. E., Thompson, C. A., & Furlong, E. E. (2010). Early development of spatial-numeric associations: Evidence from spatial and quantitative performance of preschoolers. *Developmental Science, 13*, 761–771.

Pellegrino, J. W., Hunt, E. B., Abate, R., & Farr, S. (1987). A computer-based test battery for the assessment of static and dynamic spatial reasoning abilities. *Behavior Research Methods, Instruments, and Computers, 19*(2), 231–236.

Pinhas, M., & Fischer, M. H. (2008). Mental movements without magnitude? A study of spatial biases in symbolic arithmetic. *Cognition, 109*(3), 408–415.

Protzko, J. (2015). The environment in raising early intelligence: A meta-analysis of the fadeout effect. *Intelligence, 53*, 202–210.

Qin, S., Cho, S., Chen, T., Rosenberg-Lee, M., Geary, D. C., & Menon, V. (2014). Hippocampal-neocortical functional reorganization underlies children's cognitive development. *Nature Neuroscience, 17*(9), 1263–1269.

Ramani, G. B., & Siegler, R. S. (2008). Promoting broad and stable improvements in low-income children's numerical knowledge through playing number board games. *Child Development, 79*, 375–394.

Ramani, G. B., Siegler, R. S., & Hitti, A. (2012). Taking it to the classroom: Number board games as a small group learning activity. *Journal of Educational Psychology, 104*, 661–672.

Ramful, A., Lowrie, T., & Logan, T. (2017). Measurement of spatial ability: Construction and validation of the spatial reasoning instrument for middle school students. *Journal of Psychoeducational Assessment*, 1–19. https://doi.org/10.1177/073428216659207. Advance online publication.

Rugani, R., Vallortigara, G., Priftis, K., & Regolin, L. (2015). Number-space mapping in the newborn chick resembles humans' mental number line. *Science, 347*, 534–536.

Sala, G., Bolognese, M., & Gobet, F. (2017). Spatial training and mathematics: The moderating effect of handedness. In *Proceedings of the 39th Annual Meeting of the Cognitive Science Society* (pp. 3039–3044). London, UK: Cognitive Science Society.

Sasanguie, D., Göbel, S. M., Moll, K., Smets, K., & Reynvoet, B. (2013). Approximate number sense, symbolic number processing, or number–space mappings: What underlies mathematics achievement? *Journal of Experimental Child Psychology, 114*(3), 418–431.

Schneider, M., Grabner, R. H., & Paetsch, J. (2009). Mental number line, number line estimation, and mathematical achievement: Their interrelations in Grades 5 and 6. *Journal of Educational Psychology, 101*, 359–372.

Shepard, R. N., & Metzler, J. (1971). Mental rotation of three-dimensional objects. *Science, 171*(3972), 701–703.

Siegler, R. S., & Booth, J. L. (2004). Development of numerical estimation in young children. *Child Development, 75*, 428–444.

Siegler, R. S., & Opfer, J. E. (2003). The development of numerical estimation: Evidence for multiple representations of numerical quantity. *Psychological Science, 14*, 237–243.

Siegler, R. S., & Ramani, G. B. (2008). Playing board games promotes low-income children's numerical development. *Developmental Science, 11*, 655–661.

Siegler, R. S., & Ramani, G. B. (2009). Playing linear number board games—but not circular ones—improves low-income preschoolers' numerical understanding. *Journal of Educational Psychology, 101*, 545–560.

Siegler, R. S., Thompson, C. A., & Opfer, J. E. (2009). The logarithmic-to-linear shift: One learning sequence, many tasks, many time scales. *Mind, Brain, and Education, 2*, 143–150.

Slusser, E. B., Santiago, R. T., & Barth, H. C. (2013). Developmental change in numerical estimation. *Journal of Experimental Psychology: General, 142*, 193–208.

Starr, A., Libertus, M. E., & Brannon, E. M. (2013). Number sense in infancy predicts mathematical abilities in childhood. *Proceedings of the National Academy of Sciences, 110*(45), 18116–18120.

Thompson, J. M., Nuerk, H. C., Moeller, K., & Kadosh, R. C. (2013). The link between mental rotation ability and basic numerical representations. *Acta Psychologica, 144*(2), 324–331.

Thurstone, T. G. (1974). *PMA readiness level.* Chicago, IL: Science Research Associates.

Uttal, D. H., & Cohen, C. A. (2012). 4 spatial thinking and STEM Education: When, why, and how? *Psychology of Learning and Motivation-advances in Research and Theory, 57,* 147.

Verdine, B. N., Golinkoff, R. M., Hirsh-Pasek, K., & Newcombe, N. S. (2017). Links between spatial and mathematical skills across the preschool years. *Monographs of the Society for Research in Child Development, 82*(1).

Verdine, B. N., Golinkoff, R. M., Hirsh-Pasek, K., Newcombe, N. S., Filipowicz, A. T., & Chang, A. (2014). Deconstructing building blocks: Preschoolers' spatial assembly performance relates to early mathematical skills. *Child Development, 85*(3), 1062–1076. https://doi.org/10.1111/cdev.12165.

Verdine, B. N., Irwin, C. M., Golinkoff, R. M., & Hirsh-Pasek, K. (2014). Contributions of executive function and spatial skills to preschool mathematics achievement. *Journal of Experimental Child Psychology, 126,* 37–51. https://doi.org/10.1016/j.jecp.2014.02.012.

Viarouge, A., Hubbard, E. M., Dehaene, S., & Sackur, J. (2010). Number line compression and the illusory perception of random numbers. *Experimental Psychology, 57,* 446–454.

Vosniadou, S., & Brewer, W. F. (1994). Mental models of the day/night cycle. *Cognitive Science, 18*(1), 123–183.

Wai, J., Lubinski, D., & Benbow, C. P. (2009). Spatial ability for STEM domains: Aligning over 50 years of cumulative psychological knowledge solidifies its importance. *Journal of Educational Psychology, 101*(4), 817.

Wechsler, D. (2009). *Wechsler individual achievement test* (3rd ed.). San Antonio, TX: NCS Pearson.

Wolfgang, C. H., Stannard, L. L., & Jones, I. (2001). Block play performance among preschoolers as a predictor of later school achievement in mathematics. *Journal of Research in Childhood Education, 15*(2), 173–180. https://doi.org/10.1080/02568540109594958.

Woodcock, R. W., Mather, N., McGrew, K. S., & Schrank, F. A. (2001). *Woodcock-Johnson III tests of cognitive abilities.* Itasca, IL: Riverside Publishing Company.

Woodcock, R. W., McGrew, K. S., & Mather, N. (2001). *Woodcock-Johnson test of achievement.* Rolling Meadows IL: Riverside Publishing.

Wood, G., Willmes, K., & Nuerk, H.-C., & Fischer, M. H. (2008). On the cognitive link between space and number: A meta-analysis of the SNARC effect. *Psychology Science, 50*(4), 489–525.

Zacks, J. M., Willmes, K., Nuerk, H.-C., & Fischer, M. H. (2008). Neuroimaging studies of mental rotation: A meta-analysis and review. *Journal of Cognitive Neuroscience, 20*(1), 1–19.

Zago, L., Pesenti, M., Mellet, E., Crivello, F., Mazoyer, B., & Tzourio-Mazoyer, N. (2001). Neural correlates of simple and complex mental calculation. *Neuroimage*, *13*(2), 314–327.

Zimmermann, W., & Cunningham, S. (1991). Editor's introduction: What is mathematical visualization. In *Visualization in teaching and learning mathematics*. pp. 1–7.

Neurocognitive Evidence for Spatial Contributions to Numerical Cognition

André Knops

Humboldt-Universität zu Berlin, Berlin, Germany

Numerical cognition involves the joint interplay of various cognitive capacities and domains ranging from language and executive functions to different memory systems. Here, I focus on the contribution of spatial processes to numerical cognition. To understand the cognitive and neural processes underlying numerical competencies, we need to identify and characterize the individual core processes that contribute to this multifaceted capacity. Although different theoretical models have been proposed, most researchers would agree on the distinction of the perception and representation of simple quantities that can be extracted from different formats (symbolic vs. nonsymbolic), arithmetic fact retrieval from long-term memory (e.g., multiplication tables), and procedural and approximate solving of arithmetic problems whose solutions cannot be retrieved from long-term memory (e.g., $56 - 27 = ?$). One of the most surprising insights from numerical cognition research over the past years has been the finding that each of these three facets is subject to influences from spatial processes.

This chapter is organized into three sections. (1) The Spatial Layout of the Mental Number Line will elaborate on the idea that the mental representation of numerical magnitude is spatially organized. The most commonly used metaphor to characterize the numerical magnitude representation is the mental number line (MNL). According to this idea, numerical quantity is mentally represented along a spatially oriented line with smaller numbers left from larger numbers (at least in Western societies). (2) The Organization of Arithmetic Facts in Memory describes the influence that the spatially organized mental magnitude representation exerts on retrieval of arithmetic facts from long-term memory (e.g., when retrieving the result for problems such as 2×6). During the first years of schooling, many Western and Asian curricula require the memorization of multiplication tables (e.g., $1 \times 6 = 6, 2 \times 6 = 12$, etc.). As a consequence, these arithmetic facts can later be retrieved from long-term memory without requiring any arduous computational procedures. These problems are stored in a highly interconnected associative network (see, for example, Ashcraft, 1987, pp. 302–338; Campbell, 1995; Galfano, Penolazzi, Vervaeck, Angrilli, & Umilta, 2009; Verguts & Fias, 2005). In this associative network, each operand representation is thought to be connected with the set of its multiples (e.g., the operand 7 is associated with the multiples 14, 21, 28, etc.). How a spatial organization of semantic content in memory may contribute to arithmetic fact retrieval will be addressed in The Organization of Arithmetic Facts in Memory section. (3) Mental Arithmetic as a Prime Example of Neuronal Recycling introduces the idea that mental arithmetic, which is the effortful and strategy-based solution of problems that cannot be retrieved from long-term memory, builds from foundational concepts (space, time, and number) by progressively co-opting cortical areas whose prior organization fits with the cultural need. In particular, parietal cortex hosts coordinate transformation processes that help to transform information from different reference frames to guide action and navigation. For example, to compute the position of an object in space with respect to the head position, the sensory system combines eye-centered information with the information about the eyes' posture (Beck, Latham, & Pouget, 2011). This process involves computational principles that may lend themselves to determining the outcome of an arithmetic problem by combining the positional information of the operands on the MNL. I present evidence for the idea that mental arithmetic "recycles" the operational principles of cortical circuits that evolved for spatial processes.

THE SPATIAL LAYOUT OF THE MENTAL NUMBER LINE

The mental representation of numerical magnitude is often described with the metaphor of an MNL where smaller numbers are located left from larger numbers. When perceiving numerals, a position on the MNL is activated. However, because of the analog nature of the MNL and the ubiquitous presence of noise in the cognitive system, neighboring positions (i.e., numbers) receive some coactivation as well. The closer two numerals are on the MNL the more the activation overlaps, rendering magnitude comparisons (which number is larger) increasingly difficult with decreasing numerical distance between to-be-compared numbers. Additionally, the MNL is compressed, meaning that the distances between neighboring numbers decrease as numerical magnitude increases (logarithmic model; Dehaene, 2003). Many researchers understand the spatial numerical association of response codes (SNARC) as evidence supporting a spatially organized mental magnitude representation. The SNARC effect describes the finding that in various tasks (e.g., parity or magnitude judgment) smaller numbers are associated with faster left-sided responses compared with right-sided responses, while a corresponding association exists between right-sided responses and larger numbers (Dehaene, Bossini, & Giraux, 1993). According to the most commonly held interpretation, this association is because of the correspondence between a given number's position on the MNL and the response side. Because larger numbers are represented right from smaller numbers, they are associated with right-sided responses. A similar congruency effect has been observed in the evaluation of numerical distances between adjacent numbers in a triplet (Koten, Lonnemann, Willmes, & Knops, 2011). Participants are faster when judging the side with the smaller numerical interval in a triplet such as 53__62_____98 compared with a triplet of the form 53_____62__98 because of the (task-irrelevant) congruency between numerical and spatial distances that separate numbers. Using functional magnetic resonance imaging (fMRI), we also demonstrated that this interference is located on a central semantic processing instance rather than reflecting mere response conflicts. Recently, a vertical SNARC effect has been reported implying an association of smaller numerical values with "down" and larger numerical values with "up" (Hartmann, Gashaj, Stahnke, & Mast, 2014). Larger numbers also facilitate the detection of stimuli in the upper visual field, while smaller numbers facilitate the detection of stimuli at the bottom of the visual field (Pecher, Van Dantzig, Boot, Zanolie, & Huber, 2010). Similarly, a "sagittal" SNARC effect has been reported in Japanese participants with smaller numbers eliciting faster responses close to the body and larger numbers with responses further away from the body (Ito & Hatta, 2004). Although these findings are experimentally less robust compared with the classical SNARC, a vivid discussion emerged whether vertical or horizontal number–space associations would prevail (Gertner, Henik, Reznik, & Cohen Kadosh, 2013; Wiemers, Bekkering, & Lindemann, 2014; Winter, Matlock, Shaki, & Fischer, 2015). This discussion is inspired by the embodied cognition idea, which holds that cognitive concepts are grounded in the sensorimotor experiences with the external world. Accordingly, several origins of the previously described number–space mappings have been discussed. Reading direction has been shown to modulate the horizontal SNARC effect (Shaki & Fischer, 2008). However, the sagittal SNARC (smaller numbers closer to the body than larger numbers) stands in contrast with the top-down reading direction in Japanese (Ito & Hatta, 2004). Graphical experiences such as scientific graphs, the thermometer, or the arrangements of floor numbers in elevators have been discussed as the origin for an association of the vertical dimension with numerical magnitude (Hubbard, Piazza, Pinel, & Dehaene, 2005). When pouring water into a glass or adding things to a pile, more is associated with up and hence this natural correlation between verticality and quantity is learnt during development and gives rise to the vertical number–space mapping (Winter et al., 2015). Another line of debate concerns the question whether or not these congruity effects predominantly reflect a temporally stable spatial organization of the internal magnitude representation or a volatile ad hoc mapping of numbers to space according to current task demands and experimental setting (Dollman & Levine, 2016; Ginsburg, van Dijck, Previtali, Fias, & Gevers, 2014; Santens & Gevers, 2008; van Dijck & Fias, 2011). While the jury is still out on the temporal prevalence of spatial–numerical associations, a recent report of a spatial association of larger numerosities (i.e., the number of items in a set) and the right side of space in newly hatched chicks provides compelling evidence for a culture- and education-free mapping that emerges spontaneously (Rugani, Vallortigara, Priftis, & Regolin, 2015).

More compelling evidence for a spatial organization of the MNL comes from the attentional SNARC. Fischer, Castel, Dodd, and Pratt (2003) discovered that noninformative central numerical cues induce attentional shifts: In accordance with a spatially organized MNL, small and large numbers facilitated detection of targets in the left and right visual hemifields, respectively. The effect has been extended to neuroscientific measures such as modulations of neural activity in occipital cortex as measured with fMRI (Goffaux, Martin, Dormal, Goebel, & Schiltz, 2012) or increased P1 and P3 amplitudes in EEG (Salillas, El Yagoubi, & Semenza, 2008). It is also reflected in attention-related components of the event-related potentials (ERPs) early directing attention negativity EDAN and anterior directing attention negativity ADAN; (Ranzini, Dehaene, Piazza, & Hubbard, 2009). Recent evidence suggests that these ERP correlates of number-induced attentional cueing are not restricted to mere detection tasks but generalize to color discrimination (Schuller, Hoffmann, Goffaux, & Schiltz, 2015). A similar attentional bias has been observed to influence the luminosity judgments of vertically arranged luminance gradients (2 bars that gradually changed from black to white in opposite left-to-right direction). In this grayscale task, smaller (larger) numbers oriented attention to the left (right) and correspondingly biased the luminosity judgments (Nicholls, Loftus, & Gevers, 2008). Two recent studies substantiate the interaction between spatial attention and a spatially oriented numerical magnitude representation, showing that voluntary and reflexive eye movements (both associated by corresponding attentional shifts) modulate the processing of numerical magnitude (Ranzini et al., 2015; Ranzini, Lisi, & Zorzi, 2016). These results suggest that a given number elicits attentional shifts according to its position on a left-to-right–oriented MNL, irrespective of congruency with lateralized responses as in the SNARC paradigm. Likewise, the attentional system seems to exert substantial influence on the processing of numerical magnitude.

A spatial organization of numerical quantity has also been observed at the neural level. A recent study combined ultrahigh field imaging with a predictive coding approach (Harvey et al. 2013) and revealed a topographic organization of small (1–7) nonsymbolic numerosities in human posterior parietal cortex. In particular, voxels exhibited a clear numerosity preference and exhibited a systematic spatial arrangement from smaller to larger numerosities as one moves from medial to lateral posterior superior parietal cortex. Posterior superior parietal cortex has also been demonstrated to contain topographic information of individuated objects that the visual system can flexibly exploit for the current goals. This attention-grabbing mechanism appears to provide relevant information for visual short-term memory, object tracking, grasping, or the enumeration of sets of objects (Knops, 2016; Knops, Piazza, Sengupta, Eger, & Melcher, 2014) and may take the form of a priority map. In nonhuman primates, too, a labeled line coding of numerosities has been observed where single neurons in ventral intraparietal cortex act as numerosity filters, showing maximal firing rate for a given preferred numerosity and monotonically decreasing firing rates as numerical distance between perceived numerosity and preferred numerosity increases (Nieder & Miller, 2004; Nieder, 2016). At the neural and behavioral level, evidence shows that with increasing numerical magnitude, the overlap between neighboring numbers increases, either because of compression of the underlying scale in combination with constant width of activation (logarithmic model; Dehaene, 2003) or because of increasing variability and a linearly spaced underlying scale (linear model; Gallistel and Gelman, 1992).

Together, these studies converge on the notion that numerical quantity is represented along a spatial continuum, both neurally and mentally. The exact orientation and layout of this representation may be subject to cultural and bodily experiences but shows a certain preference for placing smaller numbers left from larger numbers. The next section describes the impact the spatial organization of the MNL may have on the retrieval of arithmetic facts from long-term memory.

THE ORGANIZATION OF ARITHMETIC FACTS IN MEMORY

Healthy adults solve simple multiplications (i.e., problems with one-digit operands) mostly by retrieving the solution from a dedicated memory system, which takes the form of a highly interconnected associative network (see, for example, Ashcraft, 1987, pp. 302–338; Campbell, 1995; Verguts & Fias, 2005). On the presentation of an arithmetic problem (e.g., 7×3), both the correct result (i.e., the product, 21 in this example) and the associated multiples ([14, 28, 35,...] for the operand 7; [6, 9, 12,...] for the operand 3) are activated because of an activation spreading inside the associative network from the operand representations to the result representations (Galfano, Mazza, Angrilli, & Umilta, 2004; Galfano et al., 2009; Galfano, Rusconi, & Umilta, 2003; Niedeggen, Rösler, & Jost, 1999). In addition to the association between operand and result nodes, it is assumed that result nodes are also associated with each other. In other words, result nodes can spread activation among each other. The retrieval process is assumed to be driven by this activation spreading (among operand nodes and result nodes) within the associative network. Following the activation spreading, the result node with the highest activation (i.e., the product if the problem is solved correctly) is selected as the solution of the multiplication.

The interacting neighbors model (Verguts & Fias, 2005) assumes that results are represented in a componential fashion in two different structures, following the syntax of the base-10 system (e.g., the number 21 means 2 decades and 1 unit). For example, given the operand pair 3×7, the result 21 is represented by the coactivation of the decade number 2 and the unit number 1. During the retrieval process, activation is spread to both decades and units. Consistency of decade and unit digits of the results is a second core feature of this model. When activation spreads to the result nodes, decade and unit representations can cooperate or compete between neighboring results. Neighbors are defined as the subset of problems coactivated during the retrieval process and within a distance of ±2 from the considered problem. Namely, given the problem $N \times M$, the eight coactivated neighbors are $(N \pm 1) \times M$, $(N \pm 2) \times M$, $N \times (M \pm 1)$, and $N \times (M \pm 2)$. Cooperation occurs when (consistent) neighbors activate the same decade or unit, whereas competition occurs when (inconsistent) neighbors activate a different decade or unit.

A key question for those models refers to the quality of the representation underlying the nodes in the associative memory network. In particular, the question is whether the nodes are purely abstract (e.g., do not carry semantic magnitude information), and if so, how the referred to numbers relate to the semantic magnitude code. Early neuropsychological studies suggest that the entries in the semantic network of arithmetic facts are to a certain extent independent from the semantic magnitude code (Dehaene & Cohen, 1997). The network interference model (Campbell, 1995) assumes that (1) during the retrieval process both a symbolic code and a magnitude representation are activated; (2) both representations contribute to the activation spreading; and (3) the problem size effect originates form the compressed metric of the ANS. The interacting neighbors model, however, remains mute to the question at what level cooperation or competition occurs—a semantic or syntactic level.

An earlier EEG study provided compelling evidence for a semantic influence on the arithmetic fact network. In a verification task, the amplitude of the N400, a component indexing the degree to which a semantic expectation is violated, varied as a function of the numerical distance between incorrect but related results (Niedeggen et al., 1999). For example, the N400 had a larger amplitude when participants judged a problem such as $4 \times 6 = 30$ (related because 30 is part of the 6 series) compared with the—equally incorrect and related but numerically more distant—problem $4 \times 6 = 42$. Hence, the numerical distance between the semantic codes of these numbers influences the amount of coactivation in the associative network for arithmetic facts.

In a recent behavioral study, an asymmetric interference effect during a result verification task was observed (Didino, Knops, Vespignani, & Kornpetpanee, 2015). Participants were asked to evaluate the correctness of arithmetic problems (e.g., $8 \times 4 = 32$). The proposed incorrect result could either be a below multiple (i.e., smaller entry in one of the operands' tables; e.g., $8 \times 4 = 24$), an above multiple (e.g., larger entry in one of the operands' tables; e.g., $8 \times 4 = 40$), a below neutral ($8 \times 4 = 25$), or an above neutral ($8 \times 4 = 39$). Above multiples were rejected significantly slower than below multiples, while neutral results did not show this effect. For example, participants were slower when rejecting $8 \times 4 = 40$ (above) compared with rejecting $8 \times 4 = 24$ (below). This asymmetric interference effect increased with problem size and is indicative of a functional connection between the MNL and the associative memory network for arithmetic facts. It suggests that the compressed metric of the MNL influences the activation spreading in the associative memory network for arithmetic facts. Alternatively, it may indicate that the network itself reflects the semantic relationship between nodes, for example, by a variable strength of associations between entries.

Together, these results suggest that arithmetic fact retrieval is influenced by the characteristics of the underlying numerical metric. In particular, the amount of overlap between two entries on the MNL determines the degree to which entries in a given number table get coactivated on presentation of a given problem. The overlap on the MNL, in turn, depends on the numerical distance and the compression of the MNL.

MENTAL ARITHMETIC AS A PRIME EXAMPLE OF NEURONAL RECYCLING

Sometime, solutions to numerical problems cannot be simply retrieved from long-term memory. In this case we need to engage in more effortful arithmetic procedures or approximation of the outcome. In the third section of this chapter, I present the idea that this may involve the contribution of cortical processes that have evolutionarily evolved for spatial processing. The ability to use symbolic mathematics is one of the key cultural achievements of mankind. It allows building computers, flying to the moon, or calculating the money needed to buy a house. With respect to the time scale of our evolutionary development, the breathtaking speed of cultural inventions during the last ~5000 years since the first written traces of arithmetic appeared in Mesopotamia is tremendous. Although modern humans have benefited from a massive increase in brain volume, the evolutionary time frame was not sufficient to develop dedicated brain regions for mathematics. Hence, it has been proposed that during development, the neural implementation of mathematical concepts and operations is accomplished by occupying a "cortical niche" in the brain. In evolutionary biology, the term "exaptation" describes how a given system that emerged for a particular function is co-opted to subsequently serve another function. In cognitive neuroscience, related concepts of reuse have been proposed. *Neuronal recycling* (Dehaene & Cohen, 2007) assumes that cultural inventions are neurally implemented by co-opting cortical circuits that have evolved for evolutionary older functions. Neuronal recycling has been discussed as a basic mechanism that underlies functions in a variety of domains such as language, numerical reasoning, or social cognition (Parkinson, Liu, & Wheatley, 2014; Parkinson & Wheatley, 2013). Similarly, Anderson (2010) coined the concept as *massive redeployment*. The main difference between neuronal recycling and massive redeployment is the temporal scope of the theory (ontogeny and phylogeny, respectively). While neuronal recycling refers to the acquisition of cultural skills during life span development, the massive redeployment account is "about the evolutionary emergence of the functional organization of the brain" (Anderson, 2010). The neural recycling hypothesis holds that a given brain circuit is co-opted by cultural functions that have not (yet) developed dedicated brain areas. High-level cultural functions (e.g., reading and mathematics) build from foundational concepts (e.g., face processing and space and number, respectively) by progressively co-opting cortical areas whose prior organization fits with the cultural need. The functional scope of the co-opted brain region is enlarged.

How can we identify the involvement of a particular neural system X and its original associated process x in a given process y? The following criteria have been formulated with respect to the idea of neuronal recycling (Dehaene & Cohen, 2007, p. 385):

1. Variability in the cerebral representation of a cultural invention should be limited.
2. Cultural variability should also be limited.
3. The speed and ease of cultural acquisition in children should be predictable based on the complexity of the cortical remapping required.
4. Although acculturation often leads to massive cognitive gains, it might be possible to identify small losses in perceptual and cognitive abilities due to competition of the new cultural ability with the evolutionarily older function in relevant cortical regions.

Applying these criteria, Dehaene and Cohen identified arithmetic as one candidate domain for the concept of cultural recycling. As described in Section 1, numbers activate circumscribed areas in the horizontal aspect of the intraparietal sulcus, irrespective of format or cultural background of the participants. Parietal cortex can be considered one of the key cortical structures dedicated to the transformation of visuospatial information in the course of synchronizing action and perception. Amalric and Dehaene (2016) further demonstrate that expert mathematicians recruit largely nonverbal areas in parietal cortex when evaluating mathematical expressions, even in the absence of numbers in the stimuli. This provides further evidence for a privileged role of spatial processing circuits in mathematical thinking. The question then becomes what particular spatial functions are recruited during numerical cognition? In Section 1, I have argued that the mental magnitude representation presumably takes the form of a spatially organized line. Here, I will argue that the spatial system also contributes to the transformation of numerical content during the course of mental arithmetic that goes beyond arithmetic fact retrieval.

A number of spatial processes have been associated with arithmetic performance. Mental rotation abilities—a prime example of analog spatial transformations—correlate with numerical abilities (see chapters by Newcombe and Lourenco) but not with executive functions (Thompson, Nuerk, Moeller, and Cohen Kadosh, 2013). Visual attention helps understanding algebraic expressions (e.g., $a \times b + c \times d$) by structuring elements in adequate groups (Marghetis, Landy, and Goldstone, 2016). In adults, dual-task paradigms have shown that both visuospatial working memory and phonological working memory share processing resources with various arithmetic operations such as multiplication and subtraction (Cavdaroglu & Knops, 2016). In children, visuospatial working memory has been found to be more important for arithmetic problem solving than the phonological working memory or the central executive (Ashkenazi, Rosenberg-Lee, Metcalfe, Swigart, & Menon, 2013; Metcalfe, Ashkenazi,

Rosenberg-Lee, & Menon, 2013). These results imply that transformation and maintenance of spatial information share cognitive resources with mental arithmetic. However, these associations remain correlational and may be due to domain-general factors such as general intelligence in case of mental rotation and unspecific dual-task costs in case of the interference effects. To take this one step further and demonstrate that mental arithmetic recycles spatial circuits, we can deploy neuroimaging to test the following hypotheses that can be developed from the neuronal recycling framework.

If a given area X underlies process x and is "recycled" to also serve process y,

P1. the functional signature of area X should be reflected in processes x and y. That is, the behavioral characteristics in terms of cognitive biases and effects in the co-opting domain (y) should reflect behavioral characteristics of the processes initially associated with area X. The involvement of X in y might lead to particular cognitive biases.

P2. we should be able to identify, at the neural level, similar contributions of area X in both processes x and y. This entails
    a. common, overlapping activity and
    b. similar spatial pattern of activation in both contexts x and y (see also Mather, Cacioppo, & Kanwisher, 2013). Going beyond previous research, this stipulates that cortical circuits exhibit sufficiently stable activation patterns across tasks and domains.

At the neural level, mental arithmetic may be instantiated in the functional interactions between areas along the intraparietal sulcus and posterior superior parietal areas (Hubbard et al., 2005). Knops et al. tested this hypothesis. Specifically, we tested prediction P2, stipulating that neuronal recycling should lead to a stable pattern of activation in a given area in the context of a given cognitive process (e.g., attentional shifts) that can be identified in the context of a different process (e.g., mental arithmetic). The study involved a saccade task (participants fixated a stimulus on screen that horizontally changed position every 500–1000 ms) to induce shifts of spatial attention, as well as a symbolic and nonsymbolic calculation paradigm. Targeting the putative homolog of the lateral intraparietal area (LIP) in monkeys, we employed a multivariate classifier to activation data from posterior superior parietal lobe (PSPL) (voxels shown in Fig. 11.1C) to distinguish between leftward and rightward saccades. Without further training, this classifier then successfully differentiated between addition and subtraction trials from the calculation task. This generalization was observed with numbers presented either as Arabic symbols or as nonsymbolic sets of dots, which implies shared cognitive processes between both formats. Results are depicted in Fig. 11.1.

The observed generalization implies that mental arithmetic superimposes on a parietal circuitry originally evolved for spatial coding. Our results

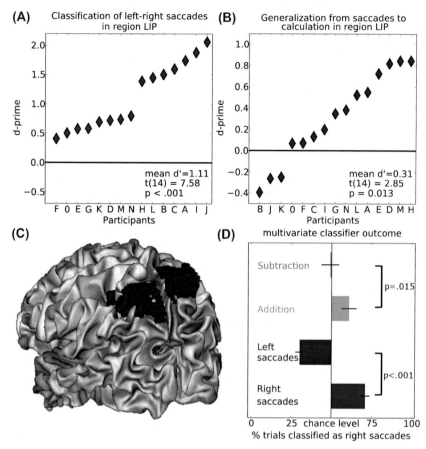

FIGURE 11.1   (A) Classification performance (denoted in d', where larger values indicate better performance) for each participant in the saccades task (participants are sorted according to d'). (B) Classification performance (d') per participant for generalization of the classifier trained on left/right saccades to subtraction/addition trials. (C) Voxel clusters in left and right PSPL region that resulted from the saccade localizer task and served as region of interest for the classifier, rendered on the white matter/gray matter boundary. (D) Percentages of trials classified as right saccades for subtraction (orange), addition (light blue), and left and right saccades (red and dark blue, respectively). *LIP*, lateral intraparietal area; *PSPL*, posterior superior parietal lobe. *From Knops, A., Thirion, B., Hubbard, E. M., Michel, V., & Dehaene, S. (2009). Recruitment of an area involved in eye movements during mental arithmetic. Science, 324(5934), 1583–1585. Reprinted with permission from AAAS.*

confirm the above hypothesis that mental calculation can be likened to a spatial shift along an MNL. In a certain sense, when a Western participant calculates 18+5, the activation moves "rightward" on the MNL from 18 to 23. This spatial shift recycles neural circuitry in PSPL shared with those involved in updating spatial information during saccadic eye movements and therefore confirms prediction P2. The PSPL area, perhaps because of its capacity for vector addition during eye movement computation (Pouget, Deneve, & Duhamel, 2002), appears to have a connectivity or internal structure that meets the functional requirements to successfully engage in mental arithmetic.

It should be emphasized that this type of multivariate analysis takes into account the spatial structure of the activity associated with a given cognitive process. Therein it goes substantially beyond the mere report of overlapping activity. Rather, it establishes a substantial similarity in the relative weight by which different cortical circuits in a given region of interest contribute to the cognitive processes at hand. Fig. 11.2 exemplifies the theoretical relationship between the spatial structures of two cognitive processes that are associated with above-threshold activity in a given cortical area.

Addition and subtraction have recently been demonstrated to be subject to systematic biases that may have their origin in the neuronal recycling.

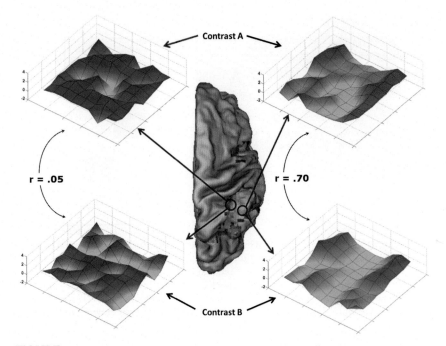

FIGURE 11.2 Overlapping activation patterns from two different contrasts sampled from 81 voxels in two regions of interest (ROIs; *red circles*) in the right hemisphere. Top row depicts a proxy for neural activity changes in functional magnetic resonance imaging (fMRI) (e.g., parameter estimates) in contrast A (e.g., saccades > fixation), and bottom row shows a proxy for neural activity changes in fMRI in contrast B (e.g., subtraction > control). Both contrasts show overlap in right parietal cortex along the intraparietal sulcus. The spatial patterns associated with contrasts A and B from the ROI on the left, however, are completely different (r = 0.05). This indicates that the voxels in this ROI are commonly involved in both processes but in a different manner. On the right we see the inverse situation, where the spatial patterns associated with both processes show high resemblance to each other (r = 0.70).

McCrink, Dehaene, and Dehaene-Lambertz (2007) investigated adult participants' ability to solve nonsymbolic addition and subtraction problems. In their task, participants were presented with nonsymbolic addition and subtraction problems using clouds of dots. They showed that for additions and subtractions, both the mean number chosen by the participants and the variability of these chosen numbers increased with the correct outcome. However, the values chosen by the participants were not centered on the correct result but were influenced by the arithmetic operation. In addition problems, the estimated outcome was larger than the actual outcome, while it was smaller than the actual outcome in subtraction problems. McCrink et al. (2007) argued that this bias showed similarity to a perceptual phenomenon called "representational momentum" (Freyd & Finke, 1984). When they watch a moving object suddenly disappear, participants tend to misjudge its final position and report a position displaced in the direction of the movement (Halpern & Kelly, 1993; Hubbard, 2014; Kerzel, 2003). Analogously, McCrink et al. described their finding as an "operational momentum" (OM) because the misjudgment was related to the arithmetic operation carried out and suggested that subjects were moving "too far" on the number line.

Despite the growing number of studies demonstrating OM in different settings (Klein, Huber, Nuerk, & Moeller, 2014; Pinhas & Fischer, 2008), the underlying mechanisms of the OM effect are currently debated. Three major hypotheses vie with one another. First, it has been suggested that the OM effect reflects the outcome of a simple heuristic that would associate different arithmetic operations with expectations concerning the numerical relationship between the outcome and the operation. For addition, this heuristic would predict that the result should be larger than either of both operands. For subtraction, the heuristic would predict outcomes that are smaller than the first operand. The same heuristic would hold for other arithmetic operations such as multiplication and subtraction. Especially the observed OM in 9-month-olds who most likely have not yet developed a full-fledged spatial representation of numerical magnitude supports this account (McCrink & Wynn, 2009). A second hypothesis assumes that OM results from the flawed logarithmic compression and decompression into a linear metric during the arithmetic process ("compression hypothesis"). As described earlier, the MNL is assumed to be logarithmically compressed (Dehaene, 2001; Nieder & Dehaene, 2009). This compression needs to be "undone" during the arithmetic process. The process that links compressed and uncompressed scales with one another may be flawed. In the extreme variant of the compression hypothesis, addition and subtraction operate on compressed values, i.e., the decompression fails entirely. This would result in massive over- and underestimations because for any two numbers the sum of their logs is the log of their product (i.e., $\log(a) + \log(b) = \log(a \times b)$). An equivalent underestimation would result for subtraction because the difference of the logs is the log of a quotient (i.e., $\log(a) - \log(b) = \log(a/b)$). According to this hypothesis, the functional characteristics of the ANS such as its individual level of compression or its precision should account for the individual variability in OM. Knops, Dehaene, Berteletti, and Zorzi (2014) tested this assumption in a simple psychophysical model. While we were able to account for the overall performance in approximate mental additions and subtractions, we were not able to fully model the OM bias on the basis of basic psychophysical parameters that characterize the ANS. These results speak against the idea that the OM is due to basic characteristics of the mental magnitude representation.

Finally, spatial accounts have been proposed to account for OM. According to the spatial competition hypothesis (Pinhas & Fischer, 2008; Pinhas, Shaki, & Fischer, 2015), OM is the result of the "competing spatial biases invoked by the operands, the operation sign, and the result of an arithmetic problem" (Pinhas et al., 2015, p. 997). Operands and results activate their respective positions on the MNL, which "compete for responses" (Pinhas & Fischer, 2008, p. 413). For subtractions, for example, the result of a given problem may be located between the operands (e.g., $7 - 2 = 5$) or to the left of both operands (e.g., $7 - 4 = 3$). Compared with problems involving zero as a second operand, these competing biases mitigate the observed bias toward the result that is more pronounced when the second operand does not additionally compete for responses. Consequently, Pinhas and Fischer (2008) observed the largest OM bias with zero problems. Recently, these authors observed an inverse OM when reversing the line labels such that the right end of the line was labeled with 0 and the left end with 10 (Pinhas et al., 2015), providing support for their spatial competition bias and underlining the observation that number-to-space mappings are highly flexible (Bächtold, Baumuller, & Brugger, 1998). According to a second spatial account, the OM reflects systematic biases from the deployment of the coordinate transformation system in parietal cortex, which also mediates attentional shifts in space (Knops, Dehaene et al., 2014; Knops, Thirion, Hubbard, Michel, & Dehaene, 2009; Knops, Viarouge, & Dehaene, 2009; Knops, Zitzmann, & McCrink, 2013). According to this approach, approximate mental arithmetic is mediated by a dynamic interaction between positional codes on the MNL and an attentional system that shifts the spatial focus to the left or right.

What does the OM tell about the contribution of the spatial system to mental arithmetic? To answer this question, we need to summarize some crucial findings. First, OM was observed with both symbolic and nonsymbolic additions and subtractions, as first demonstrated by Knops, Viarouge et al. (2009). This may be interpreted as evidence for partly shared cognitive systems and procedures in symbolic and nonsymbolic approximate calculation, in line with prediction P1. Second, OM was observed only with nonsymbolic multiplication and division, while no OM was observed during symbolic multiplication and division (Katz & Knops, 2014). In line with prediction P1, this mirrors the distinction between arithmetic fact retrieval and procedural arithmetic, with the latter being more susceptible to spatial influences than the former. Third, the above neuroimaging results imply that spatial attention plays an important role during simple arithmetic and provides evidence for prediction P2. This also renders the spatial attention account more plausible.

A number of recent findings are in line with the idea that shifts of spatial attention substantially contribute to mental arithmetic (Masson & Pesenti, 2016). Horizontal shifts of attention are associated with solving addition and subtraction problems but not with multiplication (Mathieu, Gourjon, Couderc, Thevenot, & Prado, 2016). This association has also been demonstrated with optokinetic stimulation (OKS), a technique to orient spatial attention in space. Masson, Pesenti, and Dormal (2016) showed that rightward OKS facilitates the solving of complex addition problems that contained a carry operation (i.e., the sum of the units is larger than 10), while no effect of OKS on subtraction was observed. A straightforward prediction from these results is that patients suffering from attentional deficits should be specifically impaired in mental arithmetic. Indeed, a recent study showed that patients suffering from unilateral spatial neglect made more errors when asked to solve subtraction problems in the two-digit number range compared with patients without spatial neglect and healthy controls (Dormal, Schuller, Nihoul, Pesenti, & Andres, 2014). This is particularly interesting because it goes beyond previous studies that were merely correlational and establishes a causal contribution of spatial attention to mental arithmetic. Spatial attention may be necessary to locate the result on a spatial numerical magnitude representation—the MNL. In line with this assumption, Klein et al. (2014) suggest that the OM results from a "first rough spatial anticipation followed by an evaluation/correction process." Others suggest that the association between spatial attention and numbers during mental arithmetic may be because of a semantic association between the arithmetic operator and space (Hartmann, Mast, & Fischer, 2015). However, this may well be conceived of as the result of a learned association between the operator and the induced changes in numerical magnitude. That is, during the acquisition of arithmetic principles, participants may have learned that the plus sign is generally associated with an increase of numerical magnitude and that the minus sign is generally associated with a decrease of numerical magnitude.

Taken together, the cognitive and neural mechanisms leading to OM are not yet fully understood. However, mounting evidence suggests that OM may reflect the influence of a spatial coordinate transformation system in posterior parietal cortex. Thereby, it may be understood as a prime example to further test the idea of neuronal recycling in the domain of numerical cognition.

CONCLUSIONS

Here, I delineated the role that spatial processes may play during numerical cognition. Both the representation of numerical magnitude and the associative network storing arithmetic facts exhibit a spatial organization that systematically influences performance in various settings and paradigms. Mounting evidence also suggests that the cortical circuits involved in the transformation of spatial coordinates for eye movements are associated with mathematical operations. Patient data imply a necessary contribution of spatial attention to mental arithmetic.

One question that arises is why neuronal recycling for numerical cognition should be limited to the spatial systems in parietal cortex. Beyond the spatial representations that are tied to a particular part of the body (i.e., centered on the head, eye, or hand), an impressive body of evidence suggests the existence of a second type of spatial frameworks in rodents, which codes spatial information in an allocentric reference frame (for review, see Hartley, Lever, Burgess, and O'Keefe, 2014). Located in hippocampal and surrounding parahippocampal areas, this system comprises neurons that code for an animal's position in absolute coordinates (place cells) or have multiple firing fields that tessellate the environment with a regular pattern, resembling a grid (grid cells). Originally associated with navigation in space, a recent fMRI study in humans revealed that brain regions where similar grid cells had originally been found during spatial navigation tasks (notably ventral medial prefrontal cortex and entorhinal cortex) represent abstract conceptual knowledge in a similar gridlike pattern (Constantinescu, O'Reilly, & Behrens, 2016). Together with other findings showing that the hippocampal system codes for elapsed time (Kraus et al., 2015; Kraus, Robinson, White, Eichenbaum, & Hasselmo, 2013), these findings suggest that a second, allocentric coding scheme can be exploited to code for abstract concepts and navigation therein. It appears natural to ask whether these codes lend themselves to arithmetic operations. For example, the effects of semantic distance on the recall of arithmetic facts from long-term memory (Didino et al., 2015; Niedeggen et al., 1999) may in fact be due to an allocentric spatial organization of arithmetic fact knowledge, mediated by the same cortical structures that give rise to the representation of conceptual knowledge (Constantinescu et al., 2016). Alternatively, an allocentric representation of arithmetic facts may interact with the egocentric representation of numerical magnitude in parietal areas. This may have important educational implications because the development of educational measures to improve mathematical performance should take into account whether humans preferably code numerical information in allocentric or egocentric reference frames.

A second open question is whether hand-related cortical circuits, too, may be involved in the interaction between space, number, and attention (Hubbard et al., 2005). Similar mechanisms as those involved in eye movement guide hand movements such as reaching and grasping, which rely on parietal circuits located at the posterior and anterior end of the intraparietal sulcus, respectively. However, this hypothesis has not been systematically explored, and research has mainly focused on the role of finger gnosis in counting and whole number understanding

(Andres, Michaux, & Pesenti, 2012; Michaux, Masson, Pesenti, & Andres, 2013; Noel, 2005; Wasner, Moeller, Fischer, & Nuerk, 2014).

I delineated the ways spatial processes contribute to numerical cognition and how this can be understood as an example of neuronal recycling. This is not to say that numerical cognition is entirely a spatial process. Numerical cognition also recruits a number of other cognitive domains, including verbal and domain-general processes, which are described in this book. However, it is also clear that numerical cognition can no longer be understood as operating on abstract symbolic information. The current proposal also goes beyond what embodied cognition approaches postulate. Embodied cognition frameworks hinge on two crucial assumptions. First, semantic knowledge is supposed to be associated with the sensorimotor experience during its acquisition. For example, the concept of "ball" entails the sensorimotor representations that are activated when throwing or kicking. Second, recall of semantic knowledge is mediated by a reenactment that involves the same cortical structures as during acquisition (e.g., throwing or kicking for "ball"). Despite its great appeal, the empirical support for these claims has been called into question recently. Caramazza, Anzellotti, Strnad, and Lingnau (2014) argue that most embodied cognition studies erroneously take overlapping activity in areas that may be only remotely associated with a particular primary perceptual or motor process as evidence in support of the embodiment approach. However, as shown earlier, overlapping activity is a necessary but not a sufficient condition in this context. The recycling framework allows the development of clear and testable hypotheses that go beyond merely overlapping activation patterns in associative cortices and that have already been shown to account for a number of empirical findings. This may form the foundation for achieving a more thorough understanding of the cognitive core processes that allow humans to successfully engage in numerical cognition.

References

Amalric, M., & Dehaene, S. (2016). Origins of the brain networks for advanced mathematics in expert mathematicians. *Proceedings of the National Academy of Sciences of the United States of America, 113*(18), 4909–4917. https://doi.org/10.1073/pnas.1603205113.

Anderson, M. L. (2010). Neural reuse: A fundamental organizational principle of the brain. *The Behavioral and Brain Sciences, 33*(4), 245–266. https://doi.org/10.1017/S0140525X10000853. Discussion 266–313.

Andres, M., Michaux, N., & Pesenti, M. (2012). Common substrate for mental arithmetic and finger representation in the parietal cortex. *Neuroimage, 62*(3), 1520–1528. https://doi.org/10.1016/j.neuroimage.2012.05.047.

Ashcraft, M. H. (1987). Children's knowledge of simple arithmetic: A developmental model and simulation. In C. J. Brainerd, R. Kail, & J. Bisanz (Eds.), *Formal methods in developmental research*. New-York: Springer Verlag.

Ashkenazi, S., Rosenberg-Lee, M., Metcalfe, A. W., Swigart, A. G., & Menon, V. (2013). Visuo-spatial working memory is an important source of domain-general vulnerability in the development of arithmetic cognition. *Neuropsychologia*, *51*(11), 2305–2317. https://doi.org/10.1016/j.neuropsychologia.2013.06.031.

Bächtold, D., Baumuller, M., & Brugger, P. (1998). Stimulus-response compatibility in representational space. *Neuropsychologia*, *36*(8), 731–735.

Beck, J. M., Latham, P. E., & Pouget, A. (2011). Marginalization in neural circuits with divisive normalization. *Journal of Neuroscience*, *31*(43), 15310–15319.

Campbell, J. I. (1995). Mechanisms of simple addition and multiplication: A modified network-interference theory and simulation. *Mathematical Cognition*, *1*(2), 121–164.

Caramazza, A., Anzellotti, S., Strnad, L., & Lingnau, A. (2014). Embodied cognition and mirror neurons: A critical assessment. *Annual Review of Neuroscience*, *37*, 1–15. https://doi.org/10.1146/annurev-neuro-071013-013950.

Cavdaroglu, S., & Knops, A. (2016). Mental subtraction and multiplication recruit both phonological and visuospatial resources: Evidence from a symmetric dual-task design. *Psychological Research*, *80*(4), 608–624. https://doi.org/10.10 07/s00426-015-0667-8.

Constantinescu, A. O., O'Reilly, J. X., & Behrens, T. E. (2016). Organizing conceptual knowledge in humans with a gridlike code. *Science*, *352*(6292), 1464–1468. https://doi.org/10.1126/science.aaf0941.

Dehaene, S. (2001). Subtracting pigeons: Logarithmic or linear? *Psychological Science*, *12*(3), 244–246 Discussion 247.

Dehaene, S. (2003). The neural basis of the Weber-Fechner law: A logarithmic mental number line. *Trends in Cognitive Sciences*, *7*(4), 145–147.

Dehaene, S., Bossini, S., & Giraux, P. (1993). The mental representation of parity and number magnitude. *Journal of Experimental Psychology: General*, *122*(3), 371–396. https://doi.org/10.1037/0096-3445.122.3.371.

Dehaene, S., & Cohen, L. (1997). Cerebral pathways for calculation: Double dissociation between rote verbal and quantitative knowledge of arithmetic. *Cortex*, *33*(2), 219–250.

Dehaene, S., & Cohen, L. (2007). Cultural recycling of cortical maps. *Neuron*, *56*(2), 384–398. https://doi.org/10.1016/j.neuron.2007.10.004.

Didino, D., Knops, A., Vespignani, F., & Kornpetpanee, S. (2015). Asymmetric activation spreading in the multiplication associative network due to asymmetric overlap between numerosities semantic representations? *Cognition*, *141*, 1–8. https://doi.org/10.1016/j.cognition.2015.04.002.

Dollman, J., & Levine, W. H. (2016). Rapid communication the mental number line dominates alternative, explicit coding of number magnitude. *Quarterly Journal of Experimental Psychology (Hove)*, *69*(3), 403–409. https://doi.org/10.1080/17470 218.2015.1101146.

Dormal, V., Schuller, A. M., Nihoul, J., Pesenti, M., & Andres, M. (2014). Causal role of spatial attention in arithmetic problem solving: Evidence from left unilateral neglect. *Neuropsychologia*, *60*, 1–9. https://doi.org/10.1016/j.neurop sychologia.2014.05.007.

Fischer, M. H., Castel, A. D., Dodd, M. D., & Pratt, J. (2003). Perceiving numbers causes spatial shifts of attention. *Nature Neuroscience*, *6*(6), 555–556.

Freyd, J. J., & Finke, R. A. (1984). Representational momentum. *Journal of Experimental Psychology-Learning, Memory and Cognition*, *10*(1), 126–132. https://doi.org/10.1037/0278-7393.10.1.126.

Galfano, G., Mazza, V., Angrilli, A., & Umilta, C. (2004). Electrophysiological correlates of stimulus-driven multiplication facts retrieval. *Neuropsychologia, 42*(10),
1370–1382. https://doi.org/10.1016/j.neuropsychologia.2004.02.010.

Galfano, G., Penolazzi, B., Vervaeck, I., Angrilli, A., & Umilta, C. (2009). Event-related
brain potentials uncover activation dynamics in the lexicon of multiplication
facts. *Cortex, 45*(10), 1167–1177. https://doi.org/10.1016/j.cortex.2008.09.003.

Galfano, G., Rusconi, E., & Umilta, C. (2003). Automatic activation of multiplication facts: Evidence from the nodes adjacent to the product. *Quarterly Journal
of Experimental Psychology A, 56*(1), 31–61. https://doi.org/10.1080/0272
4980244000332.

Gallistel, C. R., & Gelman, R. (1992). Preverbal and verbal counting and computation. *Cognition, 44*(1–2), 43–74. https://doi.org/10.1016/0010-0277(92)90050-R.

Gertner, L., Henik, A., Reznik, D., & Cohen Kadosh, R. (2013). Implications of
number-space synesthesia on the automaticity of numerical processing. *Cortex,
49*(5), 1352–1362. https://doi.org/10.1016/j.cortex.2012.03.019.

Ginsburg, V., van Dijck, J. P., Previtali, P., Fias, W., & Gevers, W. (2014). The
impact of verbal working memory on number-space associations. *Journal
of Experimental Psychology-Learning, Memory and Cognition, 40*(4), 976–986.
https://doi.org/10.1037/a0036378.

Goffaux, V., Martin, R., Dormal, G., Goebel, R., & Schiltz, C. (2012). Attentional
shifts induced by uninformative number symbols modulate neural
activity in human occipital cortex. *Neuropsychologia, 50*(14), 3419–3428.
https://doi.org/10.1016/j.neuropsychologia.2012.09.046.

Halpern, A. R., & Kelly, M. H. (1993). Memory biases in left versus right implied
motion. *Journal of Experimental Psychology-Learning, Memory and Cognition,
19*(2), 471–484.

Hartley, T., Lever, C., Burgess, N., & O'Keefe, J. (2014). Space in the brain: how the
hippocampal formation supports spatial cognition. *Philos Trans R Soc Lond B
Biol Sci, 369*(1635): 20120510.

Hartmann, M., Gashaj, V., Stahnke, A., & Mast, F. W. (2014). There is more than
"more is up": Hand and foot responses reverse the vertical association of
number magnitudes. *Journal of Experimental Psychology. Human Perception and
Performance, 40*(4), 1401–1414. https://doi.org/10.1037/a0036686.

Hartmann, M., Mast, F. W., & Fischer, M. H. (2015). Spatial biases during mental arithmetic: Evidence from eye movements on a blank screen. *Frontiers in
Psychology, 6*, 12. https://doi.org/10.3389/fpsyg.2015.00012.

Harvey, B. M., et al. (2013). Topographic Representation of Numerosity in the
Human Parietal Cortex. *Science, 341*(6150), 1123–1126.

Hubbard, T. L. (2014). Forms of momentum across space: Representational, operational, and attentional. *Psychonomic Bulletin and Review, 21*(6), 1371–1403.
https://doi.org/10.3758/s13423-014-0624-3.

Hubbard, E. M., Piazza, M., Pinel, P., & Dehaene, S. (2005). Interactions between
number and space in parietal cortex. *Nature Reviews Neuroscience, 6*(6), 435–448.
https://doi.org/10.1038/nrn1684.

Ito, Y., & Hatta, T. (2004). Spatial structure of quantitative representation of
numbers: Evidence from the SNARC effect. *Memory and Cognition, 32*(4),
662–673.

Katz, C., & Knops, A. (2014). Operational momentum in multiplication and division? *PLoS One, 9*(8), e104777. https://doi.org/10.1371/journal.pone.0104777.

Kerzel, D. (2003). Attention maintains mental extrapolation of target position: Irrelevant distractors eliminate forward displacement after implied motion. *Cognition, 88*(1), 109–131. https://doi.org/10.1016/s0010-0277(03) 00018-0.

Klein, E., Huber, S., Nuerk, H. C., & Moeller, K. (2014). Operational momentum affects eye fixation behaviour. *Quarterly Journal of Experimental Psychology (Hove), 67*(8), 1614–1625. https://doi.org/10.1080/17470218.201 4.902976.

Knops, A. (2016). Probing the neural correlates of number processing. *The Neuroscientist.* https://doi.org/10.1177/1073858416650153.

Knops, A., Dehaene, S., Berteletti, I., & Zorzi, M. (2014a). Can approximate mental calculation account for operational momentum in addition and subtraction? *Quarterly Journal of Experimental Psychology (Hove), 67*(8), 1541–1556. https://doi.org/10.1080/17470218.2014.890234.

Knops, A., Piazza, M., Sengupta, R., Eger, E., & Melcher, D. (2014b). A shared, flexible neural map architecture reflects capacity limits in both visual short-term memory and enumeration. *The Journal of Neuroscience, 34*(30), 9857–9866. https://doi.org/10.1523/JNEUROSCI.2758-13.2014.

Knops, A., Thirion, B., Hubbard, E. M., Michel, V., & Dehaene, S. (2009). Recruitment of an area involved in eye movements during mental arithmetic. *Science, 324*(5934), 1583–1585. https://doi.org/10.1126/science.1171599.

Knops, A., Viarouge, A., & Dehaene, S. (2009b). Dynamic representations underlying symbolic and nonsymbolic calculation: Evidence from the operational momentum effect. *Attention, Perception and Psychophysics, 71*(4), 803–821. https://doi.org/10.3758/APP.71.4.803.

Knops, A., Zitzmann, S., & McCrink, K. (2013). Examining the presence and determinants of operational momentum in childhood. *Frontiers in Psychology, 4*, 325. https://doi.org/10.3389/fpsyg.2013.00325.

Koten, J. W., Jr., Lonnemann, J., Willmes, K., & Knops, A. (2011). Micro and macro pattern analyses of FMRI data support both early and late interaction of numerical and spatial information. *Frontiers in Human Neuroscience, 5*, 115. https://doi.org/10.3389/fnhum.2011.00115.

Kraus, B. J., Brandon, M. P., Robinson, R. J., 2nd, Connerney, M. A., Hasselmo, M. E., & Eichenbaum, H. (2015). During running in place, grid cells integrate elapsed time and distance run. *Neuron, 88*(3), 578–589. https://doi.org/10.1016/j.neuron.2015.09.031.

Kraus, B. J., Robinson, R. J., 2nd, White, J. A., Eichenbaum, H., & Hasselmo, M. E. (2013). Hippocampal "time cells": Time versus path integration. *Neuron, 78*(6), 1090–1101. https://doi.org/10.1016/j.neuron.2013.04.015.

Marghetis, T., Landy, D. & Goldstone, R.L. Cogn. Research (2016) 1: 25. https:// doi.org/10.1186/s41235-016-0020-9.

Masson, N., & Pesenti, M. (2016). Interference of lateralized distractors on arithmetic problem solving: A functional role for attention shifts in mental calculation. *Psychological Research-Psychologische Forschung, 80*(4), 640–651. https://doi.org/10.1007/s00426-015-0668-7.

Masson, N., Pesenti, M., & Dormal, V. (2016). Impact of optokinetic stimulation on mental arithmetic. *Psychological Research.* https://doi.org/10.1007/s00426-016-0784-z.

Mather, M., Cacioppo, J. T., & Kanwisher, N. (2013). How fMRI can inform cognitive theories. *Perspectives on Psychological Science, 8*(1), 108–113. https://doi.org/10.1177/1745691612469037.

Mathieu, R., Gourjon, A., Couderc, A., Thevenot, C., & Prado, J. (2016). Running the number line: Rapid shifts of attention in single-digit arithmetic. *Cognition, 146*, 229–239. https://doi.org/10.1016/j.cognition.2015.10.002.

McCrink, K., Dehaene, S., & Dehaene-Lambertz, G. (2007). Moving along the number line: Operational momentum in nonsymbolic arithmetic. *Perception and Psychophysics, 69*(8), 1324–1333.

McCrink, K., & Wynn, K. (2009). Operational momentum in large-number addition and subtraction by 9-month-olds. *Journal of Experimental Child Psychology, 103*(4), 400–408. https://doi.org/10.1016/j.jecp.2009.01.013.

Metcalfe, A. W., Ashkenazi, S., Rosenberg-Lee, M., & Menon, V. (2013). Fractionating the neural correlates of individual working memory components underlying arithmetic problem solving skills in children. *Developmental Cognitive Neuroscience, 6*, 162–175. https://doi.org/10.1016/j.dcn.2013.10.001.

Michaux, N., Masson, N., Pesenti, M., & Andres, M. (2013). Selective interference of finger movements on basic addition and subtraction problem solving. *Experimental Psychology, 60*(3), 197–205. https://doi.org/10.1027/1618-3169/a000188.

Nicholls, M. E., Loftus, A. M., & Gevers, W. (2008). Look, no hands: A perceptual task shows that number magnitude induces shifts of attention. *Psychonomic Bulletin and Review, 15*(2), 413–418.

Niedeggen, M., Rösler, F., & Jost, K. (1999). Processing of incongruous mental calculation problems: Evidence for an arithmetic N400 effect. *Psychophysiology, 36*(3), 307–324.

Nieder, A. (2016). The neuronal code for number. *Nature Reviews Neuroscience, 17*(6), 366–382. https://doi.org/10.1038/nrn.2016.40.

Nieder, A., & Dehaene, S. (2009). Representation of number in the brain. *Annual Review of Neuroscience, 32*, 185–208. https://doi.org/10.1146/annurev.neuro.05 1508.135550.

Nieder, A., & Miller, E. K. (2004). A parieto-frontal network for visual numerical information in the monkey. *Proceedings of the National Academy of Sciences of the United States of America, 101*(19), 7457–7462.

Noel, M. P. (2005). Finger gnosia: A predictor of numerical abilities in children? *Child Neuropsychology, 11*(5), 413–430. https://doi.org/10.1080/09297 040590951550.

Parkinson, C., Liu, S., & Wheatley, T. (2014). A common cortical metric for spatial, temporal, and social distance. *The Journal of Neuroscience, 34*(5), 1979–1987. https://doi.org/10.1523/JNEUROSCI.2159-13.2014.

Parkinson, C., & Wheatley, T. (2013). Old cortex, new contexts: Re-purposing spatial perception for social cognition. *Frontiers in Human Neuroscience, 7*, 645. https://doi.org/10.3389/fnhum.2013.00645.

Pecher, D., Van Dantzig, S., Boot, I., Zanolie, K., & Huber, D. E. (2010). Congruency between word position and meaning is caused by task-induced spatial attention. *Frontiers in Psychology, 1*, 30. https://doi.org/10.3389/fpsyg.2010.00030.

Pinhas, M., & Fischer, M. H. (2008). Mental movements without magnitude? A study of spatial biases in symbolic arithmetic. *Cognition, 109*(3), 408–415.

Pinhas, M., Shaki, S., & Fischer, M. H. (2015). Addition goes where the big numbers are: Evidence for a reversed operational momentum effect. *Psychonomic Bulletin and Review, 22*(4), 993–1000. https://doi.org/10.3758/s13423-014-0786-7

Pouget, A., Deneve, S., & Duhamel, J. R. (2002). A computational perspective on the neural basis of multisensory spatial representations. *Nature Reviews Neuroscience, 3*(9), 741–747. https://doi.org/10.1038/Nrn914.

Ranzini, M., Dehaene, S., Piazza, M., & Hubbard, E. M. (2009). Neural mechanisms of attentional shifts due to irrelevant spatial and numerical cues. *Neuropsychologia, 47*(12), 2615–2624.

Ranzini, M., Lisi, M., Blini, E., Pitteri, M., Treccani, B., Priftis, K., et al. (2015). Larger, smaller, odd or even? Task-specific effects of optokinetic stimulation on the mental number space. *Journal of Cognitive Psychology, 27*(4), 459–470. https://doi.org/10.1080/20445911.2014.941847.

Ranzini, M., Lisi, M., & Zorzi, M. (2016). Voluntary eye movements direct attention on the mental number space. *Psychological Research, 80*(3), 389–398. https://doi.org/10.1007/s00426-015-0741-2.

Rugani, R., Vallortigara, G., Priftis, K., & Regolin, L. (2015). Animal cognition. Number-space mapping in the newborn chick resembles humans' mental number line. *Science, 347*(6221), 534–536. https://doi.org/10.1126/science.aaa1379.

Salillas, E., El Yagoubi, R., & Semenza, C. (2008). Sensory and cognitive processes of shifts of spatial attention induced by numbers: An ERP study. *Cortex, 44*(4), 406–413. https://doi.org/10.1016/j.cortex.2007.08.006.

Santens, S., & Gevers, W. (2008). The SNARC effect does not imply a mental number line. *Cognition, 108*(1), 263–270.

Schuller, A. M., Hoffmann, D., Goffaux, V., & Schiltz, C. (2015). Shifts of spatial attention cued by irrelevant numbers: Electrophysiological evidence from a target discrimination task. *Journal of Cognitive Psychology, 27*(4), 442–458. https://doi.org/10.1080/20445911.2014.946419.

Shaki, S., & Fischer, M. H. (2008). Reading space into numbers: A cross-linguistic comparison of the SNARC effect. *Cognition, 108*(2), 590–599. https://doi.org/10.1016/j.cognition.2008.04.001.

Thompson, J. M., Nuerk, H. C., Moeller, K., & Cohen Kadosh, R. (2013). The link between mental rotation ability and basic numerical representations. *Acta Psychol (Amst), 144*(2), 324–331.

van Dijck, J. P., & Fias, W. (2011). A working memory account for spatial-numerical associations. *Cognition, 119*(1), 114–119. https://doi.org/10.1016/j.cognition.2010.12.013.

Verguts, T., & Fias, W. (2005). Interacting neighbors: A connectionist model of retrieval in single-digit multiplication. *Memory and Cognition, 33*(1), 1–16.

Wasner, M., Moeller, K., Fischer, M. H., & Nuerk, H. C. (2014). Aspects of situated cognition in embodied numerosity: The case of finger counting. *Cognitive Processing, 15*(3), 317–328. https://doi.org/10.1007/s10339-014-0599-z.

Wiemers, M., Bekkering, H., & Lindemann, O. (2014). Spatial interferences in mental arithmetic: Evidence from the motion-arithmetic compatibility effect. *Quarterly Journal of Experimental Psychology (Hove), 67*(8), 1557–1570. https://doi.org/10.1080/17470218.2014.889180.

Winter, B., Matlock, T., Shaki, S., & Fischer, M. H. (2015). Mental number space in three dimensions. *Neuroscience and Biobehavioral Reviews, 57*, 209–219. https://doi.org/10.1016/j.neubiorev.2015.09.005.

12

Which Space for Numbers?

Wim Fias[1], Mario Bonato[1,2]

[1]Ghent University, Ghent, Belgium; [2]University of Padova, Padova, Italy

Numerical and mathematical cognition are closely intertwined with spatial cognition. In many everyday life contexts, numbers are often spatially represented in spatial configurations as on a line (on keyboards), on a square (on cell phones), or on a circle (as on analog clocks). A strong link with space is also present for mathematics. In the strongest case, math is about space, as in geometry. In less strong situations space serves as a useful analogy (e.g., multiplication can be conceived as surface computation) and can help in understanding and specifying abstract notions. Notably, the procedures supporting the cognitive functions that are involved in solving mathematical tasks are often spatially characterized (e.g., borrowing, column-based addition), and many physical contexts couple space and quantity (graphs and real objects such as abacus, etc.).

Accordingly, the link between numerical and mathematical cognition on the one hand and spatial cognition on the other hand has been intensively studied, from various angles (neuroscience, development, education, etc.)

Heterogeneity of Function in Numerical Cognition
https://doi.org/10.1016/B978-0-12-811529-9.00012-1

and with a focus on different levels, ranging from basic number represen- tations, over the role of basic support functions, to conceptual levels of mathematics understanding. The domains of number, math, and space are so vast and heterogeneous that every empirical investigation addressing these interactions necessarily can only cover partial aspects. This hetero- geneity is positively mirrored in the three contributions to this book sec- tion. While reporting studies, which are very relevant to the number–space domain, each of the chapters uses a different perspective. Interestingly, the complementarity between these perspectives offers a broad, original, mul- tidomain view on how spatial processing might contribute to numerical and mathematical cognition.

THE MENTAL NUMBER LINE AS A BASIC SPATIALLY DEFINED NUMBER REPRESENTATION

The chapter by Knops reviews the neurofunctional evidence about the brain areas that are functionally involved in basic number processing and calculation. The argumentation starts from the "classic" number–space interactions that are obtained using experimental psychology methods and that are generally interpreted as reflecting basic number magnitude representations that take the form of a mental number line. Perhaps the strongest conceptual claim of the chapter is that this inherently spatially coded magnitude information forms the basis for more advanced levels of mathematical knowledge. The author argues that neural systems that are originally developed for spatial processing are "neuronally recycled" to support mental arithmetic. This is especially the case for addition and subtraction that are accompanied with shifts of spatial attention along the mental number line. In the following paragraphs, we further elaborate on these ideas proposed by Knops.

Besides the subjective experience of number–space couplings, the sci- entific idea that numbers are spatially coded has a long tradition. It goes back to Galton (1880) who described people who vividly experience visual number forms with pronounced spatial configurations. Dehaene, Bossini, and Giraux (1993) were able to demonstrate that spatial coding of numbers is not restricted to some individuals (as in Galton) but that it is a general characteristic of how people mentally represent numbers, even when they are not aware of it. In that seminal paper, Dehaene et al. (1993) found, on the basis of reaction times, that numbers are systematically associated to a directionally defined spatial code as a function of their magnitude (small left, large right). The effect has been named the SNARC effect, an acronym for spatial–numerical association of response codes. This has fed the idea that a mental number line is the core number representation system from which more complex forms of mathematical knowledge and skills develop.

Knops reviewed many important basic findings in this realm starting from single number representations and extending to mathematical operations. The modulatory role for these effects is attributed to spatial attention as a crucial mental process operating on this mental number line representation. The findings from spatial cueing and similar tasks can unveil, under certain circumstances, a spontaneous association between numerical magnitude and side of target presentation, which is surprising because it is present even in the absence of spatial connotation for the response keys.

As briefly mentioned by Knops, crucial evidence supporting a key role for spatial attention in manipulating numerical magnitudes comes from the performance shown by those right-hemisphere brain-damaged patients who show impairments in left hemispace processing because of hemispatial neglect (Zorzi, Priftis, & Umiltà, 2002). When asked which number lies in the middle of a certain range (e.g., 1–9), these patients misplace their judgment toward larger numbers, as if they are ignoring the leftmost part of a left-to-right oriented interval. Intriguingly, when asked to perform a magnitude comparison task, they show impairments, which are not evident when performing a parity judgment on the same stimuli (Zorzi et al., 2012). The two biases do not seem to be directly related (van Dijck, Gevers, Lafosse, & Fias, 2012). Although the analogy with clinical behavior in line bisection has been questioned (Doricchi, Guariglia, Gasparini, & Tomaiuolo, 2005), there is little or no doubt that the performance shown by neglect patients unveils that some sort of spatial coding is attributed to numbers. Recently Bonato, Saj, and Vuilleumier (2016) showed that neglect also affects lateralized responses to a story-like list of events, therefore suggesting that the spatial coding is a general feature characterizing ordered sequences.

Interestingly, attentional effects in healthy participants were also found to extend to the mental arithmetic domain. Knops, Thirion, Hubbard, Michel, and Dehaene (2009) found that a classifier trained to distinguish the neural patterns of left and rightward oculomotor shifts was also able to distinguish whether a person had added or subtracted two numbers. The performance of left neglect patients, who present normal addition and a selective deficit for subtraction, corroborates the possibility of a mediating role of spatial attention also within the arithmetic domain (Dormal, Schuller, Nihoul, Pesenti, & Andres, 2014). Converging evidence therefore supports the idea that the mental number line forms the core of mental arithmetic and ultimately of more complex forms of math.

There is no doubt that the mental number line term is particularly effective in immediately conveying the idea that a close relationship between number and space exists. This effectiveness, however, sometimes can become counterproductive when it does not conceptually prompt the possibility that the way humans spatialize numbers might be more complex than that. There are several reasons to take a critical stance and therefore highlight, 25 years after its first formulation, some effects, which are not fully explained by the idea of a long-term mental number line playing a central role in numerical and mathematical abilities. For instance a multiplicity of reference frames (horizontal, vertical, sagittal, for review see Winter, Matlock, Shaki, & Fischer, 2015) has been reported to be associated to number. While this reinforces the idea of a close link with space, it questions whether the link necessarily has to be linear and suggests that, even assuming a core system which is long term, the number–space associations might then well (also) be influenced by context-based short-term mapping (van Dijck & Fias, 2011), rather than reflection of availability of multiple reference frames in long-term memory. This possibility seems to be confirmed by how quickly and effectively alternative spatial mappings shape the number–space association (e.g., as a clockface, Bachtold, Baumuller, & Brugger, 1998). It has been recently shown that both a long- and a short-term magnitude can coexist (Ginsburg & Gevers, 2015). Again, although all these effects do confirm the tight link between space and number at the same time, they question whether the linear relationship that the mental number line suggests is so strong and unequivocal to be the basis for all sorts of mathematics. For instance, in the case of multiplication and division there is no direct link with a line, not to mention other forms of more complex math (algebra, set theory, etc.). Also from an empirical point of view there is no clear and unequivocal link between the SNARC effect and mathematical skill (Cipora & Nuerk, 2013), and evidence suggests that at least some (e.g., SNARC) spationumerical effects do not seem to occur in expert participants (Cipora et al., 2016). Finally, the developmental trajectory of the mental number line remains to be specified. A mental number line–like effect has been reported in chicks (Rugani, Vallortigara, Priftis, & Regolin, 2015), suggesting a phylogenetic origin. Similar evidence has been described in monkeys (Adachi, 2014; Gazes et al., 2017). However, this evidence should be taken carefully because other explanations, for instance, hemispheric asymmetries in chicks, cannot be fully ruled out (Núñez & Fias, 2017).

The idea of a mental number line also needs a theory that defines *how* the basic mental number line representation would scaffold the development of more complex forms of mathematical cognition. Only then we can understand its capacity and limitations. Maybe here the neuronal recycling idea can provide an additional source of information. If we increase our understanding of the original functions of a certain region and if we can argue on theoretical grounds how these functions support certain aspects of mathematical cognition, then we can come to a more detailed picture. The fact that brain regions that are involved in oculomotor programming are implied in mental arithmetic, as demonstrated by Knops et al. (2009), may be a good starting point. It shows that spatial processing mechanisms are involved in exploring a mental representation that is spatially defined. Whether this mental space reflects a long-term mental number line or a temporary workspace (Abrahamse et al., 2016, 2017) remains to be determined.

A DEEPER UNDERSTANDING OF NUMBERS: NUMERICAL AND SPATIAL PROPORTIONS

An important step in transcending the spatial code of spatial number representations to more complex forms of math is to realize that numbers are conceptually more elaborated than pure number magnitude. It is about understanding properties of a number system that goes beyond more versus less objects and that can perhaps be better characterized as a system for quantity. This is exactly the topic of the contribution of Newcombe et al. In an effort to come to grips with such a more advanced form of mathematical knowledge, Newcombe et al. focus on proportions, which are notoriously difficult to master by children. The chapter opens with a statement, which is hard to disconfirm, namely that in most of everyday situations "more" means "larger." Consequently, we should not therefore be surprised about the fact that our cognitive system, particularly during its development, has difficulty in completely tearing apart these two aspects. On top of that, formal education initially prioritizes counting and integer-based skills, which obviously puts a lot of emphasis on extensive coding and discrete representations of number. This unbalanced treatment of extensive discrete numbers versus intensive processing of continuous magnitude constitutes a meaningful explanation for the poor performance found in situations that require intensive number processing as when a ratio must be processed, and larger numbers used as a denominator in fractions can refer to smaller quantities. Understanding proportions requires the mapping from extensive to continuous intensive information. Clearly, only if a child comes to grips with an intensive coding/representation of continuous magnitude, it can fully master the elaborated conceptual meaning of the number systems.

Newcombe et al. also interestingly contrast theories suggesting an innate sense for numerosity from those positing a general quantity system. Theories explicitly suggesting a core role for (sensorimotor, in the case of ATOM) experience highlight that the coupling across magnitudes is adaptive and shaped by repeated interactions with the environment, maybe mediated by statistical learning (Walsh, 2003). In other words, these theories—like those assuming innate and specific modules for magnitude processing—suggest the presence of a module for extensive magnitude processing. Formal teaching and environmental influences are, within these frameworks, not considered to fine-tune the system but rather to trigger a shift toward a different functioning modality—the extensive one. In turn, this functioning modality would make proportional reasoning more difficult because based on a system, which has been "forgotten." In such a way, they provide a convincing explanation as to why proportional reasoning seems intrinsically difficult for humans. The improvement in performance found when using natural frequencies rather than proportions is the complementary explanation of the same phenomenon, often investigated in number-based reasoning (Hoffrage & Gigerenzer, 1998).

Crucially for the link between numbers and space, Newcombe et al. point to structural similarity between number tasks and spatial tasks. Both can be used in an extensive and an intensive way. Developmentally extensive precedes the intensive, so the development from extensive to intensive is necessary in both, and integration of and translation between extensive and intensive is required. Newcombe et al. not only point out that number and space are structurally similar in this respect but also discuss some results that show that spatial skill is predictive of mathematical skill, which suggests a mutually dependent codevelopment.

Although there is still an important aspect of representation involved, the focus on proportions also puts more weight on the mental operations that need to be performed on the representations, i.e., the mapping procedures. An example of this is the number line estimation task (Siegler & Opfer, 2003). In this procedure, a number (e.g., 4) has to be placed on a line in a position that is determined by the values present at its extremities (for instance, toward the left of the midpoint if the extremes are 1 on the left and 9 on the right). This task has often been considered as directly revealing the underlying number representations. Specifically, the fact that children initially assign more space to the mapping of small numbers and compress larger numbers in the remaining space has been taken as evidence that mental number line representation develops from a nonlinear compressed representation to a linearized version. Contrary to this view, Newcombe et al. emphasize that the number line task is not necessarily a reflection of the number representation. Rather, it can demonstrate the ability of mapping discrete extensive to continuous intensive information. In the absence of an understanding of the continuous intensive properties of numbers (e.g., 3 is one quarter of 12), children resort to a counting-based strategy, in which they use extensive space for the small numbers and then have to "squeeze" the remaining larger numbers on the remaining part of the line. Hence, it would be the adequacy of the proportional reasoning processes that determines the accuracy of the mapping, without assuming anything about the nonlinearity or linearity of the underlying number representations (Rouder & Geary, 2014).

The case of fractions is no doubt of crucial importance for understanding how a difficulty within the numerical domain can become problematic for everyday performance. Complementing the description by Newcombe et al., we note that the first study performed in adults (Bonato, Fabbri, Umiltà, & Zorzi, 2007) suggests that university students, when presented with fractions inviting them to focus on the denominator (e.g., 1/3 is larger or smaller than 1/5), reliably present a whole number bias, which is not dissimilar from the one presented by kids. Strikingly, the same pattern of biased processing and difficulty in disengaging from the quantity expressed by the denominator has also been found when testing students of engineering and persists when using more complex ranges of stimuli. This means that, whatever be the reason for the difficulties in fraction processing, it is not abolished, but only circumvented, by formal mathematical training even after several years of formal schooling. In a way, this finding can be interpreted as supporting the idea that formal mathematical education can have "detrimental" effects in processing numbers, which are other than integers.

The structural similarity between number processing and space/quantity processing, together with the related development of spatial and mathematical thinking, raises a very important question regarding the origin of this relation. Why are both related? Which are the most important spatial skills for this link? What are the underlying cognitive mechanisms involved in proportional reasoning in the number and/or space domain? These are very important questions, especially when considering that performance on the number line estimation task has been shown over several studies to be a robust predictor of mathematical skill (Booth & Siegler, 2006, 2008).

THE IMPORTANCE OF ACTIVE PROCESSING

Knops focused mainly on number representations. Newcombe et al. approached the relationship between number and space from the perspective of conceptual understanding, thereby touching on the importance of cognitive processes, however, without fully elaborating what these processes would be and how they would relate. Lourenco et al. put full emphasis on the nature of the spatial processes that might be related to mathematical cognition. Fully realizing that spatial and mathematical cognition are not monolithic, Lourenco et al. zoom in and narrow down to what they feel is a crucial developmental factor for many mathematical tasks, namely the ability to mentally visualize and manipulate spatial information—without a priori excluding the possibility that also, other spatial skills such as navigation, could play a role in mathematical competence.

There is a considerable body of evidence that has demonstrated a link between visuospatial and mathematical reasoning. Given that many of these studies are correlational, Lourenco et al. rightly highlight the importance of understanding the directionality of the relation between math and space. Whereas the bulk of the existing studies has reasoned that good spatial skills provide a basis for the development of good mathematical skills, the opposite direction is worth considering: mathematical knowledge and insight might facilitate mastering spatial knowledge and skill. Despite the large number of studies on the relation between mathematical and spatial reasoning, only relatively few have used cross-lagged longitudinal studies or intervention studies, which of course constitute the strongest way to distinguish between the two possibilities. Lourenco et al. provide a detailed overview of these studies, most of which concentrate on the effect of spatial training on mathematical proficiency. The studies they mention include those focusing on mental rotation and those related to spatial positioning. Over studies, the results are contradictory because the effects, when present, have never been fully and independently replicated. Clearly, a lot of work is ahead of us to precisely point out how and under what conditions spatial reasoning can have an impact on the development of mathematical skill (or vice versa). A number of factors must be taken into account in future studies. First, neither mathematical nor spatial skills are monolithic. Some types of spatial processing may contribute to some mathematical tasks but not to others. Similarly, some types of spatial processing may not at all be related to mathematical processing, and some types of mathematical processing may be unrelated to spatial processing. Second, the importance of the relationship may well differ through development. Third, one has to take into account that the relation between space and math is not unidirectional, but that both may cross-fertilize, which might lead to nonlinear, dynamic, patterns of development. Fourth, one has to consider the possibility that spatial skill has not a direct impact on mathematical skill. It has to be taken into account that the possibility exists that intervening cognitive skills determine the relationship between spatial and mathematical reasoning. Taken together, the evidence so far available allows to safely conclude that spatial skill and mathematical proficiency are indeed related. However, their relationship is way more complex than a simple one-to-one linear connection. To understand this relationship in all its complexity, it is necessary to have a theory constraining it to its underlying dimensions and mechanisms. We point out an analogy between this relationship and the hotly debated issue of whether mathematical achievement can be predicted or even anticipated by measuring individual performance in nonsymbolic quantity processing (Halberda, Mazzocco, & Feigenson, 2008 vs. Gilmore et al., 2013).

Lourenco et al. propose some relevant theoretical frameworks for the future. They suggest a role for mental models, which have been proposed as a theoretical framework for cognitive reasoning skills in the past. However, from a mechanistic explanatory point of view the mental models framework remains rather general. We argue that models on working memory may bring in specific and testable predictions and might be helpful to operationalize the link between space and number. Interestingly, recent work has shown that many of the phenomena that have been interpreted as being a reflection of the mental number line (for instance, the SNARC effect) are not restricted to number tasks but have also been observed in tasks not involving numbers. In those cases the spatial effects were not generated by numbers but by the serial position of items that were kept in working memory. More specifically, as could be derived from reaction times, items from the beginning of the memorized sequence were associated with left and from the end of the sequence with right. This accords with memorized sequences taking the form of a spatially defined frame. The most prototypical frame is of course line-oriented according to writing direction, but depending on the situation other spatial frames can be adopted to allocate memorized items (Abrahamse et al., 2016, 2018). Depending on the spatial framework, additional properties can be incorporated in the memorized information. For instance, an essential spatial property of a circular presentation is that it has no beginning or end. By using such a reference frame to assign memory items, the property that all items have an equal amount of equipotent neighbors can be implemented in the relational organization of the memorized items. One can also speculate that higher levels of mathematical proficiency can be attained with more developed and more refined levels of spatial frames for storing memories. The fact that mental rotation is highly predictive for mathematical skill (as described by Lourenco et al.) is in line with such a view, as the possibility to flexibly manipulate items in spatial frames is not an interesting epiphenomenon but rather an essential process.

In sum, although the intrinsic coupling between number and space has been empirically demonstrated and theoretically confirmed, it is clear that the coupling is not a simple linear relation between two basic cognitive systems. To fully understand its nature, the complexity of the relationship has to be acknowledged and empirical phenomena at a neural and cognitive level, including a developmental perspective, need to be further documented in interaction with the development of theoretical frameworks and models that reveal the underlying dimensions. We believe that the chapters presented in the book have pointed out a number of useful trajectories along which the field can further develop.

References

Abrahamse, E., van Dijck, J.-P., & Fias, W. (2016). How does working memory enable number-induced spatial biases? *Frontiers in Psychology, 7*, 977.

Abrahamse, E., van Dijck, J.-P., & Fias, W. (2017). Grounding verbal working memory: The case of serial order. *Current Directions in Psychological Science, 26*(5), 429–433.

Adachi, I. (March 18, 2014). Spontaneous spatial mapping of learned sequence in chimpanzees: Evidence for a SNARC-like effect. *PLoS One, 9*, e90373.

Bachtold, D., Baumuller, M., & Brugger, P. (1998). Stimulus response compatibility in representational space. *Neuropsychologia, 36*, 731–735.

Bonato, M., Fabbri, S., Umiltà, C., & Zorzi, M. (2007). The mental representation of numerical fractions: Real or integer? *Journal of Experimental Psychology: Human Perception and Performance, 33*, 1410–1419.

Bonato, M., Saj, A., & Vuilleumier, P. (2016). Hemispatial neglect shows that "before" is "left". *Neural Plasticity, Vol. 2016*, 1–11.

Booth, J. L., & Siegler, R. S. (2006). Developmental and individual differences in pure numerical estimation. *Developmental Psychology, 42*, 189–201.

Booth, J. L., & Siegler, R. S. (2008). Numerical magnitude representations influence arithmetic learning. *Child Development, 79*, 1016–1031.

Cipora, K., Hohol, M., Nuerk, H. C., Willmes, K., Brożek, B., Kucharzyk, B., et al. (2016). Professional mathematicians differ from controls in their spatial-numerical associations. *Psychological Research, 80*, 710–726.

Cipora, K., & Nuerk, H. C. (2013). Is the SNARC effect related to the level of mathematics? No systematic relationship observed despite more power, more repetitions, and more direct assessment of arithmetic skill. *Quarterly Journal of Experimental Psychology, 66*, 1974–1991.

Dehaene, S., Bossini, S., & Giraux, P. (1993). The mental representation of parity and number magnitude. *Journal of Experimental Psychology: General, 122*, 371–396.

Doricchi, F., Guariglia, P., Gasperini, M. & Tomaiuolo, F. (2005). Dissociation between physical and mental number line bisection in right hemisphere brain damage. *Nature Neuroscience, 8*, 1663–1665.

Dormal, V., Schuller, A. M., Nihoul, J., Pesenti, M., & Andres, M. (2014). Causal role of spatial attention in arithmetic problem solving: Evidence from left unilateral neglect. *Neuropsychologia, 60*, 1–9.

Galton, F. (1880). Visualised numerals. *Nature, 22*, 494–495.

Gazes, R. P., Diamond, R. F. L., Hope, J. M., Caillaud, D., Stoinski, T. S., & Hampton, R. R. (2017). Spatial representation of magnitude in gorillas and orangutans. *Cognition, 168*, 312–319.

Gilmore, C., Attridge, N., Clayton, S., Cragg, L., Johnson, S., Marlow, N., et al. (June 13, 2013). Individual differences in inhibitory control, not non-verbal number acuity, correlate with mathematics achievement. *PLoS One, 8*(6), e67374.

Ginsburg, V., & Gevers, W. (2015). Spatial coding of ordinal information in short- and long-term memory. *Frontiers in Human Neuroscience, 9*, 1–10.

Halberda, J., Mazzocco, M. M. M., & Feigenson, L. (2008). Individual differences in non-verbal number acuity correlate with maths achievement. *Nature, 455*, 665.

Hoffrage, U., & Gigerenzer, G. (1998). Using natural frequencies to improve diagnostic inferences. *Academic Medicine, 73*(5), 538–540.

Knops, A., Thirion, B., Hubbard, E. M., Michel, V., & Dehaene, S. (2009). Recruitment of an area involved in eye movements during mental arithmetic. *Science, 324*(5934), 1583–1585. https://doi.org/10.1126/science.1171599.

Núñez, R., & Fias, W. (2017). Ancestral mental number lines: What is the evidence? *Cognitive Science, 41*(8), 2262–2266. https://doi.org/10.1111/cogs.12296.

Rouder, J. N., & Geary, D. C. (2014). Children's cognitive representation of the mathematical number line. *Developmental Science, 17*, 525–536.

Rugani, R., Vallortigara, G., Priftis, K., & Regolin, L. (2015). Number-space mapping in the newborn chick resembles humans' mental number line. *Science, 347*, 534–536.

Siegler, R. S., & Opfer, J. E. (2003). The development of numerical estimation: Evidence for multiple representations of numerical quantity. *Psychological Science, 14*, 237–243.

van Dijck, J. P., & Fias, W. (2011). A working memory account for spatial-numerical associations. *Cognition, 119*, 114–119.

van Dijck, J. P., Gevers, W., Lafosse, C., & Fias, W. (2012). The heterogeneous nature of number-space interactions. *Frontiers in Human Neuroscience, 10*(5), 182.

Walsh, V. (2003). A theory of magnitude: Common cortical metrics of time, space and quantity. *Trends in Cognitive Science, 7*, 483–488.

Winter, B., Matlock, T., Shaki, S., & Fischer, M. H. (2015). Mental number space in three dimensions. *Neuroscience Biobehavioural Reviews, 57*, 209–219.

Zorzi, M., Bonato, M., Treccani, B., Scalambrin, G., Marenzi, R., & Priftis, K. (2012). Neglect impairs explicit processing of the mental number line. *Frontiers in Human Neuroscience, 6*, 125.

Zorzi, M., Priftis, K., & Umiltà, C. (2002). Neglect disrupts the mental number line. *Nature, 417*, 138–139.

EXECUTIVE FUNCTIONS

13

Cognitive Interferences and Their Development in the Context of Numerical Tasks: Review and Implications

Liat Goldfarb

University of Haifa, Haifa, Israel

In everyday life people often encounter situations in which they must direct their behavior to fit their intentions, but irrelevant information interferes with this process. To study interferences by irrelevant information, the experimenter often creates a conflict situation in which the participant must respond to only a single stimulus or to one aspect of the stimulus. In these situations the participant needs to focus on the target (a stimulus or an aspect of a stimulus) and ignore the rest of the display.

Heterogeneity of Function in Numerical Cognition
https://doi.org/10.1016/B978-0-12-811529-9.00013-3

245

The Stroop task is one example of such a task. In 1935, Stroop asked participants to name the color of a color word or a color patch. Naming the color of an incongruent color word (e.g., the word "blue" written in red ink) was found to take longer than naming the color of a colored patch or any other neutral stimulus. This finding is known as the interference effect. Later on, this finding was extended by showing that it takes less time to name the color of a congruent color word (e.g., the word "blue" written in blue ink) than to name the color of a neutral stimulus. This finding is known as the facilitation effect (Dalrymple-Alford & Budayr, 1966). Both the interference and the facilitation effects are part of the Stroop effect (the difference in reaction time (RT) between incongruent and congruent stimuli) and they indicate that participants processed the irrelevant word, although they intended to name the color. In other words, the facilitation effect, the interference effect, and the Stroop effect all suggest that word naming interferes with the relevant task—i.e., name the color and ignore the word (for a review, see MacLeod, 1991). Similar findings can be found in other variations of the Stroop task and in other conflicting tasks. In all cases, these effects reflect the failure to attend to only one stimulus or dimension and to control the information provided by irrelevant stimuli or dimensions.

The magnitude of the different congruency effects found in such tasks is determined by the strength of at least two processes: automatic processing of irrelevant dimensions (e.g., the reading skill in the case of the Stroop task) and cognitive control, often also called conflict resolution (e.g., Braver, 2012; Goldfarb & Henik, 2007; MacLeod, 1991). The more the irrelevant dimension is activated and processed, the more interference will be observed. On the other hand, the better the operation of the control mechanism, the better it can resolve the conflict between the relevant and irrelevant dimensions and the less interference is observed. For example, in the case of the Stroop task, the more the word is activated, the larger the Stroop effect observed, but the more the control mechanism is activated, the smaller the Stroop effect observed. An important characteristic of the control mechanism is that its recruiting is flexible. Usually, inhibition cannot intentionally completely prevent a process from occurring. It can slowdown a process, reduce its strength, or overcome it as time goes by (e.g., Goldfarb, Aisenberg, & Henik, 2011; MacLeod, 2007, pp. 3–23).

The purpose of this paper is to examine data from the numerical cognition literature in the context of interference and control. The paper will first map some potential interferences that might exist in different numerical tasks and then it will examine the connection between observed inference in a numerical task and the development of the numerical skill.

INTERFERENCES IN NUMERICAL COGNITION

The perception of numbers and quantities and the process of solving simple or complex arithmetic problems frequently involve different aspects of interference. The numerical Stroop effect is an example of an interfering task in the numerical domain. Here, instead of word and color dimensions, the two dimensions that "pit" against each other are number and size or number and quantity. In the number-size congruency task, two digits that are different in size and numerical value are presented on the screen. In some of the trials, the numerical value and the physical size are congruent (e.g., 3 8) and in other trials they are incongruent (3 8). The common finding is that RT for deciding which digit is numerically or physically larger is faster for congruently sized numerical pairs than for incongruent pairs (e.g., Besner & Coltheart, 1979; Henik & Tzelgov, 1982; Tzelgov, Meyer, & Henik, 1992). Although both the numerical and the physical task demonstrate a situation in which interference from the irrelevant dimension is observed, the source of the interference is different in each task. While in the numerical task the attended dimension is the digit and the interfering dimension is the size, in the physical task the interference is caused by the numerical dimension that needs to be ignored and the size-congruency effect suggests that automatic numerical processing has occurred.

In another version of the numerical Stroop task, the counting Stroop, a digit or a number word is presented on the screen several times. In some of the trials the number is congruent with the number of times the digit or the word appears (e.g., "three" appears three times) and in other trials they are incongruent (e.g., "three" appears twice). Again the common finding is that RT for perceiving the quantity is faster for congruent number trials than for incongruent trials (e.g., Bush, Whalen, Shin, & Rauch, 2006).

Another interference that relates to quantity perception appears during the process of perceiving subset quantities. The perception of number of items in a subset reflects a situation in which one has to perceive the number of subset items within the total. The perception of subset numbers is necessary in everyday life, when objects usually appear as part of an overall group of items. We rarely need to simply enumerate the number of "things" (items whose identity is irrelevant), rather we enumerate predefined subset items from among a total number of items. When a child at a party decides to approach a table that contains a substantial amount of cookies, the child needs to enumerate the number of cookies among other items on the table such as spoons or flowers. We (Goldfarb & Levy, 2013; Goldfarb & Treisman, 2013) previously suggested that perceiving the number of items in a subset is qualitatively different than perceiving the total number of items. We suggested that the perception of number of subset occurs via an attentional path and it is an effortful process even for small subset numbers within the subitizing range. At an early perception stage, features in our surroundings such as size, color, and shape are perceived in their special feature maps rapidly, simultaneously, and without utilizing attentional resources (e.g., Treisman & Gelade, 1980; Treisman & Schmidt, 1982). However, to individuate identical items, so these can be counted, a representation must be created in which each individuated location is bound to the specific identity. This is in contrast to perception of the total number of items, which only requires knowledge about the location of items (and not their other features).

In line with this suggestion, we (Goldfarb; Levy, 2013) found that the perception of subset quantity involves an interference by the number of distractors (the items that do not need to be enumerated). In two experiments, participants were asked to count the number of targets (Xs) while ignoring distractors (Os). The distractors were either few or many. If counting a subset depends on prior binding between each possible location and its shape, then it was assumed that the RT for counting subset target items will be faster in a display with few distractors than in one with many distractors. On the other hand the number of distractors should not interfere with the perception of the targets' number if the number of a certain target can be directly pulled out of the scenery or of a relevant feature map such as a shape map (e.g., Huang, Treisman, & Pashler, 2007; Wolfe, 1994). Overall, the results indicated that irrelevant items do interfere and RT for counting subset target items becomes slower as the amount of distractors increases.

Other examples of interference in numerical tasks can be found in arithmetic tasks. Arithmetic is a branch of mathematics that deals with numbers and their addition, subtraction, multiplication, and division. Arithmetic processing can occur automatically, without a specific instruction to perform the task (LeFevre, Bisanz, & Mrkonjic, 1988; LeFevre & Kulak, 1994, Sklar et al., 2012). When performing arithmetic tasks, different types of unrelated information also have the potential to interfere. This can be resulted in an incorrect retrieve or in a slowdown of the processes of retrieval. For example, when we try to solve the arithmetic problem $3 \times 6 = 18$, numbers that are adjacent to the correct solution (e.g., 17) can interfere. Adjacent answers in the same times table can also interfere in a multiplication task. For example, in the case of the arithmetic problem $3 \times 6 = 18$, the number 21, which is the result of the arithmetic problem 3×7, can cause an interference. In addition, we can also observe interferences by the results of arithmetic problems that involve the relevant digits but also other irrelevant operations. Meaning that in the case of $3 \times 6 = 18$, the number 9 can interfere with the correct answer because it is the result of the irrelevant problem $3 + 6$ (e.g., Campbell, 1987; Stazyk, Ashcraft, & Hamann, 1982).

In complex arithmetic problems with more than two addends, we can also observe interferences from the intermediate sum. Complex arithmetic that involves three or more addends has specific cognitive demands such as the need to compute, hold, and manipulate the intermediate sum. It has been suggested that in these kinds of calculations, the intermediate sum might be temporarily stored in the working memory (De Stefano & LeFevre, 2004). In Abramovich and Goldfarb (2015) we examined interferences that involve intermediate sums. In this experiment participants were presented with three addends (e.g., 4, 2, 9). Then they were asked to perform two tasks: (a) calculate the sum of these addends (e.g., identify that the sum is 15) and (b) identify whether a certain digit was one of the addends in the problem displayed on the screen (e.g., identify that only the digits 4, 2, and 9 appeared on the screen as addends of the problem). RT and error rate for detecting that certain digits were not displayed (task b) were measured in two conditions of interest. In the first condition the absent digit was the intermediate sum (e.g., participants were supposed to detect that 6, which is the intermediate sum of 4 and 2, was not an addend in the addition problem). In the second condition, the absent digit was a neutral digit (e.g., participants were supposed to detect that 7 was not an addend in the addition problem). The results revealed an interference effect in which it was hard to identify that the digit representing the intermediate sum was *not* actually one of the operands, relative to a neutral digit. In a second experiment we further examined whether the intermediate sum is activated automatically when a task does not require calculation. In this experiment participants were presented with a prime of an addition problem followed by a target number. The task was to determine whether a target number is odd or even, while ignoring the addition problem in the prime. In the three addend addition problem, the target could be either congruent with the intermediate sum of the problem (e.g., prime: $8+3+4$ and target: 11) or incongruent (e.g., prime: $8+3+4$ and target: 6). The results suggested that the intermediate sum of the addition problem in the prime was activated automatically and facilitated the identification of the target.

THE MODULATION OF NUMERICAL INTERFERENCE AS A FUNCTION OF SKILL OR ABILITY

The previous part reviewed different types of interference in different numerical tasks. To deepen our understanding of numerical interference, the next part will examine how numerical interferences develop and how they are modulated by groups with different numerical skills or abilities.

Low-Skilled Groups Demonstrate Reduced Interferences

Theories that deal with the general development of a skill or ability suggest that, at the beginning, when a new skill is learned its level of automatic activation is low, and hence it will hardly interfere when it is not task relevant. However, as practice increases, the skilled dimension becomes more automatically activated and therefore it will interfere when it should be ignored (e.g., Hasher & Zacks, 1979; Logan, 1985, 1988). For example, an individual who has just begun to learn how to read in a new language will hardly experience interference by that language. That is, low proficiency in reading will result in no interference when one needs to ignore the written word in tasks such as the color-word Stroop task. As reading skill increases and therefore becomes more automatic, interference will become stronger. The notion that as practice increases, the skilled dimension will interfere more when it should be ignored can be grounded in studies that conduct controlled practice.

For example, MacLeod and Dunbar (1988) conducted experiments with four phases: the experiment started with a baseline phase, in which participants had to name four familiar colors of a square shape. Then there was the shape name practice phase, in which four white shapes appeared on the screen and the participants learned to give the shapes new names—the color names green, pink, orange, or blue. After those phases participants preformed two tests. In one test, participants had to name the colors of the shapes. This test had three conditions: a control—a color square, a congruent color—the "new name" of the shape was congruent with its color, or an incongruent color—the "new name" of the shape was incongruent with its color. In a second test, participants had to name the "new name" of the shapes. This test as well had three conditions: control—the shapes appeared in white ink, a congruent shape—the "new name" of the shape was congruent with its color, or an incongruent shape—the "new name" of the shape was incongruent with its color. This method was applied in several experiments that differed in the number of practiced trials in phase 2: in Experiment 1 participants practiced up to 2h (576 trials in 2days), in Experiment 2 they practiced up to 5h (2304 trials over 5days), and in Experiment 3 up to 20h (10,656 trials over 20days). The results of these experiments showed that in Experiment 1, with a little practice of up to 2h with giving unfamiliar shapes new color names (phase 2), color naming interfered with shape naming, but shape naming did not interfere with color naming. In Experiment 2, with more practice of up to 5h, the magnitude of interference from the shape name dimension to the color name dimension was equal to the magnitude of interference from the color name dimension to the shape name dimension. In Experiment 3, with a significant amount of practice of up to 20h, it was the shape naming that interfered with the color naming but the color naming did not interfere with the shape naming.

Similarly, a control practiced in the context of numerical tasks has been performed by Tzelgov, Yehene, Kotler, and Alon (2000). They trained participants to decide which of two arbitrary figures represents a larger magnitude. They used nine arbitrary figures that each had a magnitude corresponding to digits 1–9. However, this association was unknown to the participants. After a training stage of six sessions that each lasted about 1 h, the participants performed a form of the Stroop size congruity task. Now the arbitrary figures were presented in different sizes, which could be either congruent or incongruent with the new learned magnitude. Interestingly, when participants were asked to perform the physical comparison task and to decide which figure in the pair was physically larger, a size congruity effect appeared. This means that after training in which a magnitude is associated with arbitrary figures, this magnitude can cause interference even when it is irrelevant for the task.

The notion that as practice increases, the skilled dimension will interfere more when it should be ignored can also be evidenced in studies that examined populations with low abilities or skills as these studies usually suggest that those populations have smaller interferences. For example, in the case of the regular color-word Stroop task, unbalanced bilingual participants with low proficiency in a second language have smaller interference when the word in the Stroop task appears in that language (e.g., Mägiste, 1984; Okuniewska, 2007).

Specifically in the numerical cognition literature, in arithmetic tasks, low-skilled groups generally have a slower and less activated network of associations, when automatic activation of two addends is measured (Jackson & Coney, 2007; LeFevre & Kulak, 1994).

Likewise, Rotem and Henik (2015a, 2015b) examined the development of different interferences in multiplication problems. In Rotem and Henik (2015b), they examined the performance of typically developing children from the second, third, fourth, and sixth grades, as well as of children with a mathematics learning disability (MLD) in the sixth and eighth grades. Interestingly, they found that the interference resulting from operands that shared a timetable row was fully achieved in the third grade among typically developing children but only in the eighth grade among children with MLD. In another study, Rotem and Henik (2015a) examined the previous timetable row interference in the context of the distance interference, in which unrelated false results interfere more when they are close to the correct result than when they are distant. These effects were examined in different multiplication problems, ranging from easy (e.g., with operands smaller than 5, such as 2×3) to difficult (such as 7×9). They found that, in typically developing children, interference from the same timetable and from the false result distance increases with age (from the second to sixth grade), spreading from easy to difficult problems. In the MLD group, sixth grade children did not demonstrate these effects, while eighth grade children demonstrated them only in the less difficult problems.

Similar results can be found in the numerical Stroop task with students with developmental dyscalculia (DD), which is a specific learning disability affecting the acquisition of numerical skills (e.g., Butterworth, 2008,

Landerl, Bevan, & Butterworth, 2004; Shalev & Gross-Tsur, 2001) The physical comparison task within the numerical Stroop task can provide a good indication of the automatic interference by the numerical dimension because in this case the numerical dimension is not relevant. In line with the notion that a population with low skills/abilities will demonstrate reduced interference, it has been documented that students with DD usually demonstrate a reduced size-congruency effect in the physical comparison task (incongruent trials–congruent trials), and within this effect the component of the facilitation effect (neutral trials–congruent trial) is reduced and even eliminated in this group (e.g., Ashkenazi, Rubinsten & Henik, 2009; Rubinsten & Henik, 2005, 2006). A reduction or an elimination of the facilitation effect was also observed in acquired dyscalculia (Ashkenazi, Henik, Ifergane, & Shelef, 2008), as well as after the function of the right intraparietal sulcus, which has been documented as being involved in number processing, was disturbed by transcranial magnetic stimuli (Cohen Kadosh et al., 2007).

The Relationship Between the Development of a Skill and the Observed Interference can Be Described as an Inverted u Curve in a Wide Range of Skill Levels

Although it has been documented that the interference increases with the numerical skill/ability level, a closer look at the literature that examines a wide range of development levels, from low to extreme top, might reveal that the real relationship between the two factors can be more complex. In fact, when we look at a wide range of skills/abilities, we can also find in the literature some indications for a relationship between the two, which is better described as an inverted u shape than a simple linear relationship. Accordingly, at the beginning low-numerical skills or abilities are related to small interferences. As the numerical skill level increases, it interferes more. However, from a certain point of high proficiency, the observed interference by the skilled dimension might begin to decrease, creating a somewhat inverted u-shape pattern.

For example, we (Abramovich & Goldfarb, 2018) recently conducted a study in which we examined, in different levels of arithmetic skills: low, average, and high, the magnitude of the intermediate sum interference as we found in Abramovich and Goldfarb (2015). Remember that in that study, we found that when normal students perform a summation of three addends, the intermediate sum interferes to such a degree that participants tend to confuse the intermediate sum with the actual addends in the problem compared with a neutral stimuli.

In the new study, we found that when examining this effect for different mathematical skill levels, different behavior patterns are revealed. As expected, the average skill group demonstrated an interference effect, as shown in the previous study. On the other hand, the low-skilled group demonstrated a reduced effect in both RT and error rate. In fact, in both measures the intermediate sum effect failed to reach a significant level in the low-skill groups. This fit the notion that if the activation level of the intermediate sum is weaker for a low-skill group, then this activation will not cause strong interference. On the other hand, because the average skill group gains enough proficiency, it will result in strong activation of the intermediate sum. However, interestingly, when a wider range of skill/ability is examined and the high-proficiency group is also taken into account, the result suggests that this group also demonstrates a reduced effect in both RT and error rate compared with the average group. In fact, the pattern of results of the high-proficiency group is similar to that of the low-skilled group.

A similar indication of the inverted u pattern between proficiency level and numerical interference might also be evident in the size congruity effect in the physical comparison of the size congruity task. In two studies (Girelli, Lucangeli, & Butterworth, 2000; Rubinsten, Henik, Berger, & Shahar-Shalev, 2002), participants in different age groups preformed the physical comparison task.

Study by Girelli et al. (2000) included four age groups: first graders, third graders, fifth graders, and university students. At first grade, there was no interference by the number dimension in the form of the congruity effect in the physical task. The effect began to emerge in the third grade and was significant in the fifth grade. Within the congruity effect, the component of the facilitation effect (neutral congruent) showed a similar pattern. This, of course, fits the notion that the intrusion of the irrelevant dimension increases with the skill/ability level. However, it is worth noting that the linear shape connecting the magnitude of the facilitation effect and the skill/ability level changed when a wider range of development was taken into account and was extended to include university students as well. Although this was not specifically tested by the authors, it appears that when a wider range is taken into account, the relationship seems to actually fit the inverted u-shape pattern. It seems that while in school-aged children the facilitation effect increases with age, among university students the effect moves in the opposite direction—it starts to decline. In fact the facilitation effect failed to reach a significant level in the university students group.

In the second study (Rubinsten et al., 2002) that examined the size congruity effect in the physical comparison task, five different age groups were included: beginning of first grade, end of first grade, third grade, fifth grade, and university students. They found that the size congruity effect began to appear at the end of the first grade. Most interestingly, a glance at the pattern of the components of the facilitation effect suggests that this effect began to appear only in the third grade, increased by the fifth grade, and decreased again among undergraduate students. In the university student group, the facilitation effect was almost half the size as among the fifth grade students. (However, because this complex pattern was not originally studied by the authors these contrasts were not specifically examined in the original study, so it is not clear if the differences indeed reached significance.) Another interesting study that was performed in the context of skill/ability and the size congruity task examined the physical comparison between second and third grade students with three skill levels: low, average, and high (Heine et al., 2010). Here, the researchers examined the reverse distance effect in the incongruent condition. This effect reflects a situation where when a physical comparison task is performed, the interference in the incongruent condition will be larger when the distance between the digits is larger (i.e., 3 8) than when it is smaller (e.g., 3 4). This effect can be attributed to greater interference by the irrelevant numbers when the digits are more vividly separated (as in the larger distance condition). Interestingly, the study found that this reverse distance effect was significant only in the low and average group but not in the high-skill group. This again suggests a decline in the interference level in the high end of the skill/ability group.

How can the inverted u-shape pattern be explained, and in particular why is a sudden decrease in numerical interference observed in high-skill/ability groups? To answer this question we should remember that the observed interference is the outcome of two contrasting processes: (1) the more the irrelevant dimension is activated, the more interference will be observed and (2) the more the control mechanism is activated, the less interference will be observed. The first process increases with practice and skill level. However, if the high-skilled group also has high control abilities then a decrease in interference will be observed in this group (see Fig. 13.1). But, why would the high-proficiency group have a better control mechanism?

One possible answer is that in the case of numerosity proficiency, high control abilities are part of abilities that are required for being proficient in the numerical domain. Findings emerging across different studies in numerical cognition suggest that there is a relationship between inhibition or control abilities and numerical skills (e.g., see extensive review in Cragg & Gilmore, 2014). Numerous studies conducted on a variety of age groups found numerical and arithmetic abilities to be strongly linked to executive function (EF) abilities, which include shifting, inhibition, and working memory (Bull & Scerif, 2001; Daneman & Merikle, 1996; De Stefano & LeFevre, 2004; Merkley, Thompson, & Scerif, 2016; Passolunghi & Siegel, 2001; van der Sluis, de Jong, & van der Leij, 2004). Espy et al. (2004) even found that inhibitory control significantly predicted arithmetic skills in preschool children. Numerous findings also suggest that it is associated with deficiencies in various aspects of EF. For example, Passolunghi and Siegel (2001) suggested that students with DD had a general deficit in working memory accompanied by a difficulty in the inhibitory

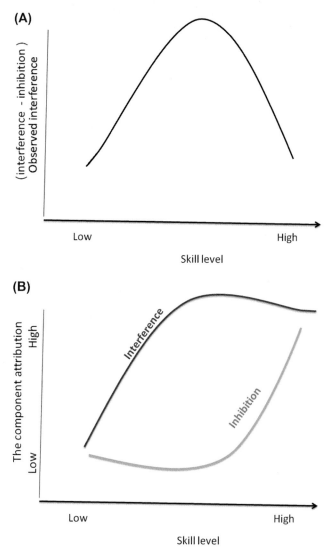

FIGURE 13.1   A possible connection between the development of a skill and the observed interference. In (A) the *black curve* represents an inverted u-shaped relationship between the development of a skill and the observed interference, which is the result of the interference level minus the inhibition level. In (B) the *red and green curves* represent the activation level of these components: interference and inhibition and the attribution of these components to the observed interference as a function of skill level. The interference by the irrelevant dimension increases as one becomes more skilled (*red curve*). In addition, higher-inhibition abilities relate to higher-skill levels (*green curve*).

mechanisms required for the inhibition of information that is no longer relevant. Similarly, Zhang and Wu (2011), who used a battery of inhibition tasks, found that children with DD had a deficit in inhibitory processes. Thus, the literature might suggest a causal link in which having good EFs and among them inhibitory functions leads to better numerical performance (e.g., Szűcs, Devine, Soltesz, Nobes, & Gabriel, 2013; Heine et al., 2010).

Specifically, in the studies described earlier, in which the high-skilled group had a reduced numerical interference, this reduction can be explained by the general high control and EF abilities associated with this group. For example, in the study that examined the reverse distance effect in the incongruent condition in second and third grade students with three skill levels, the high-skilled group differed from the other groups not only by their numerical abilities but also by their high score on intelligence and working memory tests.

A similar explanation can also be applied to the other two studies described earlier (Girelli et al., 2000; Rubinsten et al., 2002), which examined the size congruency and the facilitation effects in the physical comparison task in different age groups. It is clear that different age groups (ranging from young children to university students) differ from one another not only by their skill level but also by the different level of maturation of cognitive components, such as general EF abilities (e.g., Diamond, 2013; Espy, 1997; Hughes, 1998; Siegel & Ryan, 1989). Hence, overall, if control and EF abilities lead to better numerical skills/abilities, it is not surprising that the high-proficiency group has a better control and that a reduced interference is observed in that group.

Although this explanation seems a logical causal explanation with a one-directional influence (i.e., high inhibition ability leads to high numerical skill level), I would like to suggest that an alternative opposite causality that can explain the decline of interference in the high-proficiency group might also exist. Accordingly, high numerical skills produce high inhibition abilities. In this option, numerical skills are similar to any other skills and as such can be grounded in a general theory describing a relationship between any general type of skill and the ability to control that skill.

In 1985, Logan suggested a theory that connects the development of a skill and the development of the control of that skill. It is well agreed that the level of skill influences the level of automaticity. However, it has also been suggested that skills also consist, in addition to automatic procedures, of metacognitive knowledge of how and when to use the procedures. Specifically, it has been suggested that as an individual becomes more skilled in a task, he/she also gains better control and inhibition abilities of that specific skill (Logan, 1985). This implies nonlinear relationships between the development of a skill and the observed interference by the skilled dimension, when the latter becomes irrelevant. At the beginning, when one learns a new skill, it will hardly interfere when it should be ignored (i.e., when the skill is not relevant). However, as skill increases, the skilled dimension becomes more automatic and therefore it will interfere when it should be ignored. Yet, according to Logan (1985), when one reaches a high level of proficiency, the individual also has better inhibition and control abilities of that specific skill. This suggests that, from a certain point of high proficiency, the observed interference by the skilled dimension begins to decrease. Evidence supporting this notion was found, for example, with regard to reading proficiency in a Hebrew–Arabic bilingual Stroop task. Participants were better at controlling and reducing the Stroop interference effect in their native language than in the second language, in which they were less skilled (e.g., Tzelgov, Henik, & Leiser, 1990).

Hence, if we apply this logic to the numerical cognition literature it is possible that as one becomes more skilled in numerical and arithmetic tasks, he/she also becomes more skilled at controlling this information when it is not relevant. How does one develop this ability? Perhaps this inhibition ability develops as a parallel ability to the skill or skilled people encounter more situations in which they need to inhibit the skill. The literature on inhibition suggests that training of inhibition improves that specific inhibition. For example, the first study that investigated practice in the Stroop task was presented by Stroop himself in 1935. In Stroop's study, participants practiced naming the color of incongruent words for 8 days. Most importantly, the results suggested that practice reduced the RT for naming incongruent color words. In fact, this reduction began to appear after a single day of training.

In the context of numerical interferences, Reisberg, Baron, and Kemler (1980) asked participants to practice a version of the numerical Stroop task in which they counted digits while ignoring the digits' meaning. In the practice stage, digits that had to be ignored appeared repeatedly. In the task stage, participants had to ignore the digits' meaning once again. It was found that practice increases the ability to ignore the meaning of digits that appeared in the practice phase but not the meaning of digits that did not appear in the practice phase. These results imply that with a short practice stage in which numbers are inhibited, at least a specific control over numerical information can be learned. Hence, in everyday life, people who are high skilled might be more involved in number processing; hence they might also encounter more situations in which they have to control numbers. It is possible that practice specifically directed at controlling numbers improves the ability to inhibit numbers when they are not task relevant.

EPILOGUE

To sum up, this chapter reviews studies that suggest that various numerical tasks involve interference. In the number-size Stroop task, numerical value and physical size interfere with each other. In the counting Stroop, the quantity of the number and the numerical value interfere with each other. In the subset counting, interference by the number of items that do not need to be enumerated is observed. In arithmetic tasks, multiple irrelevant activations may interfere, such as numbers that are adjacent to the correct solution, adjacent answers in the same timetable, solutions that are the result of arithmetic problems that involve the relevant digits but with irrelevant operations. Similarly, in complex arithmetic problems with more than two addends, the intermediate sum may also interfere.

Although these interferences were reviewed in a laboratory context, they are also relevant outside the lab. For instance, in everyday activities people constantly perform arithmetic tasks, such as paying bills or calculating change. Hence, the different types of interferences reviewed previously in the context of arithmetic tasks are also relevant in real life. For example, when one invites 7 guests for a barbecue and each needs 3 paper cups, the host might mistakenly buy 24 cups because this is the product of 8 times 3. Similarly, interferences that appear in the perception of subset numbers are also relevant for everyday life. For example, when a host counts the number of bottles of wine he/she places on the table, he/she also experiences interference by the number of cups that are on the table. Interference in numerical Stroop tasks may also mimic interferences that people encounter in everyday life. For example, when one checks the number of digits in a telephone number that was given to confirm that none of the digits is missing, the identity of the digit can interfere with this task. Likewise, when one needs to dial a telephone number written on a piece of paper, the dialer must ignore the font size of the digits because they can potentially cause interference.

The second part of the chapter focuses on the relationship between the level of skill and the observed interference and suggests that when a wide range of skill levels is examined, from low to extreme top, the relationship between the two factors can be described as inverse u shaped. This means that, from a certain point of high proficiency, the observed interference might begin to decrease. Understanding interference in the context of skill level can clarify the situations in which each conflict is at its peak and specify the populations that are most at risk of encountering irrelevant interference. Two types of explanation were offered for this relationship. The first suggests that high inhibition ability leads to high numerical skill level, whereas the second suggests that high numerical skill produces high inhibition abilities. While a longitudinal study might distinguish between these options, it is possible that both factors contribute to the inverted u-shaped relationship.

It is clear that numerical tasks have number specific components and mechanisms and that numerical cognition theories usually revolve around these specific aspects. However, because numerical tasks also involve a variety of irrelevant interferences, theories regarding inhibition and automatic activation can help us rethink the findings concerning numerical abilities and widen our understanding in this field. These theories can help motivate new theoretical perspectives about numerical cognition and clarify the nature and symptoms of a wide range of numerical skills/abilities.

References

Abramovich, Y., & Goldfarb, L. (2015). Activation of the intermediate sum in intentional and automatic calculations. *Frontiers in Psychology*. https://doi.org/10.3389/fpsyg.2015.01512.

Abramovich, Y., & Goldfarb, L. (2018). *The activation of the intermediate sum as a function of numerical skills* (Submitted for publication).

Ashkenazi, S., Henik, A., Ifergane, G., & Shelef, I. (2008). Basic numerical processing in left intraparietal sulcus (IPS) acalculia. *Cortex, 44*, 439–448.

Ashkenazi, S., Rubinsten, O., & Henik, A. (2009). Attention, automaticity and developmental dyscalculia. *Neuropsychology, 23*, 535–540.

Besner, D., & Coltheart, M. (1979). Ideographic and alphabetic processing in skilled reading of English. *Neuropsychologia, 17*, 467–472.

Braver, T. S. (2012). The variable nature of cognitive control: A dual mechanisms framework. *Trends in Cognitive Sciences, 16*, 106–113.

Bull, R., & Scerif, G. (2001). Executive functioning as a predictor of children's mathematics ability: Inhibition, switching, and working memory. *Developmental Neuropsychology, 19*, 273–293.

Bush, G., Whalen, P. J., Shin, L. M., & Rauch, S. L. (2006). The counting stroop: A cognitive interference task. *Nature Protocols, 1*, 230–233.

Butterworth, B. (2008). Developmental dyscalculia. In J. Warner-Rogers, & J. Reed (Eds.), *Handbook of mathematical cognition*.

Campbell, J. I. D. (1987). Network interference and mental multiplication. *Journal of Experimental Psychology: Learning, Memory, and Cognition, 13*, 109–123.

Cohen Kadosh, R., Cohen Kadosh, K., Schuhmann, T., Kaas, A., Goebel, R., Henik, A., et al. (2007). Virtual dyscalculia induced by parietal-lobe TMS impairs automatic magnitude processing. *Current Biology, 17*, 689–693.

Cragg, L., & Gilmore, C. (2014). Skills underlying mathematics: The role of executive function in the development of mathematics proficiency. *Trends in Neuroscience and Education, 3*(2), 63–68.

Dalrymple-Alford, E. C., & Budayr, B. (1966). Examination of some aspects of the Stroop color-word test. *Perceptual and Motor Skills, 23*, 1211–1214.

Daneman, M., & Merikle, P. M. (1996). Working memory and language comprehension: A meta-analysis. *Psychonomic Bulletin and Review, 3*, 422–433.

De Stefano, D., & LeFevre, J. (2004). The role of working memory in mental arithmetic. *European Journal of Cognitive Psychology, 16*, 353–386.

Diamond, A. (2013). Executive functions. *Annual Review of Psychology, 64*, 135–168. https://doi.org/10.1146/annurev-psych-113011-143750.

Espy, K. A. (1997). The Shape School: Assessing executive function in preschool children. *Developmental Neuropsychology, 13*, 495–499.

Espy, K. A., McDiarmid, M. M., Cwik, M, F. Stalnto, M. M., Hamby, A., & Senn, T. F. (2004). Impaired neuropsychological functioning in lead-exposed children. *Developmental Neuropsychology, 26*, 465–486.

Girelli, L., Lucangeli, D., & Butterworth, B. (2000). The development of automaticity in accessing number magnitude. *Journal of Experimental Child Psychology, 76*, 104–122.

Goldfarb, L., Aisenberg, D., & Henik, A. (2011). Think the thought, walk the walk—social priming reduces the Stroop effect. *Cognition, 118*, 193–200.

Goldfarb, L., & Henik, A. (2007). Evidence for task conflict in the Stroop effect. *Journal of Experimental Psychology: Human Perception and Performance, 33*, 1170–1176.

Goldfarb, L., & Levy, S. (2013). Counting within the subitizing range: The effect of number of distractors on the perception of subset items. *PLoS One*. https://doi.org/10.1371/journal.pone.0074152.

Goldfarb, L., & Treisman, A. (2013). An evidence for the object file synchrony. *Psychological Science, 24*, 266–271.

Hasher, L., & Zacks, R. T. (1979). Automatic and effortful processes in memory. *Journal of Exceptional Psychology, 108*, 356–388.

Heine, A., Tamm, S., De Smedt, B., Schneider, M., Thaler, V., Torbeyns, J., et al. (2010). The numerical Stroop effect in primary school children: A comparison of low, normal and high achievers. *Child Neuropsychology, 16*(5), 461–477.

Henik, A., & Tzelgov, J. (1982). Is three greater than five: The relation between physical and semantic size in comparison tasks. *Memory and Cognition, 10*, 389–395.

Huang, L., Treisman, A., & Pashler, H. (2007). Characterizing the limits of human visual awareness. *Science, 317*, 823–825.

Hughes, C. (1998). Executive function in preschoolers: Links with theory of mind and verbal ability. *British Journal of Developmental Psychology, 16*(9), 233–253.

Jackson, N., & Coney, J. (2007). Simple arithmetic processing: Individual differences in automaticity. *European Journal of Cognitive Psychology, 19*, 141–160.

Landerl, K., Bevan, A., & Butterworth, B. (2004). Developmental dyscalculia and basic numerical capacities: A study of 8-9-year-old students. *Cognition, 93*, 99–125.

LeFevre, J., Bisanz, J., & Mrkonjic, L. (1988). Cognitive arithmetic: Evidence for obligatory activation of arithmetic facts. *Memory and Cognition, 16*, 45–53.

LeFevre, J., & Kulak, A. G. (1994). Individual differences in the obligatory activation of addition facts. *Memory and Cognition, 22*, 188–200.

Logan, G. D. (1985). Skill and automaticity: Relations, implications and future directions. *Canadian Journal of Psychology, 39*, 367–386.

Logan, G. D. (1988). Toward an instance theory of automatization. *Psychological Review, 95*, 492–527.

MacLeod, C. M. (1991). Half a century of research on the Stroop effect: An integrative review. *Psychological Bulletin, 109*, 163–203.

MacLeod, C. M. (2007). The concept of inhibition in cognition. In D. S. Gorfein, & C. M. MacLeod (Eds.), *Inhibition in cognition*. Washington, DC: American Psychological Association.

MacLeod, C. M., & Dunbar, K. (1988). Training and Stroop-like interference: Evidence for a continuum of automaticity. *Journal of Experimental Psychology: Learning, Memory, and Cognition, 14*, 126–135.

Mägiste, E. (1984). Stroop tasks and dichotic translation: The development of interference patterns in bilinguals. *Journal of Experimental Psychology: Learning, Memory, and Cognition*, 10, 304–315.

Merkley, R., Thompson, J., & Scerif, G. (2016). Of huge mice and tiny elephants: Exploring the relationship between inhibitory processes and preschool math skills. *Frontiers in Psychology*, 6, 1–14.

Okuniewska, H. (2007). Impact of second language proficiency on the bilingual PolishEnglish Stroop task. *Psychology of Language and Communication*, 11, 49–63.

Passolunghi, M. C., & Siegel, L. S. (2001). Short-term memory, working memory, and inhibitory control in children with difficulties in arithmetic problem solving. *Journal of Experimental Child Psychology*, 80, 44–57.

Reisberg, D., Baron, J., & Kemler, D. (1980). Overcoming Stroop interference: The effects of practice on distractor potency. *Journal of Experimental Psychology: Human Perception and Performance*, 6, 140–150.

Rotem, A., & Henik, A. (2015a). Development of product relatedness and distance effects in typical achievers and in children with mathematics learning disability. *Journal of Learning Disabilities*, 48, 577–592.

Rotem, A., & Henik, A. (2015b). Sensitivity to general and specific numerical features in typical achievers and children with mathematics learning disability. *Quarterly Journal of Experimental Psychology*, 68, 2291–2303.

Rubinsten, O., & Henik, A. (2005). Automatic activation of internal magnitudes: A study of developmental dyscalculia. *Neuropsychology*, 19, 641–648.

Rubinsten, O., & Henik, A. (2006). Double dissociation of functions in developmental dyslexia and dyscalculia. *Journal of Educational Psychology*, 98, 854–867.

Rubinsten, O., Henik, A., Berger, A., & Shahar-Shalev, S. (2002). The development of internal representations of magnitude and their association with Arabic numerals. *Journal of Experiment Child Psychology*, 81, 74–92.

Shalev, R. S., & Gross-Tsur, V. (2001). Developmental dyscalculia (DD). *Pediatric Neurology*, 24, 337–342.

Siegel, L. S., & Ryan, E. B. (1989). The development of working memory in normally achieving and subtypes of learning disabled children. *Child Development*, 60(4), 973–980.

Sklar, A. Y., Levy, N., Goldstein, A., Mandel, R., Maril, A., & Hassin, R. R. (2012). Reading and doing arithmetic nonconsciously. *Proceedings of the National Academy of Sciences*, 109, 19614–19619.

van der Sluis, S., de Jong, P. F., & van der Leij, A. (2004). Inhibition and shifting in children with learning deficits in arithmetic and reading. *Journal of Experimental Child Psychology*, 87, 239–266.

Stazyk, E. H., Ashcraft, M. H., & Hamann, M. S. (1982). A network approach to mental multiplication. *Journal of Experimental Psychology: Learning, Memory, and Cognition*, 8, 320–335.

Stroop, J. R. (1935). Studies of interference in serial verbal reactions. *Journal of Experimental Psychology*, 6, 643–662.

Szűcs, D., Devine, A., Soltesz, F., Nobes, A., & Gabriel, F. C. (2013). Developmental dyscalculia is related to visuo-spatial memory and inhibition impairment. *Cortex*, 49, 2674–2688.

Treisman, A., & Gelade, G. (1980). A feature-integration theory of attention. *Cognitive Psychology*, 12, 97–136.

Treisman, A., & Schmidt, H. (1982). Illusory conjunctions in the perception of objects. Cognitive Psychology, 14, 107–141.

Tzelgov, J., Henik, A., & Leiser, D. (1990). Controlling the Stroop interference: Evidence from a bilingual task. Journal of Experimental Psychology: Learning, Memory, and Cognition, 16, 760–771.

Tzelgov, J., Meyer, J., & Henik, A. (1992). Automatic and intentional processing of numerical information. Journal of Experimental Psychology: Learning, Memory, and Cognition, 18, 166–179.

Tzelgov, J., Yehene, V., Kotler, L., & Alon, A. (2000). Automatic comparisons of artificial digits never compared: Learning linear ordering relations. Journal of Experimental Psychology: Learning, Memory, and Cognition, 26, 1–18.

Wolfe, J. M. (1994). Guided search 2.0: A revised model of visual search. Psychonomic Bulletin and Review, 1, 202–223.

Zhang, H., & Wu, H. (2011). Inhibitory ability of children with developmental dyscalculia. Journal of Huazhong University of Science and Technology – Medical Science, 31, 131–136.

The Role of Executive Function Skills in the Development of Children's Mathematical Competencies

Camilla Gilmore[1], Lucy Cragg[2]

[1]Loughborough University, Loughborough, United Kingdom;
[2]The University of Nottingham, Nottingham, United Kingdom

OUTLINE

Heterogeneity of Function in Numerical Cognition
https://doi.org/10.1016/B978-0-12-811529-9.00014-5

INTRODUCTION

Proficiency with mathematics is important for success in modern society and impacts on our health, wealth, and quality of life (Gross, Hudson, & Price, 2009; OECD, 2013; Parsons & Bynner, 2005). However, many adults do not have the numeracy skills needed for everyday life and a large proportion of children leave school without achieving the expected level of mathematics skills (Department for Business Innovation & Skills, 2011; Gross et al., 2009). Consequently, researchers have sought to understand the range of cognitive and noncognitive skills that are involved in mathematics. Over the past two decades, an increasing body of research has identified that executive functions, the set of processes that control and guide our thoughts and behavior, play an important role in mathematics achievement and learning (e.g., review papers by Bull & Lee, 2014; Cragg & Gilmore, 2014; Raghubar, Barnes, & Hecht, 2010).

Executive functions is the name given to a group of processes that allow us to respond flexibly to our environment and engage in deliberate, goal-directed thought and action. Three executive function skills have received the most research attention: *working memory*, the ability to monitor and manipulate information in mind; *inhibition*, the ability to suppress distracting information and inappropriate responses; and *shifting*, the capacity for flexible thinking and switching attention between different tasks.

In this chapter, we will review the existing evidence for the role of executive functions in mathematics achievement before considering existing and new evidence concerning the involvement of executive function skills in specific components of mathematics. We will finish by presenting recent evidence for the direct and indirect role of executive functions on children's mathematics achievement and considering the important distinction between learning and performing mathematics.

THE DEVELOPMENT OF EXECUTIVE FUNCTION

Executive function skills begin to emerge in infancy. By 9 months old, most infants show some evidence of attentional control and are able to inhibit unwanted responses and control their behavior (Diamond, 1985). However, development of executive function skills is not linear, and different subskills may emerge at different ages and have different developmental trajectories (Anderson, 2002). For example, the ability to switch flexibly between different tasks does not emerge until 3 or 4 years old (Espy, 1997) and typically develops more slowly than attentional control. Despite the early emergence of some executive functions, development is protracted and these skills are among the last cognitive abilities to mature, continuing to develop into late adolescence (Conklin, Luciana,

Hooper, & Yarger, 2007; Huizinga, Dolan, & van der Molen, 2006; Luna, Garver, Urban, Lazar, & Sweeney, 2004).

Longitudinal studies indicate that the development of executive function skills is characterized not only by improvements in performance but also by increasing differentiation in executive subskills. While studies with adolescents and adults typically find that executive functions are multifaceted and it is possible to identify separable subskills (Huizinga et al., 2006; Miyake et al., 2000), some studies with younger children identify only a single unitary executive factor or a two-factor model in which inhibition and shifting are undifferentiated (Lee, Bull, & Ho, 2013; Wiebe, Espy, & Charak, 2008).

The protracted and differentiated development of executive functions has several implications for understanding how executive functions are involved in mathematics. First, age-related differences in the role of executive function skills in mathematics may reflect either changes in the involvement of executive function skills or changes in the underlying structure of executive functions themselves. Second, tasks used to assess executive functions may draw on different underlying subskills at different ages. Finally, during the period in which executive function skills develop, the nature of the mathematics activities in which children are engaged also changes dramatically. These factors all combine to add complexity to our understanding of how executive function skills are involved in mathematics, the consequences for interpreting individual differences in mathematics performance, and the implications for supporting children's mathematics development.

EXECUTIVE FUNCTIONS AND ACADEMIC ACHIEVEMENT

Executive function skills are particularly important when individuals are dealing with novel, rather than routine, situations and activities. This is a key characteristic of learning across all academic subjects and therefore we would expect executive function skills to be an important factor in academic achievement and success in school generally. This is indeed the case, with evidence for the role of executive function skills in mathematics, reading, writing, and science outcomes (Best, Miller, & Naglieri, 2011; Nunes, Bryant, Barros, & Sylva, 2012; St Clair-Thompson & Gathercole, 2006). However, over and above the role of executive function skills in learning and academic achievement generally, it has been suggested that executive function skills are particularly important for mathematics. For example, executive function skills measured at age 5 years account for more variance in later mathematics performance than reading (Willoughby, Blair, Wirth, Greenberg, & The Family Life Project Investigators, 2012).

Moreover, there is some evidence that the role of executive function skills differs across different academic subjects. In a longitudinal study from kindergarten to grade 5, the relationship between working memory and mathematics achievement increased, whereas the relationship between working memory and reading achievement decreased (Geary, 2011). However, other studies have suggested that executive function skills have a domain-general influence on learning and achievement that does not differ across subjects (Best et al., 2011). Differing findings across these studies may reflect the specificity of the tasks selected to measure both executive function skills and academic outcomes. To uncover the importance of executive function skills for learning and achievement in mathematics and across the academic spectrum requires sensitive measures, which can pinpoint the specific ways in which executive function skills support learning and achievement.

EXECUTIVE FUNCTIONS AND MATHEMATICS ACHIEVEMENT

A wealth of evidence, largely from correlational and longitudinal studies, has demonstrated a general relationship between executive function skills and overall mathematics achievement. These studies have shown that scores on cognitive tests of executive function skills are associated with concurrent or future mathematics performance, as measured by standardized or curriculum-based mathematics tests. Such studies have largely focused on the role of working memory; however, inhibition and shifting have received increased attention in the past few years.

Working Memory and Mathematics Achievement

The majority of studies that have explored the involvement of working memory in mathematics achievement have been based on the Baddeley and Hitch (1974, pp. 47–89) model of working memory, whereby working memory is made up of short-term stores for verbal and visuospatial information, coordinated by a central executive that allows the manipulation and storage of information at the same time. Drawing on this model, researchers have demonstrated that working memory capacity is a strong predictor of current and future mathematics achievement (Friso-van den Bos, van der Ven, Kroesbergen, & van Luit, 2013; Fuchs et al., 2010; Hecht, Torgesen, Wagner, & Rashotte, 2001; Peng, Namkung, Barnes, & Sun, 2016).

Beyond a general association between working memory and mathematics achievement, researchers have attempted to pinpoint which functions of working memory are most critical for explaining variance in mathematics achievement (e.g., see metaanalysis by Friso-van den Bos et al., 2013). This has revealed that tasks that require the simultaneous storage and manipulation of information (executive working memory) show a stronger relationship with mathematics achievement than tasks that simply measure the short-term storage of information, particularly in relation to the storage of verbal information.

Several studies have investigated whether there is a stronger relationship between verbal or visuospatial working memory and mathematics achievement. This has produced mixed findings with some studies suggesting that verbal working memory plays a greater role in mathematics achievement (e.g., Bayliss, Jarrold, Gunn, & Baddeley, 2003; Friso-van den Bos et al., 2013), while others find that visuospatial working memory is more important (e.g., Schuchardt, Maehler, & Hasselhorn, 2008; Szűcs, Devine, Soltesz, Nobes, & Gabriel, 2014). A recent metaanalysis of 110 studies found no difference in the strength of the relationship between mathematics achievement and verbal working memory, visuospatial working memory, or numerical working memory (Peng et al., 2016). One explanation for these conflicting findings is that the importance of verbal versus visuospatial working memory changes over development. Li and Geary (2013) found that verbal working memory predicted mathematics achievement in 7-year-olds but that gains in visuospatial short-term memory predicted mathematics achievement at 11 years old. These changes could reflect either developmental changes in the involvement of working memory or differences in the nature of mathematical activity (e.g., a shift from a focus on arithmetic to more advanced mathematical topics) across these years. However, few studies have explored the relationship between comprehensive measures of verbal and visuospatial working memory and mathematics achievement across development.

In a recent study, we explored whether there was a stronger relationship between verbal or visuospatial working memory and mathematics achievement in a sample of children across a wide age range (Cragg, Keeble, Richardson, Roome, & Gilmore, 2017). Groups of children aged 8–9 years, 11–12 years, 13–14 years, and young adults completed a large battery of executive function tests, including measures of verbal and visuospatial short-term and working memory. Verbal working memory was measured via a sentence span task. Participants heard a sentence with the final word missing and provided an appropriate word. After responding to a series of sentences, they were asked to recall the final word of each sentence in the series, in the correct order. Participants also completed the storage and processing elements separately. Visuospatial working memory was assessed via a complex span task. Participants saw a series of 3×3 grids each containing three symbols and they had to point to the "odd-one-out" symbol that differed from the other two. After responding to a series of grids, participants were asked to recall the position of the odd one out on each grid in the series, in the correct order. Again, participants also completed the storage and processing elements separately. Mathematics achievement was assessed using the mathematical reasoning subtest of the Wechsler Individual Achievement Test. Both verbal and visuospatial working memory performance were significant unique predictors of mathematics achievement (see Fig. 14.1), and these relationships were consistent across age groups. This suggests that the contribution of verbal and visuospatial working memory may be very similar and consistent across development, at least from middle childhood to adulthood.

In addition to the distinction between verbal and visuospatial working memory, a small number of studies have attempted to identify the exact components of working memory that contribute to mathematics achievement. Working memory measures inevitably also require the short-term storage and processing of information. When separate measures of short-term storage and processing are used alongside a complex working memory span task, in other words, when the processing component of a complex span task (e.g., sentence completion) is also measured alone, without the memory demands, a variance partitioning approach can be used to isolate the variance in performance that is associated with each element of working memory. This approach was taken in two studies by Bayliss et al. (Bayliss, Jarrold, Baddeley, Gunn, & Leigh, 2005; Bayliss et al., 2003), who found that all components of working memory play some role in mathematics achievement, but the combined storage and processing of verbal information is particularly important, at least in childhood.

We also used a variance partitioning approach to explore which specific components of working memory are important for mathematics achievement (Cragg, Keeble et al., 2017). We found that for verbal information, short-term memory, working memory, and the shared variance between short-term and working memory all accounted for significant variance in mathematics achievement. For visuospatial information, there was an additional role for processing such that short-term memory, processing, working memory, shared variance between short-term and working memory, and shared variance between short-term, processing, and working memory all accounted for significant variance in mathematics achievement. In both cases it was the shared variance between short-term and working memory that accounted for the greatest variance in mathematics achievement. This suggests that the ability to simply hold information in mind is as important for mathematics achievement as the ability to hold information while processing.

Inhibition, Shifting, and Mathematics Achievement

The role of inhibition and shifting in mathematics achievement has received less attention than working memory, although recent studies have increasingly begun to focus on these skills. The evidence is more inconsistent, with some studies finding a significant relationship

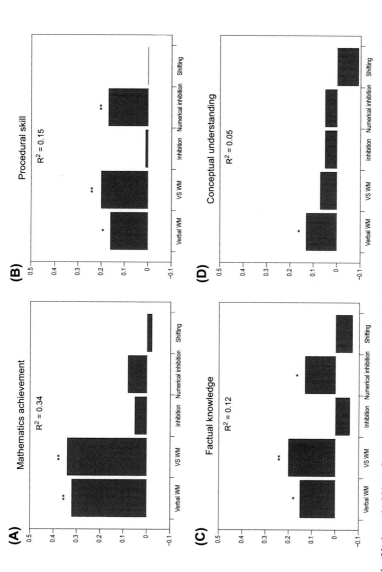

FIGURE 14.1 Variance in (A) mathematics achievement, (B) procedural skill, (C) factual knowledge, and (D) conceptual understanding explained by executive functions on mathematics achievement. Executive function skills were verbal working memory (WM), visuospatial (VS) working memory, inhibition of nonnumerical information, inhibition of numerical information, and shifting. $*P < .05$, $**P < .01$. Adapted from Cragg, L., Keeble, S., Richardson, S., Roome, H.E., & Gilmore, C. (2017). Direct and indirect influences of executive functions on mathematics achievement. Cognition, 162, 923–931, with permission of Elsevier.

between inhibition or shifting and mathematics performance in preschool and school-aged samples (Blair & Razza, 2007; Clark, Pritchard, & Woodward, 2010; Merkley, Thompson, & Scerif, 2016; St Clair-Thompson & Gathercole, 2006; Szucs, Devine, Soltesz, Nobes, & Gabriel, 2013; Yeniad, Malda, Mesman, van IJzendoorn, & Pieper, 2013), while others find no relationship (Lee et al., 2012; Monette, Bigras, & Guay, 2011; Van der Ven, Kroesbergen, Boom, & Leseman, 2012). This may be explained, in part, by shared variance with other cognitive skills. There is some evidence that inhibition is only a significant predictor of mathematical performance when shifting skills are not taken into account (Bull & Scerif, 2001; Van der Ven et al., 2012). Similarly, inhibition and shifting may be significant predictors of mathematics performance when considered alone but do not make a unique contribution when working memory (Lee et al., 2012) or intelligence (Yeniad et al., 2013) are taken into account. However Espy et al. (2004) found that inhibition was a significant predictor of mathematics after controlling for both working memory and shifting skills.

Another explanation for this inconsistent evidence, particularly concerning inhibition, is the nature of the executive function tasks. Different types of inhibition task tap into varying aspects of inhibition skill (e.g., response inhibition vs. interference control). These different forms of inhibition may not represent a single underlying construct and indeed show different developmental trajectories (Huizinga et al., 2006) and therefore may have different involvement in mathematics. Consequently, evidence for the relationship between inhibition and mathematics may depend on the specific forms of inhibition and tasks selected.

Further questioning the idea of a single, domain-general inhibitory system (Egner, 2008), there is evidence for domain specificity in the relationship with mathematics. There is some evidence that the relationship between inhibition and mathematics achievement is stronger when the inhibition task involves numerical rather than nonnumerical stimuli. Several studies have found a significant relationship between children's mathematics achievement and performance on a number-quantity Stroop task but no relationship between mathematics and color-word Stroop performance (Bull & Scerif, 2001; Navarro et al., 2011; Szucs et al., 2013), although other studies did not find this pattern (De Weerdt, Desoete, & Roeyers, 2013). We found some evidence for domain-specific effects: numerical inhibition (selecting the more numerous of two dot arrays ignoring the size of the dots), but not nonnumerical inhibition (selecting the larger animal in real life ignoring the size of the animals on the screen), was a significant predictor of mathematics achievement in children aged 8–14 years and adults (Cragg, Keeble et al., 2017). However, further research is needed to explore domain-specific effects for inhibition and in particular the extent to which this depends on the nature of the mathematics task.

A final explanation for the inconclusive evidence regarding the involvement of inhibition and shifting in children's mathematics performance is the nature of the mathematical tasks. It is difficult to pinpoint the role of inhibition and shifting in learning or performing mathematics when mathematics outcomes are measured with a general standardized or curriculum measure of mathematics. Instead, the role for these executive function skills is likely to be more specific and differs according to the mathematical activity. As we see in the following, it is therefore important to consider multiple components of mathematics to understand how and when executive function skills are involved.

MULTIPLE COMPONENTS OF MATHEMATICS

It is well established that mathematics is a complex, multicomponential skill. There are not only multiple domains, such as arithmetic, algebra, geometry, and statistics, but also each of these domains involves multiple skills. For example, when learning arithmetic, children need to learn number symbols and facts, to become proficient with different operations (addition, subtraction, etc.), to understand and apply underlying concepts and principles, to develop problem solving approaches, and to be able to apply arithmetic to real-world situations. These different components of arithmetic are not hierarchically ordered, instead there may be complex relationships among these individual skills (Dowker, 2005). Studies have shown that children may have strengths in one area despite weaknesses in another. For example, children may understand arithmetical concepts despite difficulties in performing calculations or vice versa (Canobi, 2004; Gilmore & Papadatou-Pastou, 2009).

A recent metaanalysis has explored the role of working memory across different domains of mathematics (Peng et al., 2016). This identified that working memory is most strongly related to whole number calculations and word problem solving and has a weaker relationship with geometry. However, this study only considered different domains and it is crucial to take into account different component skills when we try to identify the role of executive function skills. This is particularly true if we want to go beyond simply identifying correlations to understand precisely how executive function skills are involved in mathematics learning and performance; the role of working memory, inhibition, and shifting is likely to differ for different components. Researchers have put forward many suggestions for the ways in which different executive function skills support mathematics; for example, working memory may be important for holding interim solutions in mind during computation, inhibition may be needed to suppress unwanted number facts during retrieval, and shifting may be involved when switching between different operations and number representations (Bull & Lee, 2014; Cragg & Gilmore, 2014). However evidence for these specific mechanisms cannot be obtained from correlational studies that use general curriculum or standardized measures of overall mathematics achievement. These measures draw on a wide variety of mathematics skills, the precise constellation of which may differ from test to test, across ages even with the same test, and even from individual to individual, depending on the strategies they use. For example, a set of arithmetic problems may be solved by some children via computational strategies but by other children via retrieval. It is likely that the role of executive function skills will differ according to the strategy used, and thus the overall relationship between performance on the test and executive function performance may not be informative of the precise involvement of executive function skills. Consequently, we need to investigate how executive function skills support the learning and performance of different mathematical skills. Where possible we need to go beyond correlational designs to use experimental techniques that directly implicate specific executive function skills in specific mathematical processes.

In the following, we review existing and new evidence for the role of executive function skills in specific components of mathematics. We focus on procedural, factual, and conceptual knowledge of arithmetic as these are the skills that have received the most attention to date.

EXECUTIVE FUNCTIONS AND COMPONENTS OF ARITHMETIC

Executive Functions and Procedural Skill

Procedural knowledge of arithmetic has been defined as the ability to perform an ordered sequence of steps to solve a problem or knowing "how to" (Baroody, 2003; Hiebert & Lefevre, 1986, pp. 1–27). This involves accurately and efficiently selecting and performing appropriate operations. Executive function skills are likely to be important for procedural knowledge to represent the question, to store interim solutions or keep track of counts, to select the appropriate strategy and inhibit inappropriate ones, and to switch between operations, strategies, and notations. Evidence for the role of executive function skills and procedural skill comes largely from two types of studies: correlational or longitudinal battery studies, which focus on specific measures of procedural skill, rather than general standardized measures of achievement; and dual-task studies, which explore the online involvement of working memory while arithmetic problems are solved with procedural strategies. Evidence from these two sources will be considered in turn in the following.

Correlational studies have found a relationship between working memory and a variety of measures of procedural skill. Cowan and Powell (2014) found that a composite measure of domain-general skills, including working memory measures, was significantly related to basic calculation fluency, written arithmetic, and word problem solving. Fuchs et al. (2010) identified that central executive processes in particular were important for predicting development in multidigit arithmetic performance. Hecht, Close, and Santisi (2003) found that verbal working memory was related to fraction computation (but not conceptual understanding of fractions). Inhibition has also been implicated in procedural skills, with evidence that children with better inhibitory control made more use of the most efficient strategy to solve arithmetic problems (Lemaire & Lecacheur, 2011). Similarly, children's performance on a measure of task switching was related to procedural skills, including basic calculation and word problem solving (Andersson, 2010).

There is some evidence that executive functions may play a larger role in procedural skills for younger children than older children (Best et al., 2011; Friso-van den Bos et al., 2013); however, this comes largely from studies that use general measures of mathematics achievement, rather than specific measures of procedural skill. Over development and schooling, children's procedural skills become more automatic and they make use of different strategies. This may reduce the demands on executive function skills. However, few studies have explored the role of executive function skills on specific measures of procedural skills across development. We explored the relationship between working memory, inhibition and shifting, and procedural skill in a wide age range (Cragg, Keeble et al., 2017). Participants aged 8–9 years, 11–12 years, 13–14 years, and young adults completed a battery of executive function measures. Procedural skills were measured by response times to solve a set of single- and double-digit arithmetic problems. The problems varied across age groups to ensure that all groups would solve them via procedural strategies, rather than retrieval or guessing. The younger two age groups solved addition and subtraction problems, while the older two age groups solved addition, subtraction, multiplication, and division problems. Verbal working memory, visuospatial working memory, and inhibition of numerical information were all significant independent predictors of procedural skill and these relationships did not interact with age group. Set shifting was not related to procedural skill when entered into the same model (see Fig. 14.1). The specific components of working memory that were important for procedural skill, were further investigated via a variance partitioning approach (Bayliss et al., 2003). This revealed that verbal and visuospatial working memory and the shared variance between short-term memory and working memory, as well as visuospatial short-term memory accounted for significant variance in procedural skills. Overall the proportion of variance in procedural skills (15%) accounted for by executive function skills was less than that for overall mathematics achievement (34%).

An alternative approach to exploring the role of working memory on procedural skills is via dual-task studies. In this experimental approach, participants solve arithmetic problems both with and without a concurrent working memory load. The impact of the working memory load on performance on the arithmetic problems can shed light as to the involvement of working memory for calculation. Several studies using this methodology have identified the impact of working memory load on adults' procedural arithmetic skills. These studies have revealed that there is more interference in arithmetic performance from secondary tasks requiring central executive involvement, rather than the simple storage of information (Imbo & Vandierendonck, 2007b; Rammelaere, Stuyven, & Vandierendonck, 1999). When exploring the impact of concurrent working memory load on arithmetic procedural skills, it is important to take account of the strategies used to solve the problems. Adults and children can use a variety of strategies to solve arithmetic problems, including *retrieval* from long-term memory or procedural strategies such as *decomposition* (breaking down the problem into simpler problems, e.g., $5+7$ could be solved by $5+5=10$, $10+2=12$) or *counting*. It is essential to take account of the strategies that participants use to solve arithmetic problems in a dual-task paradigm because the involvement of working memory is likely to differ according to the strategy used. Studies have explored this by controlling the strategy that adults are required to use to solve arithmetic problems. This approach has identified that working memory tasks with a central executive component interfere with procedural strategies including counting and decomposition (Imbo & Vandierendonck, 2007b) and there is some evidence that counting strategies suffer greater impairment (Hubber, Gilmore, & Cragg, 2014).

Few studies have explored the impact of working memory load on children's procedural skills or how the effects of working memory load change across development. However, it is plausible that children rely on working memory to a greater extent when solving arithmetic problems compared with adults through the use of less-efficient strategies. Imbo and Vandierendonck (2007a) explored the impact of a working memory secondary task (continuous choice reaction time task) when children aged 10–12 years solved single-digit addition problems by counting, retrieval, or decomposition. They found that working memory load had an impact on both procedural strategies, which was greater for decomposition. However, this study did not explore whether the effects of working memory load differ across development.

We have recently explored the impact of dual-task load on arithmetic problem solving across a wide age range (Cragg, Hubber, Keeble, Richardson, & Gilmore, 2017). Children aged 9–11 years, 12–14 years, and adults solved addition problems using retrieval, decomposition, and counting under three conditions: alone (no secondary task), with a control load (with a secondary task that did not require working memory load), and with a working memory load (with a secondary task that required working memory load). The working memory load task was an n-back task involving the monitoring of either verbal or visuospatial information. The control load task involved the same responses as the working memory load task but without the need to monitor and update information. This study revealed that working memory load interfered with the performance of arithmetic whether participants solved problems via counting, decomposition, or retrieval. Surprisingly, the effect of working memory load was the same for children and adults. This suggests that the processes involved in the online performance of arithmetic are similar across development. There was some evidence that monitoring and updating verbal information interfered more than visuospatial information when counting but not when using retrieval or decomposition strategies. However, overall concurrent visuospatial load (with or without an associated working memory load) slowed arithmetic performance to a greater extent than verbal load. This study demonstrated the importance of working memory for solving arithmetic problems using procedural strategies.

Executive Functions and Factual Knowledge

The ability to retrieve the solutions to simple arithmetic problems is a focus of early mathematics instruction and an important indicator of overall mathematical performance (Geary, 2004). Several models have been proposed to account for the storage and retrieval of number fact knowledge, many of which are based on an associative network. For example, Siegler et al. proposed the *Distributions of Associations* network (Lemaire & Siegler, 1995; Siegler, 1988). According to this model, children initially solve problems using procedural strategies. Every time a problem is solved in this way, the association between the operands and the results is strengthened. With practice, the operands become more strongly associated with the correct answer. Once the association between the operands and answer exceeds a threshold, then the answer can be retrieved. This type of model can account for the types of errors typically seen when children and adults retrieve the answers to arithmetic problems. Incorrect answers are more often related to the operands than unrelated. For example, the answers to addition problems may be retrieved instead of multiplication problems (e.g., answering 30 when asked to retrieve the answer to $5+6$) or incorrect answers might be associated with one of the operands (e.g., answering 21 when asked to retrieve the answer to 7×4).

Based on this type of model of number fact retrieval, researchers have begun to consider the role of executive functions in the storage and retrieval of factual knowledge. Most attention has been paid to working memory and inhibition. It has been proposed that one role of working memory is to activate information in long-term memory (e.g., Cowan, 1999) and consequently it might be expected that working memory capacity is associated with fact retrieval. Evidence for this comes from both correlational and experimental studies.

Correlational studies have demonstrated that children with low working memory capacity are less accurate when retrieving solutions (Andersson, 2010; Geary, Hoard, Byrd-Craven, Nugent, & Numtee, 2007) and choose to make less use of retrieval than their peers with greater working memory capacity (Geary, Hoard, & Nugent, 2012). Our recent work found that both verbal and visuospatial working memory capacity were significant independent predictors of factual knowledge (correct responses to simple arithmetic problems within 3 s). This relationship was stable from age 8 years to adults (Cragg, Keeble et al., 2017). This association was further investigated using a variance partitioning approach to identify the specific components of working memory that were associated with factual knowledge. This revealed that for verbal information, short-term memory, working memory, and the shared variance between short-term and working memory all accounted for significant variance in factual knowledge. For visuospatial information, there was a greater role for processing such that short-term memory, processing, working memory, the shared variance between short-term memory and working memory, and the shared variance between short-term memory, processing, and working memory all accounted for significant variance in factual knowledge. In total, visuospatial working memory accounted for greater variance in factual knowledge than verbal working memory. This is somewhat surprising given that number facts are thought to be stored in a verbal code. It is possible that this reflects the role of visuospatial working memory in learning, rather than storing or retrieving, number facts.

Alongside evidence from these correlational studies, experimental dual-task studies have also demonstrated a role for working memory in fact retrieval. Concurrent working memory load has been found to interfere with the retrieval of arithmetic number facts by adults (Hubber et al., 2014; Imbo & Vandierendonck, 2007b), although the interference may be less than that for procedural strategies. Similarly, for children, working memory load interferes with the retrieval of number facts in comparison with either no load (Imbo & Vandierendonck, 2007a) or to a control load (Cragg, Hubber et al., 2017). Put together with the correlational findings, these studies suggest that working memory is involved not only in learning number facts but also in the online storage and retrieval of facts when children are actively problem solving.

In addition to working memory, there has been increasing interest in the role of inhibition in number fact knowledge. It has been proposed that inhibition is required to suppress the retrieval of incorrect solutions (Bull & Lee, 2014; Cragg & Gilmore, 2014). However, evidence in support of this is still somewhat scarce. Most evidence from correlational studies for the role of inhibition in specific arithmetic skills has focused on procedural skills (e.g., Lan, Legare, Ponitz, Li, & Morrison, 2011). Some evidence for the role of inhibition is starting to emerge from experimental paradigms. For example, De Visscher and Noël (2014) used an interference paradigm to explore children's susceptibility to interference in memory and found that children with poor knowledge of multiplication facts were more sensitive to interference in memory.

We explored the relationship between inhibition and factual knowledge in children aged 8–14 years and young adults (Cragg, Keeble et al., 2017). Factual knowledge was measured as the proportion of correct responses given to simple arithmetic problems within 3 s. Inhibition skill was measured with two tasks—one which involved the inhibition of numerical information and one which involved the inhibition of nonnumerical information. This revealed that inhibition of numerical information was a unique predictor of factual knowledge, over and above working memory and shifting, and that this relationship was consistent across development.

Executive Functions and Conceptual Understanding

Conceptual knowledge has been defined as understanding of the principles and relationships that underlie a domain (Hiebert & Lefevre, 1986, pp. 1–27) or *knowing why* (Baroody, 2003). It is widely recognized that good conceptual understanding is important for success in mathematics (see review by Rittle-Johnson & Schneider, 2015). Although many studies have investigated the development of conceptual understanding, little attention has been paid to the role of domain-general skills, and particularly executive functions, in conceptual understanding. However, it is plausible that executive functions play a role in both the acquisition of conceptual knowledge and the selection of conceptually based strategies. Inhibition and shifting may be involved in suppressing a prepotent procedural strategy and switching attention to identify underlying conceptual relationships. Working memory may be required to activate conceptual knowledge in long-term memory (e.g., Cowan, 1999).

The few studies that have explored the role of executive functions in conceptual understanding have produced a mixed picture. Robinson and Dubé (2013) found that children with good inhibitory control were more likely to make use of an alternative conceptually based strategy rather than computation to solve problems than children with poor inhibitory control. Similarly, Watchorn et al. (2014) found that children's conceptual understanding was associated with good attentional skills, at least for children who also had good procedural skills. Andersson (2010) found that visuospatial working memory, but not shifting, was a predictor of later conceptual understanding. However, other studies exploring the role of working memory in conceptual understanding have failed to find a relationship, at least within the domain of fractions (Hecht et al., 2003; Jordan et al., 2013).

We explored the relationship between verbal and visuospatial working memory, inhibition, and shifting with conceptual understanding in children aged 8–14 years and young adults. Conceptual understanding was measured using a task in which participants were asked to identify conceptual relationships (e.g., inversion, commutativity, associativity) between pairs of problems (based on Canobi, 2004). Across all age groups, only verbal working memory was significantly associated with conceptual understanding (see Fig. 14.1). We further explored this using a variance partitioning approach and identified that verbal working memory and the shared variance between verbal short-term and working memory were (marginally) significant predictors of conceptual understanding. Visuospatial processing (but not storage) also explained significant variance in conceptual understanding. Overall, only 5% of variance in conceptual understanding was accounted for by executive function skills (compared with 15% of procedural skills and 34% of mathematics achievement).

Overall, the role of executive functions in conceptual understanding remains unclear. One explanation for this is that it is difficult to measure conceptual understanding in isolation. Further research is needed to uncover the domain-general skills involved in both acquiring conceptual understanding and using it in problem solving. A clearer picture of the skills involved might help to explain why some children struggle to develop conceptual understanding, while other children have better conceptual understanding than expected given their procedural skills (Gilmore & Papadatou-Pastou, 2009; Watchorn et al., 2014).

DIRECT AND INDIRECT INFLUENCES OF EXECUTIVE FUNCTIONS ON MATHEMATICS ACHIEVEMENT

As described earlier, there is a large body of evidence for the involvement of executive function skills in overall mathematics achievement and some evidence for the role of executive function skills in specific components of mathematics: procedural skill, factual knowledge, and conceptual understanding. Hierarchical models of mathematics (Fuchs et al., 2010; Geary, 2004; LeFevre et al., 2010) propose that the role of executive function skills in specific components of mathematics accounts for the role of these skills in overall mathematics achievement. For example, Geary (2004) put forward a hierarchical model in which basic cognitive systems, including executive function skills, underlie conceptual and procedural knowledge of a domain, which in turn support overall competence in that domain. In other words, the role of executive function skills in conceptual and procedural knowledge fully accounts for the role of executive function skills in overall achievement. This mediating role of conceptual and procedural knowledge in explaining the relationship between executive function skills and mathematics achievement has not previously been tested.

We tested this relationship by identifying the direct and indirect effects of executive function skills on mathematics achievement in children aged 8–14 years and young adults (Cragg, Keeble et al., 2017). We first explored the relationship between procedural, conceptual, and factual knowledge and mathematics achievement using hierarchical regression. All three mathematical components were significant independent predictors of mathematics achievement, and this pattern remained stable across age. We next explored the relationship between executive function skills and mathematical components. As described in more detail in the previous sections, we found that verbal and visuospatial working memory and numerical inhibition were significant independent predictors of factual knowledge and procedural skill and that verbal working memory was a significant predictor of conceptual understanding. Nonnumerical inhibition and shifting were not significantly related to any outcomes. Again, this pattern was stable across age groups. Finally, to test the hierarchical model, we performed mediation analyses to test whether factual knowledge, procedural skill, and conceptual understanding mediated the relationship between working memory and mathematics achievement. Inhibition was not included in this model because it was not significantly associated with mathematics achievement. Two separate models were tested for verbal and visuospatial working memory separately.

These analyses revealed significant indirect effects of verbal working memory on mathematics achievement via all three component skills: factual knowledge, procedural skill, and conceptual understanding. There were also significant indirect effects of visuospatial working memory on mathematics achievement via factual knowledge and procedural skill but not conceptual understanding. In both models, however, there remained a substantial direct effect of working memory on mathematics achievement. These analyses indicate that although factual knowledge and procedural skills (and conceptual understanding) partially mediate the relationship between executive function skills and mathematics achievement, there remains a substantial role for executive function skills that remains to be explained. This direct path may represent the role of working memory in problem solving; for example, to identify the mathematical problem, construct a problem representation and select a strategy. Alternatively, this direct path may represent the role of working memory more generally in learning, which may be better captured by performance on mathematics achievement tests than on measures of specific components of arithmetic. Further work is required to better understand this as yet unexplained role played by working memory in mathematics.

CONCLUSIONS: EXECUTIVE FUNCTIONS AND LEARNING VERSUS PERFORMANCE

A large body of evidence now exists to demonstrate that executive function skills play a role in children's learning of mathematics. Over the last decade, researchers have begun to go beyond broad general relationships to understand at a more nuanced level the precise role that different executive function skills play in different components of mathematics. This endeavor has benefitted from multiple sources of evidence, including experimental and correlational studies to pinpoint mechanisms and causal relationships. With regard to children's procedural and factual knowledge, substantial progress has been made in understanding the role of executive function skills; however, further work is needed in regard to conceptual understanding.

One important distinction that requires further attention is that between performance of mathematics and learning mathematics. It is likely that executive function skills play a different role when children are learning new mathematical material compared with performing known mathematical procedures. Many of the methods currently being used to explore the role of executive functions are unable to distinguish these factors. For example, a correlation between working memory capacity and knowledge of number facts could indicate that working memory is required to store and retrieve number facts or could be a legacy from the involvement of working memory in learning these number facts in the first place. Experimental methods, for example, dual-task studies, can help to isolate the role of executive functions in the performance of mathematics. However, to reveal the role of executive functions in learning will require the use of different techniques. These might include investigating individual differences in children's ability to learn new mathematics material in school or alternatively studying learning more directly in lab-based studies (e.g., via artificial learning paradigms).

A thorough understanding of the role of executive functions in mathematics learning and performance can help us to make sense of the wide individual differences in children's success with mathematics. This in turn can help to understand why some children struggle with mathematics, identify those who are at risk of developing difficulties, and develop teaching approaches or interventions that may benefit all children. Current attempts at improving mathematics outcomes via training executive function skills have failed to show successful transfer (Melby-Lervåg & Hulme, 2013). This may be because these approaches fail to take account of the specific ways in which executive functions are involved in mathematics. A more profitable avenue to explore may be to understand and manage the executive function demands of classroom activities. As described earlier, some progress has been made in this regard, but further research, based on detailed models of mathematical cognition, is required.

References

Anderson, P. (2002). Assessment and development of executive function (EF) during childhood. *Child Neuropsychology*, *8*(2), 71–82. https://doi.org/10.1076/chin.8.2.71.8724.

Andersson, U. (2010). Skill development in different components of arithmetic and basic cognitive functions: Findings from a 3-year longitudinal study of children with different types of learning difficulties. *Journal of Educational Psychology*, *102*(1), 115–134.

Baddeley, A. D., & Hitch, G. (1974). Working memory. In G. H. Bower (Ed.), *Psychology of learning and motivation* (Vol. 8). Academic Press. Retrieved from: http://www.sciencedirect.com/science/article/pii/S0079742108604521.

Baroody, A. J. (2003). The development of adaptive expertise and flexibility: The integration of conceptual and procedural knowledge. In A. J. Baroody, & A. Dowker (Eds.), *The development of arithmetic concepts and skills: Constructive adaptive expertise*. Routledge.

Bayliss, D. M., Jarrold, C., Baddeley, A. D., Gunn, D. M., & Leigh, E. (2005). Mapping the developmental constraints on working memory span performance. *Developmental Psychology*, *41*(4), 579–597. https://doi.org/10.1037/0012-1649.41.4.579.

Bayliss, D. M., Jarrold, C., Gunn, D. M., & Baddeley, A. D. (2003). The complexities of complex span: Explaining individual differences in working memory in children and adults. *Journal of Experimental Psychology: General*, *132*(1), 71–92. https://doi.org/10.1037/0096-3445.132.1.71.

Best, J. R., Miller, P. H., & Naglieri, J. A. (2011). Relations between executive function and academic achievement from ages 5 to 17 in a large, representative national sample. *Learning and Individual Differences*, *21*(4), 327–336. https://doi.org/10.1016/j.lindif.2011.01.007.

Blair, C., & Razza, R. P. (2007). Relating effortful control, executive function, and false belief understanding to emerging math and literacy ability in kindergarten. *Child Development*, *78*(2), 647–663. https://doi.org/10.1111/j.1467-8624.2007.01019.x.

Bull, R., & Lee, K. (2014). Executive functioning and mathematics achievement. *Child Development Perspectives*, *8*(1), 36–41. https://doi.org/10.1111/cdep.12059.

Bull, R., & Scerif, G. (2001). Executive functioning as a predictor of children's mathematics ability: Inhibition, switching, and working memory. *Developmental Neuropsychology*, *19*(3), 273–293. https://doi.org/10.1207/S15326942DN1903_3.

Canobi, K. H. (2004). Individual differences in children's addition and subtraction knowledge. *Cognitive Development*, *19*(1), 81–93. https://doi.org/10.1016/j.cogdev.2003.10.001.

Clark, C. A. C., Pritchard, V. E., & Woodward, L. J. (2010). Preschool executive functioning abilities predict early mathematics achievement. *Developmental Psychology*, *46*(5), 1176–1191. https://doi.org/10.1037/a0019672.

Conklin, H. M., Luciana, M., Hooper, C. J., & Yarger, R. S. (2007). Working memory performance in typically developing children and adolescents: Behavioral evidence of protracted frontal lobe development. *Developmental Neuropsychology*, *31*(1), 103–128. https://doi.org/10.1080/87565640709336889.

Cowan, N. (1999). An embedded-processes model of working memory. In A. Miyake & P. Shah (Eds.), *Models of Working Memory* (pp. 62–101). Cambridge: University Press.

Cowan, R., & Powell, D. (2014). The contributions of domain-general and numerical factors to third-grade arithmetic skills and mathematical learning disability. *Journal of Educational Psychology*, *106*(1), 214–229. https://doi.org/10.1037/a0034097.

Cragg, L., & Gilmore, C. (2014). Skills underlying mathematics: The role of executive function in the development of mathematics proficiency. *Trends in Neuroscience and Education*, *3*(2), 63–68. https://doi.org/10.1016/j.tine.2013.12.001.

Cragg, L., Hubber, P. J., Keeble, S., Richardson, S., & Gilmore, C. (2017). When is working memory important for arithmetic? The impact of strategy and age. *PLoS ONE*, *12*(12), e0188693.

Cragg, L., Keeble, S., Richardson, S., Roome, H. E., & Gilmore, C. (2017). Direct and indirect influences of executive functions on mathematics achievement. *Cognition*, *162*, 923–931.

De Visscher, A., & Noël, M.-P. (2014). Arithmetic facts storage deficit: The hypersensitivity-to-interference in memory hypothesis. *Developmental Science*, *17*(3), 434–442. https://doi.org/10.1111/desc.12135.

De Weerdt, F., Desoete, A., & Roeyers, H. (2013). Behavioral inhibition in children with learning disabilities. *Research in Developmental Disabilities*, *34*(6), 1998–2007. https://doi.org/10.1016/j.ridd.2013.02.020.

Department for Business Innovation & Skills. (2011). *2011 Skills for life survey: Headline findings* BIS Research Paper Number 57.

Diamond, A. (1985). Development of the ability to use recall to guide action, as indicated by infants' performance on AB. *Child Development*, *56*(4), 868–883. https://doi.org/10.2307/1130099.

Dowker, A. (2005). *Individual differences in arithmetic: Implications for psychology, neuroscience and education*. Psychology Press.

Egner, T. (2008). Multiple conflict-driven control mechanisms in the human brain. *Trends in Cognitive Sciences*, *12*(10), 374–380. https://doi.org/10.1016/j.tics.2008.07.001.

Espy, K. A. (1997). The shape school: Assessing executive function in preschool children. *Developmental Neuropsychology*, *13*(4), 495–499. https://doi.org/10.1080/87565649709540690.

Espy, K. A., McDiarmid, M. M., Cwik, M. F., Stalets, M. M., Hamby, A., & Senn, T. E. (2004). The contribution of executive functions to emergent mathematic skills in preschool children. *Developmental Neuropsychology*, *26*(1), 465–486. https://doi.org/10.1207/s15326942dn2601_6.

Friso-van den Bos, I., van der Ven, S. H. G., Kroesbergen, E. H., & van Luit, J. E. H. (2013). Working memory and mathematics in primary school children: A meta-analysis. *Educational Research Review*, *10*, 29–44. https://doi.org/10.1016/j.edurev.2013.05.003.

Fuchs, L. S., Geary, D. C., Compton, D. L., Fuchs, D., Hamlett, C. L., Seethaler, P. M., et al. (2010). Do different types of school mathematics development depend on different constellations of numerical versus general cognitive abilities? *Developmental Psychology*, *46*(6), 1731–1746. https://doi.org/10.1037/a0020662.

Geary, D. C. (2004). Mathematics and learning disabilities. *Journal of Learning Disabilities*, *37*(1), 4–15. https://doi.org/10.1177/0022219404037 0010201.

Geary, D. C. (2011). Cognitive predictors of achievement growth in mathematics: A 5-year longitudinal study. *Developmental Psychology*, *47*(6), 1539–1552. https://doi.org/10.1037/a0025510.

Geary, D. C., Hoard, M. K., Byrd-Craven, J., Nugent, L., & Numtee, C. (2007). Cognitive mechanisms underlying achievement deficits in children with mathematical learning disability. *Child Development*, *78*(4), 1343–1359. https://doi.org/10.1111/j.1467-8624.2007.01069.x.

Geary, D. C., Hoard, M. K., & Nugent, L. (2012). Independent contributions of the central executive, intelligence, and in-class attentive behavior to developmental change in the strategies used to solve addition problems. *Journal of Experimental Child Psychology*, *113*(1), 49–65. https://doi.org/10.1016/j.jecp.2012.03.003.

Gilmore, C., & Papadatou-Pastou, M. (2009). Patterns of individual differences in conceptual understanding and arithmetical skill: A meta-analysis. *Mathematical Thinking and Learning*, *11*(1–2), 25–40. https://doi.org/10.1080/10986060802583923.

Gross, J., Hudson, C., & Price, D. (2009). *The long term costs of numeracy difficulties*. Every Child a Chance Trust & KPMG.

Hecht, S. A., Close, L., & Santisi, M. (2003). Sources of individual differences in fraction skills. *Journal of Experimental Child Psychology*, *86*(4), 277–302. https://doi.org/10.1016/j.jecp.2003.08.003.

Hecht, S. A., Torgesen, J. K., Wagner, R. K., & Rashotte, C. A. (2001). The relations between phonological processing abilities and emerging individual differences in mathematical computation skills: A longitudinal study from second to fifth grades. *Journal of Experimental Child Psychology*, *79*(2), 192–227. https://doi.org/10.1006/jecp.2000.2586.

Hiebert, J., & Lefevre, P. (1986). Conceptual and procedural knowledge in mathematics: An introductory analysis. In J. Hiebert (Ed.), *Conceptual and procedural knowledge: The case of mathematics* Hillsdale, NJ.

Hubber, P. J., Gilmore, C., & Cragg, L. (2014). The roles of the central executive and visuospatial storage in mental arithmetic: A comparison across strategies. *The Quarterly Journal of Experimental Psychology*, *67*(5), 936–954. https://doi.org/10.1080/17470218.2013.838590.

Huizinga, M., Dolan, C. V., & van der Molen, M. W. (2006). Age-related change in executive function: Developmental trends and a latent variable analysis. *Neuropsychologia*, *44*(11), 2017–2036. https://doi.org/10.1016/j.neuropsychologia.2006.01.010.

Imbo, I., & Vandierendonck, A. (2007a). The development of strategy use in elementary school children: Working memory and individual differences. *Journal of Experimental Child Psychology*, *96*(4), 284–309. https://doi.org/10.1016/j.jecp.2006.09.001.

Imbo, I., & Vandierendonck, A. (2007b). The role of phonological and executive working memory resources in simple arithmetic strategies. *European Journal of Cognitive Psychology*, *19*(6), 910–933. https://doi.org/10.1080/09541440601051571.

Jordan, N. C., Hansen, N., Fuchs, L. S., Siegler, R. S., Gersten, R., & Micklos, D. (2013). Developmental predictors of fraction concepts and procedures. *Journal of Experimental Child Psychology, 116*(1) 45–58, https://doi.org/10.1016/j.jecp.2013.02.001.

Lan, X., Legare, C. H., Ponitz, C. C., Li, S., & Morrison, F. J. (2011). Investigating the links between the subcomponents of executive function and academic achievement: A cross-cultural analysis of Chinese and American preschoolers. *Journal of Experimental Child Psychology, 108*(3), 677–692. https://doi.org/10.1016/j.jecp.2010.11.001.

Lee, K., Bull, R., & Ho, R. M. H. (2013). Developmental changes in executive functioning. *Child Development, 84*(6), 1933–1953. https://doi.org/10.1111/cdev.12096.

Lee, K., Ng, S. F., Pe, M. L., Ang, S. Y., Hasshim, M. N. A.M., & Bull, R. (2012). The cognitive underpinnings of emerging mathematical skills: Executive functioning, patterns, numeracy, and arithmetic: Cognitive underpinnings. *British Journal of Educational Psychology, 82*(1), 82–99. https://doi.org/10.1111/j.2044-8279.2010.02016.x.

LeFevre, J.-A., Fast, L., Skwarchuk, S.-L., Smith-Chant, B. L., Bisanz, J., Kamawar, D., et al. (2010). Pathways to mathematics: Longitudinal predictors of performance: Pathways to mathematics. *Child Development, 81*(6), 1753–1767. https://doi.org/10.1111/j.1467-8624.2010.01508.x.

Lemaire, P., & Lecacheur, M. (2011). Age-related changes in children's executive functions and strategy selection: A study in computational estimation. *Cognitive Development, 26*(3), 282–294. https://doi.org/10.1016/j.cogdev.2011.01.002.

Lemaire, P., & Siegler, R. S. (1995). Four aspects of strategic change: Contributions to children's learning of multiplication. *Journal of Experimental Psychology: General, 124*(1), 83–97. https://doi.org/10.1037/0096-3445.124.1.83.

Li, Y., & Geary, D. C. (2013). Developmental gains in visuospatial memory predict gains in mathematics achievement. *PLoS One, 8*(7), e70160. https://doi.org/10.1371/journal.pone.0070160.

Luna, B., Garver, K. E., Urban, T. A., Lazar, N. A., & Sweeney, J. A. (2004). Maturation of cognitive processes from late childhood to adulthood. *Child Development, 75*(5), 1357–1372. https://doi.org/10.1111/j.1467-8624.2004.00745.x.

Melby-Lervåg, M., & Hulme, C. (2013). Is working memory training effective? A meta-analytic review. *Developmental Psychology, 49*(2), 270–291. https://doi.org/10.1037/a0028228.

Merkley, R., Thompson, J., & Scerif, G. (2016). Of huge mice and tiny elephants: Exploring the relationship between inhibitory processes and preschool math skills. *Frontiers in Psychology, 6*. https://doi.org/10.3389/fpsyg.2015.01903.

Miyake, A., Friedman, N. P., Emerson, M. J., Witzki, A. H., Howerter, A., & Wager, T. D. (2000). The unity and diversity of executive functions and their contributions to complex 'Frontal Lobe' tasks: A latent variable analysis. *Cognitive Psychology, 41*(1), 49–100. https://doi.org/10.1006/cogp.1999.0734.

Monette, S., Bigras, M., & Guay, M.-C. (2011). The role of the executive functions in school achievement at the end of Grade 1. *Journal of Experimental Child Psychology, 109*(2), 158–173. https://doi.org/10.1016/j.jecp.2011.01.008.

Navarro, J. I., Aguilar, M., Alcalde, C., Ruiz, G., Marchena, E., & Menacho, I. (2011). Inhibitory processes, working memory, phonological awareness, naming speed, and early arithmetic achievement. *The Spanish Journal of Psychology,* 14(2), 580–588. https://doi.org/10.5209/rev_SJOP.2011.v14.n2.6.

Nunes, T., Bryant, P., Barros, R., & Sylva, K. (2012). The relative importance of two different mathematical abilities to mathematical achievement. *British Journal of Educational Psychology,* 82(1), 136–156. https://doi.org/10.1111/j.2044-8279.2011.02033.x.

OECD. (2013). *OECD skills outlook 2013.* OECD Publishing. Retrieved from: http://www.oecd-ilibrary.org/education/oecd-skills-outlook-2013_9789264204256-en.

Parsons, S., & Bynner, J. (2005). *Does numeracy matter more? National Research and Development Centre for Adult Literacry and Numeracy Research Report.* London: Institute of Education.

Peng, P., Namkung, J., Barnes, M., & Sun, C. (2016). A meta-analysis of mathematics and working memory: Moderating effects of working memory domain, type of mathematics skill, and sample characteristics. *Journal of Educational Psychology,* 108(4), 455–473. https://doi.org/10.1037/edu0000079.

Raghubar, K. P., Barnes, M. A., & Hecht, S. A. (2010). Working memory and mathematics: A review of developmental, individual difference, and cognitive approaches. *Learning and Individual Differences,* 20(2), 110–122. https://doi.org/10.1016/j.lindif.2009.10.005.

Rammelaere, S. D., Stuyven, E., & Vandierendonck, A. (1999). The contribution of working memory resources in the verification of simple mental arithmetic sums. *Psychological Research,* 62(1), 72–77. https://doi.org/10.1007/s004260050041.

Rittle-Johnson, B., & Schneider, M. (2015). Developing conceptual and procedural knowledge of mathematics. In R. Cohen Kadosh, & A. Dowker (Eds.), *Oxford handbook of numerical cognition.* Oxford, UK: Oxford University Press.

Robinson, K. M., & Dubé, A. K. (2013). Children's additive concepts: Promoting understanding and the role of inhibition. *Learning and Individual Differences,* 23, 101–107. https://doi.org/10.1016/j.lindif.2012.07.016.

Schuchardt, K., Maehler, C., & Hasselhorn, M. (2008). Working memory deficits in children with specific learning disorders. *Journal of Learning Disabilities,* 41(6), 514–523. https://doi.org/10.1177/0022219408317856.

Siegler, R. S. (1988). Strategy choice procedures and the development of multiplication skill. *Journal of Experimental Psychology: General,* 117(3), 258–275. https://doi.org/10.1037/0096-3445.117.3.258.

St Clair-Thompson, H. L. S., & Gathercole, S. E. (2006). Executive functions and achievements in school: Shifting, updating, inhibition, and working memory. *The Quarterly Journal of Experimental Psychology,* 59(4), 745–759. https://doi.org/10.1080/17470210500162854.

Szucs, D., Devine, A., Soltesz, F., Nobes, A., & Gabriel, F. (2013). Developmental dyscalculia is related to visuo-spatial memory and inhibition impairment. *Cortex,* 49(10), 2674–2688. https://doi.org/10.1016/j.cortex.2013.06.007.

Szűcs, D., Devine, A., Soltesz, F., Nobes, A., & Gabriel, F. (2014). Cognitive components of a mathematical processing network in 9-year-old children. *Developmental Science,* 17(4), 506–524. https://doi.org/10.1111/desc.12144.

Van der Ven, S. H. G., Kroesbergen, E. H., Boom, J., & Leseman, P. P. M. (2012). The development of executive functions and early mathematics: A dynamic relationship. *British Journal of Educational Psychology, 82*(1), 100–119. https://doi.org/10.1111/j.2044-8279.2011.02035.x.

Watchorn, R. P. D., Bisanz, J., Fast, L., LeFevre, J.-A., Skwarchuk, S.-L., & Smith-Chant, B. L. (2014). Development of mathematical knowledge in young children: Attentional skill and the use of inversion. *Journal of Cognition and Development, 15*(1), 161–180. https://doi.org/10.1080/15248372.2012.742899.

Wiebe, S. A., Espy, K. A., & Charak, D. (2008). Using confirmatory factor analysis to understand executive control in preschool children: I. Latent structure. *Developmental Psychology, 44*(2), 575–587. https://doi.org/10.1037/0012-1649.44.2.575.

Willoughby, M. T., Blair, C. B., Wirth, R. J., Greenberg, M., & The Family Life Project Investigators (2012). The measurement of executive function at age 5: Psychometric properties and relationship to academic achievement. *Psychological Assessment, 24*(1), 226–239. https://doi.org/10.1037/a0025361.

Yeniad, N., Malda, M., Mesman, J., van IJzendoorn, M. H., & Pieper, S. (2013). Shifting ability predicts math and reading performance in children: A meta-analytical study. *Learning and Individual Differences, 23*, 1–9. https://doi.org/10.1016/j.lindif.2012.10.004.

Systems Neuroscience of Mathematical Cognition and Learning: Basic Organization and Neural Sources of Heterogeneity in Typical and Atypical Development

Teresa Iuculano, Aarthi Padmanabhan, Vinod Menon

Stanford University, Stanford, CA, United States

Heterogeneity of Function in Numerical Cognition
https://doi.org/10.1016/B978-0-12-811529-9.00015-7

287

INTRODUCTION

Mathematical skill acquisition is hierarchical in nature, and each iteration of increased proficiency builds on knowledge of a lower-level *primitive*. For example, learning to solve arithmetical operations such as "3 + 4" requires first an understanding of what numbers mean and represent (e.g., the symbol "3" refers to the quantity of three items, which derives from the ability to attend to discrete items in the environment). Thus, all forms of mathematical cognition, from basic to complex, require proficiency in a fundamental system of "*number sense*," including elemental properties of numbers, principles of cardinality, numerosity as abstract representations of sets, and the axiomatic rules by which numerical quantity is manipulated (Dantzig, 1930; Dehaene, 1997). The brain systems supporting mathematical cognition parallel these behavioral constructs and function as a set of (partly) hierarchically organized and dynamically interacting systems. Each brain system subserves specific perceptual and cognitive processes, including visual and auditory processing, quantity processing, working memory, declarative memory, attention, and cognitive control (Fig. 15.1). Importantly, the topology of brain systems engaged during mathematical cognition varies considerably not only across individuals but also with learning and development, as individuals gain proficiency in mathematical skills.

The basic building blocks of mathematical cognition, which are learned early in a child's development, include understanding numerical magnitude and the ability to manipulate symbolic and nonsymbolic quantity ("*number sense*"). Human imaging studies using electroencephalography

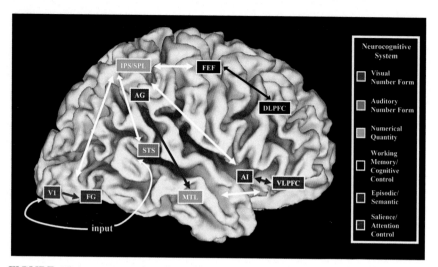

FIGURE 15.1 **Schematic circuit diagram of brain regions involved in mathematical learning and cognition.** The fusiform gyrus (FG) in the ventral temporal-occipital cortex decodes visual number form and together with the intraparietal sulcus (IPS) in the posterior parietal cortex (PPC)—which helps build visuospatial representations of numerical quantity (shown in *green boxes*)—forms the building blocks of mathematical cognition. The superior temporal sulcus (STS) aids in decoding auditory number words. Multiple parietal-frontal circuits link the IPS with working memory and cognitive control systems that include the frontal eye field (FEF), and the dorsolateral prefrontal cortex (DLPFC). These circuits facilitate visuospatial working memory for objects in space and create a hierarchy of short-term representations that allow manipulation of multiple discrete quantities over short periods (i.e., several seconds). The declarative memory system anchored in the medial temporal lobe (MTL)—and particularly the hippocampus—plays an important role in long-term memory formation and generalization beyond individual problem attributes. Finally, ventral prefrontal control circuits (shown in red) anchored in the anterior insula (AI) and ventrolateral prefrontal cortex (VLPFC) serve as flexible *hubs* for integrating information across attentional and memory systems, thereby facilitating goal-directed problem-solving and decision-making during mathematical cognition. Relative transparency for MTL indicates subsurface cortical structure. *Adapted from: Fias et al., 2013,* Trends in Neuroscience and Education © 2013 by Elsevier.

and functional magnetic resonance imaging (fMRI) have shown that these building blocks are subserved by specific and interacting brain systems (Fig. 15.1). First, visual and auditory association cortices decode the visual form (e.g., visual shape of the symbol "3") or phonological feature (e.g., the word "three") of numerical stimuli. Next, the fusiform gyrus (FG) in the higher-level visual cortex (i.e., ventral temporal–occipital cortex) plays an important role in visual object recognition and in forming relevant representations of symbolic stimuli. As the child begins to learn the use of orthographic symbols (i.e., written symbols and number words), new representations for these visual stimuli are developed in the FG (Allison, McCarthy, Nobre, Puce, & Belger, 1994; Ansari, 2008; Dehaene, Molko,

Cohen, & Wilson, 2004; Park, Hebrank, Polk, & Park, 2012; Shum et al., 2013). Simultaneously, the superior temporal sulcus (STS) and middle temporal gyrus (MTG) are involved in forming representations of number words (Thompson, Abbott, Wheaton, Syngeniotis, & Puce, 2004). The parietal attention system helps build semantic representations of quantity (Ansari, 2008) from multiple low-level *primitives* (Box 15.1), including the ability to attend to and individuate individual objects in space, and the ability to perceive the numerosity of such objects (i.e., "oneness," "three-ness"). These low-level *primitives* are processed in the posterior parietal cortex (PPC), particularly in its intraparietal sulcus (IPS) subdivision and are established many years before a child learns to process numerical symbols and number words (Hyde, Boas, Blair, & Carey, 2010). Taken together, the perceptual and cognitive brain systems that underlie *"number sense"* come on line early in development, providing a critical foundation for the acquisition of later mathematical skills.

BOX 15.1

GLOSSARY

Systems Neuroscience: refers to a subdiscipline of neuroscience and systems biology, which studies the function of brain circuits and systems. It concerns the study of brain regions and their function during a task and their interaction with other—proximal and distal—brain regions and the derived neural systems/networks.

Primitives: refer to a core set of cognitive capacities—pertinent to the intended domain—that can be utilized and integrated to develop higher-order cognitive capacities.

Schema of knowledge: refers to a cognitive construct that helps to organize categories of information and the relationship between them.

Hub: refers to a central structure within a network.

Distance effect: refers to the measurable behavior—assessed by accuracy and reaction time—which instantiates that it is harder to discriminate two sets of items as their distance decreases. It is more difficult to compare 8 versus 9—i.e., distance of 1 unit—than it is to compare 8 versus 5—i.e., distance of 3 units.

Neural distance effect: refers to the neural correlate of the *distance effect*, such that functional activation in dedicated brain regions is greater when the numerosities to be compared have a smaller distance (i.e., 8 vs. 9), compared to a larger distance (i.e., 8 vs. 5).

Proficiency in mathematics includes learning not only how to perceive and process (i.e., represent) numerical information but also to manipulate it, which requires engagement of multiple neurocognitive networks in the brain (Fig. 15.1). First, frontoposterior parietal brain circuits underlying working memory processes support the online manipulation of discrete quantities (e.g., individual objects) by creating short-term representations of such quantities and to help solve more complex problems (Metcalfe, Ashkenazi, Rosenberg-Lee, & Menon, 2013). Second, the prefrontal cortex (PFC) helps guide and maintain attention in the service of goal-directed problem-solving and cognitive control. Third, the medial temporal lobe (MTL) memory system, anchored in the hippocampus, plays an important role in long-term memory formation, and consolidation of mathematical concepts (i.e., arithmetical facts) and generalizations beyond individual problem attributes within a broader *schema of knowledge* (Box 15.1) (Davachi, 2006; Davachi, Mitchell, & Wagner, 2003). Together, these systems provide necessary cognitive scaffolding and play varying roles in mathematical skill acquisition throughout learning and development.

Considerable heterogeneity of mathematical proficiency in both children and adults is well documented, with some individuals demonstrating remarkable abilities (Cowan & Carney, 2006; Cowan & Frith, 2009; Cowan, O'Connor, & Samella, 2003; Iuculano et al., 2014; Pesenti, Seron, Samson, & Duroux, 1999; Pesenti et al., 2001) and others showing marked deficits (Butterworth, 2005, pp. 455–467; Butterworth & Kovas, 2013; Butterworth & Reigosa-Crespo, 2007, pp. 65–81; Butterworth, Varma, & Laurillard, 2011; Iuculano, Tang, Hall, & Butterworth, 2008; Kucian & von Aster, 2015; Landerl, Bevan, & Butterworth, 2004; Rousselle & Noel, 2007). This heterogeneity in mathematical abilities is likely due, in part, to variability in learning and brain plasticity throughout development. Importantly, aberrant plasticity of the brain systems described above (Fig. 15.1) is thought to underlie the pathogenesis of learning disabilities specific to mathematics or otherwise referred to as mathematical learning disabilities (MLD) (DSM-V Association, 2013). Thus, the study of typical and atypical neurocognitive development and plasticity provides a unique opportunity to help characterize sources of heterogeneity in mathematical learning.

This chapter synthesizes the extant literature on functional brain circuits underlying *"number sense"* and scaffolding of mathematical cognition and learning in children and adults, with a focus on sources of heterogeneity in typical and atypical development. We take a *systems neuroscience approach* (Box 15.1) to elucidate sources of heterogeneity arising from multiple distributed neural circuits critical for number form identification, magnitude and quantity representations, working memory, declarative memory, attention, and cognitive control (Fig. 15.1) (Arsalidou & Taylor, 2011; Fias, Menon, & Szucs, 2013; Qin et al., 2014). We conclude by highlighting directions for future research.

VENTRAL AND DORSAL VISUAL STREAMS: NEURAL BUILDING BLOCKS OF MATHEMATICAL COGNITION

Basic Organization

The cognitive building blocks of *"number sense"* are constructed from local and global functional circuits anchored in the FG and the IPS, which are considered *hub* (Box 15.1) regions within the ventral and dorsal visual streams, respectively (Figs. 15.1 and 15.2A,a,b). Specifically, these *hub* regions code perceptual and semantic representations of nonsymbolic and symbolic quantities and facilitate their dynamic manipulation in a context-dependent manner (Ansari, 2008).

The FG, which is located in the ventral temporal–occipital cortex, plays an essential role in high-level visual object recognition (Gauthier, Skudlarski, Gore, & Anderson, 2000; Gauthier, Tarr, Anderson, Skudlarski, & Gore, 1999; Goodale & Milner, 1992; Grotheer, Herrmann, & Kovacs, 2016; Holloway, Battista, Vogel, & Ansari, 2012; Shum et al., 2013; Wimmer, Ludersdorfer, Richlan, & Kronbichler, 2016) and in forming an automated representation of number forms. The IPS, which is located within the dorsal aspects of the PPC, is thought to synthesize incoming information from both visual and auditory modalities (Vogel et al., 2017). With the support of these two key neurocognitive systems, the human brain builds more complex mathematical problem-solving skills. In the following section, we describe the neurocognitive functions of IPS and FG systems and their relative roles in basic number processing.

FG regions, which are proximal to the visual word form area of the ventral temporal–occipital cortex (McCandliss, Cohen, & Dehaene, 2003; Saygin et al., 2016), show strong responses to numerical symbols in both adults and children (Ansari, 2008; Cantlon et al., 2009; Grotheer et al., 2016; Holloway et al., 2012; Shum et al., 2013) (Fig. 15.2A,b). Furthermore, intracranial electrophysiological recordings suggest that distinct subdivisions of the FG may be differentially sensitive to number stimuli compared with perceptually similar letters and nonword stimuli (Shum et al., 2013) (Fig. 15.2A,b), suggesting a unique role of the FG in perceptual decoding of numerical digits. Involvement of the FG in numerical judgment and manipulation of quantity, not just perception, is also supported by the fact that the FG, like the IPS, also exhibits *neural distance effects* (Box 15.1) during symbolic comparisons of numerical digits (Fig. 15.2A,c) (Vogel et al., 2017).

The IPS plays a crucial role in numerical quantity judgments in both adults and children (Cantlon et al., 2009; Cutini, Scatturin, Basso Moro, & Zorzi, 2014; Heine, Tamm, Wissmann, & Jacobs, 2011; Hyde & Spelke, 2009; Kucian, von Aster, Loenneker, Dietrich, & Martin, 2008; Piazza, Mechelli,

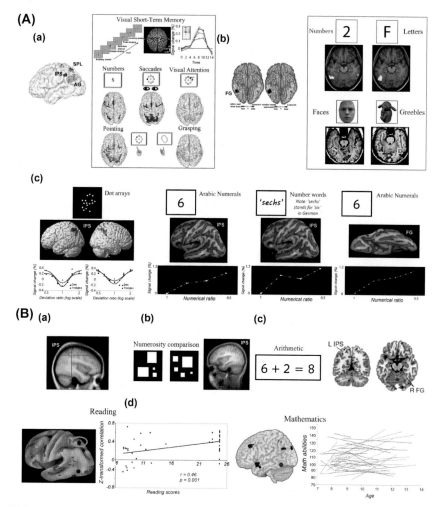

FIGURE 15.2 **Neural building blocks of mathematical cognition and learning. (A)** Basic organization. **(a)** Posterior parietal cortex. Regions of the intraparietal sulcus (IPS), superior parietal lobe (SPL), and angular gyrus (AG) implicated in numerical representations and arithmetic overlap anatomically with (*Square-top*) regions of the visual short-term memory (vSTM) system during a probe-matching (i.e., color and position) short-term memory; (*Square-bottom*) regions of the IPS that show significant functional activation for several cognitive tasks, including number comparisons, visual saccades and visual attention, pointing, and grasping; **(b)** Ventral temporal–occipital cortex. The fusiform gyrus (FG) shows preferential gamma-band responses to numerals; the FG shows significant functional activation for several stimulus categories, including numbers and letters (*Square-top*), as well as faces and complex shapes (*Square-bottom*); **(c)** Symbolic *distance effects* in the IPS and FG. **IPS**: distance/ratio effects in the bilateral parietal IPS for nonsymbolic numerical stimuli (i.e., dot arrays) (*Left*); symbolic numerical stimuli (i.e., Arabic digits) (*Middle*); "sechs" is German for "six" (*Right*). **FG**: distance/ratio effects in the FG for symbolic numerical stimuli (i.e., Arabic digits). **(B)** Heterogeneity of skills. **(a)** Reduced gray matter volume in the IPS of adolescents of low birth weight and with mathematical learning disabilities (MLD); **(b)**

Butterworth, & Price, 2002). Specifically, the IPS is engaged when a child or an adult is asked to select stimuli based on their numerosity (e.g., select the largest between two sets of stimuli—one containing 4, the other containing 5 items). Within the context of such cognitive computations, the IPS has been shown to be sensitive to the numerical distance between the nonsymbolic quantities to be compared, either when they are presented as discrete (Piazza, Izard, Pinel, Le Bihan, & Dehaene, 2004) (Fig. 15.2A,c) or analog forms (Vogel, Grabner, Schneider, Siegler, & Ansari, 2013) (i.e., the *neural distance effect*—see Box 15.1). The IPS is also strongly modulated by the numerical distance of symbolic representations and number words (Ansari, Garcia, Lucas, Hamon, & Dhital, 2005; Cohen Kadosh, Cohen Kadosh, Kaas, Henik, & Goebel, 2007; Hinton, Dymond, von Hecker, & Evans, 2010; Naccache & Dehaene, 2001) (Fig. 15.2A,c) (i.e., the *neural distance effect*—see Box 15.1). Although the IPS was initially considered solely a *"number module"* in the brain (Butterworth, 1999, 2010; Dehaene, 2003; Dehaene & Cohen, 1995), it is now widely accepted that the IPS is also critical for a broad class of related processes, including spatial attention, individuation, pointing of objects in extrapersonal space (Schaffelhofer & Scherberger, 2016; Simon, Mangin, Cohen, Le Bihan, & Dehaene, 2002), and encoding object locations into visual short-term memory (vSTM) storage (Knops, Piazza, Sengupta, Eger, & Melcher, 2014; Luck & Vogel, 2013; Todd & Marois, 2004) (Fig. 15.2A,a). This is likely due to its widespread connectivity (Vogel, Miezin, Petersen, & Schlaggar, 2012) with multiple prefrontal and parietal systems (Fig. 15.3A,a,b). Taken together, these

functional aberrancies in the IPS of 10- to 12-year-old children with MLD during a nonsymbolic numerosity comparison task; **(c)** functional aberrancies in the IPS and FG of 7- to 9-year-old children with MLD during an arithmetic task; **(d)** ventral and dorsal stream intrinsic connectivity strengthens the link between orthographic symbols and semantic representations in reading (*Left*) and math (*Right*). *Left*: significant relationship between reading scores and FG to left anterior IPS connectivity during resting. Dorsal stream regions in orange and ventral stream regions in blue. Regions of interest in black. *Right*: Functional connectivity patterns with the FG predict gains in mathematical skills over development. FG seed region in pink and functional connectivity maps in purple (including the dorsal IPS). *(A,a) Modified from: Dehaene et al., 2003*, Cognitive Neuropsychology © 2003 by Taylor & Francis; *Adapted from: Todd & Marois, 2004*, Nature © 2004 by NPG; *Adapted from: Simon et al., 2002*, Neuron © 2002 by Cell Press. *(A,b) Adapted from: Shum et al., 2013*, Journal of Neuroscience © 2013 by SfN; *Adapted from: Grotheer et al., 2016*, Journal of Neuroscience © 2016 by SfN; *Adapted from: Gauthier et al., 1999*, Nature Neuroscience © 1999 by NPG. *(A,c)* **Left**: *Adapted from: Piazza et al., 2004*, Neuron © 2004 by Elsevier; *Adapted from: Vogel et al., 2017*, Neuroimage © 2017 by Elsevier; **Far right**: *Adapted from: Vogel et al., 2017*, Neuroimage © 2017 by Elsevier. *(B,a) Modified from: Isaacs et al., 2001*, Brain © 2001 by Oxford University Press; *(B,b) Based on: Price et al., 2007*, Current Biology © 2001 by Cell Press; *(B,c) Based on: Iuculano et al., 2015*, Nature Communications © 2015 by NPG. *(B,d)* **Left**: *Modified from: Vogel et al., 2012*, Cerebral Cortex © 2015 by NPG; **Right**: *Modified from: Evans et al., 2015*, Journal of Neuroscience © 2015 by SfN.

findings suggest that the IPS builds a modal representations of quantity through neural substrates that overlap with those that serve other cognitive functions such as vSTM and attention.

Importantly, the IPS and FG do not function independently, with evidence suggesting that functional connections between dorsal and ventral streams are important for making links between visual-orthographic items and their semantic representations (Jeong & Xu, 2016; Vogel et al., 2012). Together this body of evidence suggests that both the FG and IPS and their interactions help create neural representations for symbolic form and the quantities they represent.

Heterogeneity in Typical and Atypical Development

As described in the section above, although the basic roles of the IPS and FG in *"number sense"* are well characterized in both children and adults (Ansari, 2008; Cantlon, Brannon, Carter, & Pelphrey, 2006), considerable heterogeneity exists within and across developmental groups in the function and interaction of these brain systems. With learning and development there are increased and more refined interactions between these brain systems, which gradually support proficiency and automatization of mapping between Arabic digits and quantity representations (Fan, Anderson, Davis, & Cutting, 2014; Vogel et al., 2012, 2013).

Marked heterogeneity in mathematical cognition and learning is apparent in the widespread prevalence and persistence of domain-specific learning disabilities (Butterworth et al., 2011; Shalev, Auerbach, Manor, & Gross-Tsur, 2000; Shalev, Manor, & Gross-Tsur, 2005; Szucs & Goswami, 2013) and in genetic developmental syndromes, such as Turner syndrome (TS) and Fragile X syndrome, which present significant difficulties in numerical problem-solving. Early research has consistently pointed to the IPS as a locus of dysfunction in these populations (Isaacs, Edmonds, Lucas, & Gadian, 2001; Molko et al., 2003; Price, Holloway, Rasanen, Vesterinen, & Ansari, 2007) (Fig. 15.2B,a,b,c). Specifically, lack of functional modulation during numerical judgment tasks (Price et al., 2007) and decreased gray matter density in the IPS have been reported in individuals with TS (Molko et al., 2003, 2004), Fragile X (Rivera, Menon, White, Glaser, & Reiss, 2002), as well as children (Price et al., 2007; Rotzer et al., 2008) and adolescents with MLD (Isaacs et al., 2001) (Fig. 15.2B,a,b). Furthermore, Cohen Kadosh et al. were able to induce MLD-like symptoms (i.e., reduced automaticity for processing numerical information) in a group of neurotypical adults by applying fMRI-guided transcranial magnetic stimulation over the right PPC adjoining the IPS (Kadosh et al., 2007). Functional aberrancies in the activity of the IPS and the FG have also been reported during arithmetic problem-solving tasks in MLD (Iuculano et al., 2015; Rosenberg-Lee et al., 2014) (Fig. 15.2B,c). Moreover, it has been shown that intrinsic connectivity between the FG and IPS is positively associated with mathematical skills over development (Evans et al., 2015) (Fig. 15.2B,d). This is consistent with the notion that ventral to dorsal stream connectivity is essential for the successful mapping between orthographic symbols and their semantic representations. This is a key milestone for efficient development of mathematical skills, as well as other complex and uniquely human cognitive tasks, such as reading (Vogel et al., 2012) (Fig. 15.2B,d).

As reviewed in the subsequent sections, it is increasingly clear that some forms of MLD may arise from aberrant circuits encompassing the PFC, parietal, and MTL along with the FG and IPS (Fig. 15.1). A *systems neuroscience approach* naturally leads to the proposal that heterogeneous development and outcomes in mathematical learning and cognition are the result of differences in the functional and structural architecture of these multiple functional brain systems (Fias et al., 2013; Menon, 2014).

PARIETAL-FRONTAL SYSTEMS: SHORT-TERM AND WORKING MEMORY

Basic Organization

In the previous section, we synthesized converging evidence that the IPS plays a central role in representing and manipulating numerical quantities. We also highlighted evidence that the IPS is involved in a broader class of cognitive processes that support such quantity judgments and numerical representations. These include spatial attention (Simon et al., 2002) and encoding object locations into vSTM storage (Knops et al., 2014; Luck & Vogel, 2013; Todd & Marois, 2004). For example, IPS activity is enhanced during vSTM tasks during which participants are asked to remember the spatial location of objects (Todd & Marois, 2004) (Fig. 15.2A,a). Moreover, IPS regions that are engaged during vSTM tasks overlap with regions that are activated during enumeration tasks (Knops et al., 2014), highlighting the close correspondence between these two cognitive functions. Critically, the IPS is part of an intrinsically connected dorsal frontoparietal system that includes the superior parietal lobe (SPL) and frontal eye fields (FEFs) (Corbetta & Shulman, 2002; Corbetta, Patel, & Shulman, 2008) (Fig. 15.3A,a) for online tracking of items in space.

Other cytoarchitectonic subdivisions of the parietal cortex, including the supramarginal gyrus (SMG), are more intrinsically connected

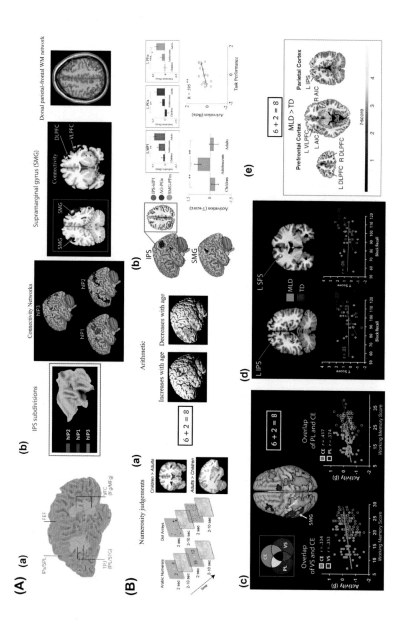

FIGURE 15.3 **Parieto-frontal working memory systems for mathematical cognition and learning. (a)** Schematic model of dorsal and ventral frontoparietal networks. Areas in blue indicate the dorsal frontoparietal network encompassing the intraparietal sulcus (IPS), superior parietal lobe (SPL), and the frontal eye field (FEF). Areas in orange indicate the ventral frontoparietal network encompassing the temporoparietal junction (TPJ), inferior parietal lobe (IPL), and superior temporal gyrus (STG), as well as ventral frontal cortex (VFC), inferior frontal gyrus (IFG, and middle frontal gyrus (MFG). The IPS–FEF network is involved in top-down control of visual processing. The TPJ–VFC network is involved in stimulus-driven control; **(b)** Parietal-circuits connectivity. **Left**: Functional connectivity of horizontal segments of the IPS subdivisions. Seed Regions of Interest (ROIs) derived from cytoarchitectonic maps for hIP1 (blue), hIP2 (red), and hIP3 (green). Colors on brain rendering surface represent voxels correlated with each seed-ROI. **Right**: Supramarginal gyrus (SMG) connectivity with dorsolateral and ventrolateral prefrontal cortices (DLPFC and VLPFC)—obtained by independent component analysis—forms the dorsal–parietal–frontal working memory (WM) network (**Far right**). **(B)** Heterogeneity of skills. **(a)** Frontoparietal shift. **Left**: Statistical comparison of 6- to 7-year-olds and adults during numerosity tasks with symbolic and nonsymbolic stimuli. Children show greater prefrontal cortex (PFC) activation compared with children (shown in red); whereas adults show greater activation in the intraparietal sulcus (IPS) compared with children (shown in green). **Right**: Brain areas where activity during arithmetic problem-solving increases or decreases with age. Increases are evident in the supramarginal gyrus (SMG) and in the ventral temporal–occipital cortex (VTOC) (shown in red); whereas decreases are seen in the lateral PFC and superior and middle temporal gyri (shown in blue); (b) Developmental trajectories of parietal circuits. **Top**: Cytoarchitectonic probabilistic labeling indicates that functional activations during arithmetic problem-solving in ventral lateral subdivisions of the IPS (hIP1), anterior angular gyrus-PGa (AG–PGa), and the posterior subdivision of the supramarginal gyrus (SMG-PFm) show linear increases from childhood to adulthood. **Bottom**: Activation in the left anterior SMG showed an inverted U-shaped profile across age groups. Greater SMG-PF activation was correlated with task performance in adolescents only; **(c)** Neural overlap between the central executive (CE) and visuospatial (VS) components of WM was observed only in the left SMG, whereas overlap between the CE and phonological loop (PL) components was observed only in the left IPS; no overlap was observed between VS and PL WM components. *Scatterplots*. ROIs characterized by overlap in correlations of activity and individual WM components show significant correlations with WM scores; **(d)** Relation between functional brain activity during arithmetic problem-solving and visuospatial working memory abilities in *a-priori*-defined frontoparietal ROIs: the IPS and the superior frontal sulcus (SFS). Typically developing (TD) children show a significant relationship between visual WM abilities and activity in these brain regions, whereas children with mathematical learning disabilities (MLD) do not; **(e)** Children with MLD show significantly greater activation than TD children in regions of the parietal-frontal network, including the bilateral DLPFC, left VLPFC, left insular cortex, and right IPS during an arithmetic problem-solving task. **(A,a)** *Modified from: Corbetta & Shulman, 2002*, Nature Reviews Neuroscience © 2002 by NPG. **(A,b) Left:** *Modified from: Uddin et al., 2010*, Cerebral Cortex © 2010 by Oxford University Press; **Right**: *Adapted from: Sridharan et al., 2008*, PNAS © 2008 by The National Academy of Sciences of the USA. **(B,a) Left:** *Modified from: Rivera et al., 2005*, Cerebral Cortex © 2005 by the Oxford University Press. **(B,b)** *Adapted from: Chang et al., 2015*, Neuroimage © 2015 by Elsevier. **(B,c)** *Modified from: Metcalfe et al., 2013*, Developmental Cognitive Neuroscience © 2013 by John Wiley & Sons, Inc. **(B,d)** *Modified from: Ashkenazi, Rosenberg-Lee et al. (2013)*, Neuropsychologia © 2013 by Elsevier. **(B,e)** *Modified from: Iuculano et al., 2015*, Nature Communications © 2015 by NPG.

to the supplementary motor area, anterior insula, and dorsolateral PFC (DLPFC) (Bressler & Menon, 2010; Sridharan, Levitin, & Menon, 2008; Uddin et al., 2010) (Fig. 15.3A,b), and support active storage and manipulation of contents in working memory (Menon, 2016). During successful mathematical learning, updating procedures such as repeated counting aid the encoding of arithmetic facts in long-term memory (Ashcraft, 1982; Groen & Parkman, 1972; Siegler & Shrager, 1984, pp. 229–293). For example, when counting on from 5 to 7 to solve the problem "5 + 2," an association is dynamically formed between the correct solution (i.e., "7") and its addends ("5" and "2"). After many repetitions, children are able to directly retrieve the answer from memory when presented with the problem (Siegler & Shrager, 1984, pp. 229–293). In this way, working memory processes support the successful encoding of information of numerical content into long-term memory. Two distinct working memory circuits, anchored in the IPS and SMG, support this process by tracking and manipulating numerical information over multiple time scales (Menon, 2016).

Heterogeneity in Typical and Atypical Development

Heterogeneity in the function of parietal–frontal working memory systems is the most apparent and insightful in the context of development (Menon, 2014). In childhood, even the most basic numerical computations require high working memory resources because mathematical problems and numerical computations need to be broken down into elemental components. It is important to note that working memory resources are still necessary in adulthood, especially when learning new concepts, or when mathematical computations become more complex (e.g., arithmetic operations with multidigits, which involve carrying and borrowing) (DeStefano & LeFevre, 2004; Imbo, Vandierendonck, & Vergauwe, 2007). Here, we focus on developmental changes at the brain level that are related to arithmetic learning during early schooling as this has been more extensively studied in the context of mathematical skill acquisition. We anticipate that this framework might translate to more complex arithmetic operations that occur later during development (see, for example, Stocco & Anderson, 2008 on the role of the prefrontal working memory system in the online manipulation of arithmetic facts during algebraic equations).

Developmentally, an anterior-to-posterior neural shift occurs as specialized regions of the PPC overtake the cognitive processes that were required by the PFC during childhood. A seminal study reported that relative to adults, children tend to engage the PPC less and the PFC more, when solving simple arithmetic problems—such as 3 + 4—reflecting higher demands on working memory processes (Rivera, Reiss, Eckert, & Menon, 2005) (Fig. 15.3B,a). This pattern of age-related engagement of the PFC extends to more elemental, nonsymbolic processing as well; while 6- and 7-year-old children engaged the bilateral inferior frontal gyrus (IFG) and adjoining insular cortex during a nonsymbolic magnitude discrimination task, young adults (24-years-olds) engage the PPC (Fig. 15.3B,a) (Cantlon et al., 2009). Other evidence suggests that children (9–11 years) show modulation of brain activity as a function of problem complexity (i.e., the *neural distance effect*) (see Box 15.1) in the PFC, including the DLPFC, precentral gyrus, and the inferior frontal gyrus, whereas young adults (19–21 years) exhibit this *neural distance effect* only in the IPS (Ansari & Dhital, 2006; Ansari et al., 2005). Taken together, these findings suggest that with learning and development, more precise representations are built within the PPC, thus "freeing up" prefrontal working memory resources.

As noted in basic organization section, the IPS and SMG are associated with distinct working memory circuits. Notably, these circuits show heterogeneous age- and ability-related patterns in their functional organization. A recent cross-sectional study with three age-cohorts (children, adolescents, and adults) found that, while engagement of the bilateral ventral IPS increased linearly with age (Fig. 15.3B,b; *top part*), a left anterior subdivision of the SMG (SMG-PF) showed an inverted U-shaped age-related profile, such that adolescents exhibited greater activation than children and young adults (Fig. 15.3B,b; *bottom part*). Furthermore, greater SMG-PF activation was associated with task performance only in adolescents (Fig. 15.3B,b, *bottom part*) and adolescents also showed greater task-related functional connectivity of the SMG-PF with ventro-temporal, anterior temporal, and prefrontal cortical regions, relative to children and adults (Chang, Metcalfe, Padmanabhan, Chen, & Menon, 2016). These results suggest that nonlinear upregulation of the SMG and its interconnected functional circuits facilitates adult-level performance in adolescents. Importantly, this study provides fine-grained evidence for the changing role of distinct working memory circuits with development.

A different view of heterogeneity in neurocognitive function comes from characterization of individual working memory components as defined by Baddeley's original model (Baddeley & Hitch, 1974). Critically, the relative roles of these individual working memory components depend on problem complexity, individual differences in learning, and stage of skill development. Recent evidence demonstrates fractionation of neurofunctional systems associated with distinct working memory components during arithmetic problem-solving in 7- to 9-year-old children.

Specifically, visuospatial working memory and the central executive components of working memory are associated with distinct patterns of brain responses (Metcalfe, Rosenberg-Lee, Ashkenazi, & Menon, 2013)

(Fig. 15.3B,c). Furthermore, visuospatial working memory is the strongest predictor of mathematical ability, compared with the other working memory components (i.e., central executive and phonological loop) (Metcalfe et al., 2013). Visuospatial working memory is also uniquely associated with increased arithmetic complexity–related responses in DLPFC and ventrolateral PFC (VLPFC) and the bilateral IPS and SMG in children (Menon, 2016). Taken together, there is evidence that distinct working memory components play unique roles in mathematical learning and are important to consider when examining developmental trajectories and sources of heterogeneity within this domain.

Atypical working memory is strongly associated with aberrancies in mathematical learning and cognition in children with MLD (Ashkenazi, Rosenberg-Lee, Metcalfe, Swigart, & Menon, 2013; Geary, Hoard, Byrd-Craven, Nugent, & Numtee, 2007). Importantly, a number of studies have reported that individuals with MLD show aberrant recruitment of prefrontal and parietal working memory systems during mathematical tasks, including arithmetic problem-solving and numerosity judgments (Butterworth et al., 2011; Davis et al., 2009; Iuculano et al., 2015; Kaufmann, Vogel, Starke, Kremser, & Schocke, 2009; Kucian & von Aster, 2015; Kucian et al., 2006, 2011; Menon, 2014; Price et al., 2007; Rosenberg-Lee et al., 2014). Evidence to date further suggests that visuospatial working memory is a specific source of vulnerability in numerical calculation deficits in children with MLD. For example, Ashkenazi, Rosenberg-Lee, et al. (2013) reported deficits in visuospatial working memory skills, in addition to deficits in arithmetic task performance in children with MLD, compared with a group of IQ- and verbal abilities-matched typically developing (TD) children (Fig. 15.3B,d). Furthermore, activation of the IPS and DLPFC and VLPFC were positively correlated with visuospatial working memory in TD children, but no such association was seen in children with MLD (Fig. 15.3B,d). Iuculano et al. (2015) also showed aberrant engagement of frontoparietal regions, including DLPFC and VLPFC, IPS, and the left insular cortex in children with MLD (Fig. 15.3B,e). These results suggest that children with MLD fail to appropriately exploit working memory resources during arithmetic problem-solving. Taken together, there is evidence to suggest that individual working memory components are engaged differently in children with MLD, further highlighting sources of heterogeneity in atypical development.

In summary, a large body of research shows that multiple working memory components and distinct frontoparietal circuits play a central and dynamic role in mathematical cognition, and that the differential engagement of these components and their associated circuits contributes to heterogeneity in both typical and atypical development.

LATERAL FRONTOTEMPORAL CORTICES: LANGUAGE-MEDIATED SYSTEMS

Basic Organization

Phonological skills are essential for successful mathematical skill acquisition, particularly during the early stages of learning (Carey, 2004). Even before formal schooling, children are introduced to number words during language acquisition. Specifically, phonological processing plays a pivotal role in learning to count and in the perceptual-to-semantic mapping from number words (i.e., "one," "two," "three," "four") to numerosities (Carey, 2004). There is evidence that phonological awareness (i.e., the ability to recognize and manipulate different word sounds) and arithmetic skills are associated with one another (De Smedt, Taylor, Archibald, & Ansari, 2010). Moreover, several behavioral studies have demonstrated a link between the ability to process linguistic information and arithmetic word problems (e.g., "John has 9 pennies. He spent 3 pennies at the store. How many pennies does he have left?"), and arithmetic computations (Fuchs, Fuchs et al, 2008; Fuchs, Powell et al, 2008; Jordan, Levine, & Huttenlocher, 1995). There is also evidence that arithmetic facts are represented as a "verbal code" in the cortex (Dehaene, Piazza, Pinel, & Cohen, 2003; Grabner et al., 2009; Prado et al., 2011). Moreover, processing of arithmetic facts engages verbal information processing regions in the STS, the MTG, and the IFG (Andres, Michaux, & Pesenti, 2012; Andres, Pelgrims, Michaux, Olivier, & Pesenti, 2011; Lee, 2000; Prado, Mutreja, & Booth, 2014; Prado et al., 2011) (Fig. 15.4A,a,b).

Heterogeneity in Typical and Atypical Development

Mathematical proficiency is dependent on successful early learning about numerical quantity, including binding of individual numbers to their names, which requires engagement and parallel development of language-related processing (Gobel, Watson, Lervag, & Hulme, 2014). Indeed, several neuroimaging studies have reported engagement of language-related brain systems during mathematical tasks in children (Evans, Flowers, Napoliello, Olulade, & Eden, 2014; Prado et al., 2011, 2014; Rosenberg-Lee et al., 2014; Zarnhofer et al., 2012). The degree of this engagement is often based on problem type (e.g., the arithmetic operation to be performed) and strategy use, highlighting a source of heterogeneity in the role of language processing on mathematical cognition. For example, Prado et al. report age-related activation increases in left lateral temporal cortex, a region typically associated with language comprehension, for multiplication, but not for subtraction (Fig. 15.4B,a). Conversely, subtraction problems showed increased activity in regions of the PPC involved in numerosity and visuospatial processing as a function of age and skills (Prado et al., 2011, 2014). These findings are in line with previous results demonstrating that multiplication relies more on verbal strategies, whereas subtraction is often solved through visuospatial strategies as reflected by the differential neural substrates engaged (Dagenbach & McCloskey, 1992; Dehaene & Cohen, 1995; Ischebeck et al., 2006). Paralleling these findings, similar results have emerged in a heterogeneous population of elementary school children, involving children with MLD. In their study, Rosenberg-Lee et al. (2014) found greater activity for addition relative to subtraction problems in the bilateral STS, middle temporal gyri (MTG), and the hippocampus, regions involved in language processing, and memory, respectively. Conversely, subtraction problems were supported by a network of posterior parietal and frontal regions (Rosenberg-Lee et al., 2014)—see also Evans et al. (2014) for a similar result.

Moreover, there is evidence that variability in strategy use modulates activation in verbal brain systems during mathematical problem-solving. Zarnhofer et al. reported that individuals who verbalize multiplication and addition operations show stronger activation in regions associated with auditory processing—in the Heschl's gyrus and Rolandic operculum—compared with individuals who simply visualize the numbers, independent of whether problems were presented as Arabic numerals or written number words (Zarnhofer et al., 2012) (Fig. 15.4B,b).

With respect to heterogeneity in mathematical abilities, children with MLD show aberrant activation in verbal information processing regions of the STS, MTG, and IFG—identified via an independent phonological localizer task (Berteletti, Prado, & Booth, 2014; Prado et al., 2011)—suggesting difficulties in verbal retrieval of problem solutions (Fig. 15.4B,c).

Furthermore, the integration of phonological and orthographic codes, anchored in the STS, MTG, and FG, respectively, might also play an important role in individual differences in the acquisition of mathematical skills, similarly to reading skills. A previous study in 7- to 9-year-olds, found that subtraction problems activated a network of posterior parietal regions, including the IPS and SMG, whereas addition problems activated a network of language-related regions in the STS and MTG (Evans et al., 2014). This same network of language-mediated areas was also active during a reading task, suggesting that verbally mediated systems underlie both arithmetic problem-solving and reading. The integration of phonological and orthographic codes has not been formally explored in mathematical cognition, however, it is plausible to predict that, at least for certain arithmetic operations (i.e., addition and multiplication), successful neural integration of these two codes at the early stages of development might predict later learning and proficiency.

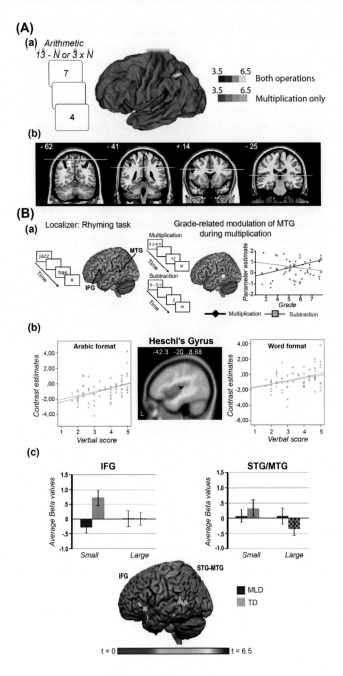

◀ FIGURE 15.4 **Language systems for mathematical knowledge. (A)** Basic organization. **(a)** Brain areas showing increased activity during mental arithmetic tasks compared to letter reading. *Left*: Schematic of the arithmetic task during which participants had to multiply the Arabic digit displayed on the screen by 3 or 4, or subtract it from 11 or 13. *Right*: brain regions of the middle temporal gyrus (MTG) and superior temporal sulcus (STS) (in blue) were more active for multiplication tasks, whereas posterior parietal regions (in orange) showed significant activation for both multiplication and subtraction tasks; **(b)** CORONAL views of brain regions showing increased functional activity during the same mental arithmetic task as **(A,a)**. Red clusters (first three from the *left*) show the parietal and frontal areas involved in subtraction and multiplication. These include the posterior superior parietal lobe (pSPL), the horizontal segment of the intraparietal sulcus (hIPS), and the inferior frontal gyrus (IFG). Blue clusters (last panel on the *right*) include perisylvian areas of the STS and MTG that showed greater activation for multiplication than subtraction. **(B)** Heterogeneity of skills. **(a)** Functional activity modulation of the MTG during multiplication, but not subtraction, in children. *Left*: Localizer task—participants decided whether two visually presented words rhymed or not. *Middle*: Arithmetic tasks—participants were asked to evaluate multiplication (*top*) or subtraction (*bottom*) math problems containing single-digit operands. *Right*: Activity in the left MTG region of interest (ROI) as a function of operation (multiplication and subtraction) and school grade; **(b)** significant correlations between peak activations in the Heschl's gyrus in the temporal lobe during mental arithmetic (multiplication and subtraction) and self-reported use of verbalization strategies for different stimulus presentation (i.e., Arabic digits and number words). *Note*: Heschl's gyrus ROI defined based on the automated anatomical labeling atlas; **(c)** Group differences in functional activation during multiplication in ROIs identified using the same rhyming task as **(B,a)**. IFG and STG–MTG clusters showed greater activation for small compared with large problems for typically developing (TD) children, but not for children with mathematical learning disabilities (MLD). *(A,a) Modified from: Andres et al., 2012, NeuroImage © 2012 by Elsevier. (A,b) Modified from: Andres et al., 2011, NeuroImage © 2011 by Elsevier. (B,a) Modified from: Prado et al., 2014, Developmental Science © 2014 by John Wiley & Sons, Inc. (B,b) Modified from: Zarnhofer et al., 2012, Behavioral and Brain Functions © 2012 by BioMed Central Ltd. (B,c) Modified from: Berteletti et al., 2014, Cortex © 2011 by Elsevier.*

This framework could also be useful for assessing the known associations of math and reading skills and comorbidity of disabilities in these two cognitive domains (Ashkenazi, Black, Abrams, Hoeft, & Menon, 2013). Indeed, a recent study demonstrated that compared with TD controls, children with developmental dyslexia and mild impairments in fact retrieval displayed greater activity in the SMG for both addition and subtraction problems (Evans et al., 2014). Conversely, the TD group engaged the SMG only for subtraction and not addition problems, suggesting that children with developmental dyslexia might use alternate nonverbal strategies for both operations to compensate for neural aberrancies in language-related areas.

Together these findings highlight the importance of these—often overlooked—verbally mediated systems in the development of mathematical cognition. The extant literature also demonstrates that it is important to consider multiple variables (i.e., type of task, type of strategy, age, comorbidity of skills) when assessing neurocognitive correlates of heterogeneity of outcomes in mathematical learning.

THE MEDIAL TEMPORAL LOBE: DECLARATIVE MEMORY

Basic Organization

The hierarchical nature of mathematical skill acquisition requires building-up associations between basic numerical concepts to more complex numerical attributes, binding old and new concepts, and accessing information about numerical attributes from long-term memory. Thus, declarative memory and associated neural circuits play a critical role in mathematical learning and cognition. These processes depend on the function of subcortical structures important for learning and memory consolidation within the MTL (Davachi, 2006; Davachi et al., 2003; Diana, Yonelinas, & Ranganath, 2007; Eichenbaum, Yonelinas, & Ranganath, 2007; Squire, 1992; Squire, Genzel, Wixted, & Morris, 2015, pp. 1–21; Squire, Stark, & Clark, 2004; Tulving, 1983) (Fig. 15.5A,a,b). Despite its critical role in learning and memory formation, the role of the hippocampus in mathematical learning has received little attention. This is likely because the critical role of this system decreases over development. For almost two decades, neuroimaging studies of mathematical cognition only assessed samples of already proficient adult participants who often do not rely on this system during simple arithmetic tasks. Here, by providing a developmental and learning perspective, we are able to shed light on the critical and unique role of the MTL memory system in the successful acquisition of mathematical skills.

Heterogeneity in Typical and Atypical Development

It is well established that during mathematical learning, inefficient procedural strategies in childhood (e.g., counting) are gradually replaced with direct retrieval of mathematical facts (Cho, Ryali, Geary, & Menon, 2011; Geary, 2011; Geary & Brown, 1991; Geary & Hoard, 2003, pp. 93–115; Qin et al., 2014). Importantly, the key role of the hippocampus in learning decreases as skills are acquired, and retrieval of numerical information is processed by neocortical systems important for mathematical cognition (Qin et al., 2014). Indeed, previous studies have demonstrated that children show increased engagement of the hippocampus relative to adolescents and adults during mathematical problem-solving (De Smedt, Holloway, & Ansari, 2011; Rivera et al., 2005). In a seminal study, Qin et al. (2014) demonstrated that children's transition from counting to memory-based retrieval strategies over a 1.2-year interval was mediated by increased hippocampal activation and increased hippocampal-neocortical connectivity (Fig. 15.5B,a,b). Critically, following an initial increase in hippocampal engagement from early to middle childhood (i.e., from 8.2 to 9.4 years), hippocampal activation decreased during adolescence and adulthood, despite further improvements in memory-based problem-solving skills. This pattern of initial increases and subsequent decreases in activation provides support for models of long-term memory consolidation, which posit that the hippocampus plays a time-limited, yet pivotal role in the early phase of knowledge acquisition (e.g., learning to associate the addend pairs 3 and 4, to the answer 7) (McClelland, McNaughton, & O'Reilly, 1995; Tse et al., 2007). Moreover, consistent with this model of developmental change, previous studies in adults have not reported reliable hippocampal engagement during basic arithmetic tasks (Dehaene et al., 2003). There is also evidence for a causal interaction between the hippocampus and the VLPFC during arithmetic problem-solving (Fig. 15.5B,c) as a function of cognitive strategy. Specifically, children with higher retrieval fluency rates show greater functional responses in the hippocampus, parahippocampal gyrus, and VLPFC (Fig. 15.5B,c). Moreover, there is evidence of a significant direct causal influence from the left VLPFC to the right hippocampus during arithmetic fact retrieval (Cho et al., 2012). Finally, differential functional activity patterns of the hippocampus are observed in 7- to 9-year-olds during arithmetic problem-solving as a function of retrieval and counting strategies (Cho et al., 2011) (Fig. 15.5B,c). Together this body of evidence suggests that the hippocampus and the PFC are engaged differently as a function of strategy in young children.

Taken together, converging evidence points to the hippocampal system and its interactions with the PFC as critical to children's early learning of arithmetic facts (Cho et al., 2011, 2012; De Smedt, Holloway, & Ansari, 2011; Qin et al., 2014), while retrieval in adolescents and adults relies predominantly on the neocortex (Dehaene, Piazza, Pinel, & Cohen, 2003; Menon, 2014).

These heterogeneous patterns of hippocampal involvement as a function of development, skill acquisition, and strategy use are consistent with the developmental cognitive model of "overlapping waves" whereby acquisition of cognitive skills is characterized by changes in the distributions of strategies over development (Siegler, 1996). In the case of mathematics, children's immature problem-solving abilities, which require breaking down numerical problems into more basic components, are slowly replaced by efficient retrieval of arithmetical facts (Cowan et al., 2011). These findings highlight the dynamic role of the hippocampus in the maturation of memory-based problem-solving strategies. Moreover, these findings demonstrate greater engagement of the hippocampus in childhood followed by decreased involvement during adolescence and adulthood.

Critically, consolidation of efficient strategy use and progressive engagement of the hippocampus during learning and development is accompanied by decreases in parietal–frontal engagement and concurrent increases in functional connectivity between hippocampal and neocortical circuits. Increases in hippocampal–PFC connectivity are significantly

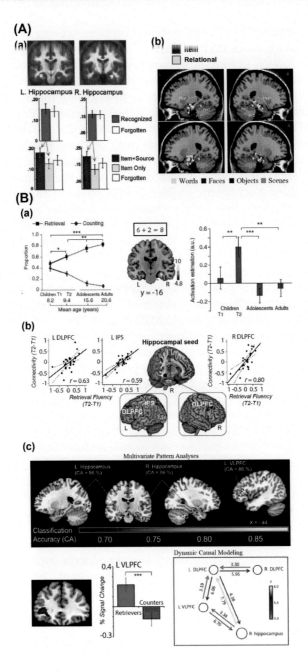

related to longitudinal improvements in retrieval fluency in early to middle childhood (Qin et al., 2014) (Fig. 15.5B,b). Together, these findings suggest that hippocampal-neocortical circuit reorganization plays a crucial role in children's shift from effortful counting to more efficient memory-based problem-solving during development. Consistent with this, children with MLD do not show hippocampal engagement during arithmetic problem-solving, suggesting that MLD, at least partially, derives from lack of hippocampal recruitment during crucial phases of mathematical learning and skill development (De Smedt et al., 2011).

Taken together, research is beginning to highlight the role of the hippocampus and its associated prefrontal-parietal circuits in consolidating basic arithmetic facts into long-term memory in children. Moreover, this body of research points to another key neurocognitive component underlying heterogeneous trajectories of typical and atypical skill development.

FIGURE 15.5 **Medial temporal lobe memory systems for mathematical knowledge. (A)** Basic organization. **(a)** Peak signal change during image encoding in bilateral hippocampi. Encoding activation is sorted according to whether the item was later recognized (*green bars*) or forgotten (*white bars*). For recognized items, both left and right hippocampi are sensitive to source memory outcome recollection (i.e., item + source > item only), suggesting that the hippocampus plays a crucial role in recollecting specific contextual details about a prior encounter (i.e., source recollection). *Red bars*: item + source; *Yellow bars*: item only; **(b)** *Top*: Metaanalysis of studies reporting medial temporal lobe (MTL) peaks for item-specific effects versus relational memory effects. Relational memory effects are predominantly seen in the bilateral hippocampus (Hipp) and the parahippocampal cortex (PhC), whereas item-specific effects are evident in the perirhinal cortex (PrC). *Bottom*: MTL peaks plotted as a function of stimulus type (i.e., words, faces, objects, scenes). **(B)** Heterogeneity of skills. **(a)** Developmental differences in hippocampal engagement. *Left*: Gradual increases in memory-based strategies and decreases in counting strategies. *Middle and Right*: Right hippocampal response during arithmetic problem-solving showing a main effect of age-group across children at different time points (Time 1 and Time 2, ~1.2 years apart), adolescents, and adults; **(b)** Hippocampal-neocortical connectivity. Longitudinal changes in hippocampal-neocortical connectivity (Y-axes) plotted against individual improvements in children's use of memory-based problem-solving strategies (X-axes); *DLPFC*, dorsolateral prefrontal cortex; *IPS*, intra parietal sulcus. **(c)** Hippocampal-prefrontal retrieval network. *Top*: Multivariate pattern analyses showing significant differences in spatial activation patterns between retrievers and counters in the bilateral hippocampus, adjoining parahippocampal gyrus, and in the left ventrolateral prefrontal cortex (VLPFC). Peak classification accuracies are shown in parentheses. *Bottom left*: The left VLPFC displays greater functional activation in 7- to 9-year-old retrievers (blue) compared with counters (red). *Bottom right*: Multivariate dynamical systems modeling of the hippocampal-prefrontal retrieval network. The left ventral and dorsal lateral aspects of the prefrontal cortex (VLPFC and DLPFC) show direct causal influence to the right hippocampus in 7- to 9-year-olds. *(A,a) Modified from: Davachi et al., 2003, PNAS © 2002 by The National Academy of Sciences of the USA. (A,b) Modified from: Davachi, 2006, Current Opinion in Neurobiology © 2006 by Elsevier. (B,a) Modified from: Qin et al., 2014, Nature Neuroscience © 2014 by NPG. (B,b) Modified from: Qin et al., 2014, Nature Neuroscience © 2014 by NPG. (B,c) Top: Modified from: Cho et al., 2011, Developmental Science © 2011 by John Wiley & Sons, Inc.; Bottom left: Modified from: Cho et al., 2011, Developmental Science © 2011 by John Wiley & Sons, Inc.; Bottom right: From: Cho et al., 2012, Journal of Cognitive Neuroscience © 2012 by MIT Press and the Cognitive Neuroscience Institute.*

THE CIRCUIT VIEW: ATTENTION AND CONTROL PROCESSES AND DYNAMIC CIRCUITS ORCHESTRATING MATHEMATICAL LEARNING

Basic Organization

The findings reviewed above demonstrate that mathematical cognition relies on interactions within and between multiple functional brain systems and circuits, including those subserving quantity processing, declarative, semantic, and working memory (e.g., Arsalidou & Taylor, 2011; Fias et al., 2013; Fig. 15.1). As noted, connectivity between the ventral and dorsal streams, and specifically from the higher-level ventral temporal–occipital visual cortices to dorsal posterior parietal cortices, helps to strengthen the link between visual percepts and their semantic representations—a key building block of numerical cognition. Although the IPS plays a key role in building representations of numerical quantity, it is also part of an intrinsically connected dorsal frontoparietal system that includes the SPL, FEFs, supplementary motor area, insula, and DLPFC (Corbetta & Shulman, 2002; Corbetta et al., 2008; Menon & Uddin, 2010; Supekar & Menon, 2012; Uddin et al., 2010) (Fig. 15.3A,a,b). This system is thought to be important for top-down goal-driven attention, including performance monitoring and manipulation of (numerical) information in working memory (Corbetta & Shulman, 2002; Corbetta et al., 2008). Furthermore, the ventral attention system, which includes the SMG, insula, and IFG, is involved in saliency processing and attentional filtering (Corbetta & Shulman, 2002; Corbetta et al., 2008; Fox, Corbetta, Snyder, Vincent, & Raichle, 2006; Menon & Uddin, 2010; Supekar & Menon, 2012) (Fig. 15.3A,a,b). Finally, memory systems anchored in the medial and anterior temporal lobe are important for encoding and retrieval of math facts (Grabner et al., 2009; Qin et al., 2014).

Critically, the flow of information within and across these multiple brain systems is regulated by flexible cognitive control systems, which facilitate the integration and manipulation of quantity and mnemonic information. These include most prominently the salience network, anchored in the anterior insula and dorsal anterior cingulate cortex, and the frontoparietal working memory network, anchored in the ventral and dorsal aspects of the lateral PFC and the SMG (Bunge, Dudukovic, Thomason, Vaidya, & Gabrieli, 2002; Cai et al., 2015; Cai, Ryali, Chen, Li, & Menon, 2014; Cole et al., 2013; Ham, Leff, de Boissezon, Joffe, & Sharp, 2013; Seeley et al., 2007; Sridharan et al., 2008) (Fig. 15.6A,a,b). Mechanistically, the implementation of cognitive control processes relies on both dynamic functional interactions within (Cai et al., 2014, 2015; Cole et al., 2013; Ham et al., 2013; Seeley et al., 2007; Sridharan et al., 2008) and across these brain systems (Bressler & Menon, 2010). Specifically, interactions between the salience

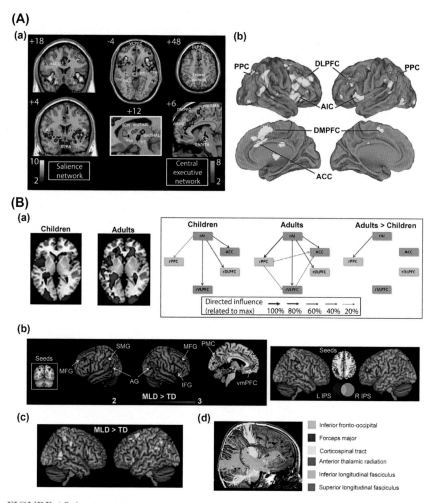

FIGURE 15.6   **Cognitive control circuits for mathematical knowledge. (A)** Basic organization. **(a)** The two core cognitive control brain networks identified using intrinsic physiological coupling during resting state fMRI. In red: the salience network (SN) important for monitoring salience of external inputs and internal brain events; in blue: the frontoparietal central executive/working memory network engaged in higher-order cognitive and attentional control tasks; **(b)** activation likelihood estimation (ALE) map of significant activated regions during an inhibitory control task. Significant clusters are evident in the anterior insular cortex (AIC), dorsolateral prefrontal cortex (DLPFC), dorsomedial prefrontal cortex (DMPFC), anterior cingulate cortex (ACC), and posterior parietal cortex (PPC). **(B)** Heterogeneity of skills. **(a)** *Left*: Brain activation in the SN and the frontoparietal central executive/working memory network during arithmetic problem-solving in children (*left* and in red) and adults (*right* and in blue). *Right*: Weaker dynamic causal interactions between cognitive control networks in children compared to adults; **(b)** *Left*: Hyperconnectivity of IPS circuits in children with mathematical learning disabilities (MLD), compared with typically developing (TD) children during an arithmetic task. *Right*: Hyperconnectivity of IPS circuits in children with MLD, compared with TD children during rest;

network and the frontoparietal working memory network help maintain attention on goal relevant numerical representations, while inhibiting irrelevant information or immature strategies. Interactions between the salience, frontoparietal, and hippocampal-frontal networks help to consolidate arithmetical representations in long-term memory.

Heterogeneity in Typical and Atypical Development

Cognitive skill acquisition during development is characterized by marked functional and structural maturation of individual brain areas (Bressler & Menon, 2010). In parallel, functional and structural brain connectivity across large-scale functional networks also undergoes significant changes with development (Bressler & Menon, 2010). Of particular interest here, connectivity between the FG and IPS increases significantly between childhood and adulthood (Vogel et al., 2012), which may underlie more efficient mapping between symbolic representations of numbers and the quantities they represent. Furthermore, intrinsic functional connectivity of the FG with the PPC and the PFC significantly predicts gains in numerical skills over development (Evans et al., 2015) (Fig. 15.2B,d). These findings illustrate how multiple functional circuits influence the development of mathematical skills and identify sources of individual differences at the brain circuit level.

The functional maturation of cognitive control systems also influences the development of mathematical skills. Despite greater PFC activation, there are decreased dynamic causal interactions between the anterior insula and the VLPFC and DLPFC in children, relative to adults (Supekar & Menon, 2012) (Fig. 15.6B,a). Moreover, weaker prefrontal control signals have been associated with worse performance on an arithmetic task with distinct pathways contributing differently to performance in children and adults (Supekar & Menon, 2012). Specifically, in children, the strength of causal signals from the anterior insula to the SMG and VLPFC significantly predicted faster reaction times; whereas in adults, an additional link from the insular cortex to the anterior cingulate cortex significantly predicted faster reaction times (Supekar & Menon, 2012). Thus, the interaction

(c) Frontoparietal regions showing greater fractional amplitude of low-frequency fluctuations (fALFF) in children with MLD, compared with TD children; (d) streamlined reconstruction of white matter pathways passing through the right temporal-parietal regions: children with MLD showed significant white matter volumetric deficits compared with TD children. *(A,a) Modified from: Seeley et al., 2007,* Journal of Neuroscience © 2007 by SfN. *(A,b) Modified from: Cai et al., 2014,* Journal of Neuroscience © 2014 by SfN. *(B,a) Modified from: Supekar & Menon, 2012,* PLoS Computational Biology © 2012 by PLoS. *(B,b)* **Left:** *Modified from: Rosenberg-Lee et al., 2014,* Developmental Science © 2014 by John Wiley & Sons, Inc.; **Right:** *Modified from: Jolles, Ashkenazi, et al. (2016),* Developmental Science © 2016 by John Wiley & Sons, Inc. *(B,c) Modified from: Jolles, Ashkenazi, et al. (2016), (B,d) Modified from: Rykhlevskaia et al., 2009,* Frontiers in Human Neuroscience © 2009 by Frontiers Media S.A.

between prefrontal cognitive control systems with frontoparietal systems for numerical manipulations represents a predominant source of heterogeneity in mathematical cognition across both development and levels of ability.

Similarly, the interaction between cognitive control systems in the ventral and dorsal aspects of the PFC with mnemonic systems anchored in the hippocampus may also be related to individual differences in mediating retrieval fluency during arithmetic problem-solving (Cho et al., 2012) (Fig. 15.5B,c). Specifically, dynamic causal modeling of fMRI data during arithmetic problem-solving has highlighted strong bidirectional causal interactions between these regions as a function of cognitive strategy for arithmetic problem-solving in elementary school children (Fig. 15.5B,c).

Both task-based and task-free fMRI studies demonstrate that connectivity of parietal-frontal circuits is impaired in children with MLD. Surprisingly, despite weaker performance, children with MLD show hyperconnectivity of the IPS with VLPFC and DLPFC and the SMG (Rosenberg-Lee et al., 2014) (Fig. 15.6B,b—*Left*). Hyperconnectivity of these circuits may arise from greater demands on working memory and cognitive control because of the need to inhibit problem-irrelevant information. This is consistent with behavioral studies showing greater intrusion errors of problem-irrelevant information in working memory during arithmetic fact retrieval in children with MLD (Barrouillet, Fayol, & Lathuliere, 1997; Geary, Hamson, & Hoard, 2000; Geary, Hoard, & Bailey, 2012). This pattern of hyperconnectivity is also apparent for intrinsic functional circuits, using "task-free" resting-state fMRI data. Compared with matched TD controls, children with MLD showed increased intrinsic functional connectivity between left and right IPS and between the IPS and DLPFC and VLPFC (Jolles, Ashkenazi, et al., 2016) (Fig. 15.6B,b—*Right*).

Taken together, converging evidence suggests that hyperconnectivity of intrinsic functional circuits associated with the IPS in children with MLD may underlie the commonly observed pattern of increased activation and task-based connectivity within these regions (Iuculano et al., 2015; Kaufmann et al., 2009a, 2009b; Kaufmann, Wood, Rubinsten, & Henik, 2011; Rosenberg-Lee et al., 2014). Notably, there is recent evidence that enhanced low-frequency fluctuations in the IPS are associated with individual differences in math abilities in children with MLD (Jolles, Ashkenazi, et al., 2016) (Fig. 15.6B,c—*Left*). Furthermore, the middle and superior frontal gyri, SMG, and superior parietal lobule also showed spontaneous regional hyperactivity and critically, hyperconnectivity with the IPS in children with MLD compared with TD children (Jolles, Ashkenazi, et al., 2016) (Fig. 15.6B,c—*Left*). These findings lead us to hypothesize that intrinsic hyperconnectivity and enhanced low-frequency fluctuations may limit flexible resource allocation and contribute to aberrant recruitment of task-related brain regions during mathematical cognition. Moreover, these results point to aberrant excitatory/inhibitory balance in the IPS in MLD. The neurotransmitters glutamate (Glx) and gamma-aminobutyric acid (GABA) play prominent roles in cortical excitability and in maintaining an optimal excitatory/inhibitory balance across brain systems and are thought to drive learning (Gupta et al., 2016; Naaijen et al., 2017; Pugh et al., 2014; Tatti, Haley, Swanson, Tselha, & Maffei, 2016). Elevated levels of Glx or decreased levels of GABA can lead to hyperexcitability and disruption of neural circuits in many neurodevelopmental and learning disorders, including dyslexia, autism, and attention-deficit hyperactivity disorder (Drenthen et al., 2016; Pugh et al., 2014; Tatti et al., 2016). Studies using magnetic resonance spectroscopy, which can provide a proxy for Glx and GABA concentrations in specific brain regions, are required to test this hypothesis in MLD.

In addition to atypical circuit function described above, structural aberrancies of gray matter volume have also been reported in MLD. These include multiple brain regions, including the parietal lobes, the ventral temporal–occipital cortices, and the anterior temporal cortices (Rykhlevskaia, Uddin, Kondos, & Menon, 2009). Individuals with MLD also show white matter deficits in the right inferior frontooccipital fasciculus, the right inferior and superior longitudinal fasciculi, the left longitudinal fasciculus, and the bilateral anterior thalamic radiation (Rykhlevskaia et al., 2009) (Fig. 15.6B,d—*Right*). Reduced integrity of white matter pathways within the right temporoparietal cortex has also been reported in MLD and has been associated with poorer performance on a standardized math task (Rykhlevskaia et al., 2009). Together, these results suggest that MLD might be reflected in gray matter structural abnormalities in a network of right temporoparietal areas and the white matter pathways associated with it. More generally, these results suggest that aberrant gray and white matter structure may drive functional abnormalities of intrinsic and task-based connectivity, resulting in inefficient communication between task-relevant brain regions. Together, these results highlight the importance of investigating local and large-scale circuit features underlying heterogeneity in typical and atypical development.

PLASTICITY IN MULTIPLE BRAIN SYSTEMS: RELATION TO LEARNING

Basic Organization

Brain plasticity is fundamental to any type of learning and a core characteristic of human cognition and development. Within this framework, learning studies are beginning to provide new insights into how mathematical knowledge is acquired, by more precisely linking learning to brain plasticity of relevant neurocognitive systems and circuits (Fig. 15.1). A popular paradigm for short-term arithmetic learning typically consists of multiple repetitions of a defined set of complex problems (i.e., training set), whereas another set is presented at a lower frequency (i.e., novel set). Training-contingent learning effects are rapid, and typically become significant after approximately eight repetitions of a problem, and remain stable over the course of the experiment (~1 week) (Ischebeck, Zamarian, Egger, Schocke, & Delazer, 2007). Comparison of brain responses between trained and novel problems is used to examine the effects of training-contingent learning; implicit in this design is the assumption that brain responses to untrained and trained problems are otherwise well matched prior to training. These studies typically reveal that training decreases responses in the PFC and IPS (Delazer et al., 2005; Ischebeck et al., 2006, 2007) (Fig. 15.7A,a,b,c), suggesting greater processing automaticity. In parallel, these studies report relative increases in activation in the left angular gyrus (AG) (Fig. 15.7A,a,b,c) (Delazer et al., 2003, 2005; Ischebeck et al., 2006, 2007; Zamarian & Delazer, 2014), although this result often reflects decreases in deactivation relative to baseline as the problems become more automatized (Wu et al., 2009).

Collectively, a handful of these studies have converged on similar results and suggest that mathematical learning is associated with not only major functional reorganization within the PPC but also other parietal-frontal circuits important for numerical reasoning (Delazer et al., 2005; Ischebeck et al., 2007; Zamarian, Ischebeck, & Delazer, 2009). More generally, these studies corroborate the evidence described in previous sections (from developmental and cross-sectional studies) regarding the pivotal contribution of all these systems to mathematical learning.

Heterogeneity in Typical and Atypical Development

Cognitive learning is shaped by brain plasticity through changes in local and large-scale functional circuits. Learning studies using continuous cohorts of participants that included TD children and children with MLD report an exceptional degree of functional and structural plasticity in the aforementioned systems (Fig. 15.1). Specifically, 2 months of a comprehensive training program emphasizing conceptual knowledge of number properties, arithmetical operations, and speeded practice of arithmetic facts (Fuchs et al., 2008, 2009, 2010, 2013; Powell, Fuchs, Fuchs, Cirino, & Fletcher, 2009) (Fig. 15.7B,a) shows significant functional and structural neuroplasticity-related effects in multiple functional brain networks important for mathematical cognition (Fig. 15.7B,a,b,c). Functional circuits associated with the IPS showed high levels of plasticity in their connectivity with the lateral PFC, the ventral temporal–occipital cortex, and the hippocampus (Jolles, Supekar, et al., 2016) (Fig. 15.7B,b). Furthermore, these connectivity changes were associated with performance gains (Fig. 15.7B,b). Conversely, functional circuits associated with other parietal regions such as the AG showed no training-related plasticity effects, highlighting the unique role of the IPS as a critical *hub* during active learning. This study extends previous

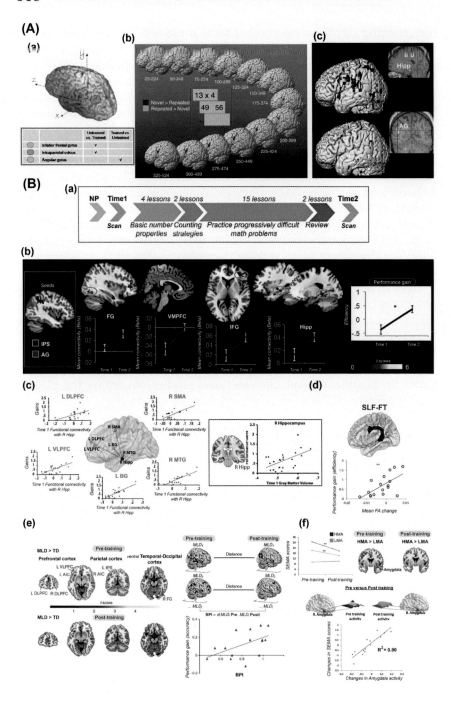

◀ FIGURE 15.7 **Cognitive learning and brain plasticity. (A)** Basic organization. **(a)** Schematic illustration of functional brain activation for untrained versus trained problems and the reverse (i.e., trained vs. untrained) during a short-term (~1 week) arithmetic training paradigm in adults. Extended activations in frontal and posterior parietal cortices, encompassing the intraparietal sulcus (IPS) are normally found for untrained versus trained problems (*orange and blue circles*). Higher activation in the lateral aspect of the posterior parietal cortex, particularly of the angular gyrus (AG), is often found on the contrast between trained and untrained problems in adults (*green circle*); **(b)** activation changes as a function of training: during task, participants had to choose the alternative solution that was closest to the actual solution of the presented multiplication problem. Rendering of moving window of 200 scans: training effects become significant starting at the 100- *to* 299-*scans* time-window. Green: repeated (trained) > novel (untrained) problems. Red: novel (untrained) > repeated (trained) problems; **(c)** *Top*: Functional activation maps for untrained versus trained multiplication problems in adults. Significant effects are evident in the IPS and the Hippocampus (Hipp); *Bottom*: Functional activation maps for trained versus untrained multiplication problems. Significant effects are seen in the AG. **(B)** Heterogeneity. **(a)** Experimental design of training studies in children. Before 2 months of cognitive training, children undergo an extensive battery of neuropsychological (NP) assessments and a scan session, which includes functional MRI (MRI), diffusion tensor imaging (DTI), resting-state fMRI (rsfMRI), and structural MRI (sMRI). 1:1 training focuses on conceptual aspects of number knowledge, effective counting strategies, and speeded practice. After training, children undergo another scanning session; **(b)** Functional IPS circuits show high level of plasticity after training. Training increased intrinsic functional connectivity between the IPS and the fusiform gyrus (FG), the ventromedial prefrontal cortex (VMPFC), the lateral prefrontal cortex encompassing the inferior frontal gyrus (IFG), the hippocampus (Hipp), and parahippocampal gyrus in a group of 7- to 9-year-old children. Intrinsic functional connectivity of the AG did not show any training-related effect except with the postcentral gyrus (not shown). Increases in IPS-functional connectivity were related to performance gains defined as efficiency gains from T1 to T2 (shown in graph); **(c)** Functional and structural hippocampal circuits show high levels of plasticity after cognitive training. *Left*: Functional connectivity of the hippocampus was correlated with improvements in arithmetic performance (i.e., gain as a function of both accuracy and RTs—efficiency measure) in response to 2-months of the same cognitive training as in **(B,a)**. Performance gains correlated with Time 1 hippocampal connectivity to the left dorsolateral prefrontal cortex (L DLPFC), left ventrolateral prefrontal cortex (L VLPFC), right supplementary motor areas (R SMA), left basal ganglia (L BG), and right middle temporal gyrus (R MTG). *Right*: Gray matter volume in the hippocampus was significantly correlated with improvements in arithmetic performance after 2 months of training; **(d)** White matter changes after training. Performance gains (i.e., efficiency) were associated with fractional anisotropy (FA) value changes in the left superior longitudinal fasciculus (SLF) linking frontal and temporal cortices (FT) in 7- to 9-year-old children; **(e)** Plasticity after training in different performance-level cohorts. *Left*: Normalization of aberrant functional brain responses in children with MLD after 2 months of cognitive training. *Top*: Before training, children with MLD showed greater functional brain activation levels compared with TD children in multiple regions of the prefrontal, parietal, and ventral temporal–occipital cortices. *Bottom*: After 2 months of training, functional brain responses in MLD children normalized to the levels of TD children. *AIC*, anterior insular cortex; *DLPFC*, dorsolateral prefrontal cortex; *FG*, fusiform gyrus; *IPS*, intraparietal sulcus; *VLPFC*, ventrolateral prefrontal cortex. *Right*: Brain plasticity index (BPI) calculated as a multivariate correlation between post- and prefunctional brain activation maps significantly correlated with performance gains (i.e., accuracy) in children with MLD; **(f)** Plasticity of emotional circuits after cognitive training. *Top left*: Mathematical training decreased scores on a math anxiety questionnaire (Scale for Early Math Anxiety—SEMA) in 7- to 9-year-old children with high levels of math anxiety (HMA). *Top right*: Mathematical

Continued

training induced functional plasticity during an arithmetic problem-solving task in children with HMA, compared to children with low levels of math anxiety (LMA). Coronal view: significant results were evident in emotional processing regions anchored in the basolateral amygdala. *Bottom:* Connectivity changes in emotion-related circuits anchored in the basolateral amygdala were significantly related to changes in SEMA scores in 7- to 9-year-olds after training. *(A,a) Modified from: Zamarian et al., 2009, Neuroscience and Biobehavioral Reviews © 2009 by Elsevier. (A,b) Modified from: Ischebeck et al., 2007, Neuroimage © 2007 by Elsevier. (A,c) Modified from: Delazer et al., 2003, Cognitive Brain Research © 2003 by Elsevier. (B,a) Adapted from: Jolles, Supekar, et al., (2016), Jolles, Wassermann, et al. (2016), Cortex © 2016 by Elsevier. (B,b) Modified from: Jolles, Supekar, et al., (2016), Cortex © 2016 by Elsevier. (B,c) Modified from: Supekar et al., 2013, PNAS © 2013 by The National Academy of Sciences of the USA. (B,d) Adapted from: Jolles, Wassermann, et al. (2016), Brain Structure and Function, © 2016 by Springer. (B,e) Modified from: Iuculano et al., 2015, Nature Communications © 2015 by NPG. (B,f) Modified from: Supekar et al., 2015, Journal of Neuroscience © 2015 by SfN.*

findings regarding connectivity features of the IPS in MLD (Rosenberg-Lee et al., 2014) (Fig. 15.6B,b) and points to cognitive training as a way to rectify functional circuit imbalances in children with MLD (see also Fig. 15.6B,c).

Neuroplasticity effects related to training-contingent learning might also differ based on an individual's math ability prior to training. In MLD, disrupted systems (Fig. 15.6B,b,c) will need to "normalize" before more efficient connectivity features can emerge, whereas in TD, increases in connectivity— accompanied by performance increases—(Fig. 15.7B,b) might reflect better and more efficient cross talk between critical brain systems.

Connectivity and structural features of other brain systems important for mathematical cognition also demonstrate plasticity as a function of training-contingent learning. In a seminal study, using a similar mathematical training paradigm to the one described above (Jolles, Supekar, et al., 2016), Supekar et al. (2013) showed that hippocampal volume prior to training was a strong predictor of performance gains in mathematical skills (Fig. 15.7B,c). Furthermore, intrinsic functional connectivity of the hippocampus with ventral and dorsal aspects of the PFC and the basal ganglia prior to training were the strongest predictors of math performance gains after training (Fig. 15.7B,c). These findings further highlight that the hippocampus plays a critical, yet underappreciated role in children's mathematical learning (Qin et al., 2014; Supekar et al., 2013). Moreover, the results of these studies suggest that quantitative measures of brain structure and organization can provide reliable and predictive markers of learning in children, and these can be more sensitive than behavioral measures (Jolles, Supekar, et al., 2016; Supekar et al., 2013).

There is also evidence of prominent changes in structural brain features as a function of the same 2-month mathematical training described above (Jolles, Wassermann, et al., 2016). Notably, individual differences in behavioral gains in mathematical skills were predicted by plasticity of the white matter tract of the left longitudinal fasciculus linking frontal and temporal cortices in 7- to 10-year-old children (Jolles, Wassermann, et al., 2016) (Fig. 15.7B,d).

Cognitive training studies not only provide insights into mechanisms of learning and brain plasticity but also, and more pragmatically, can aid in testing the efficacy of an intervention in children with MLD. Iuculano et al. (2015) used a comprehensive training protocol (Fuchs et al., 2008, 2009, 2010, 2013; Powell et al., 2009) (Fig. 15.7B,a) to test plasticity-related effects in a selected population of 7- to 9-year-old children with MLD. Results showed that in parallel with improved performance—to the level of TD controls—2 months of training resulted in dramatic and extensive functional brain changes in children with MLD and normalized functional responses to the level of TD controls (Fig. 15.7B,e). Brain plasticity effects in MLD occurred in multiple brain systems important for mathematical learning (Fig. 15.1). These include higher-level visual areas in the ventral temporal–occipital cortex that support visual form judgment and symbol recognition, posterior parietal areas that are involved in quantity representations, and prefrontal regions that support domain-general cognitive functions critical for successful learning, including attention, rule switching, and some aspects of working memory (Fig. 15.7B,e). Critically, the degree of plasticity in all these systems was significantly related to performance gains in MLD (Fig. 15.7B,e—*Right*). Consistent with previous studies in adults (Delazer et al., 2003, 2005; Ischebeck et al., 2006, 2007; Zamarian & Delazer, 2014), children with MLD showed significant reductions in widespread activation after training. This suggests that this type of training can release the burden across distributed brain systems by placing fewer demands and concurrently decreasing load on neurocognitive resources in these children. More generally, these findings suggest that a comprehensive training, one which integrates conceptual and retrieval aspects of mathematical learning, might be highly effective in eliciting significant neuroplasticity effects in MLD.

Learning does not occur in isolation; there is the growing need to assess the contribution of other factors that are often neglected in neurocognitive models but are essential in modulating relevant brain circuits and can significantly contribute to individual differences in learning and development. These include, for example, motivational, affective, and social factors. To this end, in a recent study, Supekar et al. applied the same cognitive training paradigm (Fig. 15.7B,a) (Supekar, Iuculano, Chen, & Menon, 2015) and assessed behavioral and brain changes in a group of 7- to 9-year-old children with high levels of math anxiety (Young, Wu, & Menon, 2012). Cognitive training significantly reduced high levels of math anxiety in this cohort (Fig. 15.7B,f—*Top left*). Moreover, this training remediated aberrant functional responses and connectivity in emotion-related circuits, anchored in the basolateral amygdala (Fig. 15.7B,f—*Top right*), suggesting that focused exposure to mathematical problems can reduce negative emotional response to mathematics and highlighting a key role of the amygdala in this process (Supekar et al., 2015). Critically, changes in functional circuits related to emotion processing, anchored in the basolateral amygdala, were related to reductions in anxiety with training (Fig. 15.7B,f—*Bottom*).

Taken together, cognitive learning approaches provide well-controlled settings for examining sources of individual differences in learning and brain plasticity. Moreover, building toward explanatory learning frameworks informed by cognitive and *systems neuroscience* of cognitive and affective dimensions of learning may enable us to create paradigms to reduce the burden of mathematical difficulties at different points of development and in heterogeneous groups with varying cognitive abilities.

CONCLUSIONS AND FUTURE DIRECTIONS

In this chapter, we have taken a *systems neuroscience* approach to describe the basic organization of brain processes involved in mathematical cognition and learning, and sources of heterogeneity that underlie different aspects of information processing in this domain. The perspective taken in this review is that the multilayer complexity of neurocognitive processes involved in mathematical cognition and learning is best viewed in the context of typical and atypical development. This is because of the highly dynamic role of working memory, attention, declarative and semantic memory, and cognitive control systems at different stages of proficiency and skill acquisition.

We have reviewed the key building blocks of mathematical cognition and described how they depend on core *hubs* anchored in the IPS and the FG. These regions play an essential role in the perceptual and semantic representation of quantity and help build a-modal semantic representations of numerosity by combining perceptual inputs with visuospatial *primitives* (e.g., detecting the property of "threeness" in a set of three "scanned" and "attended" items in the environment). At their core, these processes depend on the integrity of the IPS and its interactions with the FG.

Multiple PFC circuits associated with various subdivisions of the PPC, not just the IPS, facilitate access to multiple working memory circuits for additional processing and manipulation of discrete quantity. The circuit view that emerges is that perceptual and semantic representations of quantity in the ventral temporal–occipital cortex and PPC anchor mathematical cognition (e.g., the ability to understand the meaning of "three items" in a set and to make the association between such semantics and their arbitrary symbols—to know that "3" represents "three items"); while multiple prefrontal, parietal, and MTL functional brain circuits help to scaffold learning and increase capacity for problem-solving (e.g., 3 items plus 4 items results in a total of 7 items). The manner in which these circuits are engaged changes with brain maturation, cognitive development, levels of ability, and as a function of task.

There is now growing evidence to suggest that functional circuits engaged by children are not the same as those engaged by adults, who have evolved multiple strategies and schema-like knowledge for efficient learning. A number of scaffolding systems are involved during development to support the efficient acquisition of mathematical knowledge and learning. A particularly striking example is the demonstration of hippocampal-frontal and parietal-frontal circuits that are recruited in children but not in adults. The data reviewed here converge on the idea that the precise nature of this engagement is a function of developmental stage, domain knowledge, problem complexity, and individual proficiency in use of efficient problem-solving strategies.

We are still in the initial stages of understanding how functional brain circuits unfold with development and how they go awry in atypical development. It is, nevertheless, clear that the exclusive focus on activity levels in a small set of brain regions identified in highly skilled adults will likely miss important changes in network-level functional organization that accompanies learning and development. Increasingly, the focus has also shifted to multivariate analyses, as it is evident that similar levels of activation across task conditions do not necessarily imply similar kinds of information processing (Blair, Rosenberg-Lee, Tsang, Schwartz, & Menon, 2012; Prado et al., 2011; Raizada et al., 2010). These types of fine-grained analyses clearly have important implications for understanding brain mechanisms mediating the formation of unique stimulus representations and how they mature with learning and development (Ashkenazi, Rosenberg-Lee, Tenison, & Menon, 2012; Chang, Rosenberg-Lee, Metcalfe, Chen, & Menon, 2015).

Most previous normative adult and developmental studies of mathematical cognition have mainly focused on localization of activation and age-related changes, but it is becoming increasingly clear that cognition depends on interactions within and between large-scale brain systems (Bressler & Menon, 2010). New research is beginning to highlight the significant and specific changes in anterior–posterior functional connectivity that take place during time periods important for developing core competence in mathematics (Rosenberg-Lee, Barth, & Menon, 2011). A *systems neuroscience approach*, with its emphasis on networks and connectivity, rather than a pure localization approach, is better suited to further understand how even basic mathematical skills develop and are ultimately expressed in the adult brain. Moreover, analysis of network changes with learning can further clarify dynamic processing involved in mathematical learning and better address sources of individual differences in learning profiles (Bassett et al., 2011).

In sum, mathematical learning and skill acquisition require the coordination and integration of multiple cognitive processes, which rely on the engagement of short- and long-range connectivity between distributed brain systems that undergo significant changes—in terms of recruitment and organization—as a function of development and learning (Fair et al., 2008; Supekar & Menon, 2012; Supekar et al., 2010) (Box 15.2). In the many different ways we have tried to highlight in this chapter, mathematical knowledge serves as a model domain for investigating the ontogenesis of human cognitive and problem-solving skills, and explain why some individuals excel and others struggle.

BOX 15.2

FUNCTIONAL BRAIN SYSTEMS, THEIR ASSOCIATED BRAIN REGIONS, AND COGNITIVE FUNCTIONS INVOLVED IN MATHEMATICAL LEARNING AND COGNITION

Brain System	Brain Region	Cognitive Function
Ventral–dorsal visual streams	Posterior parietal cortex	*Building blocks of mathematical cognition*
	• Intraparietal sulcus	Quantity representation
	Ventral temporal–occipital cortex	
	• Fusiform gyrus	Symbols' recognition
Parietal-frontal	Posterior parietal cortex	
	• Frontal eye fields	*Working memory processing*
	• Intraparietal sulcus	Visual short-term memory and attention
	Lateral parietal cortex	
	• Supramarginal gyrus	Quantity manipulation in working memory
	• Dorsolateral prefrontal cortex	
	• Supplementary motor area	
	• Anterior insula	

Region	Structures	Functions
Lateral frontotemporal	Superior and middle temporal lobe • Superior temporal sulcus • Middle temporal gyrus	*Phonological awareness and processing* Auditory representation
	Lateral frontal lobe • Inferior frontal gyrus	Verbal information processing
Medial temporal	Hippocampus	*Arithmetic facts* Memory formation
Prefrontal	Lateral frontal lobe • Dorsolateral prefrontal cortex • Ventrolateral prefrontal cortex	*Mathematical problem-solving/reasoning* Cognitive control/executive functions
	Anterior frontal lobe • Anterior cingulate cortex • Anterior insula	Salience

References

Alllson, T., McCarthy, G., Nobre, A., Puce, A., & Belger, A. (1994). Human extrastriate visual cortex and the perception of faces, words, numbers, and colors. *Cerebral Cortex*, 4(5), 544–554. Retrieved from: http://www.ncbi.nlm.nih.gov/pubmed/7833655.

American Psychiatry Association. (2013). *Diagnostic and statistical manual of mental disorders* (5th ed.). Arlington, VA: American Psychiatry Publishing.

Andres, M., Michaux, N., & Pesenti, M. (2012). Common substrate for mental arithmetic and finger representation in the parietal cortex. *Neuroimage*, 62(3), 1520–1528. https://doi.org/10.1016/j.neuroimage.2012.05.047.

Andres, M., Pelgrims, B., Michaux, N., Olivier, E., & Pesenti, M. (2011). Role of distinct parietal areas in arithmetic: An fMRI-guided TMS study. *Neuroimage*, 54(4), 3048–3056. https://doi.org/10.1016/j.neuroimage.2010.11.009.

Ansari, D. (2008). Effects of development and enculturation on number representation in the brain. *Nature Reviews Neuroscience*, 9(4), 278–291. https://doi.org/10.1038/nrn2334.

Ansari, D., & Dhital, B. (2006). Age-related changes in the activation of the intraparietal sulcus during nonsymbolic magnitude processing: An event-related functional magnetic resonance imaging study. *Journal of Cognitive Neuroscience*, 18(11), 1820–1828. https://doi.org/10.1162/jocn.2006.18.11.1820.

Ansari, D., Garcia, N., Lucas, E., Hamon, K., & Dhital, B. (2005). Neural correlates of symbolic number processing in children and adults. *Neuroreport*, 16(16), 1769–1773. https://doi.org/10.1097/01.wnr.0000183905.23396.f1.

Arsalidou, M., & Taylor, M. J. (2011). Is 2+2=4? Meta-analyses of brain areas needed for numbers and calculations. *Neuroimage*, 54(3), 2382–2393. https://doi.org/10.1016/j.neuroimage.2010.10.009.

Ashcraft, M. H. (1982). The development of mental arithmetic: A chronometric approach. *Developmental Review*, 2(3), 213–236.

Ashkenazi, S., Black, J. M., Abrams, D. A., Hoeft, F., & Menon, V. (2013). Neurobiological underpinnings of math and reading learning disabilities. *Journal of Learning Disabilities*, 46(6), 549–569. https://doi.org/10.1177/0022219413483174.

Ashkenazi, S., Rosenberg-Lee, M., Metcalfe, A. W., Swigart, A. G., & Menon, V. (2013). Visuo-spatial working memory is an important source of domain-general vulnerability in the development of arithmetic cognition. *Neuropsychologia*, 51(11), 2305–2317. https://doi.org/10.1016/j.neuropsychologia.2013.06.031.

Ashkenazi, S., Rosenberg-Lee, M., Tenison, C., & Menon, V. (2012). Weak task-related modulation and stimulus representations during arithmetic problem solving in children with developmental dyscalculia. *Developmental Cognitive Neuroscience*, 2, S152–S166. https://doi.org/10.1016/j.dcn.2011.09.006.

Baddeley, A. D., & Hitch, G. J. (1974). Working memory. In G. A. Bower (Ed.), *Recent advances in learning and motivation*. New York: Academic Press.

Barrouillet, P., Fayol, M., & Lathuliere, E. (1997). Selecting between competitors in multiplication tasks: An explanation of the errors produced by adolescents with learning difficulties. *International Journal of Behavioral Development*, 21(2), 253–275.

Bassett, D. S., Wymbs, N. F., Porter, M. A., Mucha, P. J., Carlson, J. M., & Grafton, S. T. (2011). Dynamic reconfiguration of human brain networks during learning. *Proceedings of the National Academy of Sciences of the United States of America, 108*(18), 7641–7646. https://doi.org/10.1073/pnas.1018985108.

Berteletti, I., Prado, J., & Booth, J. R. (2014). Children with mathematical learning disability fail in recruiting verbal and numerical brain regions when solving simple multiplication problems. *Cortex, 57*, 143–155. https://doi.org/10.1016/j.cortex.2014.04.001.

Blair, K. P., Rosenberg-Lee, M., Tsang, J. M., Schwartz, D. L., & Menon, V. (2012). Beyond natural numbers: Negative number representation in parietal cortex. *Frontiers in Human Neuroscience, 6, 7.* https://doi.org/10.3389/fnhum.2012.00007.

Bressler, S. L., & Menon, V. (2010). Large-scale brain networks in cognition: Emerging methods and principles. *Trends in Cognitive Sciences, 14*(6), 277–290. pii: S1364-6613(10)00089-6 https://doi.org/10.1016/j.tics.2010.04.004.

Bunge, S. A., Dudukovic, N. M., Thomason, M. E., Vaidya, C. J., & Gabrieli, J. D. (2002). Immature frontal lobe contributions to cognitive control in children: Evidence from fMRI. *Neuron, 33*(2), 301–311.

Butterworth, B. (1999). *The mathematical brain*. London: Macmillan.

Butterworth, B. (2005). Developmental dyscalculia. In J. I. D. Campbell (Ed.), *Handbook of mathematical cognition*. Hove: Psychology Press.

Butterworth, B. (2010). Foundational numerical capacities and the origins of dyscalculia. *Trends in Cognitive Sciences, 14*(12), 534–541. https://doi.org/10.1016/j.tics.2010.09.007.

Butterworth, B., & Kovas, Y. (2013). Understanding neurocognitive developmental disorders can improve education for all. *Science, 340*, 300–305.

Butterworth, B., & Reigosa-Crespo, V. (2007). Information processing deficits in dyscalculia. In D. B. Berch, & M. M. M. Mazzocco (Eds.), *Why is math so hard for some children? The nature and origins of mathematical learning difficulties and disabilities*. Baltimore, MD, USA: Paul H Brookes Publishing Co.

Butterworth, B., Varma, S., & Laurillard, D. (2011). Dyscalculia: From brain to education. *Science, 332*(6033), 1049–1053. https://doi.org/10.1126/science.1201536.

Cai, W., Chen, T., Ryali, S., Kochalka, J., Li, C. S., & Menon, V. (2015). Causal interactions within a frontal-cingulate-parietal network during cognitive control: Convergent evidence from a multisite-multitask investigation. *Cerebral Cortex.* https://doi.org/10.1093/cercor/bhv046.

Cai, W., Ryali, S., Chen, T., Li, C. S., & Menon, V. (2014). Dissociable roles of right inferior frontal cortex and anterior insula in inhibitory control: Evidence from intrinsic and task-related functional parcellation, connectivity, and response profile analyses across multiple datasets. *Journal of Neuroscience, 34*(44), 14652–14667. https://doi.org/10.1523/JNEUROSCI.3048-14.2014.

Cantlon, J. F., Brannon, E. M., Carter, E. J., & Pelphrey, K. A. (2006). Functional imaging of numerical processing in adults and 4-y-old children. *PLoS Biology, 4*(5), e125. pii: 05-PLBI-RA-1376R3. https://doi.org/10.1371/journal.pbio.0040125.

Cantlon, J. F., Libertus, M. E., Pinel, P., Dehaene, S., Brannon, E. M., & Pelphrey, K. A. (2009). The neural development of an abstract concept of number. *Journal of Cognitive Neuroscience, 21*(11), 2217–2229. https://doi.org/10.1162/jocn.2008.21159.

Carey, S. (2004). *On the origin of concepts*. New York: Oxford University Press.

Chang, T. T., Metcalfe, A. W. S. Padmanabhan, A., Chen, T. W., & Menon, V. (2016). Heterogeneous and nonlinear development of human posterior parietal cortex function. *Neuroimage, 126*, 184–195. https://doi.org/10.1016/j.neuroimage.2015.11.053.

Chang, T. T., Rosenberg-Lee, M., Metcalfe, A. W., Chen, T., & Menon, V. (2015). Development of common neural representations for distinct numerical problems. *Neuropsychologia, 75*, 481–495. https://doi.org/10.1016/j.neuropsychologia.2015.07.005.

Cho, S., Metcalfe, A. W., Young, C. B., Ryali, S., Geary, D. C., & Menon, V. (2012). Hippocampal-prefrontal engagement and dynamic causal interactions in the maturation of children's fact retrieval. *Journal of Cognitive Neuroscience, 24*(9), 1849–1866. https://doi.org/10.1162/jocn_a_00246.

Cho, S., Ryali, S., Geary, D. C., & Menon, V. (2011). How does a child solve 7 + 8? Decoding brain activity patterns associated with counting and retrieval strategies. *Developmental Science, 14*(5), 989–1001. https://doi.org/10.1111/j.1467-7687.2011.01055.x.

Cohen Kadosh, R., Cohen Kadosh, K., Kaas, A., Henik, A., & Goebel, R. (2007). Notation-dependent and -independent representations of numbers in the parietal lobes. *Neuron, 53*(2), 307–314. https://doi.org/10.1016/j.neuron.2006.12.025.

Cole, M. W., Reynolds, J. R., Power, J. D., Repovs, G., Anticevic, A., & Braver, T. S. (2013). Multi-task connectivity reveals flexible hubs for adaptive task control. *Nature Neuroscience, 16*(9), 1348–1355. https://doi.org/10.1038/nn.3470.

Corbetta, M., Patel, G., & Shulman, G. L. (2008). The reorienting system of the human brain: From environment to theory of mind. *Neuron, 58*(3), 306–324. https://doi.org/10.1016/j.neuron.2008.04.017.

Corbetta, M., & Shulman, G. L. (2002). Control of goal-directed and stimulus-driven attention in the brain. *Nature Reviews Neuroscience, 3*(3), 201–215. https://doi.org/10.1038/nrn755.

Cowan, R., & Carney, D. P. J. (2006). Calendrical savants: Exceptionality and practice. *Cognition, 100*(2), B1–B9. https://doi.org/10.1016/j.cognition.2005.08.001.

Cowan, R., Donlan, C., Shepherd, D. L., Cole-Fletcher, R., Saxton, M., & Hurry, J. (2011). Basic calculation proficiency and mathematics achievement in elementary school children. *Journal of Educational Psychology, 103*(4), 786–803. https://doi.org/10.1037/a0024556.

Cowan, R., & Frith, C. (2009). Do calendrical savants use calculation to answer date questions? A functional magnetic resonance imaging study. *Philosophical Transactions of the Royal Society B-Biological Sciences, 364*(1522), 1417–1424. https://doi.org/10.1098/rstb.2008.0323.

Cowan, R., O'Connor, N., & Samella, K. (2003). The skills and methods of calendrical savants. *Intelligence, 31*(1), 51–65. pii: S0160-2896(02)00119-8.

Cutini, S., Scatturin, P., Basso Moro, S., & Zorzi, M. (2014). Are the neural correlates of subitizing and estimation dissociable? An fNIRS investigation. *Neuroimage, 85*(Pt 1), 391–399. https://doi.org/10.1016/j.neuroimage.2013.08.027.

Dagenbach, D., & McCloskey, M. (1992). The organization of arithmetic facts in memory: Evidence from a brain-damaged patient. *Brain and Cognition, 20*(2), 345–366.

Dantzig, T. (1930). *Number: The language of science: A critical survey written for the cultured non-mathematician.* New York, NY, USA: Macmillan.

Davachi, L. (2006). Item, context and relational episodic encoding in humans. *Current Opinion in Neurobiology, 16*(6), 693–700. https://doi.org/10.1016/j.conb.2006.10.012.

Davachi, L., Mitchell, J. P., & Wagner, A. D. (2003). Multiple routes to memory: Distinct medial temporal lobe processes build item and source memories. *Proceedings of the National Academy of Sciences of the United States of America, 100*(4), 2157–2162. https://doi.org/10.1073/pnas.0337195100.

Davis, N., Cannistraci, C. J., Rogers, B. P., Gatenby, J. C., Fuchs, L. S., Anderson, A. W., et al. (2009). Aberrant functional activation in school age children at-risk for mathematical disability: A functional imaging study of simple arithmetic skill. *Neuropsychologia, 47*(12), 2470–2479. https://doi.org/10.1016/j.neuropsychologia.2009.04.024.

De Smedt, B., Holloway, I. D., & Ansari, D. (2011). Effects of problem size and arithmetic operation on brain activation during calculation in children with varying levels of arithmetical fluency. *Neuroimage, 57*(3), 771–781.

De Smedt, B., Taylor, J., Archibald, L., & Ansari, D. (2010). How is phonological processing related to individual differences in children's arithmetic skills? *Developmental Science, 13*(3), 508–520. https://doi.org/10.1111/j.1467-7687.2009.00897.x.

Dehaene, S. (1997). *The number sense: How the mind creates mathematics.* New York: Oxford University Press.

Dehaene, S. (2003). The neural basis of the Weber-Fechner law: A logarithmic mental number line. *Trends in Cognitive Sciences, 7*(4), 145–147. Retrieved from: http://www.ncbi.nlm.nih.gov/pubmed/12691758.

Dehaene, S., & Cohen, L. (1995). Towards an anatomical and functional model of number processing. *Mathematical Cognition, 1*(1), 83–120.

Dehaene, S., Molko, N., Cohen, L., & Wilson, A. J. (2004). Arithmetic and the brain. *Current Opinion in Neurobiology, 14*(2), 218–224. https://doi.org/10.1016/j.conb.2004.03.008.

Dehaene, S., Piazza, M., Pinel, P., & Cohen, L. (2003). Three parietal circuits for number processing. *Cognitive Neuropsychology, 20*(3), 487–506.

Delazer, M., Domahs, F., Bartha, L., Brenneis, C., Lochy, A., Trieb, T., et al. (2003). Learning complex arithmetic–an fMRI study. *Brain Research Cognitive Brain Research, 18*(1), 76–88.

Delazer, M., Ischebeck, A., Domahs, F., Zamarian, L., Koppelstaetter, F., Siedentopf, C. M., et al. (2005). Learning by strategies and learning by drill–evidence from an fMRI study. *Neuroimage, 25*(3), 838–849.

DeStefano, D., & LeFevre, J. A. (2004). The role of working memory in mental arithmetic. *European Journal of Cognitive Psychology, 16*(3), 353–386.

Diana, R. A., Yonelinas, A. P., & Ranganath, C. (2007). Imaging recollection and familiarity in the medial temporal lobe: A three-component model. *Trends in Cognitive Sciences, 11*(9), 379–386.

Drenthen, G. S., Barendse, E. M., Aldenkamp, A. P., van Veenendaal, T. M., Puts, N. A., Edden, R. A., et al. (2016). Altered neurotransmitter metabolism in adolescents with high-functioning autism. *Psychiatry Research, 256*, 44–49.

Eichenbaum, H., Yonelinas, A. P., & Ranganath, C. (2007). The medial temporal lobe and recognition memory. *Annual Review of Neuroscience, 30*, 123 152.

Evans, T. M., Flowers, D. L., Napoliello, E. M., Olulade, O. A., & Eden, G. F. (2014). The functional anatomy of single-digit arithmetic in children with developmental dyslexia. *Neuroimage, 101,* 644–652. https://doi.org/10.1016/j.neuroimage.2014.07.028.

Evans, T. M., Kochalka, J., Ngoon, T. J., Wu, S. S., Qin, S. Z., Battista, C., et al. (2015). Brain structural integrity and intrinsic functional connectivity forecast 6 year longitudinal growth in children's numerical abilities. *Journal of Neuroscience, 35*(33), 11743–11750. https://doi.org/10.1523/Jneurosci.0216-15.2015.

Fair, D. A., Cohen, A. L., Dosenbach, N. U., Church, J. A., Miezin, F. M., Barch, D. M., et al. (2008). The maturing architecture of the brain's default network. *Proceedings of the National Academy of Sciences of the United States of America, 105*(10), 4028–4032. https://doi.org/10.1073/pnas.0800376105.

Fan, Q., Anderson, A. W., Davis, N., & Cutting, L. E. (2014). Structural connectivity patterns associated with the putative visual word form area and children's reading ability. *Brain Research, 1586,* 118–129. https://doi.org/10.1016/j.brainres.2014.08.050.

Fias, W., Menon, V., & Szucs (2013). Multiple components of developmental dyscalculia. *Trends in Educational Neuroscience, 2*(2), 43–47.

Fox, M. D., Corbetta, M., Snyder, A. Z., Vincent, J. L., & Raichle, M. E. (2006). Spontaneous neuronal activity distinguishes human dorsal and ventral attention systems. *Proceedings of the National Academy of Sciences of the United States of America, 103*(26), 10046–10051.

Fuchs, L. S., Fuchs, D., Stuebing, K., Fletcher, J. M., Hamlett, C. L., & Lambert, W. (2008). Problem solving and computational skill: Are they shared or distinct aspects of mathematical cognition? *Journal of Educational Psychology, 100*(1), 30–47. https://doi.org/10.1037/0022-0663.100.1.30.

Fuchs, L. S., Geary, D. C., Compton, D. L., Fuchs, D., Schatschneider, C., Hamlett, C. L., et al. (2013). Effects of first-grade number knowledge tutoring with contrasting forms of practice. *Journal of Educational Psychology, 105*(1), 58–77.

Fuchs, L. S., Powell, S. R., Hamlett, C. L., Fuchs, D., Cirino, P. T., & Fletcher, J. M. (2008). Remediating computational deficits at third grade: A randomized field trial. *Journal of Research on Educational Effectiveness, 1*(1), 2–32.

Fuchs, L. S., Powell, S. R., Seethaler, P. M., Cirino, P. T., Fletcher, J. M., Fuchs, D., et al. (2009). Remediating number combination and word problem deficits among students with mathematics difficulties: A randomized control trial. *Journal of Educational Psychology, 101*(3), 561–576. https://doi.org/10.1037/a0014701.

Fuchs, L. S., Powell, S. R., Seethaler, P. M., Fuchs, D., Hamlett, C. L., Cirino, P. T., et al. (2010). A framework for remediating number combination deficits. *Exceptional Children, 76*(2), 135–165.

Gauthier, I., Skudlarski, P., Gore, J. C., & Anderson, A. W. (2000). Expertise for cars and birds recruits brain areas involved in face recognition. *Nature Neuroscience, 3*(2), 191–197. https://doi.org/10.1038/72140.

Gauthier, I., Tarr, M. J., Anderson, A. W., Skudlarski, P., & Gore, J. C. (1999). Activation of the middle fusiform 'face area' increases with expertise in recognizing novel objects. *Nature Neuroscience, 2*(6), 568–573. https://doi.org/10.1038/9224.

Geary, D. C. (2011). Cognitive predictors of achievement growth in mathematics: A 5-year longitudinal study. *Developmental Psychology*, *47*(6), 1539–1552. https://doi.org/10.1037/a0025510.

Geary, D. C., & Brown, S. C. (1991). Cognitive addition: Strategy choice and speed-of-processing differences in gifted, normal, and mathematically disabled children. *Developmental Psychology*, *27*(3), 398–406.

Geary, D. C., Hamson, C. O., & Hoard, M. K. (2000). Numerical and arithmetical cognition: A longitudinal study of process and concept deficits in children with learning disability. *Journal of Experimental Child Psychology*, *77*(3), 236–263 Retrieved from View Full Text in PDF format (ECO).

Geary, D. C., & Hoard, M. K. (2003). Learning disabilities in basic mathematics: Deficits in memory and cognition. In J. M. Royer (Ed.), *Mathematical cognition*. Greenwich, CT: Information Age Publishing.

Geary, D. C., Hoard, M. K., & Bailey, D. H. (2012). Fact retrieval deficits in low achieving children and children with mathematical learning disability. *Journal of Learning Disabilities*, *45*(4), 291–307. https://doi.org/10.1177/0022219410392046.

Geary, D. C., Hoard, M. K., Byrd-Craven, J., Nugent, L., & Numtee, C. (2007). Cognitive mechanisms underlying achievement deficits in children with mathematical learning disability. *Child Development*, *78*(4), 1343–1359. https://doi.org/10.1111/j.1467-8624.2007.01069.x.

Gobel, S. M., Watson, S. E., Lervag, A., & Hulme, C. (2014). Children's arithmetic development it is number knowledge, not the approximate number sense, that counts. *Psychological Science*, *25*(3), 789–798. https://doi.org/10.1177/0956797613516471.

Goodale, M. A., & Milner, A. D. (1992). Separate visual pathways for perception and action. *Trends in Neurosciences*, *15*(1), 20–25.

Grabner, R. H., Ansari, D., Koschutnig, K., Reishofer, G., Ebner, F., & Neuper, C. (2009). To retrieve or to calculate? Left angular gyrus mediates the retrieval of arithmetic facts during problem solving. *Neuropsychologia*, *47*(2), 604–608.

Groen, G., & Parkman, J. (1972). Chronometric analysis of simple addition. *Psychological Review*, *329*.

Grotheer, M., Herrmann, K. H., & Kovacs, G. (2016). Neuroimaging evidence of a bilateral representation for visually presented numbers. *Journal of Neuroscience*, *36*(1), 88–97.

Gupta, S. C., Ravikrishnan, A., Liu, J., Mao, Z., Pavuluri, R., Hillman, B. G., et al. (2016). The NMDA receptor GluN2C subunit controls cortical excitatory-inhibitory balance, neuronal oscillations and cognitive function. *Scientific Reports*, *6*, 38321.

Ham, T., Leff, A., de Boissezon, X., Joffe, A., & Sharp, D. J. (2013). Cognitive control and the salience network: An investigation of error processing and effective connectivity. *Journal of Neuroscience*, *33*(16), 7091–7098. https://doi.org/10.1523/JNEUROSCI.4692-12.2013.

Heine, A., Tamm, S., Wissmann, J., & Jacobs, A. M. (2011). Electrophysiological correlates of non-symbolic numerical magnitude processing in children: Joining the dots. *Neuropsychologia*, *49*(12), 3238–3246. https://doi.org/10.1016/j.neuropsychologia.2011.07.028.

Hinton, E. C., Dymond, S., von Hecker, U., & Evans, C. J. (2010). Neural correlates of relational reasoning and the symbolic distance effect: Involvement of parietal cortex. *Neuroscience*, *168*(1), 138–148. https://doi.org/10.1016/j.neuroscience.2010.03.052.

Holloway, I. D., Battista, C., Vogel, S. E., & Ansari, D. (2012). Semantic and perceptual processing of number symbols: Evidence from a cross-linguistic fMRI adaptation study. *Journal of Cognitive Neuroscience*, 1–13.

Hyde, D. C., Boas, D. A., Blair, C., & Carey, S. (2010). Near-infrared spectroscopy shows right parietal specialization for number in pre-verbal infants. *Neuroimage, 53*(2), 647–652.

Hyde, D. C., & Spelke, E. S. (2009). All numbers are not equal: An electrophysiological investigation of small and large number representations. *Journal of Cognitive Neuroscience, 21*(6), 1039–1053. https://doi.org/10.1162/jocn.2009.21090.

Imbo, I., Vandierendonck, A., & Vergauwe, E. (2007). The role of working memory in carrying and borrowing. *Psychological Research, 71*(4), 467–483. https://doi.org/10.1007/s00426-006-0044-8.

Isaacs, E. B., Edmonds, C. J., Lucas, A., & Gadian, D. G. (2001). Calculation difficulties in children of very low birthweight: A neural correlate. *Brain, 124*(Pt 9), 1701–1707.

Ischebeck, A., Zamarian, L., Egger, K., Schocke, M., & Delazer, M. (2007). Imaging early practice effects in arithmetic. *Neuroimage, 36*(3), 993–1003.

Ischebeck, A., Zamarian, L., Siedentopf, C., Koppelstatter, F., Benke, T., Felber, S., et al. (2006). How specifically do we learn? Imaging the learning of multiplication and subtraction. *Neuroimage, 30*(4), 1365–1375. https://doi.org/10.1016/j.neuroimage.2005.11.016.

Iuculano, T., Rosenberg-Lee, M., Richardson, J., Tenison, C., Fuchs, L., Supekar, K., et al. (2015). Cognitive tutoring induces widespread neuroplasticity and remediates brain function in children with mathematical learning disabilities. *Nature Communications, 6*, 8453. https://doi.org/10.1038/ncomms9453.

Iuculano, T., Rosenberg-Lee, M., Supekar, K., Lynch, C. J., Khouzam, A., Phillips, J., et al. (2014). Brain organization underlying superior mathematical abilities in children with autism. *Biological Psychiatry, 75*(3), 223–230.

Iuculano, T., Tang, J., Hall, C. W., & Butterworth, B. (2008). Core information processing deficits in developmental dyscalculia and low numeracy. *Developmental Science, 11*(5), 669–680. https://doi.org/10.1111/j.1467-7687.2008.00716.x.

Jeong, S. K., & Xu, Y. (2016). Behaviorally relevant abstract object identity representation in the human parietal cortex. *Journal of Neuroscience, 36*(5), 1607–1619. https://doi.org/10.1523/JNEUROSCI.1016-15.2016.

Jolles, D., Ashkenazi, S., Kochalka, J., Evans, T., Richardson, J., Rosenberg-Lee, M., et al. (2016). Parietal hyper-connectivity, aberrant brain organization, and circuit-based biomarkers in children with mathematical disabilities. *Developmental Science, 19*(4), 613–631. https://doi.org/10.1111/desc.12399.

Jolles, D., Supekar, K., Richardson, J., Tenison, C., Ashkenazi, S., Rosenberg-Lee, M., et al. (2016). Reconfiguration of parietal circuits with cognitive tutoring in elementary school children. *Cortex, 83*, 231–245. https://doi.org/10.1016/j.cortex.2016.08.004.

Jolles, D., Wassermann, D., Chokhani, R., Richardson, J., Tenison, C., Bammer, R., et al. (2016). Plasticity of left perisylvian white-matter tracts is associated with individual differences in math learning. *Brain Structure and Function, 221*(3), 1337–1351. https://doi.org/10.1007/s00429-014-0975-6.

Jordan, N. C., Levine, S. C., & Huttenlocher, J. (1995). Calculation abilities in young children with different patterns of cognitive functioning. *Journal of Learning Disabilities, 28*(1), 53–64. https://doi.org/10.1177/002221949502800109.

Kadosh, R. C., Kadosh, K. C., Schuhmann, T., Kaas, A., Goebel, R., Henik, A., et al. (2007). Virtual dyscalculia induced by parietal-lobe TMS impairs automatic magnitude processing. *Current Biology, 17*(8), 689–693. https://doi.org/10.1016/j.cub.2007.02.056.

Kaufmann, L., Vogel, S. E., Starke, M., Kremser, C., & Schocke, M. (2009a). Numerical and non-numerical ordinality processing in children with and without developmental dyscalculia: Evidence from fMRI. *Cognitive Development, 24*(4), 486–494.

Kaufmann, L., Vogel, S. E., Starke, M., Kremser, C., Schocke, M., & Wood, G. (2009b). Developmental dyscalculia: Compensatory mechanisms in left intraparietal regions in response to nonsymbolic magnitudes. *Behavioral and Brain Functions, 5*(1), 35.

Kaufmann, L., Wood, G., Rubinsten, O., & Henik, A. (2011). Meta-analyses of developmental fMRI studies investigating typical and atypical trajectories of number processing and calculation. *Developmental Neuropsychology, 36*(6), 763–787.

Knops, A., Piazza, M., Sengupta, R., Eger, E., & Melcher, D. (2014). A shared, flexible neural map architecture reflects capacity limits in both visual short-term memory and enumeration. *Journal of Neuroscience, 34*(30), 9857–9866. https://doi.org/10.1523/JNEUROSCI.2758-13.2014.

Kucian, K., Grond, U., Rotzer, S., Henzi, B., Schonmann, C., Plangger, F., et al. (2011). Mental number line training in children with developmental dyscalculia. *Neuroimage, 57*(3), 782–795. https://doi.org/10.1016/j.neuroimage.2011.01.070.

Kucian, K., Loenneker, T., Dietrich, T., Dosch, M., Martin, E., & von Aster, M. (2006). Impaired neural networks for approximate calculation in dyscalculic children: A functional MRI study. *Behavioral and Brain Functions, 2*, 31. https://doi.org/10.1186/1744-9081-2-31.

Kucian, K., & von Aster, M. (2015). Developmental dyscalculia. *European Journal of Pediatrics, 174*(1), 1–13. https://doi.org/10.1007/s00431-014-2455-7.

Kucian, K., von Aster, M., Loenneker, T., Dietrich, T., & Martin, E. (2008). Development of neural networks for exact and approximate calculation: A fMRI study. *Developmental Neuropsychology, 33*(4), 447–473. https://doi.org/10.1080/87565640802101474.

Landerl, K., Bevan, A., & Butterworth, B. (2004). Developmental dyscalculia and basic numerical capacities: A study of 8-9-year-old students. *Cognition, 93*(2), 99–125. https://doi.org/10.1016/j.cognition.2003.11.004.

Lee, K. M. (2000). Cortical areas differentially involved in multiplication and subtraction: A functional magnetic resonance imaging study and correlation with a case of selective acalculia. *Annals of Neurology, 48*(4), 657–661.

Luck, S. J., & Vogel, E. K. (2013). Visual working memory capacity: From psychophysics and neurobiology to individual differences. *Trends in Cognitive Sciences, 17*(8), 391–400. https://doi.org/10.1016/j.tics.2013.06.006.

McCandliss, B. D., Cohen, L., & Dehaene, S. (2003). The visual word form area: Expertise for reading in the fusiform gyrus. *Trends in Cognitive Sciences, 7*(7), 293–299. https://doi.org/10.1016/S1364-6613(03)00134-7.

McClelland, J. L., McNaughton, B. L., & O'Reilly, R. C. (1995). Why there are complementary learning systems in the hippocampus and neocortex: Insights from the successes and failures of connectionist models of learning and memory. *Psychological Review, 102*(3), 419–457.

Menon, V. (2014). Arithmetic in child and adult brain. In R. Cohen Kadosh, & A. Dowker (Eds.), Handbook of mathematical cognition. Oxford: Oxford University Press.

Menon, V. (2016). Working memory in children's math learning and its disruption in dyscalculia. *Current Opinion in Behavioral Sciences, 10*, 125–132. https://doi.org/10.1016/j.cobeha.2016.05.014.

Menon, V., & Uddin, L. Q. (2010). Saliency, switching, attention and control: A network model of insula function. *Brain Structure and Function, 214*(5–6), 655–667.

Metcalfe, A. W., Ashkenazi, S., Rosenberg-Lee, M., & Menon, V. (2013). Fractionating the neural correlates of individual working memory components underlying arithmetic problem solving skills in children. *Developmental Cognitive Neuroscience, 6*, 162–175. https://doi.org/10.1016/j.dcn.2013.10.001.

Molko, N., Cachia, A., Riviere, D., Mangin, J. F., Bruandet, M., Le Bihan, D., et al. (2003). Functional and structural alterations of the intraparietal sulcus in a developmental dyscalculia of genetic origin. *Neuron, 40*(4), 847–858. https://doi.org/10.1016/S0896-6273(03)00670-6.

Molko, N., Cachia, A., Riviere, D., Mangin, J. F., Bruandet, M., LeBihan, D., et al. (2004). Brain anatomy in Turner syndrome: Evidence for impaired social and spatial-numerical networks. *Cerebral Cortex, 14*(8), 840–850. https://doi.org/10.1093/cercor/bhh042.

Naaijen, J., Bralten, J., Poelmans, G., Consortium, I., Glennon, J. C., Franke, B., et al. (2017). Glutamatergic and GABAergic gene sets in attention-deficit/hyperactivity disorder: Association to overlapping traits in ADHD and autism. *Translational Psychiatry, 7*(1), e999. https://doi.org/10.1038/tp.2016.273.

Naccache, L., & Dehaene, S. (2001). The priming method: Imaging unconscious repetition priming reveals an abstract representation of number in the parietal lobes. *Cerebral Cortex, 11*(10), 966–974.

Park, J., Hebrank, A., Polk, T. A., & Park, D. C. (2012). Neural dissociation of number from letter recognition and its relationship to parietal numerical processing. *Journal of Cognitive Neuroscience, 24*(1), 39–50. https://doi.org/10.1162/jocn_a_00085.

Pesenti, M., Seron, X., Samson, D., & Duroux, B. (1999). Basic and exceptional calculation abilities in a calculating prodigy: A case study. *Mathematical Cognition, 5*, 97–148.

Pesenti, M., Zago, L., Crivello, F., Mellet, E., Samson, D., Duroux, B., et al. (2001). Mental calculation in a prodigy is sustained by right prefrontal and medial temporal areas. *Nature Neuroscience, 4*(1), 103–107.

Piazza, M., Izard, V., Pinel, P., Le Bihan, D., & Dehaene, S. (2004). Tuning curves for approximate numerosity in the human intraparietal sulcus. *Neuron, 44*(3), 547–555. https://doi.org/10.1016/j.neuron.2004.10.014.

Piazza, M., Mechelli, A., Butterworth, B., & Price, C. J. (2002). Are subitizing and counting implemented as separate or functionally overlapping processes? *Neuroimage, 15*(2), 435–446. https://doi.org/10.1006/nimg.2001.0980.

Powell, S. R., Fuchs, L. S., Fuchs, D., Cirino, P. T., & Fletcher, J. M. (2009). Effects of fact retrieval tutoring on third-grade students with math difficulties with and without reading difficulties. *Learning Disabilities Research and Practice: A Publication of the Division for Learning Disabilities, Council of Exceptional Children, 24*(1), 1–11.

Prado, J., Mutreja, R., & Booth, J. R. (2014). Developmental dissociation in the neural responses to simple multiplication and subtraction problems. *Developmental Science, 17*(4), 537–552.

Prado, J., Mutreja, R., Zhang, H., Mehta, R., Desroches, A. S., Minas, J. E., et al. (2011). Distinct representations of subtraction and multiplication in the neural systems for numerosity and language. *Human Brain Mapping, 32*(11), 1932–1947.

Price, G. R., Holloway, I., Rasanen, P., Vesterinen, M., & Ansari, D. (2007). Impaired parietal magnitude processing in developmental dyscalculia. *Current Biology, 17*(24), R1042–R1043. https://doi.org/10.1016/j.cub.2007.10.013.

Pugh, K. R., Frost, S. J., Rothman, D. L., Hoeft, F., Del Tufo, S. N., Mason, G. F., et al. (2014). Glutamate and choline levels predict individual differences in reading ability in emergent readers. *Journal of Neuroscience, 34*(11), 4082–4089. https://doi.org/10.1523/JNEUROSCI.3907-13.2014.

Qin, S., Cho, S., Chen, T., Rosenberg-Lee, M., Geary, D. C., & Menon, V. (2014). Hippocampal-neocortical functional reorganization underlies children's cognitive development. *Nature Neuroscience, 17*(9), 1263–1269.

Raizada, R. D. S., Tsao, F. M., Liu, H. M., Holloway, I. D., Ansari, D., & Kuhl, P. K. (2010). Linking brain-wide multivoxel activation patterns to behaviour: Examples from language and math. *Neuroimage, 51*(1), 462–471. https://doi.org/10.1016/j.neuroimage.2010.01.080.

Rivera, S. M., Menon, V., White, C. D., Glaser, B., & Reiss, A. L. (2002). Functional brain activation during arithmetic processing in females with fragile X Syndrome is related to FMRI protein expression. *Human Brain Mapping, 16*(4), 206–218. https://doi.org/10.1002/hbm.10048.

Rivera, S. M., Reiss, A. L., Eckert, M. A., & Menon, V. (2005). Developmental changes in mental arithmetic: Evidence for increased functional specialization in the left inferior parietal cortex. *Cerebral Cortex, 15*(11), 1779–1790. https://doi.org/10.1093/cercor/bhi055.

Rosenberg-Lee, M., Ashkenazi, S., Chen, T., Young, C. B., Geary, D. C., & Menon, V. (2014). Brain hyper-connectivity and operation-specific deficits during arithmetic problem solving in children with developmental dyscalculia. *Developmental Science, 18*(3), 351–372.

Rosenberg-Lee, M., Barth, M., & Menon, V. (2011). What difference does a year of schooling make? Maturation of brain response and connectivity between 2nd and 3rd grades during arithmetic problem solving. *Neuroimage, 57*(3), 796–808. https://doi.org/10.1016/j.neuroimage.2011.05.013.

Rotzer, S., Kucian, K., Martin, E., von Aster, M., Klaver, P., & Loenneker, T. (2008). Optimized voxel-based morphometry in children with developmental dyscalculia. *Neuroimage, 39*(1), 417–422. https://doi.org/10.1016/j.neuroimage.2007.08.045.

Rousselle, L., & Noel, M. P. (2007). Basic numerical skills in children with mathematics learning disabilities: A comparison of symbolic vs non-symbolic number magnitude processing. *Cognition, 102*(3), 361–395. https://doi.org/10.1016/j.cognition.2006.01.005.

Rykhlevskaia, E., Uddin, L. Q., Kondos, L., & Menon, V. (2009). Neuroanatomical correlates of developmental dyscalculia: Combined evidence from morphometry and tractography. *Frontiers in Human Neuroscience, 3*, 51. https://doi.org/10.3389/neuro.09.051.2009.

Saygin, Z. M., Osher, D. E., Norton, E. S., Youssoufian, D. A., Beach, S. D., Feather, J., et al. (2016). Connectivity precedes function in the development of the visual word form area. *Nature Neuroscience, 19*(9), 1250–1255. https://doi.org/10.1038/nn.4354.

Schaffelhofer, S., & Scherberger, H. (2016). Object vision to hand action in macaque parietal, premotor, and motor cortices. *Elife, 5*. https://doi.org/10.7554/eLife.15278.

Seeley, W. W., Menon, V., Schatzberg, A. F., Keller, J., Glover, G. H., Kenna, H., et al. (2007). Dissociable intrinsic connectivity networks for salience processing and executive control. *Journal of Neuroscience, 27*(9), 2349–2356. https://doi.org/10.1523/JNEUROSCI.5587-06.2007.

Shalev, R., Auerbach, J., Manor, O., & Gross-Tsur, V. (2000). Developmental dyscalculia: Prevalence and prognosis. *European Child and Adolescent Psychiatry, 9*(0), S58–S64. https://doi.org/10.1007/s007870070009.

Shalev, R. S., Manor, O., & Gross-Tsur, V. (2005). Developmental dyscalculia: A prospective six-year follow-up. *Developmental Medicine and Child Neurology, 47*(2), 121–125. https://doi.org/10.1017/S0012162205000216.

Shum, J., Hermes, D., Foster, B. L., Dastjerdi, M., Rangarajan, V., Winawer, J., et al. (2013). A brain area for visual numerals. *Journal of Neuroscience, 33*(16), 6709–6715. https://doi.org/10.1523/JNEUROSCI.4558-12.2013.

Siegler, R. S. (1996). *Emerging minds: The process of change in children's thinking.* New York: Oxford University Press.

Siegler, R. S., & Shrager, J. (1984). Strategy choice in addition and subtraction: How do children know what to do? In C. Sophian (Ed.), *Origins of cognitive skills.* Hillsdale, NJ: Erlbaum.

Simon, O., Mangin, J. F., Cohen, L., Le Bihan, D., & Dehaene, S. (2002). Topographical layout of hand, eye, calculation, and language-related areas in the human parietal lobe. *Neuron, 33*(3), 475–487.

Squire, L. R. (1992). Memory and the hippocampus – a synthesis from findings with rats, monkeys, and humans. *Psychological Review, 99*(2), 195–231. https://doi.org/10.1037/0033-295x.99.2.195.

Squire, L. R., Genzel, L., Wixted, J. T., & Morris, R. G. (2015). Memory consolidation. In E. Kandel, Y. Dudai, & M. Mayford (Eds.), *Perspectives in biology: Learning and memory.* Cold Spring Harbor Laboratory Press.

Squire, L. R., Stark, C. E. L., & Clark, R. E. (2004). The medial temporal lobe. *Annual Review of Neuroscience, 27*, 279–306.

Sridharan, D., Levitin, D. J., & Menon, V. (2008). A critical role for the right fronto-insular cortex in switching between central-executive and default-mode networks. *Proceedings of the National Academy of Sciences of the United States of America, 105*(34), 12569–12574. https://doi.org/10.1073/pnas.0800005105.

Stocco, A., & Anderson, J. R. (2008). Endogenous control and task representation: An fMRI study in algebraic problem-solving. *Journal of Cognitive Neuroscience, 20*(7), 1300–1314. https://doi.org/10.1162/jocn.2008.20089.

Supekar, K., Iuculano, T., Chen, L., Menon, V., (2015). Remediation of childhood math anxiety and associated neural circuits through cognitive tutoring. *Journal of Neuroscience, 35*(36), 12574–12583.

Supekar, K., & Menon, V. (2012). Developmental maturation of dynamic causal control signals in higher-order cognition: A neurocognitive network model. *PLoS Computational Biology, 8*(2):e1002374.

Supekar, K., Swigart, A. G., Tenison, C., Jolles, D. D., Rosenberg-Lee, M., Fuchs, L., et al. (2013). Neural predictors of individual differences in response to math tutoring in primary-grade school children. *Proceedings of the National Academy of Sciences of the United States of America, 110*(20), 8230–8235. https://doi.org/10.1073/pnas.1222154110.

Supekar, K., Uddin, L. Q., Prater, K., Amin, H., Greicius, M. D., & Menon, V. (2010). Development of functional and structural connectivity within the default mode network in young children. *Neuroimage, 52*(1), 290–301. https://doi.org/10.1016/j.neuroimage.2010.04.009.

Szucs, D., & Goswami, U. (2013). Developmental dyscalculia: Fresh perspectives. *Trends in Neuroscience and Education, 2*(2), 33–37.

Tatti, R., Haley, M. S., Swanson, O. K., Tselha, T., & Maffei, A. (2016). Neurophysiology and regulation of the balance between excitation and inhibition in neocortical circuits. *Biological Psychiatry.* https://doi.org/10.1016/j.biopsych.2016.09.017.

Thompson, J. C., Abbott, D. F., Wheaton, K. J., Syngeniotis, A., & Puce, A. (2004). Digit representation is more than just hand waving. *Cognitive Brain Research, 21*(3), 412–417. https://doi.org/10.1016/j.cogbrainres.2004.07.001.

Todd, J. J., & Marois, R. (2004). Capacity limit of visual short-term memory in human posterior parietal cortex. *Nature, 428*(6984), 751–754. https://doi.org/10.1038/nature02466.

Tse, D., Langston, R. F., Kakeyama, M., Bethus, I., Spooner, P. A., Wood, E. R., et al. (2007). Schemas and memory consolidation. *Science, 316*(5821), 76–82. https://doi.org/10.1126/science.1135935.

Tulving, E. (1983). *Elements of episodic memory.* New York: Oxford University Press.

Uddin, L. Q., Supekar, K., Amin, H., Rykhlevskaia, E., Nguyen, D. A., Greicius, M. D., et al. (2010). Dissociable connectivity within human angular gyrus and intraparietal sulcus: Evidence from functional and structural connectivity. *Cerebral Cortex, 20*(11), 2636–2646. pii: bhq011. https://doi.org/10.1093/cercor/bhq011.

Vogel, S. E., Goffin, C., Bohnenberger, J., Koschutnig, K., Reishofer, G., Grabner, R. H., et al. (2017). The left intraparietal sulcus adapts to symbolic number in both the visual and auditory modalities: Evidence from fMRI. *Neuroimage, 153,* 16–27.

Vogel, S. E., Grabner, R. H., Schneider, M., Siegler, R. S., & Ansari, D. (2013). Overlapping and distinct brain regions involved in estimating the spatial position of numerical and non-numerical magnitudes: An fMRI study. *Neuropsychologia, 51*(5), 979–989.

Vogel, A. C., Miezin, F. M., Petersen, S. E., & Schlaggar, B. L. (2012). The putative visual word form area is functionally connected to the dorsal attention network. *Cerebral Cortex, 22*(3), 537–549. https://doi.org/10.1093/cercor/bhr100.

Wimmer, H., Ludersdorfer, P., Richlan, F., & Kronbichler, M. (2016). Visual experience shapes orthographic representations in the visual word form area. *Psychological Science, 27*(9), 1240–1248. https://doi.org/10.1177/0956797616657319.

Wu, S., Chang, T. T., Majid, A., Caspers, S., Eickhoff, S. B., & Menon, V. (2009). Functional heterogeneity of inferior parietal cortex during mathematical cognition assessed with cytoarchitectonic probability maps. *Cerebral Cortex, 19*(12), 2930–2945.

Young, C. B., Wu, S. S., & Menon, V. (2012). The neurodevelopmental basis of math anxiety. *Psychological Science, 23*(9), 492–501. https://doi.org/10.1177/095 6797611429134.

Zamarian, L., & Delazer, M. (2014). Arithmetic learning in adults – evidence from brain imaging. In R. Cohen Kadosh, & A. Dowker (Eds.), *The Oxford handbook of numerical cognition.* Oxford, UK: Oxford University Press.

Zamarian, L., Ischebeck, A., & Delazer, M. (2009). Neuroscience of learning arithmetic–evidence from brain imaging studies. *Neuroscience and Biobehavioral Reviews, 33*(6), 909–925. https://doi.org/10.1016/j.neubiorev.2009.03.005.

Zarnhofer, S., Braunstein, V., Ebner, F., Koschutnig, K., Neuper, C., Reishofer, G., et al. (2012). The Influence of verbalization on the pattern of cortical activation during mental arithmetic. *Behavioral and Brain Functions, 8*, 13. https://doi.org/10.1186/1744-9081-8-13.

(How) Are Executive Functions Actually Related to Arithmetic Abilities?

Kim Archambeau, Wim Gevers

Center for Research in Cognition and Neurosciences, ULB Neuroscience Institute, Université Libre de Bruxelles, Brussels, Belgium

OUTLINE

INTRODUCTION

Arithmetic abilities are required when solving problems such as "3×4" or "$24 + 33$" and are important for many everyday life situations: for instance, to make a recipe or to pay at the store. Over the past two decades, research identified many cognitive factors underlying arithmetic abilities (e.g., Cragg, Keeble, Richardson, Roome, & Gilmore, 2017; Halberda, Mazzocco, & Feigenson, 2008; Jordan, Glutting, & Ramineni, 2010). These factors can be classified as domain-specific (i.e., processes relevant in the numerical domain only) or domain-general (i.e., processes also relevant in other fields) processes. One fundamental domain-specific factor contributing to arithmetic abilities is the approximate number system (ANS). The ANS would represent a primitive system, shared by several animal species, which allows representing large numerosities in an approximate way. This system is often considered as a building block of our numerical development in general (Butterworth, 2005) and of more advanced arithmetic abilities. Some studies have shown that participants who are better to estimate or compare large numerosities are also better in arithmetic abilities (e.g., Halberda et al., 2008; Jordan et al., 2010; Libertus, Feigenson, & Halberda, 2011; Park & Brannon, 2013). More recently, the existence and/or the nature of the ANS (Gebuis, Cohen Kadosh, & Gevers, 2016; Leibovich, Katzin, Harel, & Henik, 2017) and its involvement in arithmetic abilities (Sasanguie, Göbel, Moll, Smets, & Reynvoet, 2013) became matter of intense debate.

In this section, several chapters focused on the domain-general factors that have an association with arithmetic abilities. In Chapter 15, Teresa Iuculano, Aarthi Padmanabhan, and Vinod Menon provided us with a state of the art on how neural networks and neural connectivity shape mathematical cognition throughout development. The importance of this chapter lies in demonstrating that domain-general neural systems involved in mathematical cognition change with learning, development, and training. As such, the acquisition of mathematical skill can be seen as a good example of how individual differences occur in human cognitive and problem solving skills. This is highly related to the topic described in Chapter 14 of Camilla Gilmore and Lucy Cragg. Based on the literature and listing some new empirical findings, they described how executive functions relate to mathematics achievements. Importantly, they approach both executive functions and mathematics performance not as a unitary construct but as a combination of different functions and processes. Finally, in Chapter 13, Liat Goldfarb dealt with the question of how executive functions (i.e., interference control or inhibition) are related to both numerical and arithmetic tasks. The main suggestion here is that the amount of interference varies with the level of mathematical proficiency in more complex ways than originally believed (i.e., inverse U-shaped rather than linear).

At this point, the aim is to complement these chapters with a critical review of the involvement of the different executive functions in arithmetic abilities. Executive functions have been proposed to play an important role in mathematics achievement (e.g., Bull, Espy, & Wiebe, 2008; Bull & Scerif, 2001; Cragg et al., 2017; Friso-van den Bos, van der Ven, Kroesbergen, & van Luit, 2013; Van der Ven, Kroesbergen, Boom, & Leseman, 2012). Executive functions represent a set of control processes that monitor and regulate our thoughts and actions. Three core executive functions are typically distinguished: cognitive flexibility, updating, and inhibition (Miyake & Friedman, 2012; Miyake et al., 2000). In the present closing chapter, we will therefore focus on the interaction between these domain-general factors and mathematics performance (here operational-ized as arithmetic achievements). Somewhat surprisingly, we realized that important pieces of information that would enable to directly relate these executive functions to mathematics performance are currently lacking. We outline these lacunas in the literature in the hope that we can provide interested researchers with concrete ideas on how to push forward this fascinating line of research.

THE INVOLVEMENT OF EXECUTIVE FUNCTIONS IN ARITHMETIC ABILITIES

Cognitive Flexibility

Cognitive flexibility (also referred to as "shifting") refers to our abil-ity to switch between different mental sets, tasks, or strategies (Diamond, 2013; Miyake & Friedman, 2012). In the laboratory, cognitive flexibility is typically investigated using task-switching paradigms (for a review, see Kiesel et al., 2010; Vandierendonck, Liefooghe, & Verbruggen, 2010). In this paradigm, participants are required to alternate between two or more tasks. Switching from one task to another task produces a certain cogni-tive cost. This cost is measured by the "switch cost" representing the dif-ference of performance (reaction times and/or error rate) between task switches and task repetitions (Jersild, 1927; Spector & Biederman, 1976; Vandierendonck et al., 2010). Two different types of switch costs can be identified: global and local switch costs. The global switch cost[1] refers to the difference in performance between pure blocks (i.e., block including the repetition of one single task; AAAA or BBBB) and mixed blocks (i.e., block including the alternation between two tasks; ABABAB). In contrast, local switch costs correspond to the specific difference between task-repetition trials and task-switch trials in mixed blocks. More specifically, local switch costs are measured by comparing the performance in AA and BB transitions (task-repetition trials) with the performance in BA and AB transitions (task-switch trials) in a mixed block such as AABBAABB (e.g., Kiesel et al., 2010; Kray & Lindenberger, 2000; Mayr, 2001; Vandierendonck et al., 2010). To measure cognitive flexibility, local switch costs are cur-rently preferred above global switch costs because the global switch cost is also influenced by a difference in working memory load between both blocks (Kiesel et al., 2010; Vandierendonck et al., 2010). Finally, an asym-metrical switch cost is typically observed in task-switching paradigms when the two tasks involve unequal levels of difficulty. That is, the switch

---

[1] The global switch cost is sometimes termed "mixing cost" or "general switch cost" in the literature.

cost is larger when switching from a difficult task to an easier task than the opposite, resulting in higher switch costs for the easy task (e.g., Monsell, Yeung, & Azuma, 2000; Wylie & Allport, 2000).

In the numerical domain, a lot of research investigated the relation between cognitive flexibility and mathematical performance in children (see chapter of Gilmore and Cragg). Here it is assumed that cognitive flexibility is needed in mathematical performance to support the switch between different operations like, for instance, the switch between addition and subtraction. It has also been assumed that flexibility is needed to switch between different strategies, for example, to switch between retrieval, decomposition, or transformation strategies in arithmetic problem solving (e.g., Bull & Lee, 2014; Bull & Scerif, 2001; Toll, Van der Ven, Kroesbergen, & Van Luit, 2011). For a more specific view on the role of flexibility in switching between strategies on consecutive trials, we refer the interested reader to Chapter 7.

We agree with this literature that solving a problem like "3+4−2" unequivocally implies a switch between arithmetic operations. However, the actual cognitive cost associated with this switch is unclear. Is the relation between the switch cost and the arithmetic operation the same depending on the type of transition made? For example, has the switch cost the same value when switching between addition and subtraction as when switching between addition and multiplication? Somewhat surprisingly, to the best of our knowledge, such information is currently lacking. Consequently, the question of exactly how flexibility relates to arithmetic performance remains largely unanswered.

Researchers with an interest in cognitive flexibility occasionally used arithmetic operations to examine features of task switching (e.g., Baddeley, Chincotta, & Adlam, 2001; Ellefson, Shapiro, & Chater, 2006; Jersild, 1927; Rubinstein, Meyer, & Evans, 2001). For example, Ellefson et al. (2006) used additions and subtractions to investigate the developmental changes of the asymmetrical switch cost. Given that solving additions is easier than solving subtractions, higher global and local switch costs were expected for additions compared with subtractions. Surprisingly, Ellefson et al. (2006) observed a different pattern of results in children as observed in young adults. As expected, children showed asymmetrical switch costs with larger switching costs for additions than for subtractions (i.e., the switch cost is more important when switching from subtractions to additions than the opposite). Young adults, on the other hand, exhibited global and local switch costs without any asymmetry. Apparently, this developmental difference was specific to arithmetic operations as it was not observed when the same participants switched between matching figures by color or shape. Here, both children and young adults showed the typical asymmetrical switch costs. To explain this pattern of results, Ellefson et al. (2006) suggested that the level of task familiarity changes throughout development for arithmetic operations, possibly influencing the switch cost (e.g., Meuter & Allport, 1999; Yeung & Monsell, 2003). Contrary to children, young adults have more experience and practice with additions and subtractions, making both these operations highly familiar, resulting in the absence of the asymmetrical switch cost (Ellefson et al., 2006).

Alternatively, researchers with an interest in numerical cognition did use the task-switching paradigm to examine the relation across arithmetic operations (e.g., in what way do different arithmetic operations interfere or facilitate each other; see next section) (e.g., Miller & Paredes, 1990; Zbrodoff & Logan, 1986). For instance, Miller and Paredes (1990) explored the interference between multiplications and additions via the task-switching paradigm. Participants solved arithmetic problems in pure blocks (containing only additions or only multiplications) and in mixed blocks (switching between additions and multiplications). A global switch cost was observed: additions and multiplications were solved faster in pure blocks than in mixed blocks. Another interesting pattern emerged. In pure blocks, additions were solved faster than multiplications. In mixed blocks, however, the reverse pattern was observed with faster multiplications than additions. A developmental explanation was provided. Developmentally, additions are learned earlier than multiplications. Because addition and multiplication networks are interrelated in memory, the earlier learned additions would need to be inhibited to prevent interference with the learning of the multiplications (e.g., inhibiting 5 as an answer when learning 2×3). This inhibition would persist into adulthood when both networks have to be activated for successful task performance such as mixed blocks (Miller & Paredes, 1990). Campbell and Arbuthnott (2010) more closely investigated the nature of the switch cost mixing additions and multiplications. Doing so, they replicated the results observed by Miller and Paredes (1990) mixing additions and multiplications and finding stronger global switch cost for additions than for multiplications. They argued that this finding is not because of the order of learning arithmetic operations but to the effect of asymmetrical switch costs observed in task switching. Given that additions are generally solved faster and with fewer errors than multiplications (e.g., Campbell & Arbuthnott, 2010; Campbell & Xue, 2001; Campbell, 1994), a higher switch cost for additions just reflects the more important cost for the easier task when switching involves tasks of different difficulties (Campbell & Arbuthnott, 2010).

Although a relation is often assumed between flexibility and arithmetic abilities, a review of the literature somewhat surprisingly demonstrated that this relation is not firmly empirically established. There is an important lack of studies directly addressing the question of the switch between arithmetic operations (but see Campbell & Arbuthnott, 2010), making it difficult to draw strong conclusions. Based on the aforementioned studies, the value of the switch cost between arithmetic operations seems to be influenced by the type of arithmetic operation (multiplication, addition, subtraction, division). However, to better understand the role of asymmetric switch costs, arithmetic tasks could be complemented with independent measures of the difficulty of each arithmetic operation separately. In addition, because the switch cost seems to be affected by task familiarity, different patterns of results can be obtained through development (e.g., Ellefson et al., 2006). Another outstanding issue is whether switch costs associated with arithmetic operations are completely confounded with switch costs between other types of information. Does a person presenting a large cost when switching between additions and subtractions also present a large cost when switching between other dimensions (e.g., color–shape). The observation that young adults demonstrated a different pattern of results for arithmetic as for "color–shape" switches (Ellefson et al., 2006) may be a first indication that switching between arithmetic processes is domain specific rather than domain general. If this would be the case, how would the local switch cost in arithmetic and nonarithmetic domains predict more general performances in mathematics? As outlined in the following, the question of domain specificity is also raised concerning the relation between arithmetic operations and the executive function inhibition (e.g., Gilmore and Cragg, this issue).

Inhibition

Inhibition can be conceived of as a combination of several different functions rather than a unitary construct (Dempster, 1993, pp. 3–27; Friedman & Miyake, 2004; Harnishfeger, 1995, pp. 175–204; Nigg, 2000). Furthermore, several taxonomies of inhibition exist with different conceptual distinctions (for further details, see Friedman & Miyake, 2004). For the current purposes, we define inhibition as our ability to suppress dominant responses, distractors, or irrelevant information that may disrupt our cognitive processing (e.g., Diamond, 2013; Nigg, 2000). Inhibition plays a role in the resolution or reduction of interference (e.g., Friedman & Miyake, 2004; Nigg, 2000). The term interference refers to the competition between multiple information and/or responses leading to a decrease in performance (Harnishfeger, 1995, pp. 175–204). It is to solve this interference that inhibition is assumed to be mainly involved in arithmetic abilities and especially in arithmetic facts (e.g., Bull & Lee, 2014; Cragg et al., 2017).

Arithmetic facts correspond to simple calculations for which the correct answer is retrieved from memory rather than calculated. Operands and answers of arithmetic facts are stored in interconnected associative networks in long-term memory (e.g., Ashcraft, 1992; Campbell, 1995; Verguts & Fias, 2005). In this memory representation, the presentation of a problem (e.g., 6×4) would activate the correct solution (24) together with other network-related solutions such as 28, causing interference during memory retrieval (e.g., Campbell & Graham, 1985). Strong empirical support exists for such concurrent coactivation of multiple answers creating interference. In production tasks, most of the errors produced by adults and children in multiplication solving are operand-related errors (e.g., $6 \times 7 = 48$) (e.g., Campbell & Graham, 1985; Censabella & Noël, 2004; Miller & Paredes, 1990; Siegler, 1988; Sokol, McCloskey, Cohen, & Aliminosa, 1991). Similar observations were made using multiplication verification tasks (e.g., Koshmider & Ashcraft, 1991; Niedeggen & Rösler, 1999; Stazyk, Ashcraft, & Hamann, 1982; Zamarian et al., 2007). In this type of task, participants have to decide whether a proposed solution of an arithmetic problem is correct or not. Besides the correct answer, two different types of negative trials (incorrect numerical solutions) are presented to the participants: operand-related (e.g., $6 \times 4 = 28$) and no operand-related solutions (e.g., $6 \times 4 = 26$). Participants are slower and make more errors when rejecting operand-related solutions rather than no operand-related solutions. This decrease in performance has been interpreted as support for the simultaneous activation of multiple arithmetic facts. The interference effect seems to emerge rapidly after children start to learn their multiplication tables and is also observed in children with mathematics learning disabilities (Rotem & Henik, 2015).

Similarly, an interference effect is also observed across different arithmetic operations and more specifically between multiplication and addition. The proportion of errors and reaction times increase when the result is incorrect but associated with the correct solution of the other operation (e.g., $3 + 4 = 12$) compared with incorrect results but unrelated to the other operation (e.g., $3 + 4 = 11$) (e.g., Megías & Macizo, 2015, 2016; Megías, Macizo, & Herrera, 2015; Winkelman & Schmidt, 1974; Zbrodoff & Logan, 1986). It should be noted that an interference effect is not observed between all arithmetic operations. Rather to the contrary, addition and subtraction as well as multiplication and division show facilitation effects (e.g., Campbell & Agnew, 2009; Campbell, Fuchs-Lacelle, & Phenix, 2006; De Brauwer & Fias, 2011; for further details about the facilitation effect, see the chapter of Goldfarb). Using division and multiplication, De Brauwer and Fias (2011) found that practicing one operation facilitated performance in the other operation (for similar results with addition and subtraction, see Campbell & Agnew, 2009). There is still an ongoing debate to determine whether this effect is attributed to a mediation strategy (e.g., solving a division by reference to the corresponding multiplication) or to a common representation in memory.

Recently, several studies provided empirical support for the existence of inhibitory processes in arithmetic facts solving (Campbell & Dowd, 2012; Campbell & Thompson, 2012; Megías & Macizo, 2015; Megías et al., 2015; but see Censabella & Noël, 2004 for an alternative explanation). Campbell and Thompson (2012) investigated retrieval-induced forgetting, a phenomenon often considered as a hallmark of inhibition (Campbell & Thompson, 2012; see also Anderson, 2003). Retrieval-induced forgetting occurs when the practice of one piece of information impairs the retrieval of other unpracticed but related information (Campbell & Thompson, 2012). In this study, participants performed a practice phase during which simple multiplication problems had to be solved (e.g., $2 \times 7 = ?$). During a subsequent test phase, participants solved two types of additions. Some additions were counterparts of the practiced multiplications (e.g., $2 + 7 = ?$), while the other additions were unpracticed multiplication counterparts (e.g., $3 + 5 = ?$). Retrieval-induced forgetting of addition facts was observed. Practicing multiplication impaired the performance of additions whose multiplication counterparts were practiced relative to additions whose multiplication counterparts were unpracticed. These results are explained assuming that during the practice phase, the activated addition solutions competing with the multiplication solutions are inhibited. Additional time is then needed to reactivate these addition solutions during the test phase (Campbell & Thompson, 2012; see also Campbell & Dowd, 2012; Campbell & Phenix, 2009). More direct evidence that concurrently activated arithmetic facts are inhibited during facts retrieval was provided by Megías et al. (2015). They investigated retrieval-induced forgetting on a trial-by-trial basis rather than after repeated practice blocks using the negative priming task. Negative priming is often used in studies on bilingualism to examine the inhibitory mechanisms involved in the selection of a language (e.g., Macizo, Bajo, & Cruz Martín, 2010). The idea of negative priming is that more time is needed to reactivate information that was inhibited on a previous trial. In the study of Megías et al. (2015), participants performed a verification task having to decide whether an addition was associated with its correct solution or not. The critical manipulation was in the succession of the previous (trial n-1) to the current (trial n) trial. On trial n-1, the presented solution of an addition could be correct or wrong. When wrong, the presented solution could be related to the multiplication operation (e.g., $3 + 2 = 6$) or be unrelated (e.g., $3 + 2 = 7$). On n-1 the expected interference effect across operations was observed. Reaction times were longer and more errors were made to the related compared with the unrelated solutions. Interestingly, also a negative priming effect was observed. Participants needed more time and more errors were made on trial n, if an addition was shown with a correct solution (e.g., $4 + 2 = 6$), but this solution was associated with the multiplication solution on trial n-1 (e.g., $3 + 2 = 6$). Megías et al. (2015) concluded that the multiplication solution presented on n-1 was inhibited to select the correct response on trial n. To overcome this inhibition on trial n, more time was needed to reactivate it (Megías et al., 2015; see also Megías & Macizo, 2016). A similar pattern of results was observed in children aged 10–11 years and 12–13 years (Megías & Macizo, 2015).

Direct retrieval from long-term memory is not the only way to solve arithmetic problems. For more complex (e.g., $36 + 49$) and/or less familiar arithmetic problems, a succession of procedural processes (i.e., holding and manipulation of information in mind) can be performed to solve it. Inhibition is also proposed to play a role in such procedural processes (e.g., Cragg et al., 2017; Gilmore, Keeble, Richardson, & Cragg, 2015). For instance, Abramovich and Goldfab (2015; see also the chapter of Goldfarb) investigated the interference from intermediate sums in arithmetic problems comprising three addends (e.g., $3 + 2 + 4$). In this experiment, additions with three addends were presented. Then, participants first had to decide if a proposed solution was the correct final solution to the addition (e.g., 10). In a final step, participants had to indicate whether the proposed solution was one of the three addends (e.g., 3, 2, or 4) in the addition problem. When presented with a wrong solution, participants were slower to respond when presented with the intermediate sum (e.g., 5) than when an unrelated control digit (e.g., 7) was presented. This result indicates that the intermediate sum is activated and interferes with the actual addends of the arithmetic problem. Moreover, this intermediate sum remains activated even after the final solution has been processed (Abramovich & Goldfarb, 2015). While this study indicates that an intermediate sum is activated, it remains unclear how the activation of this intermediate sum interferes with the final solution of the addition. Therefore, future studies are required to better understand the impact of this activation on complex problem solving. The role of inhibition in procedural processes was also addressed by Lemaire and Lecacheur (2011). These authors examined the role of executive functions in the strategies that children use to solve complex arithmetic problems. The results showed that children with better inhibitory processes were better to select the most efficient strategy to solve a given problem (Lemaire & Lecacheur, 2011).

In agreement with the involvement of inhibitory process in both procedural and factual skills, Cragg et al. (2017; see also the chapter of Gilmore and Cragg) found that inhibition was a significant predictor of both skills. In addition, this relation appeared to be stable from 8 years through to adulthood. However, this contradicts earlier results reported by Gilmore et al. (2015). In this study, no relation was found between inhibitory and factual skills in both children and young adults. Furthermore, inhibition was a significant predictor of procedural skills in children but not in adults (Gilmore et al., 2015). In addition, other studies relating inhibition functions to mathematics achievement (including arithmetic abilities) reported such inconsistent findings (e.g., Lee et al., 2012; Monette, Bigras, & Guay, 2011; Rose, Feldman, & Jankowski, 2011; Van der Ven et al., 2012). In these cases, however, methodological differences can largely explain the inconsistent findings (e.g., Gilmore et al., 2015). As outlined earlier, inhibition is considered as a combination of several different functions rather than a unitary construct (Dempster, 1993, pp. 3–27; Friedman & Miyake, 2004; Harnichfoger, 1995, pp. 175–204; Nigg, 2000). Therefore, inhibition tasks can tap into different aspects of the inhibitory processes (Friedman & Miyake, 2004) having different involvements in arithmetic abilities (for a similar proposal, see the chapter of Gilmore and Cragg). Importantly, however, methodological differences cannot explain the inconsistencies observed between the study of Cragg et al. (2017) and the study of Gilmore et al. (2015). Both these studies used exactly the same tasks to measure inhibition and to measure factual and procedural skills. How can these conflicting results then be explained?

First, some inconsistencies could be explained by a lack of power. Fewer participants were included in the study of Gilmore et al. (2015). Combined with a possible weak relation between inhibition and arithmetic abilities, this could lead to an absence of significant results in this study. Second, a change in the measure of central tendency (mean vs. median) for procedural skills could also explain some observed differences in the relation between procedural skills and inhibition. Finally, to avoid ceiling effects, children performed both one- and two-digit calculations in both additions and subtractions. Similarly, young adults performed all different operations, including divisions. Although this is a perfect legitimate reason, it may also have caused participants to use a mixture of retrieval, procedural, and conceptual skills. However, this is not an optimal choice if one wants to look at specific relations between inhibition and retrieval and between inhibition and procedural/conceptual skills. A possible solution could be to use only single-digit numbers in addition and multiplication (in both younger children and young adults) together with adapted response deadlines to avoid possible ceiling effects and/or a measure of reaction times across the different groups.

Updating

The function of updating is defined as our ability to monitor and replace working memory content by newer more relevant information (Miyake et al., 2000; see also Morris & Jones, 1990). The function of updating is closely related to certain measures of working memory (e.g., complex span tasks) (e.g., Schmiedek, Hildebrandt, Lövdén, Wilhelm, & Lindenberger, 2009). In the domain of numerical cognition, the terms working memory and updating are not consistently used to cover the same meaning (for a similar proposal, see Friso-van den Bos et al., 2013). Some studies considered working memory and updating to be fully identical (e.g., Cragg & Gilmore, 2014; Cragg et al., 2017). Other studies referred to working memory and updating as two different concepts but continued to use complex span tasks as a measure of updating (e.g., Bull & Lee, 2014; see also Friso-van den Bos et al., 2013). These conceptual inconsistencies make it difficult to get a clear understanding of the specific contribution of updating functions in arithmetic abilities. We decided to also report on studies investigating working memory (i.e., storage and processing of information, also referring to central executive functions), even though it is clear that the results of these studies might involve more than the specific function of updating only (i.e., selective replacement of information). The involvement of working memory in arithmetic abilities has been investigated using dual-task approaches in adults and using correlational studies examining the relation between working memory span and/or updating tasks and mathematics achievements in children (LeFevre, DeStefano, Coleman, & Shanahan, 2005, pp. 361–378; for a review, see Raghubar, Barnes, & Hecht, 2010).

The contribution of working memory in simple and complex arithmetic problems has been addressed by studies using a dual-task approach (e.g., De Rammelaere, Stuyven, & Vandierendonck, 1999, 2001; FÜrst & Hitch, 2000; Hecht, 2002; Imbo & Vandierendonck, 2007b; Imbo, Vandierendonck, & De Rammelaere, 2007; Imbo, Vandierendonck, & Vergauwe, 2007; Lemaire, Abdi, & Fayol, 1996; Seitz & Schumann-Hengsteler, 2000, 2002; but for a review see DeStefano & LeFevre, 2004). Most of these studies took the multicomponent model of working memory of Baddeley (Baddeley, 1986, 1996; Baddeley & Hitch, 1974; see also the chapter of Camos) as research framework. This approach allows to isolate the roles of the different working memory components (i.e., phonological loop, visuospatial sketchpad, and central executive) in a task of interest (e.g., solving arithmetic problem). More specifically, it involves the performance of a primary task of interest while simultaneously performing a secondary task. The nature of the secondary task is chosen to represent one of the different components of Baddeley's working memory model. For instance, random letter generation is used to tax the central executive component. Performance when the primary and the secondary task are combined is compared with performance when the primary task is completed alone. A decrease in performance in combined tasks condition is then interpreted to support the involvement of the working memory component in the primary task processing (LeFevre et al., 2005, pp. 361–378).

Using this approach, the central executive component has been shown to play a role in simple arithmetic problems. A load on the central executive component disrupts the performance on simple problem solving across all arithmetic operations (e.g., Ashcraft, Donley, Halas, & Vakali, 1992, pp. 301–329; De Rammelaere et al., 1999, 2001; Imbo & Vandierendonck, 2007a, 2007b; Lemaire et al., 1996). In addition, Imbo and Vandierendonck (2007a, 2007b) examined the involvement of the central executive component across different strategies (i.e., retrieval vs. procedural) used to solve simple arithmetic problems. The authors observed that both retrieval and procedural strategies were affected by a central executive load. For the direct retrieval strategy, it was proposed that the central executive component was involved because it is needed to activate arithmetic facts from long-term memory (Imbo & Vandierendonck, 2007a, 2007b; see also, e.g., Lemaire et al., 1996; Seitz & Schumann-Hengsteler, 2000, 2002 and the chapter of Gilmore and Cragg). For the procedural strategy, it was proposed that the central executive component was involved to maintain intermediate results and to execute carry operations (Imbo & Vandierendonck, 2007a, 2007b). How the central executive is involved in calculations requiring several steps of processing has been more closely investigated with complex arithmetic problem solving tasks. Here, it was repeatedly observed that the central executive component plays an important role in the carrying operation with complex addition and complex multiplication (e.g., Fürst & Hitch, 2000; Imbo, Vandierendonck, & De Rammelaere, 2007; Imbo, Vandierendonck, & Vergauwe, 2007). More specifically, complex arithmetic problems with carry (e.g., 154 + 328) are more affected under working memory load than complex arithmetic problems with no carry (e.g., 154 + 325). Moreover, this effect grew even larger when more carry operations (e.g., complex arithmetic problems with two carries, such as "154 + 368") had to be performed (e.g., Imbo, Vandierendonck, & De Rammelaere, 2007; Imbo, Vandierendonck, & Vergauwe, 2007). The same pattern of results was found with borrow operations (e.g., "63 – 17") in complex subtractions (Imbo, Vandierendonck, & Vergauwe, 2007). Furthermore, using arithmetic problems comprising three or four addends, Imbo, Vandierendonck, and De Rammelaere (2007) also observed that the decline of performance under central executive load is more important with higher values of the carry (e.g., complex arithmetic problems with a carry of "value 2," such as "156 + 328 + 227"). Therefore, it seems that the recruitment of central executive component increases as a function of the number of carry or borrow operations and of the value that has to be carried. All these studies provide with convincing evidence that carry and borrow manipulations are mediated by the central executive.

Other support for the involvement of working memory in arithmetic abilities comes from a growing number of studies examining the relation between working memory and mathematics achievement in children (e.g., Agostino, Johnson, & Pascual-Leone, 2010; Andersson, 2010; Bull et al., 2008; Bull & Scerif, 2001; Gathercole, Alloway, Willis, & Adams, 2006; Jenks, van Lieshout, & de Moor, 2012; Monette et al., 2011; Rasmussen & Bisanz, 2005; van der Sluis, de Jong, & van der Leij, 2007; Van der Ven et al., 2012; for a review, see Bull & Lee, 2014). Such studies can differ according to (among others) children's age, population (typical or atypical), the type of mathematics measures, and the type of working memory task used. On the one hand, studies that have taken the multicomponent model of Baddeley as research framework essentially used working memory span tasks. On the other hand, studies that have taken executive functions as research framework use working memory span tasks, updating tasks (e.g., n-back task), or both. Regardless of these differences, working memory is frequently found to be strongly related to mathematical skills (Bull & Lee, 2014). In a recent metaanalysis, Friso-van den Bos et al. (2013) examined the strength of the associations between mathematics achievement with the following components: phonological loop, visuospatial sketchpad, cognitive flexibility, inhibition, visuospatial, and verbal working memory[2] (measured by both working memory span tasks and updating tasks). The results confirmed a positive and significant relation between mathematics performance and each component under investigation. Moreover, verbal working memory showed the strongest correlation with mathematics achievement. Furthermore, the second aim of this metaanalysis was to determine which factors (e.g., tasks, sample characteristics) lead to variation in this association. The type of working memory tasks (working memory span vs. updating task) was included among these factors. Interestingly, the results indicated that the type of task moderated the relation of mathematics performance with both visuospatial and verbal working memory tasks. In other words, correlations differed depending on whether the task required the replacement of information (updating function) or the storage and manipulation of information (working memory span). Interestingly, this finding suggests that even if the concepts of working memory and updating are closely related, they could play different roles in mathematics performance (for a similar proposal, see Friso-van den Bos et al., 2013).

Working memory and updating functions seem closely related but not identical. A variety of different tasks (working memory and updating tasks) have been used to measure the involvement of updating in arithmetic abilities. In addition, while updating tasks rely more on the updating process than on working memory functions, they also require and measure working memory abilities (Ecker, Lewandowsky, & Oberauer, 2014). This makes it very difficult to isolate the specific contribution of updating in arithmetic abilities. Recently, Ecker et al. (2014) provided with a possible solution to disentangle working memory from updating processes. They provided with a method to both conceptually and theoretically disentangle updating from maintenance and processing in working memory. The results demonstrated that the measure of updating did not covary with working memory capacity (Ecker et al., 2014). Applied to arithmetic abilities, this method could be the first to provide us with a clear view on the relation between specific updating processes and arithmetic abilities. Future studies could investigate whether this more specific measures of updating obtained with this method is predictive for mathematics

---

[2] Visuospatial and verbal working memory are referred to as visuospatial and verbal updating in the metaanalysis.

achievements. At the same time, one could also investigate what is most predictive for mathematics achievements: this specific measure of updating or maintenance/processing components of working memory. Finally, because children with mathematics disabilities are often associated with lower working memory capacity (e.g., Swanson & Jerman, 2006; for a review see Raghubar et al., 2010), it could also be interesting to examine if the more specific updating measure is impaired in such children.

CONCLUSION

The aim of this closing chapter was to discuss the chapters in this section and to review the literature on the relation between executive functions (i.e., cognitive flexibility, inhibition, and updating) and arithmetic abilities. The question we tried to answer is *how* executive functions relate to arithmetic achievements. Working through the literature, it became apparent that the relation between executive functions and arithmetic achievements is often assumed rather than empirically demonstrated. As outlined across the three chapters of this section, a possible reason for this is that both arithmetic achievements and executive functions are complex multidimensional concepts influenced by several factors, including training, learning, expertise, and development.

The review of the literature taught us that the relation between cognitive flexibility and arithmetic achievements is by far the least investigated. The cost associated with the switch between arithmetic operations and how it is related to mathematics performance remains largely unanswered. Good empirical support does exist for the involvement of inhibitory processes in arithmetic facts solving. It is well documented that inhibition serves to resolve interference to select the correct answer. However, less is known about this relation with procedural skills. Finally, the updating function is frequently observed to be strongly related to mathematical skills. At the same time, the function of updating is often confounded with the working memory construct. Therefore, new empirical work based on proposed methods is needed to understand the exact role of this function in arithmetic abilities.

References

Abramovich, Y., & Goldfarb, L. (2015). Activation of the intermediate sum in intentional and automatic calculations. *Frontiers in Psychology*, 6, 1512. https://doi.org/10.3389/fpsyg.2015.01512.

Agostino, A., Johnson, J., & Pascual-Leone, J. (2010). Executive functions underlying multiplicative reasoning: Problem type matters. *Journal of Experimental Child Psychology*, 105(4), 286–305. https://doi.org/10.1016/j.jecp.2009.09.006.

Anderson, M. (2003). Rethinking interference theory: Executive control and the mechanisms of forgetting. *Journal of Memory and Language, 49*(4), 415–445. https://doi.org/10.1016/j.jml.2003.08.006.

Andersson, U. (2010). Skill development in different components of arithmetic and basic cognitive functions: Findings from a 3-year longitudinal study of children with different types of learning difficulties. *Journal of Educational Psychology, 102*(1), 115–134. https://doi.org/10.1037/a0016838.

Ashcraft, M. H. (1992). Cognitive arithmetic: A review of data and theory. *Cognition, 44*(1–2), 75–106. https://doi.org/10.1016/0010-0277(92)90051-I.

Ashcraft, M. H., Donley, R. D., Halas, M. A., & Vakali, M. (1992). *Chapter 8 working memory, automaticity, and problem difficulty. https://doi.org/10.1016/S0166-4115(08) 60890-0.*

Baddeley, A. D. (1986). *Working memory.* Oxford: Clarendon Press.

Baddeley, A. (1996). Exploring the central executive. *The Quarterly Journal of Experimental Psychology Section A, 49*(1), 5–28. https://doi.org/10.1080/713755608.

Baddeley, A., Chincotta, D., & Adlam, A. (2001). Working memory and the control of action: Evidence from task switching. *Journal of Experimental Psychology: General, 130*(4), 641–657. https://doi.org/10.1037/0096-3445.130.4.641.

Baddeley, A. D., & Hitch, G. (1974). Working memory. *Psychology of Learning and Motivation, 8*, 47–89. https://doi.org/10.1016/S0079-7421(08)60452-1.

Bull, R., Espy, K. A., & Wiebe, S. A. (2008). Short-term memory, working memory, and executive functioning in preschoolers: Longitudinal predictors of mathematical achievement at age 7 years. *Developmental Neuropsychology, 33*(3), 205–228. https://doi.org/10.1080/87565640801982312.

Bull, R., & Lee, K. (2014). Executive functioning and mathematics achievement. *Child Development Perspectives, 8*(1), 36–41. https://doi.org/10.1111/cdep.12059.

Bull, R., & Scerif, G. (2001). Executive functioning as a predictor of children's mathematics ability: Inhibition, switching, and working memory. *Developmental Neuropsychology, 19*(3), 273–293. https://doi.org/10.1207/S15326942DN1903_3.

Butterworth, B. (2005). The development of arithmetical abilities. *Journal of Child Psychology and Psychiatry, 46*(1), 3–18. https://doi.org/10.1111/j.1469-7610. 2004.00374.x.

Campbell, J. I. D. (1994). Architectures for numerical cognition. *Cognition, 53*(1), 1-44.

Campbell, J. I. D. (1995). Mechanisms of simple addition and multiplication: A modified network-interference theory and simulation. *Mathematical Cognition, 1*(2), 121–164.

Campbell, J. I. D., & Agnew, H. (2009). Retrieval savings with nonidentical elements: The case of simple addition and subtraction. *Psychonomic Bulletin and Review, 16*(5), 938–944. https://doi.org/10.3758/PBR.16.5.938.

Campbell, J. I. D., & Arbuthnott, K. D. (2010). Effects of mixing and cueing simple addition and multiplication. *European Journal of Cognitive Psychology, 22*(3), 422–442. https://doi.org/10.1080/09541440902903629.

Campbell, J. I. D., & Dowd, R. R. (2012). Interoperation transfer in Chinese–English bilinguals' arithmetic. *Psychonomic Bulletin and Review, 19*(5), 948–954. https://doi.org/10.3758/s13423-012-0277-z.

Campbell, J. I. D., Fuchs-Lacelle, S., & Phenix, T. L. (2006). Identical elements model of arithmetic memory: Extension to addition and subtraction. *Memory and Cognition, 34*(3), 633–647. https://doi.org/10.3758/BF03193585.

Campbell, J. I. D., & Graham, D. J. (1985). Mental multiplication skill: Structure, process, and acquisition. *Canadian Journal of Psychology/Revue Canadienne de Psychologie, 39*(2), 338–366. https://doi.org/10.1037/h0080065.

Campbell, J. I. D., & Phenix, T. L. (2009). Target strength and retrieval-induced forgetting in semantic recall. *Memory and Cognition, 37*(1), 65–72. https://doi.org/10.3758/MC.37.1.65.

Campbell, J. I. D., & Thompson, V. A. (2012). Retrieval-induced forgetting of arithmetic facts. *Journal of Experimental Psychology: Learning, Memory, and Cognition, 38*(1), 118–129. https://doi.org/10.1037/a0025056.

Campbell, J. I. D., & Xue, Q. (2001). Cognitive arithmetic across cultures. *Journal of Experimental Psychology: General, 130*(2), 299–315. https://doi.org/10.1037/0096-3445.130.2.299.

Censabella, S., & Noël, M.-P. (2004). Interference in arithmetic facts: Are active suppression processes involved when performing simple mental arithmetic? *Cahiers de Psychologie Cognitive, 22*(6), 635–671.

Cragg, L., & Gilmore, C. (2014). Skills underlying mathematics: The role of executive function in the development of mathematics proficiency. *Trends in Neuroscience and Education, 3*(2), 63–68. https://doi.org/10.1016/J.TINE.2013.12.001.

Cragg, L., Keeble, S., Richardson, S., Roome, H. E., & Gilmore, C. (2017). Direct and indirect influences of executive functions on mathematics achievement. *Cognition, 162*, 12–26. https://doi.org/10.1016/j.cognition.2017.01.014.

De Brauwer, J., & Fias, W. (2011). The representation of multiplication and division facts in memory. *Experimental Psychology, 58*(4), 312–323. https://doi.org/10.1027/1618-3169/a000098.

De Rammelaere, S., Stuyven, E., & Vandierendonck, A. (1999). The contribution of working memory resources in the verification of simple mental arithmetic sums. *Psychological Research, 62*(1), 72–77. https://doi.org/10.1007/s004260050041.

De Rammelaere, S., Stuyven, E., & Vandierendonck, A. (2001). Verifying simple arithmetic sums and products: Are the phonological loop and the central executive involved? *Memory and Cognition, 29*(2), 267–273. https://doi.org/10.3758/BF03194920.

Dempster, F. N. (1993). Resistance to Interference: Developmental changes in a basic processing mechanism. In *Emerging themes in cognitive development*. New York, NY: Springer New York. https://doi.org/10.1007/978-1-4613-9220-0_1.

DeStefano, D., & LeFevre, J. (2004). The role of working memory in mental arithmetic. *European Journal of Cognitive Psychology, 16*(3), 353–386. https://doi.org/10.1080/09541440244000328.

Diamond, A. (2013). Executive functions. *Annual Review of Psychology, 64*(1), 135–168. https://doi.org/10.1146/annurev-psych-113011-143750.

Ecker, U. K. H., Lewandowsky, S., & Oberauer, K. (2014). Removal of information from working memory: A specific updating process. *Journal of Memory and Language, 74*, 77–90. https://doi.org/10.1016/J.JML.2013.09.003.

Ellefson, M. R., Shapiro, L. R., & Chater, N. (2006). Asymmetrical switch costs in children. *Cognitive Development, 21*(2), 108–130. https://doi.org/10.1016/j.cogdev.2006.01.002.

Friedman, N. P., & Miyake, A. (2004). The relations among inhibition and interference control functions: A latent-variable analysis. *Journal of Experimental Psychology: General, 133*(1), 101–135. https://doi.org/10.1037/0096-3445.133.1.101.

Friso-van den Bos, I., van der Ven, S. H. G., Kroesbergen, E. H., & van Luit, J. E. H. (2013). Working memory and mathematics in primary school children: A meta-analysis. *Educational Research Review*, 10, 29–44. https://doi.org/10.1016/J.EDUREV.2013.05.003.

FÜrst, A. J., & Hitch, G. J. (2000). Separate roles for executive and phonological components of working memory in mental arithmetic. *Memory and Cognition*, 28(5), 774–782. https://doi.org/10.3758/BF03198412.

Gathercole, S. E., Alloway, T. P., Willis, C., & Adams, A.-M. (2006). Working memory in children with reading disabilities. *Journal of Experimental Child Psychology*, 93(3), 265–281. https://doi.org/10.1016/j.jecp.2005.08.003.

Gebuis, T., Cohen Kadosh, R., & Gevers, W. (2016). Sensory-integration system rather than approximate number system underlies numerosity processing: A critical review. *Acta Psychologica*, 171, 17–35. https://doi.org/10.1016/j.actpsy.2016.09.003.

Gilmore, C., Keeble, S., Richardson, S., & Cragg, L. (2015). The role of cognitive inhibition in different components of arithmetic. *ZDM*, 47(5), 771–782. https://doi.org/10.1007/s11858-014-0659-y.

Halberda, J., Mazzocco, M. M. M., & Feigenson, L. (2008). Individual differences in non-verbal number acuity correlate with maths achievement. *Nature*, 455(7213), 665–668. https://doi.org/10.1038/nature07246.

Harnishfeger, K. K. (1995). The development of cognitive inhibition: Theories, definitions, and research evidence. In F. N. Dempster, & C. J. Brainerd (Eds.), *Interference and inhibition in cognition*. San Diego: Academic Press.

Hecht, S. A. (2002). Counting on working memory in simple arithmetic when counting is used for problem solving. *Memory and Cognition*, 30(3), 447–455. https://doi.org/10.3758/BF03194945.

Imbo, I., & Vandierendonck, A. (2007a). Do multiplication and division strategies rely on executive and phonological working memory resources? *Memory and Cognition*, 35(7), 1759–1771. https://doi.org/10.3758/BF03193508.

Imbo, I., & Vandierendonck, A. (2007b). The role of phonological and executive working memory resources in simple arithmetic strategies. *European Journal of Cognitive Psychology*, 19(6), 910–933. https://doi.org/10.1080/09541440601051571.

Imbo, I., Vandierendonck, A., & De Rammelaere, S. (2007). The role of working memory in the carry operation of mental arithmetic: Number and value of the carry. *The Quarterly Journal of Experimental Psychology*, 60(5), 708–731. https://doi.org/10.1080/17470210600762447.

Imbo, I., Vandierendonck, A., & Vergauwe, E. (2007). The role of working memory in carrying and borrowing. *Psychological Research*, 71(4), 467–483. https://doi.org/10.1007/s00426-006-0044-8.

Jenks, K. M., van Lieshout, E. C. D.M., & de Moor, J. M. H. (2012). Cognitive correlates of mathematical achievement in children with cerebral palsy and typically developing children. *British Journal of Educational Psychology*, 82(1), 120–135. https://doi.org/10.1111/j.2044-8279.2011.02034.x.

Jersild, A. T. (1927). Mental set and shift. *Archives of Psychology*, 14(89), 81.

Jordan, N. C., Glutting, J., & Ramineni, C. (2010). The importance of number sense to mathematics achievement in first and third grades. *Learning and Individual Differences*, 20(2), 82–88. https://doi.org/10.1016/j.lindif.2009.07.004.

Kiesel, A., Steinhauser, M., Wendt, M., Falkenstein, M., Jost, K., Philipp, A. M., et al. (2010). Control and interference in task switching—a review. *Psychological Bulletin, 136*(5), 849–874. https://doi.org/10.1037/a0019842.

Koshmider, J. W., & Ashcraft, M. H. (1991). The development of children's mental multiplication skills. *Journal of Experimental Child Psychology, 51*(1), 53–89. https://doi.org/10.1016/0022-0965(91)90077-6.

Kray, J., & Lindenberger, U. (2000). Adult age differences in task switching. *Psychology and Aging, 15*(1), 126–147. https://doi.org/10.1037/0882-7974.15.1.126.

Lee, K., Ng, S. F., Pe, M. L., Ang, S. Y., Hasshim, M. N. A.M., & Bull, R. (2012). The cognitive underpinnings of emerging mathematical skills: Executive functioning, patterns, numeracy, and arithmetic. *British Journal of Educational Psychology, 82*(1), 82–99. https://doi.org/10.1111/j.2044-8279.2010.02016.x.

LeFevre, J., DeStefano, D., Coleman, B., & Shanahan, T. (2005). Mathematical cognition and working memory. In J. I. D. Campbell (Ed.), *Handbook of mathematical cognition*. New York: Psychology Press.

Leibovich, T., Katzin, N., Harel, M., & Henik, A. (2017). From "sense of number" to "sense of magnitude" – the role of continuous magnitudes in numerical cognition. *Behavioral and Brain Sciences, 40*, 1–62. https://doi.org/10.1017/S0140525X16000960.

Lemaire, P., Abdi, H., & Fayol, M. (1996). The role of working memory resources in simple cognitive arithmetic. *European Journal of Cognitive Psychology, 8*(1), 73–104. https://doi.org/10.1080/095414496383211.

Lemaire, P., & Lecacheur, M. (2011). Age-related changes in children's executive functions and strategy selection: A study in computational estimation. *Cognitive Development*. https://doi.org/10.1016/j.cogdev.2011.01.002.

Libertus, M. E., Feigenson, L., & Halberda, J. (2011). Preschool acuity of the approximate number system correlates with school math ability. *Developmental Science, 14*(6), 1292–1300. https://doi.org/10.1111/j.1467-7687.2011.01080.x.

Macizo, P., Bajo, T., & Cruz Martín, M. (2010). Inhibitory processes in bilingual language comprehension: Evidence from Spanish–English interlexical homographs. *Journal of Memory and Language, 63*(2), 232–244. https://doi.org/10.1016/j.jml.2010.04.002.

Mayr, U. (2001). Age differences in the selection of mental sets: The role of inhibition, stimulus ambiguity, and response-set overlap. *Psychology and Aging, 16*(1), 96–109. https://doi.org/10.1037/0882-7974.16.1.96.

Megías, P., & Macizo, P. (2015). Simple arithmetic development in school age: The coactivation and selection of arithmetic facts. *Journal of Experimental Child Psychology, 138*, 88–105. https://doi.org/10.1016/j.jecp.2015.04.010.

Megías, P., & Macizo, P. (2016). Activation and selection of arithmetic facts: The role of numerical format. *Memory and Cognition, 44*(2), 350–364. https://doi.org/10.3758/s13421-015-0559-6.

Megías, P., Macizo, P., & Herrera, A. (2015). Simple arithmetic: Evidence of an inhibitory mechanism to select arithmetic facts. *Psychological Research, 79*(5), 773–784. https://doi.org/10.1007/s00426-014-0603-3.

Meuter, R. F. I., & Allport, A. (1999). Bilingual language switching in naming: Asymmetrical costs of language selection. *Journal of Memory and Language, 40*(1), 25–40. https://doi.org/10.1006/jmla.1998.2602.

Miller, K. F., & Paredes, D. R. (1990). Starting to add worse: Effects of learning to multiply on children's addition. *Cognition, 37*(3), 213–242.

Miyake, A., & Friedman, N. P. (2012). The nature and organization of individual differences in executive functions. *Current Directions in Psychological Science, 21*(1), 8–14. https://doi.org/10.1177/0963721411429458.

Miyake, A., Friedman, N. P., Emerson, M. J., Witzki, A. H., Howerter, A., & Wager, T. D. (2000). The unity and diversity of executive functions and their contributions to complex "Frontal Lobe" tasks: A latent variable analysis. *Cognitive Psychology, 41*(1), 49–100. https://doi.org/10.1006/cogp.1999.0734.

Monette, S., Bigras, M., & Guay, M.-C. (2011). The role of the executive functions in school achievement at the end of Grade 1. *Journal of Experimental Child Psychology, 109*(2), 158–173. https://doi.org/10.1016/j.jecp.2011.01.008.

Monsell, S., Yeung, N., & Azuma, R. (2000). Reconfiguration of task-set: Is it easier to switch to the weaker task? *Psychological Research, 63*(3–4), 250–264. https://doi.org/10.1007/s004269900005.

Morris, N., & Jones, D. M. (1990). Memory updating in working memory: The role of the central executive. *British Journal of Psychology, 81*(2), 111–121. https://doi.org/10.1111/j.2044-8295.1990.tb02349.x.

Niedeggen, M., & Rösler, F. (1999). N400 effects reflect activation spread during retrieval of arithmetic facts. *Psychological Science, 10*(3), 271–276. https://doi.org/10.1111/1467-9280.00149.

Nigg, J. T. (2000). On inhibition/disinhibition in developmental psychopathology: Views from cognitive and personality psychology and a working inhibition taxonomy. *Psychological Bulletin, 126*(2), 220–246. https://doi.org/10.1037/0033-2909.126.2.220.

Park, J., & Brannon, E. M. (2013). Training the approximate number system improves math proficiency. *Psychological Science, 24*(10), 2013–2019. https://doi.org/10.1177/0956797613482944.

Raghubar, K. P., Barnes, M. A., & Hecht, S. A. (2010). Working memory and mathematics: A review of developmental, individual difference, and cognitive approaches. *Learning and Individual Differences, 20*(2), 110–122. https://doi.org/10.1016/j.lindif.2009.10.005.

Rasmussen, C., & Bisanz, J. (2005). Representation and working memory in early arithmetic. *Journal of Experimental Child Psychology, 91*(2), 137–157. https://doi.org/10.1016/j.jecp.2005.01.004.

Rose, S. A., Feldman, J. F., & Jankowski, J. J. (2011). Modeling a cascade of effects: The role of speed and executive functioning in preterm/full-term differences in academic achievement. *Developmental Science, 14*(5), 1161–1175. https://doi.org/10.1111/j.1467-7687.2011.01068.x.

Rotem, A., & Henik, A. (2015). Development of product relatedness and distance effects in typical achievers and in children with mathematics learning disability. *Journal of Learning Disabilities, 48*, 577–592. https://doi.org/10.1177/0022219413520182.

Rubinstein, J. S., Meyer, D. E., & Evans, J. E. (2001). Executive control of cognitive processes in task switching. *Journal of Experimental Psychology: Human Perception and Performance, 27*(4), 763–797. https://doi.org/10.1037/0096-1523.27.4.763.

Sasanguie, D., Göbel, S. M., Moll, K., Smets, K., & Reynvoet, B. (2013). Approximate number sense, symbolic number processing, or number–space mappings: What underlies mathematics achievement? *Journal of Experimental Child Psychology*, *114*(3), 418–431. https://doi.org/10.1016/j.jecp.2012.10.012.

Schmiedek, F., Hildebrandt, A., Lövdén, M., Wilhelm, O., & Lindenberger, U. (2009). Complex span versus updating tasks of working memory: The gap is not that deep. *Journal of Experimental Psychology: Learning, Memory, and Cognition*, *35*(4), 1089–1096. https://doi.org/10.1037/a0015730.

Seitz, K., & Schumann-Hengsteler, R. (2000). Mental multiplication and working memory. *European Journal of Cognitive Psychology*, *12*(4), 552–570. https://doi.org/10.1080/095414400750050231.

Seitz, K., & Schumann-Hengsteler, R. (2002). Phonological loop and central executive processes in mental addition and multiplication. *Psychologische Beiträge*, *44*(2), 275–302.

Siegler, R. S. (1988). Strategy choice procedures and the development of multiplication skill. *Journal of Experimental Psychology: General*, *117*(3), 258–275. https://doi.org/10.1037/0096-3445.117.3.258.

Sokol, S. M., McCloskey, M., Cohen, N. J., & Aliminosa, D. (1991). Cognitive representations and processes in arithmetic: Inferences from the performance of brain-damaged subjects. *Journal of Experimental Psychology: Learning, Memory, and Cognition*, *17*(3), 355–376. https://doi.org/10.1037/0278-7393.17.3.355.

Spector, A., & Biederman, I. (1976). Mental set and mental shift revisited. *The American Journal of Psychology*, *89*(4), 669. https://doi.org/10.2307/1421465.

Stazyk, E. H., Ashcraft, M. H., & Hamann, M. S. (1982). A network approach to mental multiplication. *Journal of Experimental Psychology: Learning, Memory, and Cognition*, *8*(4), 320–335. https://doi.org/10.1037/0278-7393.8.4.320.

Swanson, H. L., & Jerman, O. (2006). Math disabilities: A selective meta-analysis of the literature. *Review of Educational Research*, *76*(2), 249–274. https://doi.org/10.3102/00346543076002249.

Toll, S. W. M., Van der Ven, S. H. G., Kroesbergen, E. H., & Van Luit, J. E. H. (2011). Executive functions as predictors of math learning disabilities. *Journal of Learning Disabilities*, *44*(6), 521–532. https://doi.org/10.1177/0022219410387302.

van der Sluis, S., de Jong, P. F., & van der Leij, A. (2007). Executive functioning in children, and its relations with reasoning, reading, and arithmetic. *Intelligence*, *35*(5), 427–449. https://doi.org/10.1016/j.intell.2006.09.001.

Van der Ven, S. H. G., Kroesbergen, E. H., Boom, J., & Leseman, P. P. M. (2012). The development of executive functions and early mathematics: A dynamic relationship. *British Journal of Educational Psychology*, *82*(1), 100–119. https://doi.org/10.1111/j.2044-8279.2011.02035.x.

Vandierendonck, A., Liefooghe, B., & Verbruggen, F. (2010). Task switching: Interplay of reconfiguration and interference control. *Psychological Bulletin*, *136*(4), 601–626. https://doi.org/10.1037/a0019791.

Verguts, T., & Fias, W. (2005). Interacting neighbors: A connectionist model of retrieval in single-digit multiplication. *Memory and Cognition*, *33*(1), 1–16. https://doi.org/10.3758/BF03195293.

Winkelman, J. H., & Schmidt, J. (1974). Associative confusions in mental arithmetic. *Journal of Experimental Psychology*, *102*(4), 734–736. https://doi.org/10.1037/h0036103.

Wylie, G., & Allport, A. (2000). Task switching and the measurement of "switch costs". *Psychological Research, 63*(3–4), 212–233. https://doi.org/10.1007/s004269900003.

Yeung, N., & Monsell, S. (2003). Switching between tasks of unequal familiarity: The role of stimulus-attribute and response-set selection. *Journal of Experimental Psychology: Human Perception and Performance, 29*(2), 455–469. https://doi.org/10.1037/0096-1523.29.2.455.

Zamarian, L., Stadelmann, E., Nürk, H.-C., Gamboz, N., Marksteiner, J., & Delazer, M. (2007). Effects of age and mild cognitive impairment on direct and indirect access to arithmetic knowledge. *Neuropsychologia, 45*(7), 1511–1521. https://doi.org/10.1016/j.neuropsychologia.2006.11.012.

Zbrodoff, N. J., & Logan, G. D. (1986). On the autonomy of mental processes: A case study of arithmetic. *Journal of Experimental Psychology: General, 115*(2), 118–130. https://doi.org/10.1037/0096-3445.115.2.118.

MEMORY

Numerical Cognition and Memory(ies)

Pierre Barrouillet

University of Geneva, Geneva, Switzerland

Numerical cognition, the ensemble of mental states and mechanisms by which numbers are used to represent, organize, understand, predict, and transform our environment and ourselves, encompasses such a variety of processes that it involves almost all the structures and functions of the mind. Thus, it is not only a key domain of study for cognitive psychology and neurosciences to understand how animal and human organisms adapt themselves to their environment but it also constitutes a privileged access for understanding mind structure and functioning. Ranging from mechanisms often described as automatic (digit identification, subitizing, activation of arithmetic facts) to the most controlled activities (word or algebra problem solving, multidigit operations, manipulation of fractions), numerical cognition involves different mnemonic systems. The maintenance and transformation of information required by complex activities such as word problem solving (Thevenot & Barrouillet, 2015, pp. 158–179) or fraction arithmetic (Lortie-Forgues, Tian, & Siegler, 2015) obviously solicit short-term or working memory (WM). However, numerical cognition also involves the acquisition and use of a variety of skills and strategies, stored in procedural long-term memory (LTM), to manipulate and process knowledge about numbers stored in declarative LTM. Key questions about the role of mnemonic systems in numerical cognition concern the mechanisms by which this knowledge might be acquired if it is not innate, the way numbers and numerical facts are organized in and retrieved from LTM, and the impact on number processing of constraints on retrieval from LTM and limitations in WM functioning.

This chapter addresses the question of how WM and LTM interact in numerical cognition. Among the possible ways of interaction, I focus in the following pages on the question of a possible and often evoked trade-off between working memory–demanding activities and direct retrieval from LTM (e.g., Logan, 1988). Are some of the complex, highly controlled and resource-demanding numerical activities progressively replaced, through practice, by the fast and automatic retrieval of answers from LTM? Such a trade-off, if it exists, presupposes some functional and structural relationships between the two mnemonic systems. In other words, thinking about the interactions between WM and LTM in numerical cognition requires to hypothesize a cognitive architecture in which the connections between the two systems are made explicit. A first section in this chapter describes some examples of such a cognitive architecture. A second section reviews a series of studies that aimed at investigating the impact that WM limitations might have on various numerical activities. As we will see, what these studies have in common is the idea that the constraints on numerical cognition resulting from WM limitations can be alleviated by substituting fast and automated direct retrievals from LTM to working memory–based and consequently slow and resource-demanding controlled processes. In a third section, I review empirical evidence suggesting that retrieval from LTM might not be the panacea for alleviating the intrinsic limitations from which the human cognitive system suffers when processing numbers. Finally, recent studies are presented that suggest that direct retrieval from LTM might neither be the inescapable end point of the repeated use of procedural strategies in arithmetic problem solving nor necessarily the fastest cognitive process that humans have at their disposal when dealing with numbers.

MNEMONIC SYSTEMS IN COGNITIVE ARCHITECTURES FOR NUMERICAL COGNITION: SOME EXAMPLES

One of the first attempts to delineate a cognitive architecture for numerical cognition was proposed by Siegler and Shrager (1984, pp. 229–293) in their distributions of associations model. This model was designed to account for addition and subtraction solving in children, with the aim to show how adaptive strategy choices could result from simple cognitive processes without any recourse to something like an executive processor. The two main parts of the model are a representation of knowledge about particular problems and a repertoire of strategies that operate on these representations to produce answers, these answers reshaping, in turn, the representations from which they originate. The representation of knowledge consists of associations between problems (e.g., $5 + 3$) and possible answers, either correct or incorrect. For example, $5 + 3$ would not only be associated with 8 but also with 5, 6, 7, and 9. Each problem could be characterized by the way these associations with possible answers are distributed. Problems for which a single answer concentrates most of the associative strength, ordinarily the correct answer, are assumed to present a peaked distribution, whereas problems for which the associative strength is distributed among several answers present a flat distribution in which no answer clearly emerges.

The strategies that operate on these representations are assumed to follow three distinct phases. The first would correspond to an attempt to directly retrieve an answer. The subject would set a confidence criterion, which determines the minimal strength that the association between the problem and a given answer must reach for retrieving this answer, as well as a search length criterion, which determines the number of retrieval attempts before moving to another strategy. If retrieval fails, either because no association reached the confidence criterion or because the successive attempts did not return any answer within the limit of the search length, children would try to elaborate a representation of the problem, for example, by putting up fingers, from which the answer could be read out. If this strategy fails, children would eventually rely on an algorithmic strategy such as counting the fingers that have been put up to represent each operand. Importantly, the distributions of associations model is also a learning model, which learns by doing. Whatever the strategy used to produce an answer, its production modifies the representation of knowledge by strengthening its association with the problem, in such a way that problems frequently solved in a correct way inescapably lead to a peaked distribution and, at the end, to their solving through retrieval of the strengthened answer.

This model incorporates several characteristics that are relevant for our purpose. The associations that constitute the knowledge about problems are most certainly held in some associative network stored in declarative LTM, whereas the repertoire of strategies can be seen as pertaining to a procedural LTM. WM is lacking, though it can be assumed that the production of answers through algorithmic strategies takes place in such a system.

The relationships between these mnemonic systems are more explicitly detailed in cognitive architectures inspired from the adaptive control of thought—rational (ACT-R) model (Anderson, 1993; Anderson & Lebière, 1998; Anderson et al., 2004), such as the ADAPT model for number transcoding (Barrouillet, Camos, Perruchet, & Seron, 2004). This model describes how numbers are transcoded from their verbal to their Arabic form. The model assumes that word number are stored as a verbal string that a parser segments into a series of meaningful units. As illustrated in Fig. 17.1, the parsing of the verbal input is driven by knowledge stored in LTM; the parser identifying segments that correspond either to entries in a lexicon that associates verbal forms to their Arabic correspondent (e.g., "six" is associated with 6, and "twenty-five" with 25) or to separators (e.g., "hundred" in "two hundred twenty-five"). These segments are encoded and sequentially sent to WM where they are processed by a production system made of conditions–actions procedures. A procedure is triggered when the current content of WM matches its conditions of application. During transcoding, WM can contain representations of the units

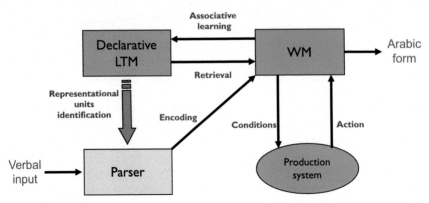

FIGURE 17.1   A cognitive architecture for the ADAPT model (described by Barrouillet et al., 2004). Number words are encoded as verbal strings segmented by a parser into units sent to working memory (WM) where they are processed by conditions–actions rules that read and transform the content of WM to produce the Arabic form of the number. The content of WM creates representational units stored through associative learning in long-term memory (LTM) from which they can be retrieved by subsequent transcoding activities.

isolated by the parser, digital forms retrieved from LTM, and the ordered chain made of digits and empty slots constructed by the production rules that have already been applied (e.g., after having processed "twelve" and "thousand," this chain would contain digits and slots to be filled such as 12 _ _ _). In the same way as producing answers modifies knowledge in the distributions of associations model, the functioning of ADAPT modifies the content of LTM. Chain of digits constructed by the production system are associated in LTM with their corresponding verbal strings. These associations that represent new representational units are reinforced at each transcoding of the same verbal string, resulting in the same chain of digits in such a way that frequently encountered verbal forms are eventually no longer processed by the production system but directly retrieved from LTM.

The ADAPT model illustrates the relationships existing between WM and LTM; the former holding the information retrieved from the latter, while the different elements simultaneously held in WM are deposited, through a process of associative learning, in LTM where they create new chunks of declarative knowledge (Anderson, 1993; Cowan, 1988, 2005; Logan, 1988). These new associations, when sufficiently reinforced, would circumvent the need to resort to algorithmic strategies for producing answers that can henceforth be retrieved. As we will see in this chapter, this transition from algorithmic computing to direct retrieval has often been assumed to be one of the main mechanisms by which the constraints resulting from WM limitations could be overcome in a variety of numerical processes. However, these WM limitations affect numerical activities in different ways depending on the nature of the processes and the type of representations they involve.

WORKING MEMORY IN NUMERICAL COGNITION

WM is usually defined as a limited-capacity system devoted to the temporary storage of a small number of short-lived items of information for ongoing cognitive processes (Baddeley, 2007). In its most popular account, known as the multicomponent model, WM is thought to consist of a capacity-limited central executive in charge of complex processing and of the control and coordination of domain-specific slave systems for storage among which the phonological loop for verbal information and the visuospatial sketchpad for visuospatial information have received the most detailed descriptions (Baddeley, 1986; Baddeley & Logie, 1999, pp. 28–61). Despite the upsurge in theoretical models for WM in the last decades (Barrouillet and Camos, 2015; Cowan, 2005; Engle, 2002; Oberauer, 2009; Unsworth & Engle, 2007) and some evolution of the multicomponent model itself (Baddeley, 2000, 2003, 2007), the tripartite structure hypothesized by

Baddeley (1986) has oriented most of the research on the role of WM in numerical cognition. This section does not aim to exhaustively review the abundant literature that has explored the relationships between WM and numerical cognition,[1] but instead to illustrate how WM limitations constrain a variety numerical activities and, more importantly, how the recourse to alternative memory systems can alleviate these constraints. I will, in turn, evoke counting, transcoding, and arithmetic problem solving.

Working Memory for Counting

Counting sets of objects is one of the most basic and early mastered numerical skills (Kaufman, Lord, Reese, & Volkmann, 1949) for which it has been suggested that human beings would be endowed with innate principles (Gelman & Gallistel, 1978). To the best of my knowledge, the first attempt to investigate counting within the framework of WM is due to Logie and Baddeley (1987). They suggested that counting involves three types of processing: the perception of the items to be counted, the access to the counting sequence in LTM, and the short-term storage of a running total, to which might be added the vocal or subvocal articulation of numbers in the counting sequence. This analysis pointed toward a prominent role of the phonological loop in articulating numbers and storing the running total. Consequently, Logie and Baddeley explored the role of the phonological loop by hindering its operation through concurrent articulatory suppression, asking adult participants to continuously repeat the word *the* while counting arrays of 1–25 dots. To discard the possibility of an effect of articulatory suppression simply due to the requirement to carry out a

---

[1] One can consult the excellent review by Raghubar, Barnes, and Hecht (2010) that follows a previous reviews by LeFevre, DeStefano, Coleman, and Shanahan (2005) and DeStefano and LeFevre (2004). Examining the relationship between WM and math from different approaches (experimental investigations, studies in children with math disabilities, longitudinal studies), Raghubar et al. conclude that WM is related with mathematical performance in adults, typically developing children, and children suffering from math disabilities. A more recent metaanalysis by Peng, Namkung, Barnes, and Sun (2016) indicates that the medium correlation between math and WM is about 0.35 with a 95% confidence interval from 0.32 to 0.37. The stronger relations are observed with whole-number calculations (including one-digit number operations) and word problem solving, but even basic number knowledge significantly correlates with WM. Geometry is the domain exhibiting the weakest relation. Quite surprisingly, the relation between math and WM does not seem to be modulated when considering different domain-specific WM systems such as verbal, numerical, or visuospatial WM. Peng et al. conclude from this observation that the relation between WM and mathematics is driven by a domain-general central executive. Interestingly, correlations are stronger in individuals with mathematical difficulties, especially when associated with other cognitive deficits.

concurrent secondary task, they added a condition in which counting was performed while tapping the table with the hand independently of counting. The results revealed that articulatory suppression had a significant effect on counting over and above any disruptive effect of a secondary task like tapping. However, Logie and Baddeley noted that subjects' counting abilities were not totally disrupted. Of course, counting was slower and more prone to errors under articulatory suppression, but this slowing down was far from being dramatic (9.6 s instead of 8 s to count 25 dots) and the mean absolute error surprisingly low (about only one unit for the largest arrays). In a subsequent experiment, they verified that unattended speech, known to automatically enter the phonological loop and disrupt its functioning (Colle & Welsh, 1976; Salamé & Baddeley, 1982), had very little effect.

These findings did not perfectly fit the phonological loop hypothesis. Logie and Baddeley (1987) reasoned that if the phonological loop was in charge of registering the running total, articulatory suppression would completely prevent this and lead to very large errors, something that did not happen. Another system than WM evidently intervenes. Logie and Baddeley suggested that this system is LTM in which the lexical representations of numbers might be successively primed. Thus, counting could involve a representation of the running total encoded on more than one dimension, "one of these being reflected by the articulatory loop and the other being a reflection of priming effects in some other system" (Logie & Baddeley, 1987, p. 325). This other system to which Logie and Baddeley refer and that stores lexical representations concerns the declarative part of LTM. However, it might be that its procedural part is also involved, as suggested by the investigations run by Camos, Barrouillet, and Fayol (2001).

Camos et al. (2001), although sharing the same theoretical framework, analyzed the counting process in a slightly different way than Logie and Baddeley (1987). Indeed, besides saying number words, counting involves pointing at objects for keeping track of what has been counted, a process that should involve the visuospatial sketchpad. The hypothesis tested by Camos et al. (2001) was that coordinating the two actions to comply with the one-to-one correspondence principle should involve the central executive, the cost of this coordination progressively decreasing with age as WM capacity increases and the two components of the task become more and more automatized. Camos et al. reasoned that if coordinating saying and pointing involves a cost, counting should take longer than performing the slowest of its two components in isolation (i.e., reciting the number chain without pointing at objects or pointing silently at objects), a difference that should progressively decrease with age. Findings revealed the exact reverse of the expected developmental pattern. Counting took slightly longer than its slowest component in adults, but this difference disappeared in 8-year-old children, and counting time was even lower than the time needed by the slowest component in 5-year-old children. In subsequent experiments, Camos et al. tested the hypothesis that increasing the cost of saying would make its coordination with pointing more demanding. For this purpose, children and adults were asked to count using the alphabet as a number chain or adults were asked to count in a less automatized foreign language (e.g., as English for French speakers) and even in a language they just began to learn, such as Tahitian. Once more, results were surprising. Increasing the difficulty of saying did not increase the cost of coordination. On the contrary, when adults counted with a largely nonautomatized number chain such as Tahitian, counting became faster than its slowest component, a finding that echoed the facilitation observed in young children. Camos et al. concluded that coordinating the two components of counting involves a negligible cost (see Towse & Hitch, 1997, for similar conclusions). They suggested that, at an early age, the operations necessary to count objects have been compiled into automated production rules as described by the ACT-R model (Anderson, 1993), in such a way that their integrated mobilization does no longer need attentional control.

In summary, these studies show that a complex task such as counting that seemingly solicits several components of the WM system might not be strongly constrained by WM limitations. Largely automatized processes, involving LTM declarative and procedural knowledge, such as associative priming and production rules instantiation and firing, alleviate the attentional demand of counting, even at an early age. Transcoding is another numerical process in which LTM mechanisms can replace controlled activities that load WM.

Working Memory for Transcoding

Despite the formal simplicity of the written decimal system that consists of only 10 elements (0, 1, 2, 3, 4, 5, 6, 7, 8, and 9) with a single principle, positional notation, and the fact that this code is the subject of systematic tuition, transcoding numbers from their verbal to their Arabic form remains a complex task for young children. Traditional models assumed that this transcoding involves the semantic representation of quantity to which the number refers (McCloskey, Caramazza, & Basili, 1985; Power & Dal Martello, 1990), making the task especially difficult for young children when large numbers must be processed. By contrast, asemantic models such as the ADAPT model described above assume that numbers can be transcoded without accessing their meaning. Nonetheless, this does not turn transcoding into an easy task. Within this theoretical framework, transcoding requires a series of processing steps taking place in a limited-capacity WM in charge of the maintenance and transformation of the chain of digits under construction. Camos (2008) has tested the hypothesis that children's performance should depend both on their WM capacity and on the number of production rules that the ADAPT model requires for transcoding a given number.

Seven-year-old children with either high, medium, or low WM capacities were asked to transcode one- to four-digit numbers from their verbal to their Arabic format, these numbers requiring from 2 to 7 rules in ADAPT (e.g., 14 and 40 require 2 rules, whereas 3708 requires 7 rules). Results revealed that error frequency increased with the number of rules the numbers required and, more importantly, that this effect was stronger in children with low WM capacities. Qualitative analyses of the errors revealed that children with low WM capacities were not only less efficient but also suffered from a developmental delay in the acquisition of the transcoding rules.

These findings indicate that WM limitations constrain both the acquisition and the use of the rules needed for transcoding numbers. However, we mentioned earlier that ADAPT incorporates a learning mechanism by which the most frequently transcoded numbers result in the association in LTM of their verbal and Arabic forms. Verbal strings are segmented by the parser and consequently trigger the production system, insofar as these verbal strings do not correspond to a representational unit in LTM. But if such a representational unit is accessed, the Arabic form is directly retrieved. Computational simulations and empirical findings indicated that this shift from algorithmic computing to direct retrieval mainly interests two-digit and some frequent three-digit numbers (Barrouillet et al., 2004). This relative rarity is of course due to the low frequency with which large numbers (e.g., 4827 or 32,012) are transcoded. Arithmetic problem solving is a domain in which the algorithmic computing to direct retrieval shift has been assumed to be more frequent and extensively documented.

Working Memory for Additions and Subtractions

As mentioned earlier, Baddeley's (1986) multicomponent model has oriented most of the investigations of the relations between WM and numerical cognition, and this is especially true for arithmetic problem solving. For example, Raghubar, Barnes, and Hecht (2010) emphasized the fact that the role in arithmetic problem solving of the different components of WM (as described by Baddeley) depends on the strategy used by the subject to solve a given problem. Different strategies (e.g., finger counting, verbal counting, direct retrieval from LTM, use of derived facts) might involve a variety of representations and processes that most probably solicit and load WM in different ways. The study by Imbo and Vandierendonck (2007) on single-digit addition and subtraction solving in adults perfectly illustrates this point. Imbo and Vandierendonck noted that many studies had already reported that the central executive is needed to solve simple arithmetic problems (e.g., De Rammelaere, Stuyven, & Vandierendonck, 1999; Lemaire, Abdi, & Fayol, 1996), but these studies neglected the fact that individuals use several strategies to solve even the simplest problems. Thus, they used the choice/no-choice method (Siegler & Lemaire, 1997) in which participants are asked in a first phase to solve problems using whatever method they want (choice), and then to solve these problems while using a predetermined strategy (no-choice). The strategies in this latter condition were retrieval, transformation in which one of the operands is decomposed to reach an intermediary step ($8+5=8+2+3=10+3=13$) and counting (for $8+5$, counting 9, 10, 11, 12, 13). The role of WM components was investigated by loading either the central executive with a concurrent choice reaction time task or the phonological loop by asking participants to maintain a preload of five letters while solving the problems. These conditions were compared with a no-load condition. We will focus here on the results concerning additions, response time analyses revealing approximately the same effects on subtractions. The effects of executive load and phonological load affected all the strategies but were smaller for retrieval. Interestingly, loading WM affected strategy efficiency but did not impact strategy choice in the choice condition.

The interpretation of these facts seems rather straightforward, at least as far as algorithmic strategies such as transformation and counting are concerned. These strategies that require the attentional control of a multi-step process, (e.g., controlling for the number of steps in the number line for the counting procedure) and the maintenance of intermediary results (remembering that 3 must be added to 10 when decomposing 5 for $8+5$) might involve the operations of the central executive. Moreover, they also rely on the phonological loop for browsing the number line and maintaining numerical results. Things are more complex concerning retrieval. The significant effect of a phonological load was unexpected by Imbo and Vandierendonck (2007) who suggested to explain this finding by a methodological artifact, maintaining five letters being quite hard and probably requiring attention.[2] The role of the central executive and the involvement of attentional resources in the retrieval process are a matter of debate and were discussed by the authors. They argued that even if candidate answers are automatically activated (e.g., Campbell, 1995), selecting the correct one and inhibiting the others could involve executive resources (Deschuyteneer & Vandierendonck, 2005), explaining the detrimental effect of an executive load. Nonetheless, the effects of executive and phonological loads on response times were smaller for the retrieval strategy, preserving the idea that the shift from algorithmic to direct retrieval is an efficient way to shield numerical processes from the limitations inherent to the functioning of our cognitive system, making responses faster and more accurate.

---

[2] It might be noted that Vergauwe, Camos, and Barrrouillet (2014) have observed that whereas the maintenance of four letters leaves a concurrent attention-demanding task unaffected, maintaining five letters has an effect, suggesting that the maintenance of five letters exceeds the capacities of the phonological loop and requires attentional processes.

Overall, this rapid survey of the relations between of WM and numerical processes such as counting, transcoding, and arithmetic problem solving suggests that functioning frees these activities from strong cognitive constraints by substituting the access to LTM knowledge (either procedural or declarative) for controlled demanding processes. The resulting shift from algorithmic processes to direct retrieval that seems to characterize transcoding and arithmetic problem solving and that Logan (1988) theorizes as an automatization is also a shift from one mnemonic system (WM) to another (LTM). However, does the more and more frequent reliance on LTM knowledge instead of controlled algorithmic processing free numerical cognition from any constraint? Is retrieval the panacea for numerical cognition?

IS RETRIEVAL THE PANACEA FOR NUMERICAL COGNITION?

It could be concluded from the few examples selected in the previous section that retrieving information from LTM confers such a decisive adaptive advantage in terms of speed and accuracy that numerical cognition might irresistibly shift from WM to LTM processes through functioning. However, quite ironically, the structure of human memory does not seem especially appropriate for storing number-related knowledge. Dehaene (2011) in his book *The number sense* emphasizes that the associative nature of our memory makes, at the same time, not only its strength when allowing us to rapidly and effortlessly retrieve information semantically related to the task at hand, but also its weakness when it matters to keep pieces of knowledge separated from each other and immune from interference. Unfortunately, this is exactly what numerical cognition requires when number facts must be accurately retrieved within an inextricable associative network connecting a limited number of entries to a host of possible answers (the numbers from 1 to 9 to hundreds of numerical facts). Spreading activation along the connections of such a dense network makes that irrelevant and incorrect responses receive activation and creates interference.

Accordingly, several studies have provided evidence that arithmetic number facts are obligatorily activated by the mere presentation of their constituents. One of the most famous demonstrations of this phenomenon has been provided by LeFevre, Bisanz, and Mrkonjic (1988). Adult participants were presented with a pair of numbers replaced, after a variable delay, by a probe number, their task being to decide whether this probe was part of the pair that just disappeared. The critical manipulation concerned negative trials and consisted in presenting a probe that was either the sum of the two numbers presented or an unrelated number (e.g., either 6 or 7 for the pair 4–2). LeFevre et al. reasoned that the task requires the comparison between the probe and a set of activated elements in memory.

A probe matching a highly activated number calls for a positive response. However, if the sum obligatorily receives some activation from the two presented numbers through spreading activation, the rejection of a probe that matches this sum requires an extra processing for resisting interference and selecting the appropriate negative response. In line with this hypothesis, presenting the sum of the two numbers instead of an unrelated number slowed down the negative responses, suggesting that the activation of simple additive facts is obligatory even when addition is completely irrelevant for the task. Interestingly, it appeared that interference was larger, and thus obligatory activation stronger, in individuals skilled at multidigit arithmetic (LeFevre & Kulak, 1994).

This obligatory activation of arithmetic facts creates interference even between operations, as several studies demonstrated. For example, Zbrodoff and Logan (1986) observed that in a verification multiplication task (e.g., $3 \times 4 = 12$?), associative lures corresponding to the sum of the operands ($3 \times 4 = 7$) resulted in slower rejections than nonassociative lures ($3 \times 4 = 11$). This phenomenon suggests that the operation of addition is initiated by associative lures despite its irrelevance, creating interference. However, Zbrodoff and Logan went beyond this observation and clarified the nature of the processes underpinning this obligatory activation. They wondered whether the activation of stored number facts is autonomous in the sense that (1) it could begin without intention, being entirely triggered by the presence in the environment of the appropriate stimuli (i.e., the operands), and (2) it would run to completion ballistically once it begins. The conclusion reached by Zbrodoff and Logan was that the processes underlying simple arithmetic are only partially autonomous. Indeed, the associative confusion effect proved weaker when operations were performed in pure (i.e., only additions or multiplications) than in mixed blocks and was affected by manipulations of the relative frequency of associative lures, the effect being stronger when these lures are rarer. This suggests that the activation of number facts can begin without intention but is not totally independent from intentions. Moreover, Zbrodoff and Logan's results revealed that mental arithmetic processes do not totally run to completion ballistically as they can be intentionally stopped under certain conditions. However, though being not totally autonomous, mental arithmetic processes present some degree of autonomy that makes memory for number facts especially prone to interference, as the associative confusion effect testifies.

Although it could be surmised that the occurrence of interference requires a prolonged strengthening of problem–answer association through practice, the associative confusion effect appears rapidly during learning as demonstrated by Lemaire, Fayol, and Abdi (1991) who observed it in 9- and 10-year-old children. In line with Zbrodoff and Logan's (1986) conclusions of a partial autonomy of the processes for simple mental arithmetic, Lemaire et al. showed that the associative confusion effect can be inhibited, but that this inhibition is more difficult as subjects are younger. This was established by manipulating the stimulus onset asynchrony (SOA) between the presentation of problems and associative lures. Whereas the effect was strong at short SOAs (i.e., 0 ms and 100 ms) for addition and multiplication problems, it was no longer observed in both adults and fifth graders at an SOA of 300 ms. Lemaire et al. accounted for these findings by assuming that longer delays enabled their participants to inhibit the activated irrelevant responses. The associative confusion effect was also observed in fourth graders but was still significant in this group at SOA 300 ms for both operations. Such a finding not only suggests lower inhibitory capacities in younger children but also that the associative network storing arithmetic facts can be even more prone to interference in children. A related finding was reported by Barrouillet, Fayol, and Lathulière (1997) who investigated the sources of the difficulties that adolescents with mathematical disabilities encounter in solving simple multiplication. Are these difficulties due to storage deficits, these adolescents having failed to properly learn multiplication tables, or instead to retrieval failures resulting from a difficulty to inhibit incorrect responses? The authors reasoned that if the locus of the problem was at retrieval, alleviating the need for inhibiting incorrect responses should result in better performance. Thus, they presented mathematically disabled adolescents with a task in which they had to identify the answer of difficult multiplication problems (e.g., 7×8; 9×7; 8×9; 7×6) among four numbers. In a no-interference condition, the correct answer was surrounded by distractors that did not pertain to any multiplication table (for example, for 7×6, 42 was presented along with 41, 44, and 38). In a weak-interference condition, the distractors were numbers found on tables, but not on those of the operands (e.g., 45, 32, and 40), whereas in a strong-interference condition, distractors were issued from the tables of one of the two operands (e.g., 49, 35, 38). Whereas, in a pretest, the adolescents tested gave only 56% of correct responses to the problems involved in the study, their rate of correct recognition of the answer in the no-interference condition was 82%, a rate that did not differ from that observed in the weak-interference condition (83%). However, the strong-interference condition elicited a significantly lower rate of correct responses (74%). These data showed that memorization deficits are not sufficient to explain the difficulties experienced by these adolescents who also encounter problems that are specific to the inhibition of incorrect responses associated with the operands.

Overall, research on the mnemonic mechanisms related to the activation and retrieval of numerical knowledge leads to a paradox. On the one hand, the shift from algorithmic processes that load a capacity-limited WM to the direct retrieval of answers from LTM is presented as an adaptive solution to overcome the limitations of the human cognitive system when processing numbers, and it is at the basis of several theoretical models (Ashcraft & Battaglia, 1978; Barrouillet et al., 2004; Campbell, 1994, 1995; Logan, 1988; Siegler & Shrager, 1984, pp. 229–293). On the other hand, the associative structure of human LTM has been claimed to be poorly adapted for numerical cognition (Dehaene, 2011), and an abundant literature from which we gave only few examples has documented its interfering nature (see De Visscher & Noël, 2016, for a recent review). We have seen that executive processes of inhibition (Lemaire et al., 1991) and response selection (Deschuyteneer & Vandierendonck, 2005) are involved in retrieving even simple arithmetic answers. This suggests that retrieval from LTM is far from being the panacea in circumventing the cognitive limitations that constrain numerical cognition. It has even been argued that numerical processes assumed to rely on retrieval from LTM solicit WM to the same extent as algorithmic processes. Barrouillet, Lépine, and Camos (2008) asked adult participants whose WM capacities had been previously assessed to perform a series of numerical tasks such as counting, subitizing, reading digits, and solving very small additions with operands from 1 to 4. As it could be expected, performance on an algorithmic process such as counting as well as the slope between counting times and the size of the to-be-counted array significantly correlated with WM capacities ($r = -0.294$ and -0.325, respectively). This means that individuals with higher WM capacities are more efficient in implementing the counting procedure and in monitoring its progress and outcome. However, reading digits, a process that indisputably relies on retrieval, and subitizing, for which retrieval has been hypothesized (Mandler & Shebo, 1982), exhibited correlations in the same range with WM capacity ($r = -0.307$ and -0.263 respectively). A comparable correlation ($r = -0.385$) was observed with the response times for very small addition problems for which there is a large consensus that they are solved through direct retrieval (LeFevre, Sadesky, & Bisanz, 1996). In the same way, the size effect on these small additions correlated with WM capacities ($r = -0.354$).

These latter findings question the distinction usually made between algorithmic processes and retrieval, the former being assumed to be cognitively demanding and WM dependent, whereas the latter would be largely autonomous and would underpin automaticity (Klapp, Boches, Trabert, & Logan, 1991; Logan, 1988). It is, for example, surprising that the size effect that affected the solving of very small additions correlated in the same way with WM capacities as the slope of the counting process, this slope reflecting the efficiency of the step-by-step process of moving from one item to be counted to the next. If arithmetic fact retrieval consists in accessing chunks of declarative knowledge associating problems with their answer, one can wonder why the speed of this process is affected by the size of the numbers stored in these chunks, and why this size effect correlates with WM capacities. It might be that the numerical processes subsumed under the generic description of "retrieval" are more complex and diverse than expected.

WHAT IS EXACTLY RETRIEVED FROM LONG-TERM MEMORY IN "RETRIEVAL" PROCESSES?

I have already evoked the mechanism by which the distributions of associations model (Siegler & Shrager, 1984, pp. 229–293) accounts for the shift from algorithmic processes to direct retrieval in addition problem solving. Algorithmic computing would create and reinforce problem–answer associations in LTM, the probability of their retrieval depending on associative strength. The probability of retrieving an answer is high when the strength of its association with the problem surpasses that of any other answer and a subjective confidence criterion. Because any production of an answer, whatever the strategy used, reinforces its association with the problem, the model necessarily converges toward the more and more frequent use of retrieval. A similar learning mechanism is assumed by the instance theory of automaticity (Logan, 1988). Because children spontaneously develop a variety of counting strategies for solving simple additions long before any systematic tuition, this learning mechanism would explain why, from the age 10 onward, the answer of small additions is retrieved from LTM (Ashcraft & Battaglia, 1978; LeFevre et al., 1996; see Zbrodoff & Logan, 2005, pp. 331–345, for a review).

A straightforward prediction of the retrieval model is that the effects related with the size of the operands, which characterize step-by-step counting strategies, should disappear with the transition from algorithmic computing to direct retrieval (Klapp et al., 1991). Indeed, there is a priori no reason that the time needed for accessing a piece of knowledge would depend on the size of the numbers it contains (e.g., reading times for digits do not systematically increase with the size of the corresponding number, Barrouillet et al., 2008). Notwithstanding, the problem-size effect, i.e., the increase in solution time with the size of the operands, is among the most striking phenomena affecting addition solving as it has been observed in virtually all the studies (Zbrodoff & Logan, 2005, pp. 331–345; Fig. 17.2). Groen and Parkman (1972) suggested that it resulted from the sporadic recourse to slower nonretrieval strategies, an intuition corroborated by LeFevre et al. (1996) who established that nonretrieval strategies were more frequently used with larger operands. However, reaction time (RTs) increase with the size of the operands even in those trials that are reported by the participants as retrieved. This size effect in "retrieved" problems is more difficult to account for. It has been suggested that retrieval latencies could reflect acquisition history. Because large problems are more often solved through algorithmic computing in the course of development, the resulting associations would be less accessible (LeFevre et al., 1996). It has also been hypothesized that small problems are more frequently encountered and solved, resulting in stronger and more rapidly retrieved memory traces (Ashcraft & Guillaume, 2009, pp. 121–151; Hamann & Ashcraft, 1986). Structural properties of the problems have also been advocated. Ashcraft and Battaglia (1978) supposed that the size effect results from

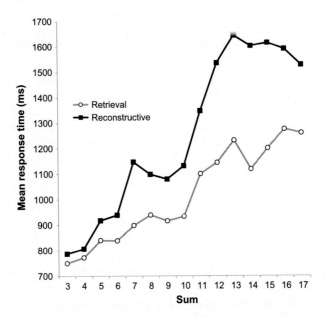

FIGURE 17.2   The size effect in simple addition problem solving. Mean response times as a function of the sum in additions with operands from 1 to 9 for problems solved through retrieval or through reconstructive (algorithmic) strategies. *Data from Uittenhove, K., Thevenot, C., Barrouillet, P. (2016). Fast automated counting procedures in addition problem solving: When are they used and why are they mistaken for retrieval? Cognition, 146, 289–303. Used with permission by Elsevier.*

a search in a 10×10 tabular representation of the 100 basic addition facts. Beginning at 0-0, the search would progress along the rows and columns of the table until reaching the cell corresponding to the problem at hand, large operands corresponding to longer moves in the table (see Widaman, Geary, Cormier, & Little, 1989, for a related idea). Others have suggested that the effect results from interference, with large problems suffering from more interference (Zbrodoff, 1995; Zbrodoff & Logan, 2005, pp. 331–345). The diversity of these explanations is indicative of the difficulties encountered by current cognitive psychology to accommodate the retrieval hypothesis with an effect that is, instead, rather suggestive of the use of some step-by-step algorithmic process. This is probably for discarding this alternative account that Zbrodoff and Logan (2005, pp. 331–345) emphasized the fact that the increase in RTs with the size of the operands is not strictly monotonic. In the same vein, Ashcraft (1992, p. 80) stated "the term problem-size effect is now generally considered as a misnomer" because the effect would not be due to the size of the operands per se, but to some variable that coincidentally covaries with this size, such as frequency, history of acquisition, or spatial position in a table.

However, is the problem-size effect really not monotonic and is it due to the coincidental covariation of some effectual factor with the size of the operands? Answering this question requires to collect RTs in a large set of participants solving several times each simple addition. This is what Uittenhove, Thevenot, and Barrouillet (2016) did in a sample of 92 adults who solved 6 times each of the 81 simple additions with operands from 1 to 9. To analyze the size effect in RTs for "retrieved" problems, the authors focused on small additions (i.e., with a sum ≤10), participants reporting the use of reconstructive strategies in more than 50% of the large additions. Moreover, the analyses were conducted on a subset of 51 participants who reported having retrieved 98% of these small additions. These "retrieved" small problems nonetheless exhibited a strong size effect with a correlation between mean RTs and the size of their sum of 0.63 ($P<.001$), RTs ranging from 720 ms for $2+1$ to 994 ms for $3+7$. However, the analysis of this problem-size effect revealed a striking and unexpected finding: RTs increased monotonically with the sum up to 7 and then plateaued from 7 to 10. This surprising pattern of size effect, which does not fit with any of the explanations reviewed above, finds its solution when considering different types of small additions. After having set aside *tie* problems (i.e., $n+n$ problems such as $2+2$) that are known since Groen and Parkman (1972) for being processed in a particular way, Uittenhove et al. first isolated what they called *very small* additions in which both operands do not exceed 4. These problems had already been studied in isolation by Barrouillet and Thevenot (2013) because of the relevance of the number 4 for numerical cognition, which corresponds to the upper limit of the subitizing range (Kaufman et al., 1949), of the object-file representations (Kahneman, Treisman, & Gibbs, 1992) and of the focus of attention (Cowan, 2001). These problems exhibited in Barrouillet and Thevenot (2013) a specific size effect that Uittenhove et al. intended to replicate and explain. They were compared with the remaining small additions with at least one operand larger than 4, what they called *medium small* additions, from which they removed the problems involving 1 (i.e., $n+1$ or $1+n$ with $n>4$, called the $n+1$ problems), as they are known to be solved using some specific rule. In a quite counterintuitive way, the problem-size effect mainly concentrated in the *very small* problems (increment in mean RT of 47 ms per unit, Fig. 17.3), whereas there was no size effect at all on *medium small* problems (size-related slope of −5 ms, Fig. 17.4) and only a small effect on *ties* and $n+1$ problems (slopes of 8 and 7 ms respectively). This pattern of size effect that does not fit with any of the explanations reviewed above suggests that *very small* additions are solved in a way that differs from the other small additions, the systematic increase in RTs with the size of both operands already observed by Barrouillet and Thevenot (2013) evoking some step-by-step process. However, its speed (increment of 47 ms per step) and the verbal reports of retrieval are incompatible with

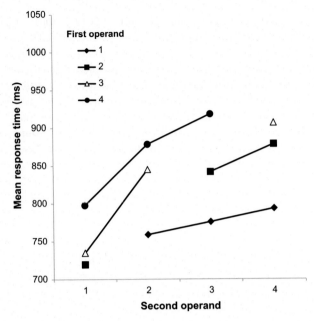

FIGURE 17.3   The size effect for the *very small* additions (operands from 1 to 4) in Uittenhove et al. (2016). The figure shows how RTs increase with the magnitude of both operands. Data from a subsample of 51 participants who reported having retrieved 100% of these problems. *Used with permission by Elsevier from Uittenhove, K., Thevenot, C., Barrouillet, P. (2016). Fast automated counting procedures in addition problem solving: When are they used and why are they mistaken for retrieval?* Cognition, 146, 289–303.

the use of consciously controlled algorithmic strategies based on counting. Thus, how are these very small additions solved?

Extending a hypothesis already put forward by Baroody (1983), Uittenhove et al. suggested that very small additions, which are the most frequently used and the first to be practiced in development, are solved through a rapid sequential procedure that reaches automatization through practice, hence its speed even though its duration remains determined by the size of the operands. Newell (1990) in his state, operator and result (SOAR) model has provided a fine-grained description of this type of automated and fast cognitive process he calls decision cycle. A decision cycle corresponds to the smallest deliberate act and the smallest unit of serial operations because it involves repeated accessing of knowledge from LTM through the firing of productions. These productions that relate conditions to actions are very fast with a duration of the order of tens of milliseconds and constitute the lowest level of cognition. The number of productions involved in the decision cycle determines its duration, which is of the order of hundreds of milliseconds. Interestingly, decision cycles are involuntary, automatic, running to quiescence, and delivering their

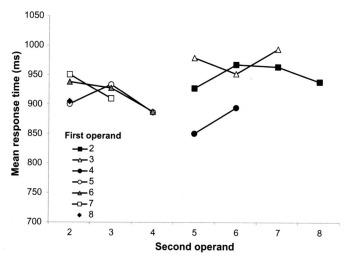

FIGURE 17.4   Mean RTs for the *medium small* additions (sum ≤ 10 with at least one oper-
and larger than 4, see text) in Uittenhove et al. (2016). This pattern can be compared with
the effect of the operands magnitude on RTs for the *very small* additions in Fig. 17.3, no clear
pattern emerging here for the *medium small* problems. Data from a subsample of 51 participants who reported having retrieved 98% of these problems. *Used with permission by Elsevier
from Uittenhove, K., Thevenot, C., Barrouillet, P. (2016). Fast automated counting procedures in
addition problem solving: When are they used and why are they mistaken for retrieval? Cognition,
146, 289–303.*

response, while the subject remains unaware of their process. These characteristics perfectly fit what is usually considered as retrieval in arithmetic cognition, with the fast access to an answer automatically activated
by the operands, the process being so fast that it remains unconscious,
individuals experiencing at the phenomenological level "retrieval" from
LTM. Nonetheless, the decision cycle hypothesis allows to understand
why the time course of this "retrieval" is affected by the size of the operands because this size determines the number of productions involved.

Uittenhove et al. (2016) suggested that the automated procedure for additions could take the form of a decision cycle in which both operands are successively represented in WM in an analogical way capturing their meaning.
Each token of these representations would in turn trigger a *next-token-next-
value* production that would access the number chain in LTM for tagging this
token with the next value retrieved and move to the next token, the process
running until all the tokens of both operands have been tagged (Fig. 17.5).
Running to quiescence, the decision cycle would deliver the last number
accessed as a response. This would account for the fact that the time needed
to produce this response depends on the sum of the operands that determines
the number of firings of the production and for the size-related slope which
is in the range of duration of a production cycle (i.e., tens of milliseconds,

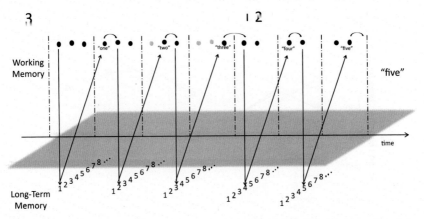

FIGURE 17.5 Illustration of the successive steps of the automated procedure for solving very small additions (here 3 + 2) hypothesized by Uittenhove et al. (2016). The first operand is encoded in working memory (WM) in an analogical format triggering the *next-token-next-value* production. Each space between two successive vertical dashed lines corresponds to a state of WM and to a production cycle in which the production moves to the next token and accesses the next value in the number chain stored in long-term memory (LTM), tags the token with this value, moves to the next token, and so on. When the first operand (3) has been totally processed in this way, the second operand (2) is encoded and processed in turn. The value retrieved from LTM for tagging the last token ("five") determines the response. Each production cycle has a duration of the order of tens of milliseconds, the entire process corresponding to a decision cycle, which runs automatically to quiescence.

here 47 ms). This would also explain why the use of this automated procedure is limited to operands up to 4, which is the maximum number of items that can be simultaneously represented in a single focus of attention (Cowan, 2001). Finally, the process is so fast that individuals remain unconscious of its successive steps and mistake it for retrieval, hence their verbal report.

The fact that medium small additions did not exhibit any size effect at all is, of course, totally compatible with their solving by direct retrieval from LTM. Within a race model in which different processes run in parallel and compete for determining performance (e.g., Logan, 1988), this leads to suppose that the automated procedure described above would be faster than the retrieval process itself, winning the race as long as it can be used (i.e., for additions with operands up to 4). For other small additions, retrieval would be the fastest process and win the race. Uittenhove et al. noted that in the subset of 51 frequent retrievers, the mean RT for the *medium small* additions (928 ms) was very close to the time needed to solve the slowest of the *very small* additions (i.e., 3 + 4 and 4 + 3, mean RT of 912 ms), suggesting that the process used to solve the former was slower than that used to solve the latter, something that corroborates the race model hypothesis. Overall, a careful analysis of the RTs in addition problem solving leads to temperate the scientific consensus on a shift from algorithmic computing to direct retrieval for the most frequent problems. What had been previously overlooked is that the smallest problems are the only problems exhibiting a systematic size effect, whereas the disappearance of this effect, which should be the signature of retrieval (Klapp et al., 1991), is observed on larger problems that are less frequent, more prone to interference and practiced later in development.

CONCLUSION

Numerical cognition is among the most remarkable achievements of human intelligence. It involves such a variety of processes, from subitizing to the most complex mathematical demonstrations that it solicits almost all the structures and functions of human mind including its different mnemonic systems. This is why the title of this chapter refers to these different memories. As I tried to demonstrate here, cognitive psychologists have often suggested that one of the most efficient ways human mind has developed for overcoming its inherent limitations in number processing is to exchange controlled and demanding processes that load WM for retrieval of knowledge from LTM. There is no doubt that moving from one mnemonic system to the other makes functioning smoother and easier. However, the algorithmic computing—direct retrieval shift that is assumed to characterize automation, development, and expertise acquisition might be more complex than expected, even in its simplest and universally admitted forms, such as simple addition problem solving. Indeed, a detailed analysis of adult performance makes hardly tenable the hypothesis that the most frequently encountered operations are eventually solved by retrieving their answer from memory. Instead, prolonged practice seems to result, in some cases, in the storage and retrieval of problem–answer associations, but in others in the automation of the algorithmic process itself, when a simple one-greater rule is not used for n + 1 problems. This variety in the trajectories taken by the different memories in their involvement in numerical cognition testifies from the high degree of adaptability of the cognitive system when confronted to complex activities such as number processing.

References

Anderson, J. R. (1993). *Rules of the mind*. Hillsdale, NJ: Lawrence Erlbaum Associates.

Anderson, J. R., Bothell, D., Byrne, M. D., Douglass, S., Lebiere, C., & Qin, Y. (2004). An Integrated Theory of Mind. *Psychological Review, 111*, 1036–1060.

Anderson, J. R., & Lebière, C. (1998). *The atomic components of thought*. Mawhaw, NJ: Lawrence Erlbaum Associates.

Ashcraft, M. H. (1992) Cognitive arithmetic: A review of data and theory. *Cognition*, *44*, 75–106.

Ashcraft, M. H., & Battaglia, J. (1978). Cognitive arithmetic: Evidence for retrieval and decision processes in mental addition. *Journal of Experimental Psychology: Human Learning and Memory, 4*, 527–538.

Ashcraft, M. H., & Guillaume, M. M. (2009). Mathematical cognition and the problem size effect. In B. Ross (Ed.). B. Ross (Ed.), *The psychology of learning and motivation: (Vol. 51)*. Burlington: Academic Press.

Baddeley, A. D. (1986). *Working memory*. Oxford: Clarendon Press.

Baddeley, A. D. (2000). The episodic buffer: A new component of working memory? *Trends in Cognitive Sciences, 4*, 417–423.

Baddeley, A. D. (2003). Working memory: Looking back and looking forward. *Nature Reviews Neuroscience, 4*, 829–839.

Baddeley, A. D. (2007). *Working memory, thought, and action*. Oxford: Oxford University Press.

Baddeley, A. D., & Logie, R. H. (1999). Working memory: The multiple-component model. In A. Miyake, & P. Shah (Eds.), *Models of working memory: Mechanisms of active maintenance and executive control*. Cambridge: Cambridge University Press.

Baroody, A. J. (1983). The development of procedural knowledge: An alternative explanation for chronometric trends of mental arithmetic. *Developmental Review, 3*, 225–230.

Barrouillet, P., Camos, V., Perruchet, P., & Seron, X. (2004). A developmental asemantic procedural transcoding (ADAPT) model: From verbal to Arabic numerals. *Psychological Review, 111*, 368–394.

Barrouillet, P., & Camos, V. (2015). *Working memory: Loss and reconstruction*. Hove, UK: Psychology Press.

Barrouillet, P., Fayol, M., & Lathulière, E. (1997). Difficulties in selecting between competitors when solving elementary multiplication tasks: An explanation of the errors produced by adolescents with learning difficulties. *International Journal of Behavioral Development, 21*, 253–275.

Barrouillet, P., Lépine, R., & Camos, V. (2008). Is the influence of working memory capacity on high level cognition mediated by complexity or resource-dependent elementary processes? *Psychonomic Bulletin and Review, 15*, 528–534.

Barrouillet, P., & Thevenot, C. (2013). On the problem size effect on small additions: Can we really discard any counting-based account? *Cognition, 128*, 35–44.

Camos, V., Barrouillet, P., & Fayol, M. (2001). Does the coordination of verbal and motor information explain the development of counting in children? *Journal of Experimental Child Psychology, 78*, 240–262.

Camos, V. (2008). Low working memory capacity impedes both efficiency and learning of number transcoding in children. *Journal of Experimental Child Psychology, 99*, 37–57.

Campbell, J. I. D. (1994). Architectures for numerical cognition. *Cognition, 53*, 1–44.

Campbell, J. I. D. (1995). Mechanisms of simple addition and multiplication: A modified network-interference theory and simulation. *Mathematical Cognition, 1*, 121–164.

Colle, H. A., & Welsh, A. (1976). Acoustic masking in primary memory. *Journal of Verbal Learning and Verbal Behavior, 15*, 17–32.

Cowan, N. (1988). Evolving conceptions of memory storage, selective attention, and their natural constraints within the human information processing system. *Psychological Bulletin, 104,* 163–191.

Cowan, N. (2001). The magical number 4 in short-term memory: A reconsideration of mental storage capacity. *Behavioral and Brain Sciences, 24,* 87–185.

Cowan, N. (2005). *Working memory capacity.* Hove, East Sussex, UK: Psychology Press.

Dehaene, S. (2011). *The number sense.* Oxford: Oxford University Press.

De Rammelaere, S., Stuyven, E., & Vandierendonck, A. (1999). The contribution of working memory resources in the verification of simple arithmetic sums. *Psychological Research, 62,* 72–77.

De Visscher, A., & Noël, M.-P. (2016). Similarity interference in learning and retrieving arithmetic facts. *Progress in Brain Research, 227,* 131–158.

Deschuyteneer, M., & Vandierendonck, A. (2005). Are "input monitoring" and "response selection" involved in solving simple mental arithmetical sums? *European Journal of Cognitive Psychology, 17,* 347–370.

DeStefano, D., & LeFevre, J.-A. (2004). The role of working memory in mental arithmetic. *European Journal of Cognitive Psychology, 16,* 353–386.

Engle, R. W. (2002). Working memory capacity as executive attention. *Current Directions in Psychological Science, 11,* 19–23.

Gelman, R., & Gallistel, C. R. (1978). *The child's understanding of number.* Cambridge, MA: Harvard University Press.

Groen, G. J., & Parkman, J. M. (1972). A chronometric analysis of simple addition. *Psychological Review, 79,* 329–343.

Hamann, M. S., & Ashcraft, M. H. (1986). Textbook presentations of the basic addition facts. *Cognition and Instruction, 3,* 173–192.

Imbo, I., & Vandierendonck, A. (2007). The role of phonological and executive working memory resources in simple arithmetic strategies. *European Journal of Cognitive Psychology, 19,* 910–933.

Kahneman, D., Treisman, A., & Gibbs, B. J. (1992). The reviewing of object files: Object-specific integration of information. *Cognitive Psychology, 24,* 175–219.

Kaufman, E. L., Lord, M. W., Reese, T. W., & Volkmann, J. (1949). The discrimination of visual number. *American Journal of Psychology, 62,* 498–525.

Klapp, S. T., Boches, C. A., Trabert, M. L., & Logan, G. D. (1991). Automatizing alphabet arithmetic: II. Are there practice effects after automaticity is attained? *Journal of Experimental Psychology: Learning, Memory, and Cognition, 17,* 196–209.

LeFevre, J.-A., Bisanz, J., & Mrkonjic, L. (1988). Cognitive arithmetic: Evidence for obligatory activation of arithmetic facts. *Memory and Cognition, 16,* 45–53.

LeFevre, J.-A., & Kulak, A. G. (1994). Individual differences in the obligatory activation of addition facts. *Memory and Cognition, 22,* 188–200.

LeFevre, J., DeStefano, D., Coleman, B., & Shanahan, T. (2005). Mathematical cognition and working memory. In J. I. D. Campbell (Ed.), *The handbook of mathematical cognition* (pp. 361–378). New York: Psychology press.

LeFevre, J.-A., Sadesky, G. S., & Bisanz, J. (1996). Selection of procedures in mental addition: Reassessing the problem size effect in adults. *Journal of Experimental Psychology: Learning, Memory, and Cognition, 22,* 216–230.

Lemaire, P., Abdi, H., & Fayol, M. (1996). The role of working memory resources in simple cognitive arithmetic. *European Journal of Cognitive Psychology, 8,* 73–103.

Lemaire, P., Fayol, M., & Abdi, H. (1991). Associative confusion effect in cognitive arithmetic: Evidence for partially autonomous processes. *CPC: European Bulletin of Cognitive Psychology, 11*, 587–604.

Logan, G. D. (1988). Toward an instance theory of automatization. *Psychological Review, 95*, 492–527.

Logie, R. H., & Baddeley, A. D. (1987). Cognitive processes in counting. *Journal of Experimental Psychology: Learning, Memory, and Cognition, 13*, 310–326.

Lortie-Forgues, H., Tian, J., & Siegler, R. S. (2015). Why is learning fraction and decimal arithmetic so difficult? *Developmental Review, 38*, 201–221.

Mandler, G., & Shebo, B. J. (1982). Subitizing: An analysis of its component processes. *Journal of Experimental Psychology: General, 111*, 1–22.

McCloskey, M., Caramazza, A., & Basili, A. (1985). Cognitive mechanisms in number-processing and calculation: Evidence from dyscalculia. *Brain and Cognition, 4*, 171–196.

Newell, A. (1990). *Unified theories of cognition.* Cambridge, MA: Harvard University Press.

Oberauer. (2009). Design for a working memory. *Psychology of Learning and Motivation, 51*, 45–100.

Peng, P., Namkung, J., Barnes, M., & Sun, C. (2016). A Meta-analysis of mathematics and working memory: Moderating effects of working memory domain, type of mathematics skill, and sample characteristics. *Journal of Educational Psychology, 108*, 455–473.

Power, R. J. D., & Dal Martello, M. F. (1990). The dictation of Italian numerals. *Language and Cognitive Processes, 5*, 237–254.

Raghubar, K. P., Barnes, M. A., & Hecht, S. A. (2010). Working memory and mathematics: A review of developmental, individual difference, and cognitive approaches. *Learning and Individual Differences, 20*, 10–122.

Salamé, P., & Baddeley, A. (1982). Disruption of short-term memory by irrelevant speech: Implications for the structure of working memory. *Journal of Verbal Learning and Verbal Behavior, 21*, 150–164.

Siegler, R. S., & Lemaire, P. (1997). Older and younger adults' strategy choices in multiplication: Testing predictions of ASCM using the choice/no-choice method. *Journal of Experimental Psychology: General, 126*, 71–92.

Siegler, R. S., & Shrager, J. (1984). Strategic choices in addition and subtraction: How do children know what to do? In C. Sophian (Ed.), *Origins of cognitive skills.* Hillsdale: Erlbaum.

Thevenot, C., & Barrouillet, P. (2015). Arithmetic word problem solving and mental representations. In R. Cohen Kadosh, & A. Dowker (Eds.), *Oxford handbook of numerical cognition.* New York, NY: Oxford University Press.

Towse, J. N., & Hitch, G. N. (1997). Integrating information in object counting: A role for a central coordination process? *Cognitive Development, 12*, 393–422.

Uittenhove, K., Thevenot, C., & Barrouillet, P. (2016). Fast automated counting procedures in addition problem solving: When are they used and why are they mistaken for retrieval? *Cognition, 146*, 289–303.

Unsworth, N., & Engle, R. W. (2007). The nature of individual differences in working memory capacity: Active maintenance in primary memory and controlled search from secondary memory. *Psychological Review, 114*, 104–132.

Vergauwe, E., Camos, V., & Barrrouillet, P. (2014). The impact of storage on processing: How is information maintained in working memory? *Journal of Experimental Psychology: Learning, Memory, and Cognition, 40*, 1072–1095.

Widaman, K. F., Geary, D. C., Cormier, P., & Little, T. (1989). A componential model for mental addition. *Journal of Experimental Psychology: Learning, Memory, and Cognition, 15*(5), 898–919.

Zbrodoff, N. J. (1995). Why is 9 + 7 harder than 2 + 3? Strength and interference as explanation of the problem size effect. *Memory and Cognition, 23*, 689–700.

Zbrodoff, N. J., & Logan, G. D. (1986). On the autonomy of mental processes: A case study of arithmetic. *Journal of Experimental Psychology: General, 115*, 118–130.

Zbrodoff, N. J., & Logan, G. D. (2005). What everyone finds: The problem-size effect. In J. I. D. Campbell (Ed.), *Handbook of mathematical cognition* (pp. 331–345). New York and Hove: Psychology press.

18

Hypersensitivity-to-Interference in Memory as a Possible Cause of Difficulty in Arithmetic Facts Storing

Marie-Pascale Noël, Alice De Visscher

Université Catholique de Louvain, Louvain-la-Neuve, Belgium

Heterogeneity of Function in Numerical Cognition
https://doi.org/10.1016/B978-0-12-811529-9.00018-2

INTRODUCTION

Children spend most of their first four grades in primary school learning to solve simple arithmetic operations. The expectation is that, by the end of the fourth grade approximately, children are able to use mature strategies to solve most of the calculations and that they have stored in memory some of the answers, in particular, the products. Yet, for some children, this learning is problematic. More specifically, this is often so in the case of math learning disabilities or of dyscalculia (Garnett & Fleischner, 1983; Geary, Hoard, & Hamson, 1999).

In this chapter, we focus on multiplications because the memorization of the answers of the problem is particularly stressed in school compared with other operations. Much previous research assumed that additions were first solved through counting strategies and that, at some point, children would store the answers of at least the small additions (this idea is, however, debated; see Barrouillet & Thevenot, 2013). In contrast, children learn the meaning of multiplication by performing repeated additions but very soon they are encouraged to learn the answers by heart, as other strategies such as counting or repeated additions are too time-consuming and error prone.

Despite a number of differences between the existing theoretical models, there is a general consensus that arithmetic facts are organized in an interrelated network in long-term memory (e.g., Campbell, 1995; McCloskey & Lindemann, 1992, pp. 365–409; Verguts & Fias, 2005) and that when confronted with a multiplication problem, not only the correct answer but also other related incorrect answers are activated. Consequently, the set of associated but incorrect answers constitutes competing answers that interfere with the retrieval of the correct answer (e.g., Campbell, 1987a, 1987b, 1995; Campbell & Tarling, 1996). Based on this idea, Barrouillet, Fayol, and Lathulière (1997, see also Barrouillet's chapter in this volume) proposed that difficulties in recalling multiplication answers might be because of difficulties in inhibiting the competitors. To support this view, they presented multiple choice multiplications to adolescents with learning disabilities. The adolescents erred more often when the false answers were multiples of one of the operands (e.g., $4 \times 6 = 18$, 24, 28, or 30?) than when they were not (e.g., $4 \times 6 = 21$, 24, 25, or 27?) or when the false answers did not come from multiplication tables (e.g., $4 \times 6 = 22$, 23, 24, or 26?). However, one should be cautious when drawing conclusions from this study. First, these adolescents did not present specific math learning disability but presented global learning disability with abnormally low IQ. Second, there was no group control to ensure that their profile was specific. Following this study, some authors proposed that inhibition weakness might account for difficulties in retrieving multiplication answers from memory.

ARE DIFFICULTIES IN ARITHMETIC DUE TO INHIBITION PROBLEMS?

A series of research tested whether general math performance in math correlated with inhibition capacities and many of them found significant correlations (Bull & Scerif, 2001; Espy et al., 2004; Gilmore et al., 2013; St Clair-Thompson & Gathercole, 2006 and see Cragg & Gilmore, 2014, for a recent review). Others tested the hypothesis that children with math learning disabilities would present weaker inhibition skills than typically developing children but failed to find an inhibition deficit in these math-disabled children (e.g., Censabella & Noël, 2005; van der Sluis, de Jong, & van der Leij, 2004). But only a few studies specifically examined whether inhibition capacities are specifically related to performance in simple arithmetic. In particular, Censabella and Noël (2008) addressed this question by testing three inhibitory functions in children with or without math learning difficulties and also in children with a specific deficit in arithmetic facts. These three functions were the suppression of irrelevant information from working memory, the inhibition of proponent response, and the interference control of distractors. The first of these was assessed using the listening span task: the child listens to a series of sentences, judges whether they are true or false, and then has to recall the last word of each sentence. The authors examined the number of intrusions of not ending words in the child's recall. The inhibition of proponent response was assessed by a Stroop task. An Erikson task, in which the participant has to judge a central target surrounded by the same or different distractors, assessed resistance to the interference caused by external distractors. The authors found no significant differences of performance in any of these inhibition tasks between control children and either all the children presenting arithmetic disabilities or those with difficulties in arithmetic facts retrieval. In the same vein, Bellon, Fias, and De Smedt (2016) tested typically developing third graders and found no correlation between performance in an addition or multiplication verification task and inhibition abilities tested in numerical and color Stroop tasks.

Censabella and Noël (2008) then reasoned that most of these studies reported earlier tested inhibition capacities, i.e., the active suppression of irrelevant or competing stimuli or responses. However, interference can also arise in long-term memory when the retrieval of information is based on a process of spreading activation (Anderson, 1983). In this view, the ease of retrieving a fact is a function of the level of activation of this fact relative to the activation of other associated facts. If other associated facts are also activated, then the retrieval of the target fact will suffer interference and the retrieval will be less efficient. Following this idea, Censabella and Noël (2004) tested in adult participants whether arithmetic fact performance was more closely related to performance in a typical inhibition task (the color

Stroop task) or in an activation-based interference paradigm. The latter was assessed using the fan effect paradigm (Anderson, 1974): participants had to learn sets of sentences such as "this character is in this location." Yet one character could be associated with either a single or several locations. The number of associations between the character and the location(s) is called the fan. In this experiment, the fan was either one or three. After the learning phase, participants were asked to verify propositions such as "this character is in this location." Typically, it is found that participants are slower and less accurate when the fan is large. The assumption is that the character's name acts as a memory probe and spreads activation to associated location node(s). Yet, as the capacity of spreading activation is supposed to be limited, it leads to full activation of associated location nodes in the case of a fan of one but to reduced activation if the fan is larger. Indeed, in the case of a larger fan, the activation is divided into the different connected pathways and this leads to a reduced accessibility of the associated location concepts.

Censabella and Noël (2004) found no correlation between arithmetic abilities and performance in the Stroop task but found a significant correlation with the measure of activation-based interference: the larger the fan effect in a participant, the longer his/her response time in multiplications. The authors concluded that, in arithmetic facts models, the key determinant of performance and retrieval efficiency is the associative strength of the correct answer relative to the associative strengths of all answers. This relative associative strength entails retrieval interference: the correct answer's relative associative strength is weaker as the number of associated but incorrect answers increases. This activation-based interference does not necessarily require active inhibitory mechanisms, as the reduced associative strength of the correct answer is sufficient to account for weaker performances. This interference is a passive one because of the "overfacilitation" of competitors (Anderson & Bjork, 1994, pp. 265–325; see also Logan, 1994, pp. 189–239).

All these previous studies consider that interference could occur when the person is attempting to retrieve the correct answer of a problem and suppose that arithmetic facts have been stored in memory, where the problems are associated with the correct and incorrect related answers. However, the study of one single case showing a dramatic impairment in storing multiplication facts in memory led us to consider that some other types of interference could already play a role at the learning stage.

A CASE STUDY WITH ARITHMETIC FACT IMPAIRMENT AND HYPERSENSITIVITY-TO-INTERFERENCE IN MEMORY

DB was a 42-year-old woman previously diagnosed with developmental dyscalculia and high intellectual potential (De Visscher & Noël, 2013). When we assessed her numerical skills, we found a very specific profile of dyscalculia characterized by extremely slow answers in single-digit multiplications and to a lesser extent in additions. This extreme slowness was explained by her use of computational and finger-counting strategies, while the control subjects used retrieval. Further investigations revealed that she actually stored only very few products. Indeed, when we forced her to produce the first answer activated by the problem, by using a time-limited multiplication production task, she performed as well as control participants only for 2 out of 8 small problems (2×3 and 3×3) and gave no correct answers for the 8 large problems except for 8×8. This suggests that she only stored the answers of a very few items: some of the small products and some of the ties. Furthermore, in a table membership judgment task where participants had to decide whether the number displayed was a possible result of a multiplication of two single digits, control participants were 95% correct while she reached only 66% and based her answers mainly on the parity status of the item: she accepted all the even numbers as table facts and dismissed too many odd ones.

We therefore explored what might explain such a severe and persistent difficulty in memorizing arithmetic facts, in the context of high intelligence, good memory, and high motivation in practicing the times tables. Deep cognitive investigation showed that none of the hypotheses currently found in the literature to account for dyscalculia or difficulties in learning multiplication facts were supported. Indeed, DB had perfectly normal performance in tasks tapping the approximate number system, access to number magnitude from Arabic symbols or rote verbal memory. Eventually, we considered a new hypothesis based on hypersensitivity-to-interference in memory.

The rationale was that single-digit multiplication facts are made up of three elements: the two operands, selected from 10 possibilities (0–9), and the product, corresponding to a combination of 2 elements from these 10 possibilities. If we consider the 64 problems from 2×2 to 9×9, any given operand appears in 15 problems and several problems share the same product (for instance, 24 is the product of 8×3, 3×8, 4×6, and 6×4 and 18 is the product of 6×3, 3×6, 9×2, and 2×9). In the memory literature, it is well known that during the storage stage, memorization is hindered when the items are similar to one another. This detrimental effect of similarity has been observed in many different paradigms both in long-term (e.g., paired-associated learning such as in Hall, 1971) and short-term memory (e.g., immediate recall paradigms such as Conrad & Hull, 1964; or Conlin, Gathercole, & Adams, 2005). According to Oberauer and Kliegl (2006), any two items have a certain degree of overlap regarding their features, with more similar items sharing a larger proportion of feature units. When items have to be stored in memory, the features of item representations interact with one another, which lead to mutual interference. This interference partially degrades the memory traces, which in turn leads to slower processing and to retrieval errors. When we learn multiplication facts by rote, we have to learn distinct associations of three elements with many features in common. For instance, we have to learn that $9 \times 4 = 36$ but also that $6 \times 4 = 24$, $8 \times 4 = 32$, and so on. Considering the high overlap of features, arithmetic facts can be qualified as prone to interference. Accordingly, we proposed that a heightened sensitivity to interference could prevent the storage of arithmetic facts in long-term memory.

Using different kinds of paradigms assessing interference in working memory and in learning in long-term memory, we tested this hypothesis on the patient DB (De Visscher & Noël, 2013). In association learning tasks, DB performed perfectly when the verbal associates were composed of dissimilar items, but she had impaired performance when the associates were composed of similar items or when the associations included shared items. In this type of experiment, the sensitivity to item similarity is typically because of greater proactive interference (Underwood, 1983). The hypothesis of high susceptibility to proactive interference in DB was directly tested using the recent-probes task (originally from Monsell, 1978). In a visual version of the task, two unknown visual signs were presented for 3000 ms followed by one visual sign, and participants had to decide whether this sign was one of the two just previously seen. DB had no difficulty in accepting the sign when it was a part of the stimulus trial (positive trials) or in rejecting a sign that had not been seen before (negative trials). However, she had considerable difficulty in rejecting lures that were seen in a previous trial (negative lures, see Fig. 18.1). Thus, her difficulties were not because of memory problems per se but to a very high sensitivity to proactive interference between the previous trial's target set and the current trial's negative response (see Jonides & Nee, 2006).

This single case is particularly interesting as it described the case of an adult with very good general cognitive abilities but with a circumscribed deficit in arithmetic fact storage. We found that she suffered from hypersensitivity-to-interference in memory and we argued that this prevented her from storing very similar associations in memory, such as arithmetic facts. From this new observation, we decided to investigate whether DB was one isolated case of dyscalculia or whether this new theoretical explanation could also account for arithmetic facts deficit in other people.

HYPERSENSITIVITY-TO-INTERFERENCE IN CHILDREN WITH LOW ARITHMETIC SKILLS

The hypersensitivity-to-interference account of difficulties in storing arithmetic facts in memory was first tested in children at the age when they are supposed to build this memory network (De Visscher & Noël, 2014a), namely in fourth grade. Two groups of fourth graders were compared: one characterized by weak arithmetic fluency (i.e., ability to solve many addition or multiplication facts within a given time limit) and the

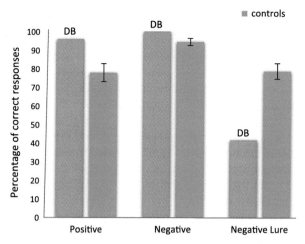

FIGURE 18.1   Percentage of correct responses for DB and the control participants in the recent probe task for the positive, negative, and negative lure trials.

other showing normal arithmetic fluency. Children from the two groups were matched for school, gender, and age. We designed a simple and child-friendly task to assess their sensitivity to proactive interference in the context of associative memory. Children were presented with pairs of pictures associating one famous cartoon character with a location (e.g., beach, shop, garage, etc.) and were asked to remember where each character was. In each task block, children successively saw three different associations of characters and locations. Directly after the three presentations, a verification phase started, offering associations of character–location that the child had to verify (pressing the "true" or "false" button). After three verifications, children were instructed that the characters were traveling and that they had to learn new associations of character–location. The next block then started with the presentation of three new character–location associations. Twenty of these blocks were presented, corresponding to a total of 60 verifications. Half of them were true and half were false. Regarding the true character–location verifications, half of the trials were low interfering because they were new items that were never seen before. The other half was high interfering because the same pictures were associated differently in the previous block, causing proactive interference. For example, in the first block Asterix was in the mountain and Marsupilami was in the shop, and in the second block Asterix was in the shop in both the learning and the verification phase. Regarding the false character–location verifications, some parts of the trials were low interfering because they were composed of one picture from the learning phase and one totally new picture. The other parts of the false verifications were high interfering either because they associated differently one character and one location that had both been encountered during the current block or because they associated differently a character and a location used in the previous block. The two groups of children performed equally well in the verification of low-interfering character–location associations. However, and in accordance with the hypersensitivity-to-interference hypothesis, children with low fluency in arithmetic performed worse than the other group in the high-interfering condition (see Fig. 18.2). Their impairment in the high-interference condition is a clear sign of an abnormal proactive interference, while their normal performance in the low-interference condition indicates that they had typical associative memory in general. These results corroborate the findings from the DB case study and support the hypothesis that hypersensitivity-to-interference in memory can prevent storage of arithmetic facts in memory in children.

HYPERSENSITIVITY-TO-INTERFERENCE AS AN EXPLANATION FOR SPECIFIC IMPAIRMENT IN ARITHMETIC FACTS

Our suggestion is that people who are very sensitive to interference in memory have great difficulties in learning multiplication facts because arithmetic facts are associations made of very similar features. This hypothesis is thus one possible explanation for one very specific difficulty

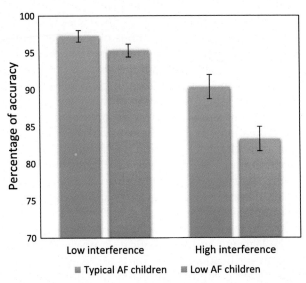

FIGURE 18.2   Percentage of accuracy for associations with low- and high interference in both typical children and children with low arithmetic fluency (AF).

that is frequently encountered in dyscalculia. Yet dyscalculia is heterogeneous and other explanations should be considered for other types of numerical disabilities. Accordingly, we conducted a study that aimed first to examine whether we could replicate the pattern of results of DB in a group of adults presenting dyscalculia and, second, to test the specificity of our hypothesis, i.e., testing whether hypersensitivity-to-interference accounts for dyscalculia characterized by a specific difficulty in learning arithmetic facts (De Visscher, Szmalec, Van der Linden, & Noël, 2015). With these aims in mind, we compared adults with no learning difficulties with two groups of adults with dyscalculia: one group who presented global math impairment and one group selectively impaired in arithmetic facts. We predicted that only the latter would show hypersensitivity-to-interference. Regarding the former group, we tested another hypothesis that has emerged in the scientific literature, namely, the hypothesis of a serial-order learning deficit.

Recently, a number of researchers have questioned the quality of ordinality processing in dyscalculia. Rubinsten and Sury (2011) found that adults with dyscalculia experienced difficulties in judging whether three stimuli (dots or digits) were correctly ordered. Kaufmann, Vogel, Starke, Kremser, and Schocke (2009) showed different brain activations between dyscalculic and control children during ordinal judgment tasks of numerical (digits) and nonnumerical (physical size of symbols) stimuli. Furthermore, Lyons and Beilock (2011) found that the ability to judge whether three Arabic digits are presented in increasing order is highly correlated with arithmetic ability. However, in all these studies, ordinality processing was confounded with magnitude processing, as researchers used digits, amounts of dots, or physical size of items, all of which have a magnitude aspect. A deficit in magnitude processing could in itself explain these results. Accordingly, we (De Visscher, Szmalec et al., 2015) selected a task of serial-order learning that would not involve any processing of magnitude. We hypothesized that a deficit at that level would already impair early numerical learning—in particular, learning the counting sequence—and would lead to global math disability.

To test these hypotheses, we selected the Hebb repetition learning paradigm. In this paradigm, participants have to recall the order of presentation of nine syllables presented sequentially in the center of the screen. After the presentation of the sequence, all syllables are randomly positioned on a virtual circle around a question mark and the participants have to click on the nine syllables in the same order as the presentation. In the Hebb learning paradigm, the sequences' order is random and always different for all the fillers, but one sequence's order is repeated throughout the experiment, without the participants being aware of it. What is typically found is that participants benefit from the repetition of the same order sequence and show an improvement over time in remembering that specific repeated sequence. In our experiment, two sets of meaningless consonant–vowel syllables were used to build three types of sequence: nonrepeated sequences (fillers), interfering repeated sequence, and noninterfering repeated sequence. So, in our paradigm, two sequences were repeated throughout the experiment. The interfering repeated sequence was composed of the same syllables as the fillers, hence increasing proactive interference. Conversely, the noninterfering repeated sequence was constructed from different syllables than those involved in the fillers (and therefore in the interfering sequence), hence reducing proactive interference. The experiment ended when the participant correctly reproduced two successive noninterfering repeated trials, with a maximum of 20 repetitions.

In line with our hypotheses, not only did the performance of participants with dyscalculia differ from the controls but also the profile of performance between the two groups with dyscalculia differed. In agreement with serial-order learning deficit, the group with global dyscalculia was poorer than the controls in retrieving the order of the fillers, thereby already revealing problems in processing order in short-term memory. Regarding long-term memory ability, as measured by the Hebb repetition effect, they needed more repetitions than the control group to learn the noninterfering repeated sequence. Moreover, after learning the noninterfering repeated sequence, we asked participants to travel within the sequence by giving the syllable that came immediately after or two places after a given syllable. They produced many more errors in this task than the controls did. Thus, although they reached the same stopping criterion indicating that they had learned the sequence, long-term consolidation of the acquired sequential information was not as efficient as that of the controls. Importantly, these difficulties were not because of a general memory problem. Indeed, they did not differ from the controls in a word list memory task that did not involve any order constraint. These results support the view of a serial-order learning deficit in this group with global math deficit.

By contrast, the profile of adults with a specific problem with arithmetic facts showed a typical Hebb effect and reached the stopping criterion with the same number of repetitions as the control participants. In addition, the quality of their memory trace for the noninterfering repeated sequence was similar to that of the controls, as evidenced by their performance in answering which syllable came immediately after or two places after a given syllable. These results support the assumption of good serial-order learning skills in this group. Contrariwise, in the interfering situation, their performance dropped drastically and they showed an interference effect while the two other groups did not (see Fig. 18.3).

In sum, these findings supplied new evidence for a link between an arithmetic facts deficit and hypersensitivity-to-interference in adults with

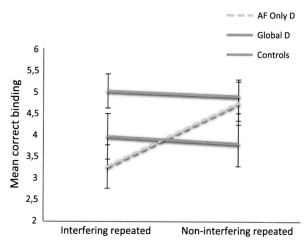

FIGURE 18.3   Mean correct binding (or correct succession of two syllables) in the Hebb paradigm for the interfering and noninterfering repeated sequences in the three groups of participants: controls, with global dyscalculia, and with specific problems in arithmetic facts. *AF*, arithmetic fluency.

arithmetic fact dyscalculia and showed its specific character as it is found in certain profiles of dyscalculia, not in all of them.

AN INDEX OF INTERFERENCE FOR EACH MULTIPLICATION PROBLEM

So far, we have shown that hypersensitivity-to-interference in memory could account for difficulties in learning arithmetic facts, mostly multiplication facts. This has been shown in a case study and in groups of children or adults performing poorly in arithmetic fact tasks. All these studies assumed that arithmetic facts interfere because of features overlap but they did not test this assumption directly.

Such a similarity-based interference in arithmetic facts was already proposed in the network interference model by Campbell (1995). According to this model, when a problem is presented, it activates a physical code and a magnitude code. The former corresponds to the physical representation of the two operands and the operation sign and it activates other problems according to the feature overlap of these operands and sign. A weight of interference is given according to the type of feature overlap (a weight for each common operand and sign). The second corresponds to the magnitude representation of the answers. Globally, the problem node activation depends on the total similarity corresponding to the sum of the feature matching of operands and sign (i.e., physical code) and magnitude similarity of the answers (magnitude code).

Campbell mainly tried to account for differences in errors and response times between the different multiplication problems in typical adults who are supposed to have built a multiplication fact network. He was mainly interested in accounting for differences between problems in the retrieval stage of the answers. Here, we (De Visscher & Noël, 2014b) were interested in understanding how children develop their arithmetic fact network to explain why some never succeed in learning these facts. Taking a learning perspective, one could consider that the first problem–answer associations that a child learns probably do not lead to great difficulties. However, learning would become more problematic if children are hypersensitive-to-interference and learn other problem–answer associations that share many features with the already stored facts. Based on this view, the order of learning arithmetic facts and the degree of overlap between each problem and the already learned problems should matter in the process of learning. Of course, the order in which problems are learned varies from one education system to another, but in general the small tables are learned before the large ones. Accordingly, we assumed that table 2 is learned before table 3, which itself is learned before table 4 and so on. Second, we assumed that learning a new multiplication fact is subject to proactive interference. More precisely, the more features of this new multiplication fact that overlaps with the previously learned ones, the weaker its memory trace will be. To calculate the degree of features overlap, we calculated the frequency of co-occurrences of digits associations between the target problem and the previously learned ones (for both the operands and the answers).

The problems learned in the first place are $2 \times 2 = 4$ and $2 \times 3 = 6$. These do not have any associations of digits in common and therefore have a level of interference of 0. The third problem learned is $2 \times 4 = 8$, which shares the association of 2 and 4 with problem $2 \times 2 = 4$. The interference level is thus 1. Let us consider a larger problem, $3 \times 9 = 27$. Its level of interference is 9. Indeed, this problem shares the association of 2 and 3 with four previously learned problems ($\underline{3} \times \underline{2} = 6$, $4 \times \underline{3} = 1\underline{2}$, $\underline{3} \times 7 = \underline{2}1$, $8 \times \underline{3} = \underline{2}4$), of 2 and 7 with two previously learned problems ($2 \times 7 = 14$, $3 \times 7 = 21$), of 2 and 9 with one other problem ($9 \times 2 = 18$), of 3 and 7 with one other problem ($3 \times 7 = 21$), and of 3 and 9 ($3 \times 3 = 9$), thus leading to a total of 9. It is important to note that the problems encountered later are not necessarily the most interfering. The problem $7 \times 7 = 49$, for instance, is learned 31st but ranked 16th in interference weight. Conversely, the problem $4 \times 8 = 32$ is learned 20th but is the most interfering problem. The measure of the features overlap between the problem and the previously learned ones is therefore a proxy aiming to measure the degree of proactive interference that a problem has undergone during the learning stage. The position of the digits in the associative structure of problem–answer is not considered because the positional value of the features has been shown to not play a role in the feature overlap model of Oberauer and Lange (2008). We intended to use the most parsimonious proxy of proactive interference, but this index could be refined on different levels. With this interference parameter, we wanted to test whether the features overlap in arithmetic fact learning impacts on performance across problems.

We first tested whether this interference parameter could explain differences of performance across multiplication problems by analyzing data published by Campbell (1997). These data were error and response times for solving each multiplication problem collected on 44 undergraduates. We considered jointly the interference parameter and an index of the problem size. Indeed, the problem size is one of the more powerful factors accounting for performance variations across problems and indicates that small problems (i.e., made of small operands) are usually solved more accurately and more rapidly than large problems (Campbell & Graham, 1985; Zbrodoff & Logan, 2005, pp. 331–346). Both the problem size and the interference parameter played a significant and unique role in accounting for differences in reaction time and accuracy across multiplication problems.

In addition, we collected data from third-grade children, fifth-grade children, and undergraduates and analyzed the error rates and response times according to the problem size and interference parameter. Regarding accuracy, the problem size explained a substantial part of the variance in performance in each of the three age groups, while the interference parameter explained variance in the third-grade group only. Regarding response time, the interference parameter was significant in all age groups and the problem size accounted for independent and significant explanatory power only in undergraduates (and was marginally significant in third and fifth graders). Concretely, the time needed to solve a multiplication increases as the level of interference increases throughout the child's development.

We were also interested in accounting for interindividual differences. Accordingly, we measured sensitivity to the interference parameter and/or sensitivity to the problem size for each individual (based on the regression slopes of the person's response times). The two factors significantly accounted for interindividual differences in multiplication response time in each of the three groups. More precisely, children or adults who were more sensitive to the interference parameter were globally slower to solve multiplications.

Coming back to our patient DB, we examined her sensitivity to the interference parameter and to the problem size. It appeared that the interference parameter but not the problem size accounted for her response time in multiplication. Moreover, DB's sensitivity to the interference parameter was larger than that of the controls. These current results corroborate the previous findings and clearly show that the abnormally slow response time of DB in multiplication was because of her hypersensitivity-to-interference weight of the problems.

Finally, we considered the data obtained in fourth-grade children who had poor versus normal arithmetic facts fluency (see De Visscher & Noël, 2014a) and examined whether the interference parameter and/or the problem size effect predicted the individual differences in multiplication in fourth grade and 1 year later when they were in fifth grade. Again, the slope of the problem size and the interference parameter were calculated for each child and then used to account for interindividual differences in accuracy and speed in multiplication. Sensitivity to the interference parameter proved to be a significant explanation of differences in multiplication accuracy, with higher sensitivity being associated with lower accuracy. Being sensitive to the problem size was less influential. Finally, only the interference parameter significantly accounted for differences in time needed for solving multiplications. When considering multiplication performance 1 year later when the children were in fifth grade, sensitivity to the interference parameter remained strongly associated with lower accuracy and longer response times in multiplications.

We then verified the assumption according to which hypersensitivity-to-interference disturbs the learning stage and consequently hampers retrieval from memory by using a time-limited multiplication task and a multiplication table membership judgment task. The multiplication production task without time limit allowed children to use different strategies, such as computing strategies or retrieval strategies. We therefore used a multiplication production task in which children had only 2 s to answer, forcing them to use a retrieval strategy. Results showed that children who were more sensitive to the interference parameter of multiplication were also those who performed worse in the time-limited multiplication task. This finding supports the idea that hypersensitivity-to-interference prevents storage of arithmetic facts in memory, hampering the retrieval of the answer to arithmetic problems.

Then, we used a multiplication table membership judgment task to distinguish between a storage versus an access deficit. Indeed, when a child can recognize whether a number belongs to a multiplication table's answers or not, it means that he/she has to a certain extent created an arithmetic facts network. Conversely, difficulties in this task reveal a storage deficit. Results showed that children who were the more sensitive to the interference parameter of multiplication were also slower in the table membership judgment task.

In summary, this interference parameter accounted for variation of performance across multiplication problems in adults (Campbell's data in 1997 and our own data) and in children (third grade and fifth grade). Sensitivity to this interference parameter also accounted for interindividual differences with a higher sensitivity being observed in children and adults who are slow in multiplication solving, in children with low arithmetic fluency, and in the patient DB. Finally, children who are more sensitive to this interference parameter are performing weaker in a time-limited multiplication task and in a table membership judgment task, suggesting that they failed to build an appropriate arithmetic fact network.

INDEX OF INTERFERENCE AND PROBLEM SIZE

As already mentioned, Campbell (1995) proposed that both a physical and a magnitude code are activated when a problem is presented. The index of interference just described is a way of assessing the interference that can be associated with the physical code. The magnitude code would be activated for the answer and would explain the problem size effect, i.e., the fact that smaller single-digit multiplications are solved more quickly and accurately than larger single-digit multiplications (e.g., 2×3 compared with 7×8). Indeed, problems would activate the magnitude representation of the answers and because the magnitude representation is increasingly compressed with larger numbers (Dehaene, 1992), the magnitude representations of large answers would be more difficult to discriminate than those of small answers. Because of their higher similarity in the magnitude code, larger problems would create more interference, which would increase the retrieval time. Together with De Smedt (De Visscher, Noël, & De Smedt, 2016), we have measured the unique influence of the physical representation and the numerical magnitude representation on the multiplication performance of 79 fourth graders. To measure the influence of the magnitude code, we presented a magnitude comparison task to the participants, in which they had to compare two numbers, which actually corresponded to two close answers of multiplication problems (e.g., 42 and 45). The underlying idea was that if they have a very precise number magnitude representation, they would perform better in this task and show less overlap of magnitude code for problems close in magnitude; according to Campbell's model, this should lead to better performance in multiplication.

The time taken to compare two close answers was indeed related to children's performance in multiplication, but it was also related to their performance in addition, subtraction, division, in a mixture of all these, as well as to performance in a general math test that included word problems, geometry, and other mathematical knowledge. The ease of processing number magnitude is therefore a factor related to all kinds of mathematical dimensions because all numerical tasks require the processing of number magnitude.

By contrast, sensitivity to the interference parameter was specifically related to performance in operations that are mainly tapping facts stored

In memory (i.e., additions and multiplications) but not to subtractions, mixed operation calculations, or to a general mathematics test. Moreover, sensitivity to this physical similarity interference uniquely explained the use of retrieval strategies. Indeed, more interfering problems were less often solved via retrieval and individuals with higher sensitivity to physical interference also had lower frequencies of retrieval use. Physical representation therefore seems to have an important and specific role in storing and retrieving arithmetic facts in long-term memory.

HYPERSENSITIVITY-TO-INTERFERENCE IN MEMORY AND CLOSE CONCEPTS

Finally, to better understand this "hypersensitivity-to-interference in memory," we tried to disentangle it from other closely related concepts. Firstly, we explored whether sensitivity to the interference parameter in multiplication reflected a general sensitivity to interference in memory or if it was specific to numbers. We measured the correlation between two measures of sensitivity to interference, the one measured in the multiplication task (sensitivity to the interference parameter) and one measured in a nonnumerical task, the associative memory task of character–location explained earlier. A positive correlation was obtained between sensitivity to the interference parameter in multiplication and sensitivity to interference in this nonnumerical associative memory task (Spearman's rho = .506).

Secondly, we wanted to determine whether this sensitivity to interference was a specific concept or whether it was confused with closely related concepts such as inhibition capacities, verbal memory, or associative memory capacities. To that end, we tested inhibition capacities with a Stroop color task, verbal memory capacities with a word list memory task and associative memory capacities with a (noninterfering) paired-associates memory task. Sensitivity to interference in the nonnumerical associative memory task did not correlate with any of these measures, showing the specificity of that concept.

In summary, the sensitivity to the interference parameter in multiplication reflects a general sensitivity to interference in memory. This sensitivity to interference in memory is a specific concept that should not be confounded with close concepts such as inhibition, memory, or associative memory.

DISCUSSION

This chapter aimed at understanding what could account for the difficulty in storing arithmetic facts in memory that is frequently encountered in children with math learning disability and considered hypersensitivity-to-interference in memory as a possible cause.

After the learning stage, arithmetic facts are stored in an interconnected network in memory, in which not only the correct response but also other associated incorrect responses are activated (e.g., Campbell, 1987a, 1987b). Accordingly, some researchers have proposed that difficulty in retrieving arithmetic facts from memory might be because of weak inhibition capacities preventing the individual from suppressing the interference of competitors (Barrouillet et al., 1997). However, this hypothesis has not received much empirical support (Bellon et al., 2016; Censabella & Noël, 2008).

More recently, we (De Visscher & Noël, 2013) considered the learning stage of arithmetic facts and proposed that arithmetic fact deficit might be caused by hypersensitivity-to-interference in memory. We argued that memorizing arithmetic facts amounts to storing associations of operands with answers and that these associations share many common features, which creates interference. Individuals who show heightened sensitivity to this kind of interference would be hampered in creating distinct memory traces for each of the arithmetic facts.

In a way, this hypothesis is congruent with previous finding of Censabella and Noël (2004) who found that the time taken to solve multiplications was correlated with adults' sensitivity to the fan effect. Indeed, several authors consider that the fan effect is due to proactive interference in memory. For instance, Hasher, Tonev, Lustig, and Zacks (2001), chap. 23, pp. 286–297, wrote, "These fan effects are widely seen as the basis for proactive interference" (p. 293, see also, Bunting, Conway, & Heitz, 2004).

This hypothesis of a hypersensitivity-to-interference was proposed to explain the profile of a patient showing a specific impairment in arithmetic facts (De Visscher & Noël, 2013) and was later supported in a group study of children (De Visscher & Noël, 2014a) and of adults (De Visscher, Berens, Keidel, Noël, & Bird, 2015; De Visscher, Szmalec et al., 2015) with difficulties in arithmetic facts. Furthermore, it was shown to be a hypothesis accounting for specific disorder in learning arithmetic facts rather than global math difficulties (De Visscher, Berens et al., 2015; De Visscher, Szmalec et al., 2015).

Subsequently, we (De Visscher & Noël, 2014b) went further in understanding how this proactive interference took place when learning multiplication facts. We created an index for each multiplication problem, aiming at assessing the amount of proactive interference each problem received from the previous learned ones. This index of interference accounted for a significant part of the variance across the different multiplication problems, beyond the well-known problem size effect. The more a problem had a high index of interference, the slower and the more error prone was the answer, and the less frequent retrieval strategy was used (De Visscher et al., 2016). Sensitivity to this index of interference also accounted for differences across individuals. Individuals with high sensitivity to this index of interference were slower to solve multiplications and used fewer retrieval strategies (De Visscher et al., 2016). This has been shown in populations of unselected children and in adults (De Visscher & Noël, 2014b). Regarding atypical arithmetic development, high sensitivity to this index of interference was characteristic of the patient DB and of children with weak arithmetic fact fluency (De Visscher & Noël, 2014b).

Sensitivity to the parameter of interference specifically accounted for interindividual differences in arithmetic tasks that mostly required recalling facts from memory and not in arithmetic operations that are usually solved through procedural strategies or performance in a global math test (De Visscher et al., 2016). In addition, in the case of atypical development, this sensitivity to the interference parameter in multiplication correlated with a measure of nonnumerical sensitivity to interference in memory. That is, when testing participants with arithmetic fact deficit, both in adults and children, a correlation between sensitivity to interference in arithmetic and in a nonnumerical domain was found. Finally, sensitivity to interference has been distinguished from the capacities to inhibit dominant responses or global memory capacities (De Visscher & Noël, 2014b). This concept thus proved to be very specific and to account for very specific difficulties in mathematics.

Several of these studies have compared individual(s) with impairment in arithmetic facts to control individuals. We do not yet know whether one needs to have a certain degree of resistance to interference in memory to be able to constitute a normal arithmetic fact network or whether there is a linear relationship between arithmetic fact storing and sensitivity to interference. Further research might also consider the consequences of becoming more sensitive to interference, due to aging or brain damage: to what extent would this affect the person's ability to retrieve answers from an already constituted arithmetic fact network or the ability to learn new arithmetic facts?

Further work should also examine what other areas of learning could suffer from hypersensitivity-to-interference and could thus be expected to be impaired along with arithmetic facts. One possible avenue to consider is dyslexia. Indeed, people with dyslexia often report problems with fact retrieval (see Göbel, 2015, pp. 680–715 for a review). Bogaerts et al. (2015) found a greater difficulty in resisting proactive interference in adults with dyslexia than in typical adults. This sensitivity to interference may perhaps account for some of the reading difficulties of people with dyslexia and for associated deficits in arithmetic. However, one needs to clarify the process by which this sensitivity to interference might account for reading problems. One possibility is to consider learning the grapheme-phoneme conversions, which may include several shared features. For instance, in French, we have proper sounds for the combinations of the letters in, on, an, ain, un, en, ou, etc.

It is interesting to note that at the brain level, the left angular gyrus is involved in both phonological and in calculation tasks (Göbel, 2015, pp. 680–715). Dysfunction of brain areas centered in and around the left angular gyrus is found in dyslexia and functional dysconnectivity is mainly found in tasks that require phonological assembly (Pugh et al., 2000). This region also proved to be sensitive to the level of interference of multiplication problems. De Visscher, Berens et al. (2015) compared the brain activation of adults while they were verifying multiplication problems. They contrasted small and large problems and, in each of these categories, problems characterized by a low- or high index of interference. They found specific modulation of brain activity according to the level of interference of the problem in the left angular gyrus with greater activation for low-interfering problems than for high-interfering problems. Further research might try to understand if there is a common underlying process that accounts for this activation in the left angular gyrus for both phonological tasks and problems according to their level of interference.

According to Ansari (2008), the angular gyrus is engaged in general mapping processes. For instance, it is considered to be involved in associating symbols with nonsymbolic events, such as written letters with sounds (Booth et al., 2004). In arithmetic, Grabner, Ansari, Koschutnig, Reishofer, and Ebner (2013) suggested that it could mediate the automatic mapping of arithmetic problems onto answers in long-term memory. As the level of activation of the angular gyrus was seen to be affected by the degree of interference of multiplication problems (De Visscher, Berens et al., 2015; De Visscher, Szmalec et al., 2015), one possibility would be to consider that this mapping to information stored in memory would vary according to the level of proactive interference in establishing the memory trace. In the particular case of multiplication, the low-interfering problems would lead to a highly automated mapping with their corresponding answers in memory compared with high-interfering problems.

All this work supports the view that similarity between arithmetic facts provokes interference already during the learning stage and that individuals who are hypersensitive-to-interference struggle in learning those facts. The next and most important question, then, is what can be done to help those individuals with hypersensitivity-to-interference learning arithmetic facts. Clinical research should find ways to improve their ability to resist interference or ways to increase the distinctiveness of the different arithmetic facts so as to decrease similarity-based interference (perhaps by adding differentiating—but irrelevant—features). Further research is needed to answer these questions.

Acknowledgments

The two authors are supported by the FSR-FNRS Fonds National de la recherche scientifique de Belgique (National Research Fund of Belgium).

References

Anderson, M. C., & Bjork, R. A. (1994). Mechanisms of inhibition in long-term memory: A new taxonomy. In D. Dagenbach, & T. H. Carr (Eds.), *Inhibitory processes in attention, memory, and language*. San Diego, CA: Academic Press.

Anderson, J. R. (1974). Retrieval of propositional information from long-term memory. *Cognitive Psychology, 6*, 451–474.

Anderson, J. R. (1983). A spreading activation theory of memory. *Journal of Verbal Learning and Verbal Behavior, 22*, 261–295.

Ansari, D. (2008). Effects of development and enculturation on number representation in the brain. *Nature Reviews Neuroscience, 9*, 278–291.

Barrouillet, P., Fayol, M., & Lathulière, E. (1997). Selecting between competitors in multiplication tasks: An explanation of the errors produced by adolescents with learning difficulties. *International Journal of Behavioral Development, 21*, 253–275.

Barrouillet, P., & Thevenot, C. (2013). On the problem-size effect in small additions: Can we really discard any counting-based account? *Cognition, 128*(1), 35–44.

Bellon, E., Fias, W., & De Smedt, B. (June 2016). Are individual differences in arithmetic fact retrieval in children related to inhibition? *Frontiers in Psychology, Vol. 7* Article 825.

Bogaerts, L., Szmalec, A., Hachmann, W. M., Page, M. P. A., Woumans, E., & Duyck, W. (2015). Increased susceptibility to proactive interference in adults with dyslexia ? *Memory, 23*, 268–277. https://doi.org/10.1080/09658211.2014.882957.

Booth, J. R., Burman, D. D., Meyer, J. R., Gitelman, D. R., Parrish, T. B., & Mesulam, M. M. (2004). Development of brain mechanisms for processing orthographic and phonologic representations. *Journal of Cognitive Neuroscience, 16*, 1234–1249.

Bull, R., & Scerif, G. (2001). Executive functioning as a predictor of children's mathematics ability: Inhibition, switching and working memory. *Developmental Neuropsychology, 19*(3), 273–293.

Bunting, M. F., Conway, A. R. A., & Heitz, R. P. (2004). Individual differences in the fan effect and working memory capacity. *Journal of Memory and Language, 51*, 604–622.

Campbell, J. I. D. (1987a). Network interference and mental multiplication. *Journal of Experimental Psychology: Learning, Memory, and Cognition, 13*(1), 109–123.

Campbell, J. I. D. (1987b). Production, verification, and priming of multiplication facts. *Memory and Cognition, 15*, 349–364.

Campbell, J. I. D. (1995). Mechanisms of simple addition and multiplication: A modified network-interference theory and simulation. *Mathematical Cognition, 1*(2), 121–164.

Campbell, J. I. D. (1997). On the relation between skilled performance of simple division and multiplication. *Journal of Experimental Psychology: Learning, Memory, and Cognition, 23*(5), 1140–1159.

Campbell, J. I. D., & Graham, D. J. (1985). Mental multiplication skill: Structure, process, and acquisition. *Canadian Journal of Psychology, 39*, 338–366.

Campbell, J. I. D., & Tarling, D. P. M. (1996). Retrieval processes in arithmetic production and verification. *Memory and Cognition, 24*, 156–172.

Censabella, S., & Noël, M. P. (2004). Interference in arithmetic facts: Are active suppression processes involved when performing simple mental arithmetic? *Cahiers de Psychologie Cognitive/Current Psychology of Cognition, 22*, 635–671.

Censabella, S., & Noël, M.-P. (2005). The inhibition of exogenous information in children with learning disabilities. *Journal of Learning Disabilities, 38*, 400–410.

Censabella, S., & Noël, M.-P. (2008). The inhibition capacities of children with mathematical disabilities. *Child Neuropsychology, 14*(1), 1–20.

Conlin, J. A., Gathercole, S. E., & Adams, J. W. (2005). Stimulus similarity decrements in children's working memory span. *Quarterly Journal of Experimental Psychology Section A – Human Experimental Psychology, 58*(8), 1434–1446.

Conrad, R., & Hull, A. J. (1964). Information, acoustic confusion and memory span. *British Journal of Psychology, 55*, 429–432.

Cragg, L., & Gilmore, C. (2014). Skills underlying mathematics: The role of executive function in the development of mathematics proficiency. *Trends in Neuroscience and Education, 3*, 63–68.

De Visscher, A., Berens, S. C., Keidel, J. L., Noël, M.-P., & Bird, C. M. (August 1, 2015). The interference effect in arithmetic fact solving: An fMRI study. *NeuroImage, 116*, 92–101. https://doi.org/10.1016/j.neuroimage.2015.04.063.

De Visscher, A., & Noël, M. P. (2013). A case study of arithmetic facts dyscalculia caused by a hypersensitivity-to-interference in memory. *Cortex, 49*(1), 50–70.

De Visscher, A., & Noël, M.-P. (2014a). Arithmetic facts storage deficit: The hypersensitivity-to-interference in memory hypothesis. *Developmental Science*, 1–9. https://doi.org/10.1111/desc.12135.

De Visscher, A., & Noël, M.-P. (2014b). The detrimental effect of interference in multiplication facts storing: Typical development and individual differences. *Journal of Experimental Psychology: General, 143*(6), 2380–2400.

De Visscher, A., Noël, M.-P., & De Smedt, B. (2016). The role of physical digit representation and numerical magnitude representation in children's multiplication fact retrieval. *Journal of Experimental Child Psychology, 152*, 41–53.

De Visscher, A., Szmalec, A., Van der Linden, L., & Noël, M.-P. (2015b). Serial-order learning impairment and hypersensitivity-to-interference. *Cognition, 144*, 38–48.

Dehaene, S. (1992). Varieties of numerical abilities. *Cognition, 44*, 1–42.

Espy, K. A., McDiarmid, M. M., Cwik, M. F., Stalets, M. M., Hamby, A., & Senn, T. E. (2004). The contribution of executive functions to emergent mathematic skills in pre-school children. *Developmental Neuropsychology, 26*(1), 465–486.

Garnett, K., & Fleischner, J. E. (1983). Automatization and basic fact performance of normal and learning disabled children. *Learning Disability Quarterly, 6*, 223–230.

Geary, D. C., Hoard, M. K., & Hamson, C. O. (1999). Numerical and arithmetical cognition: Patterns of functions and deficits in children at risk for a mathematical disability. *Journal of Experimental Child Psychology, 74*(3), 213–239.

Gilmore, C., Attridge, N., Clayton, S., Cragg, L., Johnson, S., Marlow, N., et al. (2013). Individual differences in inhibitory control, not non-verbal number acuity, correlate with mathematics achievement. *PLoS One, 8*(6), e67374.

Göbel, S. (2015). Number processing and arithmetic in children and adults with reading difficulties. In R. C. Kadosh, & A. Dowker (Eds.), *The Oxford handbook of numerical cognition*. Oxford Library of Psychology.

Grabner, R., Ansari, D., Koschutnig, K., Reishofer, G., & Ebner, F. (2013). The function of the left angular gyrus in mental arithmetic: evidence from the associative confusion effect. *Human Brain Mapping, 34*(5), 1013–1024. https://doi.org/10.1002/hbm.21489.

Hall, J. F. (1971). Formal intralist response similarity: Its role in paired-associate learning. *American Journal of Psychology*, *84*(4), 521–528.

Hasher, L., Tonev, S. T., Lustig, C., & Zacks, R. T. (2001). Inhibitory control, environmental support, and self-initiated processing in ageing. In M. Naveh-Benjamin, M. Moscovictch, & H. L. Roediger (Eds.), *Perspectives on human memory and cognitive aging*. New york: Psychology Press.

Jonides, J., & Nee, D. E. (2006). Brain mechanisms of proactive interference in working memory. *Neuroscience*, *139*(1), 181–193.

Kaufmann, L., Vogel, S. E., Starke, M., Kremser, C., & Schocke, M. (2009). Numerical and non-numerical ordinality processing in children with and without developmental dyscalculia: Evidence from fMRI. *Cognitive Development*, *24*(4), 486–494. https://doi.org/10.1016/j.cogdev.2009.09.001.

Logan, G. D. (1994). On the ability to inhibit thought and action: A users' guide to the stop signal paradigm. In D. Dagenbach, & T. H. Carr (Eds.), *Inhibitory processes in attention, memory, and language*. San Diego, CA: Academic Press.

Lyons, I. M., & Beilock, S. L. (2011). Numerical ordering ability mediates the relation between number-sense and arithmetic competence. *Cognition*, *121*(2), 256–261. https://doi.org/10.1016/j.cognition.2011.07.009.

McCloskey, M., & Lindemann, A. M. (1992). MATHNET: Preliminary results from a distributed model of arithmetic fact retrieval. In J. I. D. Campbell (Ed.), *The nature and origins of mathematical skills*. Amsterdam: Elsevier.

Monsell, S. (1978). Recency, immediate recognition, and reaction time. *Cognitive Psychology*, *10*, 465–501.

Oberauer, K., & Kliegl, R. (2006). A formal model of capacity limits in working memory. *Journal of Memory and Language*, *55*(4), 601–626.

Oberauer, K., & Lange, E. B. (2008). Interference in verbal working memory: Distinguishing similarity-based confusion, feature overwriting, and feature migration. *Journal of Memory and Language*, *58*(3), 730–745. https://doi.org/10.1016/j.jml.2007.09.006.

Pugh, K. R., Mencl, W. E., Shaywitz, B. A., Shaywitz, S. E., Fulbright, R. K., Constable, R. T., et al. (2000). The angular gyrus in developmental dyslexia: Task-specific differences in functional connectivity within posterior cortex. *Psychological Science*, *11*, 51–56.

Rubinsten, O., & Sury, D. (2011). Processing ordinality and quantity: The case of developmental dyscalculia. *PLoS One*, *6*(9), e24079. https://doi.org/10.1371/journal.pone.0024079.

St Clair-Thompson, H. L., & Gathercole, S. E. (2006). Executive functions and achievements in school: Shifting, updating, inhibition, and working memory. *Quarterly Journal of Experimental Psychology*, *59*(4), 745–759.

Underwood, B. J. (1983). *Attributes of memory*. Glenview, IL: Scott Foresman.

van der Sluis, S., de Jong, P. F., & van der Leij, A. (2004). Inhibition and shifting in children with learning deficits in arithmetic and reading. *Journal of Experimental Child Psychology*, *87*, 239–266.

Verguts, T., & Fias, W. (2005). Interacting neighbours: A connectionist model of retrieval in single-digit multiplication. *Memory and Cognition*, *36*, 1–16.

Zbrodoff, N. J., & Logan, G. D. (2005). What everyone finds: The problem-size effect. In J. I. D. Campbell (Ed.), *Handbook of mathematical cognition*. New York: Psychology Press.

CHAPTER

19

Working Memory for Serial Order and Numerical Cognition: What Kind of Association?

Steve Majerus[1,2], Lucie Attout[1]

[1]Université de Liège, Liège, Belgium; [2]Fund for Scientific Research FNRS, Brussels, Belgium

OUTLINE

NUMERICAL COGNITION: THE IMPORTANCE OF WORKING MEMORY FOR SERIAL ORDER

Many studies have explored the associations between working memory (WM) and numerical cognition by considering that WM capacity may be an important determinant of performance in numerical tasks, such as mental calculation. Indeed, in these tasks, the target numerical information (the numbers and the operators of the arithmetic problem) has to be temporarily maintained in WM and the operations that are carried out on the numbers and any intermediate operations and results. At the same time, the studies that have explored the associations between WM capacity and mental calculation show an inconsistent picture, especially as regards the involvement of WM storage abilities studies (e.g., De Smedt et al., 2009; Gathercole & Pickering, 2000a; Holmes, Adams, & Hamilton, 2008; Jarvis & Gathercole, 2003; Noël, 2009; Swanson & Kim, 2007; see Raghubar, Barnes, & Hecht, 2010, for a review). Many studies suggest an association between executive processes of WM tasks, which involve processing of temporarily stored information, and numerical cognition (e.g., Bull & Johnston, 1997; De Smedt et al., 2009; Gathercole & Pickering, 2000b; Imbo, Vandierendonck, & Vergauwe, 2007; Noël, 2009). But studies exploring the association between WM *storage* and numerical abilities have led to contradictory results, especially as regards the involvement of *verbal* temporary storage capacities (De Smedt et al., 2009; Holmes & Adams, 2006; Noël, Seron, & Trovarelli, 2004).

Those studies typically did not distinguish between item and serial-order storage capacities, two aspects which have been shown to be critical when exploring the associations between WM storage capacity and performance in other cognitive domains (Majerus, Poncelet, Greffe & Van der Linden, 2006; Martinez Perez, Majerus, & Poncelet, 2012). Item information involves the maintenance of the identity of the memoranda and their temporary activation in long-term memory bases (such as language knowledge for verbal stimuli), whereas serial-order information involves the order in which the memoranda have been presented (e.g., Nairne & Kelley, 2004). Several studies have shown that the serial-order component of WM may be particularly predictive of performance in cognitive tasks that require the processing of sequentially organized information, such as numerical information. Attout, Noël, and Majerus (2014) conducted a longitudinal study in typical developing children first tested when they were in third year kindergarten. The authors measured WM for item information by presenting a delayed nonword repetition task; on each trial, a single nonword with a consonant-vowel-consonant structure was presented to minimize serial-order retention requirements while maximizing sublexical phonological retention requirements. Serial-order WM was assessed via a serial-order reconstruction task; the children heard sequences of familiar animal names in different orders, and for each sequence, they had to reconstruct the order of the animals using cards on which the animals that had occurred (and only those) were depicted. This procedure ensured that item-processing requirements were minimized. Attout, Fias, Salmon, and Majerus (2014) and Attout, Noël et al. (2014) showed that performance on the serial-order WM but not on the item WM task predicted performance on mental calculation tasks (additions and subtractions), after control of verbal and nonverbal intelligence estimates. Furthermore, Attout and Majerus (2015) showed that children with dyscalculia present impaired performance for verbal WM tasks that maximize serial-order retention requirements, while showing no deficits for verbal WM tasks that maximize item retention requirements (see also De Visscher, Szmalec, Van Der Linden, & Noël, 2015). Attout, Salmon, and Majerus (2015) observed the same pattern of results in a subsequent study with adult participants presenting a history of developmental dyscalculia.

These results suggest a specific association between serial-order WM abilities and numerical abilities. They also allow us to clarify the inconsistent findings about verbal WM impairment in children with dyscalculia observed in previous studies, given that most of these studies used standard verbal WM tasks that confound serial-order and item retention abilities. The studies by Attout et al. indeed show that depending on the type of information that is assessed, verbal WM performance is either preserved or impaired.

However, while the studies described here indicate a close association between serial-order WM and numerical cognition, they do not directly inform us about the reasons for these associations. The most straightforward explanation would be to consider that numerical tasks, such as mental calculation tasks, require temporary sequential storage abilities for maintaining the numbers and operators of a calculation problem in correct order in WM and also for relating the successive results of the different operations and suboperations to each other. This functional link between WM and numerical tasks is supported by the results from the longitudinal study by Attout, Fias et al. (2014), and Attout, Noël et al. (2014), showing that early serial-order WM abilities predict later calculation abilities, while the reverse prediction is not observed. However, as we will see, the association between serial-order WM and numerical cognition may also reflect deeper, representational links because of the possible sharing of codes used for representing order information in the numerical and WM domains. We consider here that numbers are not only associated with codes informing about their numerical magnitude but also with codes informing about their ordinal position in the numerical chain, in line with recent studies suggesting that magnitude and ordinal processing can dissociate at behavioral and neural levels (Lyons & Beilock, 2013; Turconi, Jemel, Rossion, & Seron, 2004). Before examining further this representational account of the links between serial-order WM and numerical cognition, we need to examine how current models of WM explain the representation and maintenance of serial-order information.

WORKING MEMORY: THE NATURE OF
SERIAL-ORDER CODES

The WM literature is particularly rich as regards theoretical accounts for the representation of serial order. While differing in the precise manner serial-order representations are implemented, the vast majority of WM models currently consider distinct representational levels for item and order information. This is in line with an increasing number of empirical studies suggesting a separation of cognitive and neural processes involved in the maintenance of item versus serial-order information, such as the studies by Attout et al. discussed in the previous section (see also Hachmann et al., 2014; Henson, Hartley, Burgess, Hitch, & Flude, 2003; Martinez-Perez et al., 2012; Majerus, Poncelet, Greffe et al., 2006; Majerus et al., 2010; Majerus, Attout, Artielle, & Van der Kaa, 2015; Nairne & Kelley, 2004, and later sections of this chapter).

Models considering a separation between item and serial-order representational levels assume the existence of specific context signals for encoding serial position information. These context signals can take different forms. Some models consider that the context signal is a unidimensional primacy marker, which is most active for items occurring at the beginning of a list (e.g., Page & Norris, 1998). Other models assume more specific codes for the representation of serial order. One of these models is the Start-End Model proposed by Henson (1998). This model assumes that items are associated to nodes representing the start and the end of the list, respectively. Early items of a WM list will be strongly associated with the start node and weakly with the end node, while the reverse will be true for end-of-list items; items of the middle of a WM list will show intermediate levels of association with both the start and the end nodes. This model was one of the first positional models of serial-order WM by explicitly assuming that items of a WM list are associated with distinct codes as a function of their serial position within a list. At the same time, these positional codes are still rudimentary, as they code for only two positional dimensions: start-of-list and end-of-list.

The idea of position-specific codes for serial order has been further developed by proposing context signals that contain specific codes for each serial position. One instance of this type of model is the architecture proposed by Burgess and Hitch (1999, 2006). Burgess and Hitch proposed a mechanism by which item representations are associated with a context signal dynamically changing for each successively presented item during WM list encoding. At recall, a replay of the context signal and its successive states allows to retrieve the items associated with each state and to output the items in correct serial position. This model assumes that each serial position is represented by a specific code, although the nature of this code is not further specified. Brown, Preece, and Hulme (2000) proposed an even more explicit architecture by assuming that the contextual signals are time-based signals such as oscillators. In this type of architecture, each item is associated with the successive peaks of the oscillator, which can be compared with the advancement of the hands of a clock. The authors further assume that multiple oscillators can be in operation, oscillating at different speeds and allowing to encode serial position information simultaneously at different time scales, as needs to be the case when items are temporally grouped; in that case, the order of the items can be defined not only relative to their order in the whole memory list but also relative to their order within the temporal subgroups. In sum, Brown et al. assume that serial position information is encoded via time-based codes that have an intrinsically ordinal organization, just like numerical codes. This type of account has been recently refined by Hartley, Hurlstone, and Hitch (2016). The authors proposed an architecture containing a population of oscillators for encoding serial position information at different levels of the list, with the oscillators fitting best the temporal regularities of a stimulus sequence being chosen in a bottom-up manner, based on the amplitude modulations of the speech envelope.

A final model comes even closer to the notion of ordinal codes for the representation of serial-order information by proposing numerical rank-based ordinal codes. In their neurocomputational model, Botvinick and Watanabe (2007) proposed that numerical rank-order information is associated with each item representation as a function of the order of presentation of each item. A similar assumption characterizes the context serial order in a box model (Lewandowsky & Farrell, 2008) and the first model proposed by Burgess and Hitch (1992); these models also considered that serial order is encoded via a context signal that reflects absolute position from the start of the sequence, such as numerical rank-order information.

It should be noted that a few models do not consider a separation between item and serial-order codes but consider that serial order is encoded via item-by-item chaining mechanisms by intrinsic encoding energy levels that decrease with each new incoming item or by cues inherent to phonological item information (e.g., Botvinick & Plaut, 2006; Farrell & Lewandowsky, 2002; Jones, Madden, & Miles, 1992; Lewandowsky & Murdock, 1989). These models typically have difficulties in accounting for several critical phenomena of serial-order processing such as temporal grouping or in dealing with item repetitions (Farrell & Lewandowsky, 2002; Hartley et al., 2016; Hurlstone, Hitch, & Baddeley, 2014). Furthermore, as already mentioned, an increasing number of studies suggest that capacities for maintaining item and serial-order information can dissociate at both behavioral and neural levels (Majerus, Poncelet, Greffe et al., 2006; Majerus et al., 2010, 2015; Henson et al., 2003; Majerus et al., 2015; Marshuetz, Smith, Jonides, DeGutis, & Chenevert, 2000). Importantly, distinct variables influence item and serial-order recall, with linguistic factors such as word frequency and semantic knowledge affecting mainly item recall (Nairne & Kelley, 2004; Poirier & Saint-Aubin, 1995); conversely, articulatory suppression, temporal grouping, and rhythmic grouping exert stronger effects on serial order than on item memory performance

(Henson et al., 2003). Hence, there is currently no strong evidence in favor of serial-order WM models that consider that item and serial-order information are represented via a single, common mechanism.

This short review of the most representative serial-order WM models allows us to consider the associations observed between serial-order WM and numerical cognition in a new light. One common feature of most of the models reviewed here is the use of domain-general positional codes, which in addition, for some models, have an inherent ordinal organization, such as temporal or numerical signals. By comparing these theoretical proposals with the empirical data presented in the first section of this chapter, we can raise the hypothesis that serial-order WM and numerical cognition are not only associated because of the involvement of WM resources in numerical tasks but also that their association may be because of a common activation of ordinal and possibly numerical codes. In other words, the codes used to represent the ordinality of number information may also be those used to represent the serial positions in WM. In the next section, we will first examine the evidence in favor of domain-general codes used for representing serial-order information across different WM domains as implicitly or explicitly assumed by current theoretical models of serial-order WM. After this more general focus on the existence of domain-general serial-order codes across WM domains, we will examine more specific empirical evidence for shared ordinal codes in the numerical and the WM domains.

EVIDENCE FOR DOMAIN-GENERAL CODES FOR THE REPRESENTATION OF SERIAL ORDER IN WM

An increasing number of empirical studies are supporting the assumption of domain-general mechanisms for the representation and maintenance of serial-order information in WM. A first part of evidence stems from studies that have compared serial-order processing across verbal and visuospatial WM domains. These studies show that several hallmark effects of serial-order WM in the verbal domain can also be observed in the visuospatial domain (see Hurlstone et al., 2014, for a recent review). A first hallmark effect concerns the overall shape of the serial response curve, which is marked by a strong primacy and smaller recency effect, reflecting better recall of initial and late items as compared with middle of list items. These primacy and recency effects are very robust in verbal immediate serial recall (e.g., Lee & Estes, 1981) and have also been observed for various visuospatial material such as spatial locations (Guérard & Tremblay, 2008), visual configurations (Avons, 1998), and unknown faces (Smyth, Dennis, & Hitch, 2005; Ward, Avons, & Melling, 2005).

The type of serial-order errors observed during serial recall provide further and also more direct evidence for the similarity of the serial-order maintenance processes involved in verbal and visuospatial domains.

Of special importance here are the transposition gradients that reflect a locality constraint: serial position exchanges are more likely to occur for adjacent than for more distant items, with the likelihood of serial position exchange errors decreasing linearly with the increase of the distance between the two serial positions. This gradient is very typical for verbal immediate serial recall tasks (e.g., Burgess & Hitch, 1999; Henson, 1998) and is also observed for the reproduction of visuospatial sequences (Hurlstone & Hitch, 2015; Parmentier, Andres, Elford, & Jones, 2006; Smyth et al., 2005).

A further important characteristic of serial order is the way it can influence recall performance by the nature of its temporal organization: items whose serial positions are temporally grouped (by using shorter interitem temporal intervals for within-group items relative to items at the boundaries of two groups) lead to higher recall performance than items presented at a monotonous presentation rate. Temporal grouping effects have been studied quite extensively at the level of verbal WM (e.g., Frankish, 1985; Henson, 1998; Hitch, Burgess, Towse, & Culpin, 1996). In addition to leading to higher WM recall performance, they also induce within-group microprimacy and microrecency effects. As a corollary, temporal grouping leads to an increase of interposition errors, with serial position exchanges involving more frequently the same intragroup serial positions (e.g., the likelihood of serial position exchanges between the first item of a group and the first item of another group is higher than the likelihood of serial position exchanges between the first item of a group and the second item of another group). The temporal grouping effect and several of its consequences listed here have also been observed for the reproduction of auditory spatial and visuospatial sequences (Parmentier, Maybery, & Jones, 2004; Parmentier et al., 2006). Further evidence for cross-modal serial-order coding mechanisms stems from studies that have shown that retention of serial-order information in the verbal modality is impacted by a concurrent task involving the retention of serial-order information in the visuospatial domain, and vice versa (Depoorter & Vandierendonck, 2009; Vandierendonck, 2016). Importantly, these cross-domain interference effects were diminished when the concurrent task involved item rather than serial-order recall.

Some of these hallmark effects of serial-order coding can also be observed in other modalities than in the verbal and visuospatial domains, although these domains have been investigated less extensively. In the auditory nonverbal domain such as the musical WM domain, several studies have shown similar serial-order effects as for the verbal and visuospatial domains. A recent study by Gorin, Kowialiewski, and Majerus (2016) showed that tone sequence WM tasks probing retention of serial order but not item information lead to detrimental performance when a temporally organized, rhythmic sequence has to be reproduced during the retention interval. These findings parallel earlier findings observed for the verbal WM domain by Henson et al. (2003), where the same type of rhythmic interference task had a negative impact on serial order but not item recognition of verbal memoranda. Similarly, the impact of temporal grouping, observed in the verbal and visuospatial domains, has also been observed for musical WM tasks: Deutsch (1980) observed higher memory performance for grouped than ungrouped tone sequences in expert musicians, and Gorin, Mengal, and Majerus (submitted) observed similar findings in nonmusician participants using a serial-order recognition task for tone sequences. Finally, the same type of transposition gradients for serial-order exchange errors observed in the verbal and visuospatial WM domains also characterizes the serial-order errors observed during musical sequence recall tasks (Mathias, Pfordresher, & Palmer, 2014; Pfordresher, Palmer, & Jungers, 2007). A recent study also showed that reproduction of tactile stimulus sequences leads to the same type of serial position curve, with marked primacy and recency effects, as in the verbal WM tasks (Johnson, Shaw, & Miles, 2016).

Neuroimaging studies further support the similarity of mechanisms involved in the retention of serial-order information in verbal and visuospatial domains, as well as the specificity of serial-order coding mechanisms relative to item maintenance. Several studies showed that the encoding and recognition of serial-order information recruits a frontoparietal network centered on the right intraparietal sulcus for sequences of verbal information such as letters, words and nonwords, as well as for sequences of visual information such as sequences of unfamiliar faces (Majerus, Poncelet, Van der Linden et al., 2006; Majerus et al., 2007, 2010; Marshuetz et al., 2000; Martinez Perez, Poncelet, Salmon, & Majerus, 2015). These shared neural activity foci cannot be simply attributed to domain-general attentional control processes involved in WM (Cowan, 1995; Barrouillet, Bernardin, & Camos, 2004), given that the activations in the right frontoparietal network were specific to the serial-order conditions; cross-domain and cross-condition shared activity foci were also observed but these involved a frontoparietal network centered on the left intraparietal sulcus, which, in other studies, has indeed been associated with domain-general attentional control processes (Cowan et al., 2011; Majerus et al., 2010, 2012, 2016). It should also be noted that domain-specific activations were observed for the verbal and visuospatial WM tasks and mainly involved sensory cortices, with frontotemporal cortices for the verbal WM tasks, and fusiform cortices for visual WM tasks.

EVIDENCE FOR DOMAIN-GENERAL CODES FOR THE REPRESENTATION OF ORDINAL INFORMATION IN WM AND NUMERICAL COGNITION

The cognitive and neural similarities observed for coding of serial-order information across several WM domains presented in the preceding section and synthesized in Table 19.1 are in line with the assumption of

TABLE 19.1   Synthesis of Studies Examining the Domain Generality of Serial-Order Processing

Main Findings	Studies
Serial-order effects across working memory (WM) domains	
• Primacy and recency effects in verbal, visual, visuospatial, and tactile WM	Avons (1998), Guérard and Tremblay (2008), Johnson et al. (2016), Lee and Estes (1981), Smyth et al. (2005) and Ward et al. (2005)
• Transposition gradients in verbal, visuospatial, and musical WM	Burgess and Hitch (1999), Henson (1998), Hurlstone and Hitch (2015), Mathias et al. (2014), Palmer and Pfordresher (2003), Parmentier et al. (2006), Pfordresher et al. (2007) and Smyth et al. (2005)
• Temporal grouping effects in verbal, auditory spatial, visuospatial, and musical WM	Deutsch (1980), Frankish (1985), Henson (1998), Hitch et al. (1996) and Parmentier et al. (2004, 2006)
• Cross-domain (verbal, visuospatial, motor, musical) interference by maintenance/ reproduction of serial order	Depoorter and Vandierendonck (2009), Gorin et al. (2016), Henson et al. (2003) and Vandierendonck (2016)
Ordinal effects in WM and numerical cognition	
• Distance effects in WM and number comparison tasks	Attout, Fias et al. (2014), Attout, Noël et al. (2014), Henson et al. (2003), Holyoak (1977), Marshuetz et al. (2000), Moyer and Landauer (1967) and Turconi et al. (2006)
• Position–space associations in WM and numerical cognition	Antoine et al. (2017), Dehaene et al. (1993), Ginsburg et al. (2014), Guida et al. (2016), van Dijck and Fias (2011) and van Dijck et al. (2013)
• Independence of position–space associations in WM and numerical cognition	Ginsburg et al. (2014) and Ginsburg and Gevers (2015)
Overlapping neural substrates	
• Bilateral or right intraparietal sulcus involvement for serial-order WM in verbal and visuospatial domains and for ordinal numerical processing	Attout et al. (2015), Majerus, Poncelet, Greffe et al. (2006), Majerus, Poncelet, Van der Linden et al. (2006, 2007, 2010), Marshuetz et al. (2000, 2006), Martinez Perez et al. (2015) and Pinel et al. (2001)

specific but domain-general mechanisms for serial order, as proposed by most of the current theoretical models of serial-order WM. Importantly, some of the effects described in the preceding section extend the WM domain and also characterize the numerical domain.

A first of these effects is the distance effect in recognition WM tasks, which is the corollary of the transposition gradients that characterize the serial-ordering errors in WM recall tasks. In recognition WM tasks, participants make more errors and show slowed response times when judging items from adjacent serial positions than when judging items from more distant serial positions (Henson et al., 2003; Holyoak, 1977; Marshuetz et al., 2000). This distance effect is one of the hallmark effects also observed in the numerical literature. An important number of studies have shown that during number comparison tasks, such as magnitude judgment or ordinal judgment, number pairs that are close (such as 64–65) take more time to be judged than more distant numbers (such as 32–65) (Buckley & Gillman, 1974; Foltz, Poltrock, & Potts, 1984; Moyer & Landauer, 1967; Turconi, Campbell, & Seron, 2006). Very interestingly, these distance effects in the numerical domain seem to depend on the same neural substrates as those supporting serial-order coding in the WM domain. In several neuroimaging studies, the intraparietal sulcus has been shown to be sensitive to the distance between different numbers, with greater activation for close versus distant numbers (Pinel, Dehaene, Rivière, & LeBihan, 2001; Chochon, Cohen, van de Moortele, & Dehaene, 1999). Marshuetz, Reuter-Lorenz, Smith, Jonides, and Noll (2006) showed that the same intraparietal sulcus area is also sensitive to distance effects in serial-order WM tasks: the intraparietal sulcus is more strongly activated when participants have to compare close serial positions than when they have to compare more distant serial positions.

In a recent neuroimaging study, Attout, Fias et al. (2014) and Attout, Noël et al. (2014) directly compared ordinal number comparison and serial-order WM recognition tasks, while varying the distance effect in both tasks. In the number ordinal comparison tasks, the participants had to decide whether a number appearing on the screen occurs before or after a target number; the numerical distance between the probe number and the target number was varied. In the WM tasks, the participants had to maintain a sequence of four letters, followed by a probe display presenting two letters from the sequence, and the participants had to decide whether the two letters were presented in the same order as in the memory list; the serial position distance between the two target letters was varied. The authors observed a parametric modulation of activity in the intraparietal sulcus for the distance effects in both the numerical and serial-order WM tasks, suggesting that processing of ordinal information in WM and numerical domains is supported by similar neural substrates. It should be noted that these distance-sensitive activity foci in the intraparietal sulcus were bilateral, while previous studies associated serial-order processing in WM more specifically with the right intraparietal sulcus (Majerus et al., 2010). This bilateral activation of the intraparietal sulci could have been due to the additional involvement of attentional control processes: when judging near and distant serial positions/numbers, there is not only a difference at the level of ordinal representations but also nearer positions/numbers are harder to judge, which may recruit attentional control processes to a larger extent. As we have already noted, the left intraparietal sulcus has been associated more specifically with attentional control processes as compared with the right intraparietal sulcus in the context of WM tasks (see also, Majerus et al., 2016). Studies comparing direct magnitude and ordinal judgment in the numerical domain, thereby controlling for attentional processes equally associated with both types of judgments, have indeed observed a stronger recruitment of right hemisphere activity for the ordinal judgment tasks as compared with the magnitude judgment tasks (Turconi et al., 2004).

Further evidence for a close association between the representations involved in coding order information in the WM and numerical domains comes from a series of studies initiated by van Dijck and Fias. A hallmark effect in the numerical literature is the SNARC effect for spatial-numerical association of response codes. The SNARC effect, in the numerical domain, is characterized by faster responses with the left hand for judging small numbers (and hence positioned to the left of the number sequence) and faster responses with the right hand for judging larger numbers (positioned to the right in the number sequence) (Dehaene, Bossini, & Giraux, 1993). This effect has been explained as representing an association between numerical and spatial representations. Importantly, a very similar effect has been recently documented in the WM domain. In their seminal study, van Dijck and Fias (2011) had participants memorize sequences of words, which, during the maintenance interval, were represented for semantic judgment together with words that had not been present in the memory list; participants had to only judge those words that had been present in the list (go-trial). No explicit numerical or positional judgment had to be performed, but participants had to activate each word in WM before deciding to make a response. The authors observed similar positional effects and position–space associations as observed during the SNARC effect: participants were faster in judging the semantic category of the words with their left hand when the words came from the beginning of the memory list, and they were faster in responding with their right hand when the words were situated toward the end of the memory list. These results suggest that the serial position of stimuli in a WM list activates similar representations as those involved when processing numbers in number comparison tasks and further suggest that these representations are not only organized in an ordinal numerical manner but also in a spatial manner. This positional SNARC effect has been shown to be independent of the visual or auditory presentation of the memory list and hence cannot be explained by spatial encoding strategies only (Ginsburg, van Dijck, Previtali, Fias, & Gevers, 2014; Guida, Leroux, Lavielle-Guida, & Noël, 2016).

The importance of the spatial dimension associated with serial position codes has been further demonstrated by a follow-up study. van Dijck, Abrahamse, Majerus, and Fias (2013) had participants conduct a dot location task rather than a word or number judgment task during the retention interval of a list of numbers. The dots appeared on the left or the right side of the screen and were preceded by the presentation of numbers, some being part of the memory list and some being new; participants had to respond to the dots only if the number that preceded was from the memory list. The authors observed that participants were faster in detecting dots located on the left of the screen when the memory list item they had just seen was from an early serial position in the memory list. Conversely, they were faster in detecting dots located on the right of the screen when they had just seen a memory item stemming from late serial positions. A similar result was observed in a study by Antoine, Ranzini, Gebuis, Van Dijck, and Gevers (2017). These authors used the paradigm developed by van Dijck et al. (2013) but replaced the dot location task by a line bisection task. Antoine et al. observed that the line bisection midpoint was shifted to the right when participants were retrieving items from the end of the memory and it was shifted to the left when participants were retrieving items from the start of the memory list. These results suggest that codes involved in the representation of serial position information in WM have a spatial, ordinal organization, just like numbers, and are able to bias the focus of spatial attention (Abrahamse, van Dijck, Majerus, & Fias, 2014).

EVIDENCE AGAINST A DIRECT ASSIMILATION OF NUMERICAL ORDINAL CODES AND SERIAL-ORDER WM CODES

These findings lead us to further speculate on the nature of the codes used to represent serial-order information in WM. Some of the studies presented here suggest that their nature could be spatial (van Dijck et al., 2013; Abrahamse et al., 2014; Ginsburg et al., 2014; van Dijck & Fias, 2011). Other studies indicate a number of striking similarities between the processing of serial position information in WM and ordinal numerical information, as shown by the similar distance effects that appear when participants have to compare different serial positions in WM or different numbers in a number comparison task (Attout, Fias et al., 2014; Attout, Noël et al., 2014). This raises the possibility that numerical codes may also support the coding of serial position information in WM, in line with some of the computational models of serial-order WM described earlier in this chapter

(Botvinick & Watanabe, 2007). Intuitively, we are very frequently using numbers to label, maintain, and retrieve serial position information. Typically, when recalling information in WM tasks, we are using numbers to retrieve the items and their serial positions by saying: "This item was first, the second and the third items were …, the fourth and fifth items I don't remember any more ….".

We report here the results of two recent studies that have explored more directly the hypothesis of the intervention of numerical ordinal codes for the representation of serial-order information in WM. A first study reexplored the associations between serial-order WM and numerical abilities by taking into account general numerical arithmetic abilities, ordinal numerical judgment abilities, and magnitude numerical judgment abilities in 102 children aged 8–10 years (Attout & Majerus, 2018). The authors first confirmed the findings of previous studies by showing a specific association between serial-order WM abilities and mental arithmetic abilities after control of age, nonverbal, and verbal intelligence estimates ($r_{partial} = .30$, $P < .01$). Critically, when determining to what extent the association between serial-order WM abilities and mental arithmetic abilities was due to activation of numerical representations, the authors observed that ordinal numerical judgment abilities fully mediated the link between serial-order WM and mental arithmetic abilities. At the same time, the link between ordinal numerical and mental arithmetic abilities remained significant when testing the mediation of this effect by serial-order WM abilities. These results suggest that, at least for children aged 8–10 years, the link between serial-order WM abilities and mental arithmetic abilities reflects a shared component of ordinal processing abilities. However, this link is not necessarily specific to processing of numerical information, given that numerical magnitude judgment did not mediate the link between serial-order WM and mental arithmetic abilities. The existence of more general ordinal processing abilities being involved in serial-order WM abilities is also supported by findings from an earlier study by Attout, Fias et al. (2014) and Attout, Noël et al. (2014), showing that the cognitive and neural distance effects observed in serial-order WM and ordinal numerical judgment tasks also characterize alphabetic judgment abilities; furthermore, the correlations between the sizes of the behavioral distance effects were more robust for serial-order WM and alphabetic judgment tasks than for serial-order WM and numerical judgment tasks.

Another recent study assessed the hypothesis of numerical codes for serial-order coding in WM by preinstalling associations between numbers and stimuli to be maintained in a verbal WM task (Majerus & Oberauer, 2018). The rationale of this study was that if number codes are directly used to represent serial position information in WM, then word-digit associations may facilitate or interfere with serial position coding in WM, depending on the congruency of the digit associated to each word with the numerical rank of the word's serial position in the WM list. Participants learnt a list of 12 word-digit associations, with each digit (from 1 to 6) being associated to 2 different words. In a subsequent phase, the words were presented in memory lists, with the words occurring in serial positions whose numerical rank matched the learnt word-digit association (facilitation condition) or did not match the word-digit association (interference condition). Memory was probed either using a serial-order recognition task or an immediate serial recall task. For the serial-order recognition task, no effect of either facilitation or interference of word-digit associations on memory performance was observed. In the immediate serial recall task, an effect of interference was observed for both the interference and facilitation conditions. These results indicate that if number codes intervene during serial-order WM tasks, then this intervention is limited to the recall stage, given that no effect was observed for the serial-order recognition task. But even for the recall stage, the results need to be nuanced, given that interference effects were observed for both the facilitation and the interference conditions, indicating that, if numerical codes accompany retrieval of serial position information, they are easily disturbed by the concomitant presentation of any other numerical information.

The equivalence of numerical and WM ordinal representations has also been questioned in a series of studies following up the SNARC effect observed in WM tasks. In several studies, Ginsburg and collaborators showed that SNARC effects in WM tasks and typical numerical SNARC effects can coexist and hence result from different levels of representation. In a first study, Ginsburg et al. (2014) showed that when participants had to categorize both memorized and nonmemorized numbers during a WM retention delay, the positional SNARC effect for WM items disappeared, while a typical numerical SNARC effect emerged. They observed faster left-hand responses for small numbers and faster right-hand responses for large numbers, independently of the numbers' WM status and, for the numbers in WM, independently of their serial position in the WM list. Ginsburg et al. (2014) argued that the positional SNARC effect is because of the temporary creation of position–space bindings in WM, whereas the numerical SNARC effect is because of the long-term activation of the mental number line. They directly showed the operation of these two types of SNARC effects in a subsequent study in which participants were randomly instructed, within the same task, to classify only WM items or to classify all items: both a positional SNARC and a numerical SNARC effect were observed (Ginsburg & Gevers, 2015). These results show that numerical ordinal representations have their own existence, independently of temporary serial position–space associations created in WM.

CONCLUSIONS

Two main conclusions can be drawn from this theoretical review. First, there is an increasingly large body of research indicating that domain-general representations and processes support the representation of serial-order information in verbal, nonverbal auditory, and visuospatial WM. This is in line with the majority of current theoretical accounts of serial-order WM, which consider that serial order is processed by specific positional context signals distinct from the modality-specific mechanisms used to represent item information in WM (see also Hurlstone et al., 2014). At the same time, there remain many open questions as regards the nature of the codes used to represent serial position information in WM. Serial position effects across domains can look similar and yet stem from distinct processes. This is nicely illustrated by the multiplicity of computational models of serial-order WM models reviewed in this chapter: all the models we discussed are able to simulate most of the hallmark effects that characterize serial-order WM while substantially differing in the precise implementation of the mechanisms used for representing serial-order information. Also, overlapping neural activity has been observed in the right intraparietal sulcus for processing serial-order information in the verbal and visuospatial domains, but multivariate voxel pattern analysis (MVPA) studies still need to be conducted to determine whether the same type of neural information characterizes right intraparietal sulcus activity when processing serial-order information in the verbal and visuospatial domains (see Fias, Lammertyn, Caessens, & Orban, 2007; Zorzi, Di Bono, & Fias, 2011, for studies illustrating how univariate and MVPA analyses of intraparietal sulcus activity can lead to very different conclusions about the similarity of neural processes involved in several order processing tasks). Also, we should not exclude the possibility that both domain-general and domain-specific processes could support serial-order WM. Kalm and Norris (2014), using MVPA, showed that neural patterns in linguistic sensory cortices, dedicated to verbal item processing, can also represent serial-order information, at least as regards sublexical phonological sequence information. At the same time, Kalm and Norris showed that, within the language cortices, distinct codes were involved for representing item and serial-order information, and the neural representations identified by MVPA supported positional context signal rather than chaining models of serial-order coding.

The second important conclusion we can draw from this theoretical review is that there is a growing body of research indicating close associations between serial-order WM and numerical cognition. A number of studies have shown that there is a functional association between the two domains by highlighting the specific links that exist between verbal WM capacity for serial-order information and arithmetic abilities and by showing specific serial-order WM deficits in participants with dyscalculia or a history of dyscalculia. As we have argued at the beginning of this chapter, this association could reflect the fact that temporary sequential storage processes are necessary for actively maintaining the numbers and operators of a mental calculation problem and the sequence of operations carried out and their associated results. There are also some indications for a representational link between serial-order WM and number domains as shown by the existence of similar ordinal effects in both domains such as the distance effect or the SNARC effect. Furthermore, the same neural substrates in the intraparietal sulci support distance effects in the serial-order WM and the numerical domains, although, again, these conclusions are drawn from studies using univariate study analysis designs; the neural representational similarity of distance effects observed in WM and numerical tasks still needs to be determined using MVPA experimental designs. Further evidence for representational overlap comes from the recent study by Attout and Majerus (submitted), showing that the link between serial-order WM and mental calculation is fully mediated by ordinal numerical judgment abilities, suggesting that there are shared ordinal representations or processes in the WM and numerical domains.

At the same time, these results do not indicate that the shared ordinal representations are necessarily or exclusively numerical. There is even some evidence against a purely numerical nature of these shared representations. As we have noted, ordinal distance effects can be observed across many different domains, and Attout, Fias et al. (2014) and Attout, Noël et al. (2014) observed that the correlation between alphabetical and serial-order WM distance effects is stronger than the correlation between numerical and serial-order WM distance effects. Furthermore, the studies by Ginsburg and colleagues have shown a clear independence between numerical SNARC effects and positional SNARC effects in WM. The fact that numerical representations cannot drive the positional SNARC effect in WM is most clearly demonstrated by the fact that a numerical SNARC effect can occur on the stimuli held in WM (Ginsburg & Gevers, 2015, Experiment 2). If the positional SNARC effect in WM was because of the activation of number codes associated with the serial positions of number stimuli held in WM, then it should be difficult to obtain a typical numerical SNARC on the same numbers: the numerical value of the numbers and that of their serial position in the WM list should interfere with each other. In the study by Ginsburg and Gevers (2015), there was no hint that the numerical SNARC would have been attenuated for items activated in WM relative to non-WM items. Similarly, the study by Majerus and Oberauer provided very limited evidence for a direct involvement of number codes in the representation of serial-order information in WM.

What could the codes used to represent serial order then be like? On the basis of the empirical and theoretical studies discussed here, the codes used to represent serial-order information appear to be positional, ordinal, and in close connection to spatial and/or temporal dimensions (see also Chapter 12). The results of the positional SNARC effect and its ability to orient spatial attention support the notion of spatial markers for serial position in a left-to-right dimension. These results are also in line with some of the theoretical models of serial-order WM discussed here, such as the Start-End model (Henson, 1998), which represents serial position information via a dimension that can be easily translated in spatial terms (the start of a verbal sequence is typically represented to the left and the end of a verbal sequence list is typically represented to the right, at least for individuals used to a left-to-right writing system). Other studies suggest the role of temporal processes, such as the studies demonstrating the impact of temporal regularities on serial position coding, with temporal grouping of memoranda facilitating serial-order coding and the presentation of nonmemory rhythmic sequences interfering with serial-order recall in verbal, visuospatial, and musical modalities. We also have seen that a significant number of computational models implementing temporal codes provide robust accounts for the representation of serial-order information and are the only models able to explain temporal grouping effects (Brown et al., 2000; Hartley et al., 2016). Currently, given the available data, it is difficult to tease apart the spatial and temporal hypotheses; both types of codes could intervene simultaneously or could reflect an even more abstract type of ordinal reference frame, which would allow to represent the spatial and temporal relations between successive elements (Fias, et al., 2007; Ragni, Franzmeier, Maier, & Knauff, 2016). Neuroimaging studies highlighting right intraparietal involvement for serial-order processing in WM are compatible with both spatial and temporal hypotheses as the right parietal cortex is typically involved in both spatial and temporal tasks (Arend, Rafal, & Ward, 2011; Cabeza et al., 1997; Cazzoli, Müri, Kennard, & Rosenthal, 2015; Chechlacz, Rotshtein, & Humphreys, 2014; Corbetta & Shulman, 2011; Fias, Lammertyn, Reynvoet, Dupont, & Orban, 2003; Hubbard, Piazza, Pinel, & Dehaene, 2005; Konoike et al., 2015; Rao, Mayer, & Harrington, 2001).

Future studies need to determine the exact nature of the context signals used for representing serial-order information in WM, by manipulating simultaneously temporal and spatial dimensions and their impact on serial-order WM, to determine whether only one of the two dimensions is important for serial-order WM, whether both dimensions support independently serial-order WM, or whether both dimensions reflect a more general, abstract ordinal reference frame. Future studies also will need to determine how these dimensions interact with numerical ordinal representations and mediate the link observed between serial-order WM and numerical cognition.

Acknowledgments

This work was supported by grants F.R.S.-FNRS T.1003.15 (Fund for Scientific Research FNRS, Belgium), ARC 12/17-01-REST (French-speaking community of Belgium), and PAI-IUAP P7/11 (Belgian Science Policy).

References

Abrahamse, E., van Dijck, J. P., Majerus, S., & Fias, W. (2014). Finding the answer in space: The mental whiteboard hypothesis on serial order in working memory. *Frontiers in Human Neuroscience, 8*, 932. https://doi.org/10.3389/fnhum.2014.00932.

Antoine, S., Ranzini, M., Gebuis, T., Van Dijck, J. P., & Gevers, W. (2017). Order information in verbal working memory shifts the subjective midpoint in both the line bisection and the landmark tasks. *Quarterly Journal of Experimental Psychology, 70*(10), 1973–1983.

Arend, I., Rafal, R., & Ward, R. (2011). Temporal feature integration in the right parietal cortex. *Neuropsychologia, 49*(7), 1788–1793.

Attout, L., Fias, W., Salmon, E., & Majerus, S. (2014). Common neural substrates for ordinal representation in short-term memory, numerical and alphabetical cognition. *PLoS One, 9*, e92049. https://doi.org/92010.91371/journal.pone.0092049.

Attout, L., & Majerus, S. (2015). Working memory deficits in developmental dyscalculia: The importance of serial order. *Child Neuropsychology, 21*, 432–450.

Attout, L., & Majerus, S. (2017). *Serial Order Working Memory and Numerical Ordinal Processing Share Common Processes and Predict Arithmetic Abilities British Journal of Developmental Psychology*, in press. https://doi.org/10.1111/bjdp.1221.

Attout, L., Noël, M. P., & Majerus, S. (2014). The relationship between working memory for serial order and numerical development: A longitudinal study. *Developmental Psychology, 50*, 1667–1679.

Attout, L., Salmon, E., & Majerus, S. (2015). Working memory for serial order is dysfunctional in adults with a history of developmental dyscalculia: Evidence from behavioral and neuroimaging data. *Developmental Neuropsychology, 40*, 230–247.

Avons, S. E. (1998). Serial report and item recognition of novel visual patterns. *British Journal of Psychology, 89*, 285–308.

Barrouillet, P., Bernardin, S., & Camos, V. (2004). Time constraints and resource sharing in adults' working memory spans. *Journal of Experimental Psychology: General, 133*, 83–100.

Botvinick, M., & Plaut, D. C. (2006). Short-term memory for serial order: A recurrent neural network model. *Psychological Review, 113*, 201–233.

Botvinick, M., & Watanabe, T. (2007). From numerosity to ordinal rank: A gain-field model of serial order representation in cortical working memory. *Journal of Neuroscience, 27*, 8636–8642.

Brown, G. D. A., Preece, T., & Hulme, C. (2000). Oscillator-based memory for serial order. *Psychological Review, 107*, 127–181.

Buckley, P. B., & Gillman, C. B. (1974). Comparisons of digits and dot patterns. *Journal of Experimental Psychology, 103*, 1131–1136.

Bull, R., & Johnston, R. S. (1997). Children's arithmetical difficulties: Contributions from processing speed, item identification, and short-term memory. *Journal of Experimental Child Psychology, 65*, 1–24.

Burgess, G. C., & Hitch, G. (1992). Toward a network model of the articulatory loop. *Journal of Memory and Language, 31*, 429–460.

Burgess, N., & Hitch, G. J. (1999). Memory for serial order: A network model of the phonological loop and its timing. *Psychological Review, 106*, 551–581.

Burgess, N., & Hitch, G. J. (2006). A revised model of short-term memory and long-term learning of verbal sequences. *Journal of Memory and Language, 55*, 627–652.

Cabeza, R., Mangels, J., Nyberg, L., Habib, R., Houle, S., McIntosh, A. R., et al. (1997). Brain regions differentially involved in remembering what and when: A PET study. *Neuron, 19*, 863–870.

Cazzoli, D., Müri, R. M., Kennard, C., & Rosenthal, C. R. (2015). The role of the right posterior parietal cortex in letter migration between words. *Journal of Cognitive Neuroscience, 27,* 377–386.

Chechlacz, M., Rotshtein, P., & Humphreys, G. W. (2014). Neuronal substrates of Corsi Block span: Lesion symptom mapping analyses in relation to attentional competition and spatial bias. *Neuropsychologia, 64,* 240–251.

Chochon, F., Cohen, L., van de Moortele, P. F., & Dehaene, S. (1999). Differential contributions of the left and right inferior parietal lobules to number processing. *Journal of Cognitive Neuroscience, 11,* 617–630.

Corbetta, M., & Shulman, G. L. (2011). Spatial neglect and attention networks. *Annual Review of Neuroscience, 34,* 569–599.

Cowan, N. (1995). *Attention and memory: An integrated framework.* New York: Oxford University Press.

Cowan, N., Li, D., Moffitt, A., Becker, T. M., Martin, E. A., Saults, J. S., & Christ, S. E. (2011). A neural region of abstract working memory. *Journal of Cognitive Neuroscience, 23,* 2852–2863.

De Smedt, B., Janssen, R., Bouwens, K., Verschaffel, L., Boets, B., & Ghesquiere, P. (2009). Working memory and individual differences in mathematics achievement: A longitudinal study from first grade to second grade. *Journal of Experimental Child Psychology, 103,* 186–201.

De Visscher, A., Szmalec, A., Van Der Linden, L., & Noël, M. P. (2015). Serial-order learning impairment and hypersensitivity-to-interference in dyscalculia. *Cognition, 144,* 38–48.

Dehaene, S., Bossini, S., & Giraux, P. (1993). The mental representation of parity and number magnitude. *Journal of Experimental Psychology: General, 122,* 371–396.

Depoorter, A., & Vandierendonck, A. (2009). Evidence for modality-independent order coding in working memory. *Quarterly Journal of Experimental Psychology, 62,* 531–549.

Deutsch, D. (1980). The processing of structured and unstructured tonal sequences. *Perception and Psychophysics, 28,* 381–389.

Farrell, S., & Lewandowsky, S. (2002). An endogenous distributed model of ordering in serial recall. *Psychonomic Bulletin and Review, 9,* 59–79.

Fias, W., Lammertyn, J., Caessens, B., & Orban, G. A. (2007). Processing of abstract ordinal knowledge in the horizontal segment of the intraparietal sulcus. *The Journal of Neuroscience, 27,* 8592–8596.

Fias, W., Lammertyn, J., Reynvoet, B., Dupont, P., & Orban, G. A. (2003). Parietal representation of symbolic and nonsymbolic magnitude. *Journal of Cognitive Neuroscience, 15,* 47–56.

Foltz, G. S., Poltrock, S. E., & Potts, G. R. (1984). Mental comparison of size and magnitude: Size congruity effects. *Journal of Experimental Psychology: Learning Memory and Cognition, 10,* 442–453.

Frankish, C. R. (1985). Modality-specific grouping effects in short-term memory. *Journal of Memory and Language, 24,* 200–209.

Gathercole, S. E., & Pickering, S. J. (2000a). Assessment of working memory in six- and seven-year-old children. *Journal of Educational Psychology, 92,* 377–390.

Gathercole, S. E., & Pickering, S. J. (2000b). Working memory deficits in children with low achievements in the national curriculum at 7 years of age. *British Journal of Educational Psychology, 70)*, 177–194.

Ginsburg, V., & Gevers, W. (2015). Spatial coding of ordinal information in short- and long-term memory. *Frontiers in Human Neuroscience, 9*, 8. https://doi.org/10.3389/fnhum.2015.00008.

Ginsburg, V., van Dijck, J. P., Previtali, P., Fias, W., & Gevers, W. (2014). The impact of verbal working memory on number-space associations. *Journal of Experimental Psychology: Learning Memory and Cognition, 40*, 976–986.

Gorin, S., Kowialiewski, B., & Majerus, S. (2016). Domain-generality of timing-based serial order processes in short-term memory: New insights from musical and verbal domains. *PLoS One, 11*(12), e0168699. https://doi.org/10.1371/journal.pone.0168699.

Guérard, K., & Tremblay, S. (2008). Revisiting evidence for modularity and functional equivalence across verbal and spatial domains in memory. *Journal of Experimental Psychology: Learning Memory and Cognition, 34*, 556–569.

Guida, A., Leroux, A., Lavielle-Guida, M., & Noël, Y. (2016). A SPoARC in the dark: Spatialization in verbal immediate memory. *Cognitive Science, 40*, 2108–2121.

Hachmann, W. M., Bogaerts, L., Szmalec, A., Woumans, E., Duyck, W., & Job, R. (2014). Short-term memory for order but not for item information is impaired in developmental dyslexia. *Annals of Dyslexia, 64*, 121–136.

Hartley, T., Hurlstone, M. J., & Hitch, G. J. (2016). Effects of rhythm on memory for spoken sequences: A model and tests of its stimulus-driven mechanism. *Cognitive Psychology, 87*, 135–178.

Henson, R. N. A. (1998). Short-term memory for serial order: The start-end model. *Cognitive Psychology, 36*, 73–137.

Henson, R., Hartley, T., Burgess, N., Hitch, G., & Flude, B. (2003). Selective interference with verbal short-term memory for serial order information: A new paradigm and tests of a timing-signal hypothesis. *Quarterly Journal of Experimental Psychology, 56A*, 1307–1334.

Hitch, G. J., Burgess, N., Towse, J. N., & Culpin, V. (1996). Temporal grouping effects in immediate recall: A working memory analysis. *The Quarterly Journal of Experimental Psychology, 49A*, 116–139.

Holmes, J., & Adams, J. W. (2006). Working memory and children's mathematical skills: Implications for mathematical development and mathematics curricula. *Educational Psychology, 26*, 339–366.

Holmes, J., Adams, J. M., & Hamilton, C. (2008). The relationship between visuo-spatial sketchpad capacity and children's mathematical skills. *European Journal of Cognitive Psychology, 20*, 272–289.

Holyoak, K. J. (1977). The form of analog size information in memory. *Cognitive Psychology, 9*, 31–51.

Hubbard, E. M., Piazza, M., Pinel, P., & Dehaene, S. (2005). Interactions between number and space in parietal cortex. *Nature Reviews Neuroscience, 6*, 435–448.

Hurlstone, M. J., & Hitch, G. J. (2015). How is the serial order of a spatial sequence represented? Insights from transposition latencies. *Journal of Experimental Psychology: Learning Memory and Cognition, 41*, 295–324.

Hurlstone, M. J., Hitch, G. J., & Baddeley, A. D. (2014). Memory for serial order across domains: An overview of the literature and directions for future research. *Psychological Bulletin, 140*, 339–373.

Imbo, I., Vandierendonck, A., & Vergauwe, E. (2007). The role of working memory in carrying and borrowing. *Psychological Research, 71*, 467–483.

Jarvis, H. L., & Gathercole, S. E. (2003). Verbal and non-verbal working memory and achievements on National Curriculum tests at 11 and 14 years of age. *Educational and Child Psychology, 20*, 123–140.

Johnson, A. J., Shaw, J., & Miles, C. (2016). Tactile order memory: Evidence for sequence learning phenomena found with other stimulus types. *Journal of Cognitive Psychology, 28*, 718–725.

Jones, D., Madden, C., & Miles, C. (1992). Privileged access by irrelevant speech to short-term memory: The role of changing state. *Quarterly Journal of Experimental Psychology, 44*, 645–669.

Kalm, K., & Norris, D. (2014). The representation of order information in auditory-verbal short-term memory. *The Journal of Neuroscience, 34*, 6879–6886.

Konoike, N., Kotozaki, Y., Jeong, H., Miyazaki, A., Sakaki, K., Shinada, T., et al. (2015). Temporal and motor representation of rhythm in fronto-parietal cortical areas: An fMRI study. *PLoS One, 10*(6), e0130120. https://doi.org/10.1371/journal.pone.0130120.

Lee, C. L., & Estes, W. K. (1981). Item and order information in short-term memory: Evidence for multilevel perturbation processes. *Journal of Experimental Psychology: Human Learning and Memory, 7*, 149–169.

Lewandowsky, S., & Farrell, S. (2008). Phonological similarity in serial recall: Constraints on theories of memory. *Journal of Memory and Language, 58*, 429–448.

Lewandowsky, S., & Murdock, B. B. (1989). Memory for serial order. *Psychological Review, 96*, 25–57.

Lyons, I. M., & Beilock, S. L. (2013). Ordinality and the nature of symbolic numbers. *Journal of Neuroscience, 33*, 17052–17061.

Majerus, S., Attout, L., Artielle, M. A., & Van der Kaa, M. A. (2015). The heterogeneity of verbal short-term memory impairment in aphasia. *Neuropsychologia, 77*, 165–176.

Majerus, S., Attout, L., D'Argembeau, A., Degueldre, C., Fias, W., Maquet, P., et al. (2012). Attention supports verbal short-term memory via competition between dorsal and ventral attention networks. *Cerebral Cortex, 22*, 1086–1097.

Majerus, S., Bastin, C., Poncelet, M., Van der Linden, M., Salmon, E., Collette, F., et al. (2007). Short-term memory and the left intraparietal sulcus: Focus of attention? Further evidence from a face short-term memory paradigm. *NeuroImage, 35*, 353–367.

Majerus, S., Cowan, N., Peters, F., Van Calster, L., Phillips, C., & Schrouff, J. (2016). Cross-modal decoding of neural patterns associated with working memory: Evidence for attention-based accounts of working memory. *Cerebral Cortex, 26*, 166–179.

Majerus, S., D'Argembeau, A., Martinez, T., Belayachi, S., Van der Linden, M., Collette, F., et al. (2010). The commonality of neural networks for verbal and visual short-term memory. *Journal of Cognitive Neuroscience, 22*, 2570–2593.

Majerus, S., & Oberauer, K. (2018). *No Evidence for a Role of Numerical Codes in Serial Order Working Memory* (Manuscript in preparation).

Majerus, S., Poncelet, M., Greffe, C., & Van der Linden, M. (2006). Relations between vocabulary development and verbal short-term memory: The importance of short-term memory for serial order information. *Journal of Experimental Child Psychology, 93,* 95–119.

Majerus, S., Poncelet, M., Van der Linden, M., Albouy, G., Salmon, E., Sterpenich, V., et al. (2006). The left intraparietal sulcus and verbal short-term memory: Focus of attention or serial order? *NeuroImage, 32,* 880–891.

Marshuetz, C., Reuter-Lorenz, P. A., Smith, M.-C., Jonides, J., & Noll, D. C. (2006). Working memory for order and the parietal cortex: An event-related functional magnetic resonance imaging study. *Neuroscience, 139,* 311–316.

Marshuetz, C., Smith, E. E., Jonides, J., DeGutis, J., & Chenevert, T. L. (2000). Order information in working memory: fMRI evidence for parietal and prefrontal mechanisms. *Journal of Cognitive Neuroscience, 12,* 130–144.

Martinez Perez, T., Poncelet, M., Salmon, E., & Majerus, S. (2015). Functional alterations in order short-term memory networks in adults with dyslexia. *Developmental Neuropsychology, 40,* 407–429.

Martinez Perez, P. T., Majerus, S., & Poncelet, M. (2012). The contribution of short-term memory for serial order to early reading acquisition: Evidence from a longitudinal study. *Journal of Experimental Child Psychology, 111,* 708–723.

Mathias, B., Pfordresher, P. Q., & Palmer, C. (2014). Context and meter enhance long-range planning in music performance. *Frontiers in Human Neuroscience, 8,* 1040. https://doi.org/10.3389/fnhum.2014.01040.

Moyer, R. S., & Landauer, T. K. (1967). Time required for judgements of numerical inequality. *Nature, 215,* 1519–1520.

Nairne, J. S., & Kelley, M. R. (2004). Separating item and order information through process dissociation. *Journal of Memory and Language, 50,* 113–133.

Noël, M. P. (2009). Counting on working memory when learning to count and to add: A preschool study. *Developmental Psychology, 45,* 1630–1643.

Noël, M. P., Seron, X., & Trovarelli, F. (2004). Working memory as a predictor of addition skills and addition strategies in children. *Current Psychology of Cognition, 22,* 3–25.

Page, M. P. A., & Norris, D. (1998). The primacy model: A new model of immediate serial recall. *Psychological Review, 105,* 761–781.

Palmer, C., & Pfordresher, P. Q. (2003). Incremental Planning in Sequence Production. *Psychological Review, 110,* 683–712.

Parmentier, F. B., Andres, P., Elford, G., & Jones, D. M. (2006). Organization of visuo-spatial serial memory: Interaction of temporal order with spatial and temporal grouping. *Psychological Research, 70,* 200–217.

Parmentier, F. B., Maybery, M. T., & Jones, D. M. (2004). Temporal grouping in auditory spatial serial memory. *Psychonomic Bulletin and Review, 11,* 501–507.

Pfordresher, P. Q., Palmer, C., & Jungers, M. K. (2007). Speed, accuracy, and serial order in sequence production. *Cognitive Science, 31,* 63–98.

Pinel, P., Dehaene, S., Rivière, D., & LeBihan, D. (2001). Modulation of parietal activation by semantic distance in a number comparison task. *NeuroImage, 14,* 1013–1026.

Poirier, M., & Saint-Aubin, J. (1995). Memory for related and unrelated words: Further evidence on the influence of semantic factors in immediate serial recall. *Quarterly Journal of Experimental Psychology, 48A*, 384–404.

Raghubar, K. P., Barnes, M. A., & Hecht, S. A. (2010). Working memory and mathematics: A review of developmental, individual difference, and cognitive approaches. *Learning and Individual Differences, 20*, 110–122.

Ragni, M., Franzmeier, I., Maier, S., & Knauff, M. (2016). Uncertain relational reasoning in the parietal cortex. *Brain and Cognition, 104*, 72–81.

Rao, S. M., Mayer, A. R., & Harrington, D. L. (2001). The evolution of brain activation during temporal processing. *Nature Neuroscience, 4*, 317–323.

Smyth, M. M., Dennis, C. H., & Hitch, G. (2005). Serial position memory in the visual-spatial domain: Reconstructing sequences of unfamiliar faces. *Quarterly Journal of Experimental Psychology, 58A*, 909–930.

Swanson, H. L., & Kim, K. (2007). Working memory, short-term memory, and naming speed as predictors of children's mathematical performance. *Intelligence, 35*, 151–168.

Turconi, E., Campbell, J. I., & Seron, X. (2006). Numerical order and quantity processing in number comparison. *Cognition, 98*, 273–285.

Turconi, E., Jemel, B., Rossion, B., & Seron, X. (2004). Electrophysiological evidence for differential processing of numerical quantity and order in humans. *Cognitive Brain Research, 21*, 22–38.

van Dijck, J. P., Abrahamse, E. L., Majerus, S., & Fias, W. (2013). Spatial attention interacts with serial-order retrieval from verbal working memory. *Psychological Science, 24*, 1854–1859.

van Dijck, J. P., & Fias, W. (2011). A working memory account for spatial-numerical associations. *Cognition, 119*, 114–119.

Vandierendonck, A. (2016). Modality independence of order coding in working memory: Evidence from cross-modal order interference at recall. *Quarterly Journal of Experimental Psychology, 69*, 161–179.

Ward, G., Avons, S. E., & Melling, L. (2005). Serial position curves in short-term memory: Functional equivalence across modalities. *Memory, 13*, 308–317.

Zorzi, M., Di Bono, M. G., & Fias, W. (2011). Distinct representations of numerical and non-numerical order in the human intraparietal sulcus revealed by multivariate pattern recognition. *NeuroImage, 56*, 674–680.

Do Not Forget Memory to Understand Mathematical Cognition

Valérie Camos
Université de Fribourg, Fribourg, Switzerland

The perspective taken to write this discussion chapter is from someone working on memory and interesting by the different memory systems (long-term, short-term, and working memory), the different types of knowledge (declarative and procedural, mostly), and their interactions with attention. I always thought that mathematical cognition, more specifically numerical and arithmetic activities, is particularly suitable to study human cognition. In my view, these activities are probably more adequate than other human cognitive activities, such as language tasks, because even the most basic numerical task such as counting requires the interplay of different memory systems and different types of knowledge.

Sustaining the idea that mathematical cognition is an adequate field to examine human cognition, mathematical activities have often been used in assessment tools of basic, fundamental, and human cognitive processes. From the digit span task, which is one of the earliest component subtests included in the Wechsler intelligence scales for both adults (WAIS, Wechsler, 2008) and children (WISC, Wechsler, 2005), to the now classic (arithmetic) operation span task created by Turner and Engle (1989) that measures working memory capacity or to Oberauer, Süß, Schulze, Wilhelm, and Wittmann (2000) memory updating task, which involves the solving of simple additions and subtractions, mathematical activities were chosen as support activities to assess human cognitive functioning. This clearly shows the particular place mathematical activities have in human cognitive functioning. Indeed, contrary to other knowledge children are acquiring during childhood such as object physical properties, mathematics is a human creation. Unlike understanding object properties, the properties of mathematical objects are a creation of the human mind. Thus, one may expect that studying the acquisition of this particular knowledge will, more than any other acquisitions, inform us about the fundamental characteristics and the basic processes of the human cognitive system.

In this chapter, I will summarize and enlighten some of the points made in the previous chapters of this section. For this purpose, I will reintegrate them in one of the most central questions in the field of memory. This question touches the essence of how cognitive psychology considers the relationships between working memory and long-term memory. The debate on the existence of a separate working memory store distinct from long-term memory is still vivid, and different theoretical perspectives and models embraced the view that human cognitive functioning is based either on a dedicated working memory system distinct from long-term memory or on a unitary memory system. I will also propose some perspectives that could direct future research plans in the field of mathematical cognition.

DISTINCTION BETWEEN WORKING MEMORY AND LONG-TERM MEMORY

One of the most fundamental questions, or maybe the essential issue in the study of human memory, concerns the existence (or not) of distinct memory systems. This debate can be traced back to the early age of memory research (Ebbinghaus, 1885/1913). Based on the observation of his own behavior, Ebbinghaus distinguished an immediate memory from a more stable form of memorization. Human memory was then segregated into two distinct memory systems: a primary memory system, in which a small amount of information is held active and represents our conscious present and a secondary memory system, which embraces the vast body of knowledge stored over a lifetime (James, 1890; or for a revival of this distinction; Unsworth & Engle, 2007). Although this depiction of memory mirrors our subjective experience and has led to fruitful research (on which we will elaborate further below), an increasing number of studies have cast doubt on a strict separation between these two types of memory. There is a steady line of researchers believing in the unity of memory and proposing that a single set of principles would capture all memory phenomena, from the very immediate to the more lasting ones (e.g., Crowder, 1982; Lewandowsky, Duncan, & Brown, 2004; Melton, 1963; Nairne, 2002; Neath & Surprenant, 2003; Wickelgren, 1974). The aim of this chapter is not to discuss the evidence that could depart these two theoretical views on memory, and readers interested by the question could refer to the review by Cowan (2008, pp. 323–338). At the same time, this chapter is organized around this debate. In this first part of the chapter, we will examine how research in mathematical cognition has been influenced by the distinction of two memory systems, with a particular emphasis on the role of working memory in numerical and arithmetic tasks. The second part of this chapter will be dedicated to the unitary view on memory. Although this conception frames less research in mathematical cognition than the two-store view, more recent works in arithmetic explicitly or implicitly integrate this perspective.

Working Memory

The segregation of human memory into primary memory and secondary memory introduced by Ebbinghaus (1885/1913) and James (1890) at the end of the 19th century gained renewed interest when the information-processing approach promoted by Newell and Simon emerged. The proposal to conceive the human mind as an information processing system such as a computer calls for a structure that would have to temporarily maintain the to-be-processed information and that could also execute and control programs to process information (Tolman, 1948). The emergence of different models of short-term memory led to the replacement of the "primary"/"secondary" memory labels by "short-" and "long-term" memory labels (Atkinson & Shiffrin, 1968, pp. 89–195, 1971; Broadbent, 1958; Craik, 1971; Waugh & Norman, 1965). In these models, short-term memory is considered to be distinct from a sensory store and from long-term memory and to be a central component of human cognition. Indeed, long-term learning relies on the capacity of the short-term store to temporarily hold information until it is transferred to the long-term store.

This transfer depends on the amount of time information resides in the short-term store. However, in these models, the role of the short-term store is not limited to maintenance but is assumed to be responsible for the different ways to encode incoming information (e.g., visual, auditory, or haptic) after the first processing steps by sensory systems. The short-term store also retrieves knowledge from the long-term store or makes decisions for action. For example, in Atkinson and Shiffrin's (1971) model, short-term memory has the role to control, coordinate, and monitor the subroutines needed to acquire new information and to retrieve old ones from long-term memory. However, despite the fact that short-term memory had been frequently assigned such a role of operational or working memory, Baddeley and Hitch (1974, pp. 647–667) observed that empirical evidence supporting this hypothesis was remarkably sparse. Aiming at testing whether short-term memory can have the operational role often hypothesized, Baddeley and Hitch (1974, pp. 647–667) provided an important set of studies, which had a long-standing impact on the field of memory. They found that short-term storage had relatively little impact on concurrent processing. Based on this finding, they proposed a working memory model in which the storage and processing functions, previously attributed to the short-term store, are separated. This conception in which working memory is conceived as a storage-processing device is at the root of one of the most influential models of working memory, known as the multicomponent model (Baddeley, 1986, 2007, 2012).

The Multicomponent Model

The multiple-component framework has developed during the past 40 years and inspired an impressive amount of behavioral, developmental, neuropsychological, and neuroimaging studies (see, Baddeley, 2007, for review). This model assumes multiple domain-specific cognitive functions, each of them drawing on its own pool of resources and hence having its own capacity limit (Baddeley & Logie, 1999, pp. 28–61). Although a central executive component is in charge of the processing and control of action, distinct storage components, i.e., the phonological loop and the visuospatial sketchpad, are specialized in maintaining verbal and visuospatial information, respectively. In the last version of the multicomponent model, Baddeley (2000) introduced the idea that maintenance is not solely domain-specific, but that integrated representations, in which features pertaining to different domains or retrieved from long-term memory, are stored in an episodic buffer under the control of the central executive.

The multicomponent model inspired a large set of studies in mathematical cognition. These studies examined the respective implication of each working memory subcomponent in different mathematical tasks. As reviewed in Chapter 17, most arithmetic tasks require some temporary storage of information. As a consequence, it has been shown that the phonological loop is involved in solving simple arithmetic problems and complex arithmetic problems. It is also required in counting arrays. The visuospatial sketchpad, another subcomponent of working memory, intervenes in complex arithmetic, but only in some simple arithmetic tasks. For example, solving simple divisions imposes demand on the visuospatial sketchpad, whereas solving simple multiplications does not. Because the central executive is in charge of coordinating the subcomponents and controlling actions, its role has been quite unsurprisingly evidenced in complex arithmetic activities, which require the maintenance of intermediate results and carry operations. Although these are among the most difficult aspects of arithmetic, the involvement of the central executive is not restricted to high-level arithmetic skills. In children, even simple arithmetic involves the central executive. It is also important to note that several accounts of mathematical difficulties give a determinant role to one or several subcomponents of working memory. Moreover, in explaining dyscalculia, the implication of some domain-general factors is now recognized, and authors often consider working memory as described by the multicomponent model when referring to these factors.

The implication of the different subcomponents in mathematical cognition has been largely documented (but see Majerus's chapter for a discussion on the involvement of the phonological loop). Contrasting with this vast amount of studies, the most recently added component, i.e., the episodic buffer, did not, so far, stimulate much work. This state of affairs is rather surprising, knowing the important role Baddeley assigns to this component. The episodic buffer should be in charge of the integration of multiple codes. For example, the maintenance of series of letters with their spatial locations on screen requires the binding of verbal and visuospatial information. These integrated representations would be then stored and maintained in the episodic buffer (Allen, Baddeley, & Hitch, 2006; Langerock, Vergauwe, & Barrouillet, 2014). In the field of mathematical cognition, numbers are an excellent example of information existing in different formats. As put forward by Dehaene (1992) in his triple-code model, numbers can be represented under three different codes: a verbal, an Arabic, and an approximate number representation. Each of these representations is particularly involved in certain types of mathematical activities. Counting, arithmetic facts, and mental exact arithmetic are thought to rely on verbal number representations. The Arabic number representations are required when solving complex arithmetic problems presented in a written form, whereas the approximate number representations support quantity comparison and approximate arithmetic tasks. As a consequence, the episodic buffer should allow the maintenance of multicode number representations. To our knowledge, this aspect has not been examined yet.

Moreover, some predictions can be drawn based on recent studies on the functioning of the episodic buffer in short-term memory tasks. It was shown that the maintenance of integrated representations (like what I mentioned above, letters associated to specific locations) relies on attentional mechanisms. Thus, the maintenance of integrated representations would compete for allocation of attention with any concurrent attention-demanding task. However, it has been recently shown that the maintenance of these integrated representations is not more demanding than the maintenance of any featuring element (Langerock et al., 2014). This implies that the binding of the different features into an integrated representation is not attention-demanding (Allen et al., 2006; Langerock et al., 2014), contrary to the initial proposal made by Baddeley (2000; Repovs & Baddeley, 2006). In the number domain, the same pattern can thus be expected. The binding of different number codes should also not be attention-demanding.

Although studies that examine this particular prediction are still lacking, other findings in mathematical cognition seem to conflict with such an idea. Number codes and their manipulation have been particularly studied through transcoding tasks. Transcoding tasks require individuals to produce numbers in a representation that is different from the one given as input. The most frequently studied transcoding task is a task in which individuals are asked to write numbers in Arabic form based on their auditory–verbal presentation. This activity is still widely used in primary schools to teach the number system. In a study comparing groups of 7-year-old children differing on their working memory capacity, Camos (2008) showed that low-span children made more transcoding errors than children with high working memory capacity. Because these differences in working memory capacity are considered to reflect differences in executive attention (Engle & Kane, 2004, pp. 145–199), this finding means that attention is involved in the transformation of one number code to the other. However, the binding of these codes should not demand any attentional resources. This apparent conflict, if it were confirmed by further studies, should question how the different codes interact and are manipulated.

Another type of integrated representations is also involved in mathematical cognition. As presented in Majerus' and Fias's chapter, time, number, and space can be supported by integrated representations, which permit the maintenance of serial order.

The Coding and Maintenance of Serial Order

Among the multitude of works in the working memory field, the coding and maintenance of serial order represents a field on its own. This very flourished domain has produced numerous models to account for the ability to remember and recall sequences of information (e.g., Botvinick & Plaut, 2006; Brown, Neath, & Chater, 2007; Burgess & Hitch, 1999; Farrell & Lewandowsky, 2002; Henson, 1998; Page & Norris, 1998, and see Lewandowsky & Farrell, 2008, and Majerus' chapter, for review).

Among these proposals, some serial-order models (e.g., Burgess & Hitch, 1999; Farrell & Lewandowsky, 2002) and some more general models of working memory (e.g., Barrouillet & Camos, 2015) envision serial order as a feature of integrated representations. In a previous example, it was explained that location is one particular feature of an integrated representation of a letter in its location. Similarly, the order of an item in a sequence of different memory items would be encoded as a feature of the representation of this particular item. Within this perspective, the integrated representation should be built in working memory and maintained in an episodic buffer through some attentional mechanism of maintenance, as it was previously reported in studies examining the maintenance of verbal items in spatial locations.

Contrasting with this account of serial order, an alternative view about serial order took its inspiration in mathematical cognition and more specifically in the representation of numbers. As summarized by Fias in his chapter, this perspective relies on the role of spatial encoding of order. Inspired by the SNARC effect (i.e., spatial–numerical association of response codes; Dehaene, Bossini, & Giraux, 1993) and other evidence that numbers are spatially represented on an oriented number line, van Dijck, Fias et al. proposed that the order in which information of memory items is stored in working memory is also spatially organized, with initial items being on the left and end items located on the right of a left-to-right oriented spatial continuum (Abrahamse, van Dijck, & Fias, 2016; van Dijck, Abrahamse, Acar, Ketels, & Fias, 2014; van Dijck & Fias, 2011). This interesting proposal exemplifies how the particularities of the mathematical activities can inspire research outside the field of mathematical cognition and can renew conceptions in memory functioning. This shows how much research in mathematical cognition and in memory can beneficially influence each other. However, this is not restricted to working memory, and research in long-term memory has also important resonance with mathematical cognition.

Declarative and Procedural Knowledge in Long-Term Memory

Long-term memory has been fragmented in several different subtypes of memories. Among them, Anderson (1993) proposed to distinguish declarative from procedural memory knowledge. Procedures can act on declarative knowledge to process and transform it, to create new declarative knowledge that can be learnt (i.e., stored in long-term memory), whereas declarative knowledge can be retrieved from long-term memory and temporarily maintained to constitute the triggering conditions of procedures. This depiction provides an adequate framework to explain learning and account for the age-related changes in the achievement of some arithmetic tasks (Table 20.1; see also Prado's chapter in this book). Across childhood, the efficiency in performing arithmetic tasks increases, with

TABLE 20.1   Different Examples of Numerical and Arithmetic Tasks for Which Age-Related Improvements Depend on Changes in Efficiency and Knowledge in Declarative or Procedural Long-Term Memory (LTM)

Age-Related Increase in	Declarative LTM	Procedural LTM
Efficiency (response times, accuracy, cognitive cost)	Multiplicative facts	Additive facts
	Subitizing (as pattern recognition)	Subitizing (as counting procedure)
Knowledge (amount, diversity)	Additive facts	Counting
		Complex arithmetic
Teaching method	*Rote repetition*	*Abacus*

faster response times, lower error rates, and smaller cognitive cost in older than younger children. In parallel, the amount of knowledge children acquired on these tasks increases, as well as its diversity. These age-related increases in efficiency and knowledge are observed in both declarative and procedural long-term memory, and they are illustrated by performance patterns on some widely used arithmetic tasks. For example, multiplicative facts are stored as declarative knowledge, and children become more and more efficient in retrieving these facts from long-term memory (see Noel's chapter in this book). In the same way, it was suggested that subitizing, i.e., the fast and accurate labeling of small quantities, relies on the recognition of visuospatial patterns (Mandler & Shebo, 1982), children becoming increasingly fast at retrieving the numerical labels of these patterns. The increase in the amount of declarative knowledge has been often mentioned to explain the developmental improvement in solving simple additions. Through practice, more and more additive facts should be stored in long-term memory, and more and more simple additions should thus be solved by direct retrieval of their answer (Siegler & Shrager, 1984, pp. 229–293, see Barrouillet's chapter in this book). By contrast, complex arithmetic tasks benefit from developmental increase in the amount of procedural knowledge, which expands the range of problems children can solve. Similarly, part of the age-related improvement observed in counting relies on the acquisition of diverse procedures (Camos, 2003). Finally, in a model conceiving subitizing as a fast counting procedure (Gallistel & Gelman, 1992), age-related changes result from the increasing efficiency in the use of procedural knowledge. The idea is that the counting procedure becomes increasingly accurate and fast with age, to reach a state of ultra-automatization, i.e., in which the procedure is virtually automatic and without any cognitive cost. At this state, children have no conscious experience of counting, and they have the impression that the cardinal of the presented array pops up in their mind. Similarly, it was recently suggested that an akin fast counting procedure could account for solving some simple additions, i.e., those with the smallest operands (see Barrouillet and Prado's chapters). An increasing efficiency in performing this counting procedure would explain the developmental pattern in solving simple additions.

This brief overview shows that the dichotomy introduced in the memory literature between declarative and procedural knowledge finds strong echoes in the field of mathematical cognition. Activities, such as subitizing or solving simple additions, could be considered as relying either on declarative or on procedural long-term memory depending on theoretical accounts. It should be noted that beyond these theoretical divergences, this dichotomy reflects also differences in pedagogical methods to teach mathematics. The use of rote repetition to acquire additive or multiplicative facts endorses in classroom practices the importance of declarative knowledge in arithmetic. Alternatively, introducing the manipulation of abacus in school curricula is the recognition of the role of procedures in the acquisition of mathematical skills.

UNITARY VIEW OF MEMORY

As introduced at the beginning of this chapter, one of the most fundamental issues in the memory literature is the debate concerning the unity or dichotomy of memory systems. In the previous section, I have shown that the conception of a distinct working memory, separated from a long-term memory, had a strong influence on the study of mathematical cognition. Contrasting with the separate memory systems approach, the unitary view of memory inspired research on numerical or arithmetic activities to a much smaller degree. There is nevertheless one interesting recent example exposed in Noel's chapter.

An Interesting Example: The Multiplicative Facts

Among theories that explain short-term and long-term memory phenomena within a unitary system, Nairne's (1990) feature model served as a framework for studying the storage and retrieval of multiplication facts (De Visscher & Noel, 2014). In the feature model, representations stored in long-term memory are suffering from representation-based interferences generated by the overlap of features. In other words, two representations sharing a feature may interfere from each other, leading to impairment in storage and poorer recall performance when one tries to retrieve one or the other representation. In the particular case of the storage of multiplication facts, the similarity between the facts would provoke interference and would affect their retrieval from long-term memory. To test this idea,

De Visscher and Noel (2014) computed an index of the interference weight for each multiplication by counting the overlap of digits between multiplications, i.e., the number of digits that multiplications shared between them. In accordance with the feature model, they showed that this interference parameter substantially predicted performance, with poorer accuracy for multiplication having a high interference index (i.e., sharing many digits with other multiplications; see Noel's chapter in this book). This extensive and nice study collected evidence in adult populations and in different children age groups. Nevertheless, and despite what it tells us about the retrieval of multiplicative facts, this study also stimulates certain questions inspired by what is known about long-term memory and how the feature model describes its functioning.

Long-term memory is often described as a semantically organized system. As a consequence, it is difficult to fully understand how a parameter that does not take into account any semantic information is such a good predictor of the retrieval of information from long-term memory. Indeed, following the interference index used by De Visscher and Noel, the digit 1 in "10" should provoke the same interference as 1 in "81," although the same digit 1 has two different meanings, representing a decade and a unit, respectively. To understand how a parameter that does not take into account the meaning of digits could account for multiplication retrieval, one can propose that the storage of multiplication facts does not rely on semantic codes in long-term memory. Actually, the multiplication facts (i.e., solutions of the simple multiplications) are often learnt by rote repetition of the multiplication tables. Thus, because of this learning method, multiplication facts could be stored as verbal–phonological representations and not semantic representations. As a consequence, representation-based interference between multiplication facts should be sensitive to the phonological similarity between the facts. However, the interference index used by De Visscher and Noel is based on Arabic formats, and could not reflect such phonological interference, because the numerical system in European languages is non-transparent. The non-transparency of European languages means that there is a poor correspondence between the verbal and the Arabic formats of numbers. Thus, except if one considers that multiplication facts are stored digit by digit (e.g., 81 would be stored as two separated digits 8 and 1, and not as a whole number), this is difficult to envision why such an interference index accounts for the retrieval of multiplication facts. This is even more difficult to envision when one considers the findings from the few studies that attempted to manipulate the amount of overlapping features in memory tasks. Using working memory tasks, these studies varied the amount of phonological features shared either between verbal memory items or between verbal memory items and verbal distractors (e.g., Camos, Mora, & Oberauer, 2011; Oberauer, 2009). As expected from the feature model, an increased amount of shared features led to poorer recall performance but only when items were maintained through subvocal rehearsal. Thus, the overlap of features could account for representation-based interference only when information is stored under a phonological format.

Finally, although De Visscher and Noel's (2014) findings could be accounted for by the feature model, the fact that the retrieval of multiplication facts depends on the amount of common features shared across representations does not provide direct evidence for the unitary nature of memory. If one considers that multiplications are represented in an interconnected network of digits, the time to retrieve a multiplication fact would depend on the number of associations, with more associations between digits leading to longer response times. Indeed, activation would spread among a larger number of connections in the network, and it would take longer to reach the correct answer. This effect known as the fan effect is the characteristic of the retrieval of declarative knowledge from long-term memory (Anderson, 1974). The retrieval of multiplication facts should also exhibit a fan effect as for any declarative knowledge stored in long-term memory, and the interference index could be seen as the expression of the amount of associations between multiplications. As a consequence, the retrieval of multiplication facts should be sensitive to variation in working memory capacity. This was previously reported by Cantor and Engle (1993) when examining individual differences in a memory task in which adults had to memorize series of unrelated sentences. Individuals with lower capacity showed an exaggerated fan effect because they had less cognitive resources (as measured through working memory tasks) that can spread out in the interconnected network. Accordingly, low-span individuals should also exhibit a larger fan effect in the retrieval of multiplication facts. Similarly, children who have less working memory capacity should also display larger fan effects when solving or verifying simple multiplications. Examining how the fan effect would affect recall is of paramount importance because it should inform us on the role of interferences in memory.

The Role of Interferences

In the unitary view of memory, forgetting is often conceived as resulting from interference (see Lewandowsky et al., 2004). In the previous example, retrieving multiplication facts is impaired because representations shared some features, which results on representation-based interferences. When evoking interference, most models actually refer to this type of interferences. However, another type of interferences based on the use of common processes has been mentioned in the previous chapters.

Attout et al. suggested that the maintenance of serial order in working memory relies on the use of the numerical order (see Majerus's chapter for a discussion of these findings). As a consequence, a working memory task requiring the maintenance of order would share a process with a numerical task. This sharing would give rise to some process-based interferences and to poorer performances in the working memory task when performed concurrently with the numerical task. Another domain of mathematical cognition could also be sensitive to process-based interference. In his chapter, Barrouillet proposed that simple additions are not solved through direct retrieval of their answer from long-term memory but through some rapid counting procedure (see also Prado's chapter). Accordingly, it might be expected that simple additions would suffer from specific process-based interferences. If simple additions need counting procedure to be achieved, any counting activities should thus interfere with simple additions, something that remains to be tested. Once again, the domain of mathematical cognition could provide adequate tasks to examine these fundamental issues of the functioning of human memory.

CONCLUSION

Through this chapter, I discussed some of the issues presented in this section of the book, which considered the involvement of memory systems in mathematical cognition. More importantly, I wanted to show how the field of mathematical cognition is an important domain for anyone interested in the functioning of human memory. The diverse numerical and arithmetical activities reviewed and discussed in the previous chapters could provide a fruitful playground to examine memory.

Among these different activities, it could be surprising that little was said about the approximate number system, although it represents such a large number of studies in mathematical cognition. Since several decades, research has established that an approximate number system allows us to evaluate quantities (see Geary, Berch, & Mann Koepke, 2015, for review). The absence of discussion could be taken as a sign of clear acceptance of such a system and of agreement on the importance of this system in mathematical cognition. The acceptance of its existence seems widely shared. Ample evidence has indeed been collected showing that the approximate number system is an innate system that human beings share with many other species, such as non-human primates, dogs, birds, and even fishes. However, the importance of the approximate number system for mathematical cognition could be questioned. Indeed, mathematics is a human creation. Should we not search in the specificities of human cognition the roots of the emergence of mathematics and its development throughout human childhood? In such a quest, researchers studying mathematical cognition should not forget the role of memory. Memory allows us to temporarily maintain information, process it, learn new knowledge, and retrieve previous acquired knowledge, which, in sum, is the heart of human cognition.

Acknowledgments

The author thanks Steve Majerus for his careful reading and comments on a previous draft of this chapter.

References

Abrahamse, E., van Dijck, J. P., & Fias, W. (2016). How does working memory enable number-induced spatial biases? *Frontiers in Psychology: Cognition, 7*(977).

Allen, R. J., Baddeley, A. D., & Hitch, G. J. (2006). Is the binding of visual features in working memory resource-demanding? *Journal of Experimental Psychology: General, 135*, 298–313.

Anderson, J. R. (1974). Retrieval of propositional information from long-term memory. *Cognitive Psychology, 5*, 451–474.

Anderson, J. R. (1993). *Rules of the mind.* Hillsdale, NJ: Erlbaum.

Atkinson, R. C., & Shiffrin, R. M. (1968). Human memory: A proposed system and its control processes. In K. W. Spence, & J. T. Spence (Eds.), *The psychology of learning and motivation: Advances in research and theory* (Vol. 2). New York, NY: Academic Press.

Atkinson, R. C., & Shiffrin, R. M. (1971). The control of short-term memory. *Scientific American, 225*, 82–90.

Baddeley, A. D. (1986). *Working memory.* Oxford: Clarendon Press.

Baddeley, A. D. (2000). The episodic buffer: A new component of working memory? *Trends in Cognitive Sciences, 4*, 417–423.

Baddeley, A. D. (2007). *Working memory, thought, and action.* Oxford: Oxford University Press.

Baddeley, A. D. (2012). Working memory: Theories, models, and controversies. *Annual Review of Psychology, 63*, 1–29.

Baddeley, A. D., & Hitch, G. J. (1974). Working memory. In G. A. Bower (Ed.), *Recent advances in learning and motivation* (Vol. 8). New York, NY: Academic Press.

Baddeley, A. D., & Logie, R. H. (1999). Working memory: The multiple-component model. In A. Miyake, & P. Shah (Eds.), *Models of working memory: Mechanisms of active maintenance and executive control.* Cambridge: Cambridge University Press.

Barrouillet, P., & Camos, V. (2015). *Working memory: Loss and reconstruction.* Hove, UK: Psychology Press.

Botvinick, M. M., & Plaut, D. C. (2006). Short-term memory for serial order: A recurrent neural network model. *Psychological Review, 113*, 201–233.

Broadbent, D. E. (1958). *Perception and communication.* London: Pergamon Press.

Brown, G. D. A., Neath, I., & Chater, N. (2007). A temporal ratio model of memory. *Psychological Review, 114*, 539–576.

Burgess, N., & Hitch, G. J. (1999). Memory for serial order: A network model of the phonological loop and its timing. *Psychological Review, 106*, 551–581.

Camos, V. (2003). Counting strategies from 5 years to adulthood: Adaptation to structural features. *European Journal of Psychology of Education, 18*, 251–265.

Camos, V. (2008). Low working memory capacity impedes both efficiency and learning of number transcoding in children. *Journal of Experimental Child Psychology, 99*, 37–57.

Camos, V., Mora, G., & Oberauer, K. (2011). Adaptive choice between articulatory rehearsal and attentional refreshing in verbal working memory. *Memory and Cognition, 39*, 231–244.

Cantor, J., & Engle, R. W. (1993). Working memory capacity as long-term memory activation: An individual differences approach. *Journal of Experimental Psychology: Learning, Memory and Cognition, 19*, 1101–1114.

Cowan, N. (2008). What are the differences between long-term, short-term, and working memory? In W. S. Sossin, J.-C. Lacaille, V. F. Castellucci, & S. Belleville (Eds.), *Progress in brain research* (Vol. 169). Amsterdam: Elsevier.

Craik, F. I. M. (1971). Primary memory. *British Medical Bulletin, 27*, 232–236.

Crowder, R. G. (1982). The demise of short term memory. *Acta Psychologica, 50*, 291–323.

De Visscher, A., & Noel, M.-P. (2014). The detrimental effect of interference in multiplication facts storing: Typical development and individual differences. *Journal of Experimental Psychology: General, 143*, 2380–2400.

Dehaene. (1992). Varieties of numerical cognition. *Cognition, 44*, 1–42.

Dehaene, S., Bossini, S., & Giraux, P. (1993). The mental representation of parity and number magnitude. *Journal of Experimental Psychology: General, 122*, 371.

van Dijck, J. P., Abrahamse, E., Acar, F., Ketels, B., & Fias, W. (2014). A working memory account of the interaction between numbers and spatial attention. *Quarterly Journal of Experimental Psychology, 67*, 1500–1513.

van Dijck, J. P., & Fias, W. (2011). A working memory account for spatial-numerical associations. *Cognition, 119*, 114–119.

Ebbinghaus, H. (1885/1913). *Memory: A contribution to experimental psychology*. New York: Teachers College, Columbia University (H.A. Ruger & C.E. Bussenius, Trans.).

Engle, R. W., & Kane, M. J. (2004). Executive attention, working memory capacity, and a two-factor theory of cognitive control. In B. Ross (Ed.), *The psychology of learning and motivation. Vol. 44*). New York, NY: Elsevier.

Farrell, S., & Lewandowsky, S. (2002). An endogenous distributed model of ordering in serial recall. *Psychonomic Bulletin and Review, 9*, 59–79.

Gallistel, C., & Gelman, R. (1992). Preverbal and verbal counting and computation. *Cognition, 44*, 43–74.

Geary, D. C., Berch, D. B., & Mann Koepke, K. (2015). *Evolutionary origins and early development of number processing*. London: Academic Press.

Henson, R. A. (1998). Short-term memory for serial order: The Start–End Model. *Cognitive Psychology, 36*, 73–137.

James, W. (1890). *Principles of psychology*. New York, NY: Henry Holt.

Langerock, N., Vergauwe, E., & Barrouillet, P. (2014). The maintenance of cross-domain associations in the episodic buffer. *Journal of Experimental Psychology: Learning, Memory, and Cognition, 40*, 1096–1109.

Lewandowsky, S., Duncan, M., & Brown, G. D. A. (2004). Time does not cause forgetting in short-term serial recall. *Psychonomic Bulletin and Review, 11*, 771–790.

Lewandowsky, S., & Farrell, S. (2008). Short-term memory: New data and a model. *The Psychology of Learning and Motivation, 49*, 1–48.

Mandler, G., & Shebo, B. J. (1982). Subitizing: An analysis of its component processes. *Journal of Experimental Psychology: General, 111*, 1–22.

Melton, A. W. (1963). Implications of short-term memory for a general theory of memory. *Journal of Verbal Learning and Verbal Behavior, 2,* 1–21.

Nairne, J. S. (1990). A feature model of immediate memory. *Memory and Cognition, 18,* 251–269.

Nairne, J. S. (2002). Remembering over the short-term: The case against the standard model. *Annual Review of Psychology, 52,* 53–81.

Neath, I., & Surprenant, A. (2003). *Human memory.* Belmont, CA: Wadsworth.

Oberauer, K. (2009). Interference between storage and processing in working memory: Feature overwriting, not similarity-based competition. *Memory and Cognition, 37,* 346–357.

Oberauer, K., Süß, H.-M., Schulze, R., Wilhelm, O., & Wittmann, W. W. (2000). Working memory capacity: Facets of a cognitive ability construct. *Personality and Individual Differences, 29,* 1017–1045.

Page, M. P. A., & Norris, D. (1998). The primacy model: A new model of immediate serial recall. *Psychological Review, 105,* 761–781.

Repovs, G., & Baddeley, A. D. (2006). The multi-component model of working memory: Explorations in experimental cognitive psychology. *Neuroscience, 139,* 5–21.

Siegler, R. S., & Shrager, J. (1984). Strategic choices in addition and subtraction: How do children know what to do? In C. Sophian (Ed.), *Origins of cognitive skills.* Hillsdale, NJ: Erlbaum.

Tolman, E. C. (1948). Cognitive maps in rats and men. *Psychological Review, 55,* 189–208.

Turner, M. L., & Engle, R. W. (1989). Is working memory capacity task dependent? *Journal of Memory and Language, 28,* 127–154.

Unsworth, N., & Engle, R. W. (2007). The nature of individual differences in working memory capacity: Active maintenance in primary memory and controlled search from secondary memory. *Psychological Review, 114,* 104–132.

Waugh, N. C., & Norman, D. A. (1965). Primary memory. *Journal of Experimental Psychology, 72,* 89–104.

Wechsler, D. (2005). *Wechsler intelligence scale for children* (4th ed.). San Antonio, TX: The Psychological Corporation.

Wechsler, D. (2008). *Wechsler adult intelligence scale* (4th ed.). San Antonio, TX: Pearson.

Wickelgren, W. A. (1974). Single-trace fragility theory of memory dynamics. *Memory and Cognition, 2,* 775–780.

Index

Printed in the United States
By Bookmasters